Multimedia Networking

From Theory to Practice

This authoritative guide to multimedia networking is the first to provide a complete system design perspective based on existing international standards and state-of-the-art networking and infrastructure technologies, from theoretical analyses to practical design considerations.

The four most critical components involved in a multimedia networking system – data compression, quality of service (QoS), communication protocols, and effective digital rights management – are intensively addressed. Many real-world commercial systems and prototypes are also introduced, as are software samples and integration examples, allowing readers to understand the practical tradeoffs in the real-world design of multimedia architectures and to get hands-on experience in learning the methodologies and design procedures.

Balancing just the right amount of theory with practical design and integration knowledge, this is an ideal book for graduate students and researchers in electrical engineering and computer science, and also for practitioners in the communications and networking industry. Furthermore, it can be used as a textbook for specialized graduate-level courses on multimedia networking.

Jenq-Neng Hwang is a Professor in the Department of Electrical Engineering, University of Washington, Seattle. He has published over 240 technical papers and book chapters in the areas of image and video signal processing, computational neural networks, multimedia system integration, and networking. A Fellow of the IEEE since 2001, Professor Hwang has given numerous tutorial and keynote speeches for various international conferences as well as short courses in multimedia networking and machine learning at universities and research laboratories.

Multimedia Networking

From Theory to Practice

JENQ-NENG HWANG

University of Washington, Seattle

CAMBRIDGE UNIVERSITY PRESS
Cambridge, New York, Melbourne, Madrid, Cape Town, Singapore, São Paulo, Delhi

Cambridge University Press
The Edinburgh Building, Cambridge CB2 8RU, UK

Published in the United States of America by Cambridge University Press, New York

www.cambridge.org
Information on this title: www.cambridge.org/9780521882040

First published 2009

Printed in the United Kingdom at the University Press, Cambridge

A catalog record for this publication is available from the British Library

Library of Congress Cataloging in Publication data
Hwang, Jenq-Neng.
  Multimedia networking : from theory to practice / Jenq-Neng Hwang.
    p.   cm.
  Includes bibliographical references.
  ISBN 978-0-521-88204-0 (hardback)
  1. Multimedia systems.   2. Computer networks.   I. Title.
  QA76.575.H86 2009
  006.7–dc22
  2008053943

ISBN 978-0-521-88204-0 hardback

To my wife Ming-Ying, my daughter Jaimie, and my son Jonathan, for their endless love and support.

Contents

Preface

With the great advances in digital data compression (coding) technologies and the rapid growth in the use of IP-based Internet, along with the quick deployment of last-mile wireline and wireless broadband access, networked multimedia applications have created a tremendous impact on computing and network infrastructures. The four most critical and indispensable components involved in a multimedia networking system are: (1) data compression (source encoding) of multimedia data sources, e.g., speech, audio, image, and video; (2) quality of service (QoS) streaming architecture design issues for multimedia delivery over best-effort IP networks; (3) effective dissemination multimedia over heterogeneous IP wireless broadband networks, where the QoS is further degraded owing to the dynamic changes in end-to-end available bandwidth caused by wireless fading or shadowing and link adaptation; (4) effective digital rights management and adaptation schemes, which are needed to ensure proper intellectual property management and protection of networked multimedia content.

This book has been written to provide an in-depth understanding of these four major considerations and their critical roles in multimedia networking. More specifically, it is the first book to provide a complete system design perspective based on existing international standards and state-of-the-art networking and infrastructure technologies, from theoretical analyses to practical design considerations. The book also provides readers with learning experiences in multimedia networking by offering many development-software samples for multimedia data capturing, compression, and streaming for PC devices, as well as GUI designs for multimedia applications. The coverage of the material in this book makes it appropriate as a textbook for a one-semester or two-quarter graduate course. Moreover, owing to its balance of theoretical knowledge building and practical design integration, it can serve also as a reference guide for researchers working in this subject or as a handbook for practising engineers.

Acknowledgements

This book was created as a side product from teaching several international short courses in the past six years. Many friends invited me to offer these multimedia networking short courses, which enabled my persistent pursuit of the related theoretical knowledge and technological advances. Their constructive interactions and suggestions during my short-course teaching helped the content and outline to converge into this final version. More specifically, I am grateful to Professor LongWen Chang of National Tsing Hua University, Professor Sheng-Tzong Steve Cheng of National Cheng Kung University, Director Kyu-Ik Cho of the Korean Advanced Institute of Information Technology, Professor Char-Dir Chung of National Taiwan University, Professor Hwa-Jong Kim of Kangwon National University, Professor Ho-Youl Jung of Yeungnam University, Professor Chungnan Lee of National Sun Yat-sen University, Professor Chia-Wen Lin of National Chung Cheng University, Professor Shiqiang Yang of Tsinghua University at Beijing, and Professor Wei-Pang Yang of National Chiao Tung University. I also would like to record my deep appreciation to my current and former Ph.D. students, as well as visiting graduate students, for their productive research contributions and fruitful discussions, which have led me to a better understanding of the content presented in this book. In particular, I would like to thank Serchen Chang, Wu Hsiang Jonas Chen, Timothy Cheng, Chuan-Yu Cho, Sachin Deshpande, Hsu-Feng Hsiao, Chi-Wei Huang, ChangIck Kim, Austin Lam, Jianliang Lin, Shiang-Jiun Lin, Qiang Liu, Chung-Fu Weldon Ng, Tony Nguyen, Jing-Xin Wang, Po-Han Wu, Yi-Hsien Wang, Eric Work, Peng-Jung Wu, Tzong-Der Wu, and many others. Moreover, I would like to thank Professors Hsuan-Ting Chang and Shen-Fu Hsiao and Clement Sun for their help in proofreading my manuscript while I was writing this book.

Abbreviations

AAC	advanced audio coding
AAC-LC	AAC low-complexity profile
AAC-SSR	AAC scalable sample rate profile
AC	access category
ACAP	advanced common application platform
ADPCM	adaptive delta pulse code modulation
ADSL	asymmetric digital subscriber line
ADTE	adaptation decision taking engine
AES	advanced encryption standard
AIFS	arbitration interframe space
AIFSN	arbitration interframe space number
AIMD	additive-increase multiplicative decrease
ALM	application-level multicast
AMC	adaptive modulation coding
AODV	ad hoc on-demand distance vector
AP	access point
AR	auto-regressive
ARF	autorate fallback
ARIB	Association of Radio Industries and Business
ARK	Add Round Key
ARQ	automatic repeat request
AS	autonomous system
ASF	advanced system format
ASO	arbitrary slice ordering
ATIS	Alliance for Telecommunications Industry Solutions
ATSC	Advanced Television Systems Committee
AVC	advanced video coding
BBGDS	block-based gradient descent search
BE	best-effort
BER	bit error rate
BGP	border gateway protocol
BIC	bandwidth inference congestion control
BIFS	binary format for scenes
BSAC	bit-sliced arithmetic coding
BSS	basic service set
BST-OFDM	band segmented transmission OFDM

CA	certification authority
CABAC	context-adaptive binary arithmetic coding
CAST	Carlisle Adams and Stafford Tavares
CATV	cable TV
CAVLC	Context-Adaptive Variable Length Coding
CBR	constant bitrate
CBT	core-based tree
CCIR-601	Consultative Committee on International Radio, Recommendation 601
CD	compact disc
CD-I	CD-interactive
CDMA	code-division multiple access
CDN	content delivery network
CELP	code-excited linear prediction
CIF	common intermediate format
CLC	cross-layer congestion control
CLUT	color look-up table
CMMB	China Mobile Multimedia Broadcasting
CMS	content management system
COFDM	coded orthogonal frequency-division multiplex
COPS	common open policy service
CPE	customer premise equipment
CPU	central processing unit
CQ	custom queuing
CQICH	channel quality indication channel
CRC	cyclic redundancy check
CS-ACELP	conjugate structure – algebraic code-excited linear prediction
CSI	channel-state information
CSMA/CA	carrier sense multiple access with collision avoidance
CSMA/CD	carrier sense multiple access with collision detection
CTS	clear to send
CW	contention window
CWA	contention window adaptation
DAB	digital audio broadcasting
DAM	digital asset management
DBS	direct broadcast satellite
DCC	digital compact cassette
DCF	distributed coordination function
DCI	digital cinema initiative
DCT	discrete cosine transform
DES	data encryption standard
DFS	distributed fair scheduling
DI	digital item
DIA	digital item adaptation
DID	digital item declaration
DIDL	digital item declaration language
DiffServ	differentiated services

DIFS	DCF interframe space
DII	digital item identification
DIM	digital item method
DIMD	doubling increase multiplicative decrease
DIML	digital item method language
DIP	digital item processing
DIXO	digital item extension operation
DL	downlink
DLNA	digital living network alliance
DRM	digital rights management
DSC	digital still camera
DSCP	differentiated service code point
DSL	digital subscriber line
DTV	digital TV
DVB	digital video broadcasting
DVB-H	digital video broadcasting – handheld
DVB-T	digital video broadcasting – terrestrial
DVD	digital versatile disk
DVMRP	distance-vector multicast routing protocol
EBCOT	embedded block coding with optimized truncation
EBU	European Broadcasting Union
EDCA	enhanced distributed channel access
EDTV	enhanced definition TV
ertPS	extended-real-time polling service
ESS	extended service set
ETSI	European Telecommunication Standards Institute
EV-DO	evolution-data only
FATE	fair airtime throughput estimation
FDD	frequency-division duplex
FDDI	fiber distributed data interface
FDMA	frequency-division multiple access
FEC	forward error correction
FIFO	first-in first-out
FMO	flexible macroblock ordering
FTP	file transfer protocol
FTTH	fiber to the home
GIF	graphics interchange format
GOP	group of pictures
GPRS	general packet radio service
GSM	global system for mobile
GUI	graphical user interface
HCCA	HCF controlled channel access
HCF	hybrid coordination function
HD-DVD	high-definition digital versatile disk
HDTV	high-definition TV
HFC	hybrid fiber cable

HHI	Heinrich Hertz Institute
HSDPA	high speed downlink packet access
HSUPA	high speed uplink packet access
HTTP	hypertext transfer protocol
IANA	Internet Assigned Numbers Authority
IAPP	inter access-point protocol
ICMP	Internet control message protocol
IDEA	international data encryption algorithm
IEC	International Electrotechnical Commission
IETF	Internet engineering task force
IGMP	Internet group management protocol
IIF	IPTV interoperability forum
IMT-2000	International Mobile Telecommunications 2000
IntServ	integrated services
IP	intellectual property
IP	Internet protocol
IPMP	intellectual property management and protection
IPTV	Internet protocol TV
IPv4	Internet protocol Version 4
IPv6	Internet protocol Version 6
ISBN	International Standard Book Number
ISDB-T	integrated services digital broadcasting – terrestrial
ISDN	integrated services digital network
ISMA	International Streaming Media Alliance
ISO	International Organization for Standardization
ISP	Internet service provider
ISPP	interleaved single-pulse permutation
ISRC	International Standard Recording Code
ITS	intelligent transportation system
ITU-T	International Telecommunication Union
iTV	interactive TV
JBIG	Joint Bi-level Image experts Group
JP3D	JPEG2000 3D
JPEG	Joint Photographic Experts Group
JPIP	JPEG2000 Interactive and Progressive
JPSEC	JPEG2000 Secure
JPWL	JPEG2000 Wireless
JVT	Joint Video Team
LAN	local area network
LD-CELP	low-delay code-excited linear prediction
LLC	logical link control
LPC	linear predictive coding
LSA	link-state advertisement
LSP	line spectral pairs
LTE	long-term evolution
LTP	long-term prediction

MAC	media access control
MAF	minimum audible field
MAN	metropolitan area network
MBS	Multicast and broadcast service
Mbone	multicast backbone
MCF	medium-access coordination function
MCL	mesh connectivity layer
MCU	multipoint control unit
MD5	message digest 5
MDC	multiple description coding
MDCT	modified discrete cosine transform
MELP	mixed-excitation linear prediction
MFC	Microsoft Foundation Class
MIMO	multiple-input multiple-output
MMP	multipoint-to-multipoint
MMR	mobile multi-hop relay
MMS	Microsoft Media Server
MOS	mean opinion score
MOSPF	multicast open shortest path first
MPC	multiple-pulse coding
MPDU	MAC protocol data unit
MPE-FEC	multiprotocol encapsulated FEC
MPEG	Moving Picture Experts Group
MPLS	multiprotocol label switching
MRP	multicast routing protocol
MSDU	MAC service data unit
MSE	mean squared error
MVC	multi-view video coding
NAL	network abstraction layer
NAT	network address translation
NAV	network allocation vector
NGN	next generation network
nrtPS	non-real-time polling service
OC-N	optical carrier level N
OFDM	orthogonal frequency-division multiplex
OFDMA	OFDM access
OLSR	optimized link-state routing
OS	operating system
OSPF	open shortest path first
OTT	one-way trip time
OWD	one-way delay
P2P	peer-to-peer
PAL	phase alternating line
PARC	Palo Alto Research Center
PCF	point coordination function
PCM	pulse code modulation

PDA	personal digital assistant
PER	packet error rate
PES	packetized elementary stream
PGP	pretty good privacy
PHB	per-hop behavior
PHY	physical layer
PIFS	PCF interframe spacing
PIM	protocol-independent multicast
PIM-DM	protocol-independent multicast-dense mode
PIM-SM	protocol-independent multicast-sparse mode
PKC	public key cryptography
PKI	public key infrastructure
PLC	packet loss classification
PLR	packet loss rate
PLM	packet-pair layered multicast
PMP	point-to-multipoint
PNA	progressive network architecture
POTS	plain old telephone service
PQ	priority queuing
PSI	program specific information
PSNR	peak signal-to-noise ratio
PSTN	public-switched telephone network
QAM	quadrature amplitude modulation
QMF	quadrature mirror filter
QoE	quality of experience
QoS	quality of service
QPSK	quadrature phase-shift keying
RBAR	receiver-based autorate
RC	Rivest Cipher
RCT	reversible color transform
RDD	rights data dictionary
RDT	real data transport
RED	random early detection/discard/drop
REL	rights expression language
RIP	routing information protocol
RLC	receiver-driven layered congestion control
RLC	run-length code
RLM	receiver-driven layered multicast
ROTT	relative one-way trip time
RPE	regular pulse excitation
RPF	reverse path forwarding
RSVP	resource reservation protocol
RTCP	real-time transport control protocol
RTP	real-time transport protocol
rtPS	real-time Polling Service
RTS	request to send

RTSP	real-time streaming protocol
RTT	round-trip time
RVLC	reversible variable-length code
SAD	sum of absolute differences
SAN	storage area network
SBR	spectral band replication
SCM	superposition coded multicasting
SDM	spatial-division multiplex
SDMA	space-division multiple access
SDK	software development kit
SDP	session description protocol
SDTV	standard definition TV
SECAM	sequential color with memory
SFB	scale factor band
SHA	secure hash algorithm
SIF	source input format
SIFS	short interframe space
SIP	session initiation protocol
SKC	secret key cryptography
SLA	service-level agreement
SLTA	simulated live transfer agent
SMCC	smooth multirate multicast congestion control
SMIL	synchronized multimedia integration language
SMPTE	Society for Motion Picture and Television Engineers
SPL	sound pressure level
SRA	source rate adaptation
SSP	stream synchronization protocol
STB	set-top box
STP	short-term prediction
STS-N	SONET Telecommunications Standard level N
SVC	scalable video coding
TBTT	too busy to talk
TCP	transmission control protocol
TDAC	time-domain aliasing cancellation
TDD	time-division duplex
TDMA	time-division multiple access
T-DMB	terrestrial digital multimedia broadcasting
TFRC	TCP-friendly rate control
TIA	Telecommunication Industry Association
TOS	type of service
TPEG	Transport Protocol Experts Group
TTL	time to live
UAC	user agent client
UAS	user agent server
UDP	user datagram protocol
UED	usage environment description

UGS	unsolicited grant service
UL	uplink
UMB	ultra-mobile broadband
UMTS	universal mobile telecommunications system
URI	uniform resource identifier
URL	uniform resource locator
VBR	variable bitrate
VCD	video CD
VCEG	Video Coding Experts Group
VCL	video coding layer
VDSL	very-high-bitrate digital subscriber line
VFW	Video for Windows
VLBV	very-low-bitrate video
VO	video object
VoD	video on demand
VoIP	voice over IP
VOP	video object plane
VQ	vector quantization
VRML	virtual reality modeling language
VSB	vestigial sideband
VSELP	vector-sum-excited linear prediction
WAN	wide area networks
WCDMA	wideband CDMA
WEP	wired equivalent privacy
WFQ	weighted fair queuing
Wi-Fi	wireless fidelity
WiMAX	Worldwide Interoperability for Microwave Access
WLAN	wireless local area network
WMN	wireless mesh network
WMV	Windows Media Video
WNIC	wireless network interface card
WPAN	wireless personal area network
WRALA	weighted radio and load aware
WRED	weighted random early detection
WT	wavelet transform
WWW	World Wide Web
XML	extensible markup language
XrML	extensible rights markup language
XOR	exclusive OR

1 Introduction to multimedia networking

With the rapid paradigm shift from conventional circuit-switching telephone networks to the packet-switching, data-centric, and IP-based Internet, networked multimedia computer applications have created a tremendous impact on computing and network infrastructures. More specifically, most multimedia content providers, such as news, television, and the entertainment industry have started their own streaming infrastructures to deliver their content, either live or on-demand. Numerous multimedia networking applications have also matured in the past few years, ranging from distance learning to desktop video conferencing, instant messaging, workgroup collaboration, multimedia kiosks, entertainment, and imaging [1] [2].

1.1   Paradigm shift of digital media delivery

With the great advances of digital data compression (coding) technologies, traditional analog TV and radio broadcasting is gradually being replaced by digital broadcasting. With better resolution, better quality, and higher noise immunity, digital broadcasting can also potentially be integrated with interaction capabilities.

In the meantime, the use of IP-based Internet is growing rapidly [3], both in business and home usage. The quick deployment of last-mile broadband access, such as DSL/cable/T1 and even optical fiber (see Table 1.1), makes Internet usage even more popular [4]. One convincing example of such popularity is the global use of voice over IP (VoIP), which is replacing traditional public-switched telephone networks (PSTNs) (see Figure 1.1). Moreover, local area networks (LANs, IEEE 802.3 [5]) or wireless LANs (WLANs, also called Wi-Fi, 802.11 [6]), based on office or home networking, enable the connecting integration and content sharing of all office or home electronic appliances (e.g., computers, media centers, set-top boxes, personal digital assistants (PDAs), and smart phones). As outlined in the vision of the Digital Living Network Alliance (DLNA), a digital home should consist of a network of consumer electronics, mobile and PC devices that cooperate transparently, delivering simple, seamless interoperability so as to enhance and enrich user experiences (see Figure 1.2) [7]. Even the recent portable MP3 players (such as the Microsoft Zune, http://www.zune.net/en-US/) are equipped with Wi-Fi connections (see Figure 1.3). Wireless connections are, further, demanded outside the office or home, resulting in the fast-growing use of mobile Internet whenever people are on the move.

These phenomena reflect two societal trends on paradigm shifts: a shift from digital broadcasting to multimedia streaming over IP networks and a shift from wired Internet to wireless Internet. Digital broadcasting services (e.g., digital cable for enhanced definition TV (EDTV) and high-definition TV (HDTV) broadcasting, direct TV via direct broadcast

Table 1.1 The rapid deployment of last-mile broadband access has made Internet usage even more popular

Services/Networks	Data rates
POTS	28.8–56 kbps
ISDN	64–128 kbps
ADSL	1.544–8.448 Mbps (downlink) 16–640 kbps (uplink)
VDSL	12.96–55.2 Mbps
CATV	20–40 Mbps
OC-N/STS-N	$N \times 51.84$ Mbps
Ethernet	10 Mbps
Fast Ethernet	100 Mbps
Gigabit Ethernet	1000 Mbps
FDDI	100 Mbps
802.11b	1, 2, 5.5, and 11 Mbps
802.11a/g	6–54 Mbps

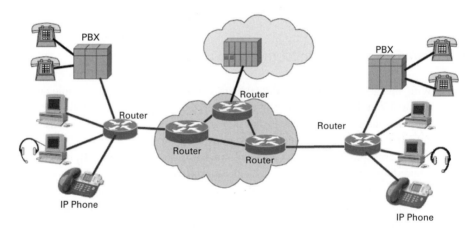

Figure 1.1 The growing use of voice over IP (VoIP) is quickly replacing the usage of traditional public-switched telephone networks (PSTNs): private branch exchanges (PBXs) are used to make connections between the internal telephones of a private business.

satellite (DBS) services [8], and digital video broadcasting (DVB) [9]) are maturing (see Table 1.2), while people also spend more time on the Internet browsing, watching video or movie by means of on-demand services, etc. These indicate that consumer preferences are changing from traditional TV or radio broadcasts to on-demand information requests, i.e., a move from "content push" to "content pull." Potentially more interactive multimedia services are taking advantage of bidirectional communication media using IP networks, as evidenced by the rapidly growing use of video blogs and media podcasting. It can be confidently predicted that soon Internet-based multimedia content will no longer be produced by traditional large-capital-based media and TV stations, because everyone can have a media station that produces multimedia content whenever and wherever they want, as long

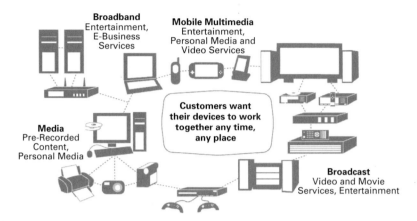

Figure 1.2 The vision of the Digital Living Network Alliance (DLNA) [7].

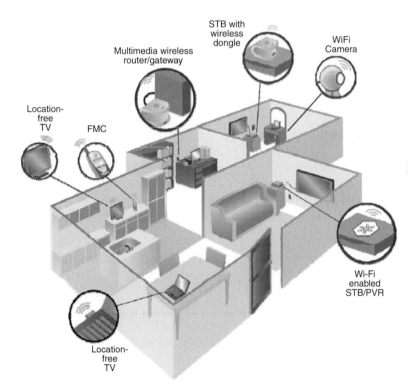

Figure 1.3 WLAN-based office or home networking enables the connecting, integration, and content sharing of all office or home electronic appliances (www.ruckuswireless.com).

as they have media-capturing devices (e.g., digital camera, camcorder, smart phone, etc.) with Internet access (see Figure 1.4). A good indication of this growing trend is the recent formation of a standardization body for TV over IP (IPTV) [10], i.e., the IPTV Interoperability Forum (IIF), which will develop ATIS (Alliance for Telecommunications Industry

Table 1.2 Digital broadcasting is maturing [11]

Region	Fixed reception standards	Mobile reception standards
Europe, India Australia, Southeast Asia	DVB-T	DVB-H
North America	ATSC	DVB-H
Japan	ISDB-T	ISDB-T one-segment
Korea	ATSC	T-DMB
China	DVB-T/T-DMB/CMMB	

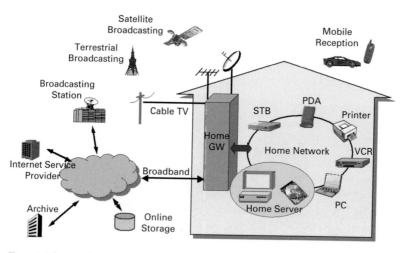

Figure 1.4 Interactive multimedia services take advantage of the bidirectional communication media of IP networks.

Solutions) standards and related technical activities that enable the interoperability, inter-connection, and implementation of IPTV systems and services, including video-on-demand and interactive TV services.

The shift from wired to wireless Internet is also coming as a strong wave (see Figure 1.5) [12] [24]. The wireless LAN (WLAN or the so-called Wi-Fi standards) technologies, IEEE 802.11a/b/g and the next generation very-high-data-rate ($>$ 200 Mbps) WLAN product IEEE 802.11n, to be approved in the near future, are being deployed everywhere with very affordable installation costs [6]. Also, almost all newly shipped computer products and more and more consumer electronics come with WLAN receivers for Internet access. Further-more wireless personal area network (WPAN) technologies, IEEE 802.15.1/3/4 (Bluetooth/ UWB/Zigbee), which span short-range data networking of computer peripherals and con-sumer electronics appliances with various bitrates, provide an easy and convenient mech-anism for sending and receiving data to and from the Internet for these end devices [14]. To provide mobility support for Internet access, cellular-based technologies such as third generation (3G) [14] [15] networking are being aggressively deployed, with increased multimedia application services from traditional telecommunication carriers. Furthermore,

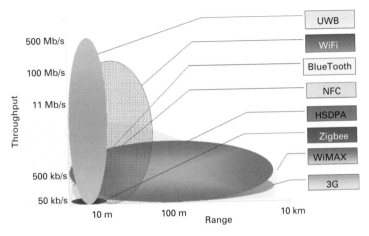

Figure 1.5 The WLAN technologies, IEEE 802.11 a/b/g/n, are being deployed everywhere with very affordable installation costs. Furthermore, the WPAN technologies, IEEE 802.15.1/3/4, provide an easy and convenient mechanism for sending or receiving data to or from the internet for these end devices [24].

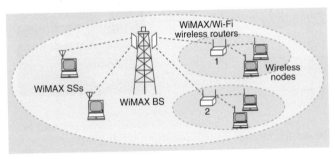

Figure 1.6 Fixed or mobile WiMAX (IEEE 802.16d/e) can serve as an effective backhaul for WLAN [23] (© IEEE 2007).

mobile wireless microwave access (WiMAX) serves as another powerful alternative to mobile Internet access from data communication carriers. Fixed or mobile WiMAX (IEEE 802.16d and 802.16e) [16] [17] can also serve as an effective backhaul for WLAN whenever this is not easily available, such as in remote areas or moving vehicles with compatible IP protocols (see Figure 1.6).

1.2 Telematics: infotainment in automobiles

Another important driving force for wireless and mobile Internet is telematics, the integrated use of telecommunications and informatics for sending, receiving, and storing information via telecommunication devices in road-traveling vehicles [18]. The telematics market is rolling out fast thanks to the growing installation in vehicles of mobile Internet access, such as the general packet radio service (GPRS) or 3G mobile access [12]. It ranges from front-seat information and entertainment (*infotainment*) such as navigation, traffic status, hand-free communication, location-aware services, etc. to back-seat infotainment, such as multimedia entertainment and gaming, Internet browsing, email access, etc. Telematics systems have also been designed for engine and mechanical monitoring, such as remote diagnosis, care data collection, safety and security, and vehicle status and location monitoring. Figure 1.7 shows an example of new vehicles equipped with 3G mobile access (www.jentro.com).

In addition to the growing installation of mobile Internet access in vehicles, it is also important to note the exponentially growing number of WLAN and WPAN installations on vehicles (see Figure 1.8). This provides a good indication of the wireless-access demand for

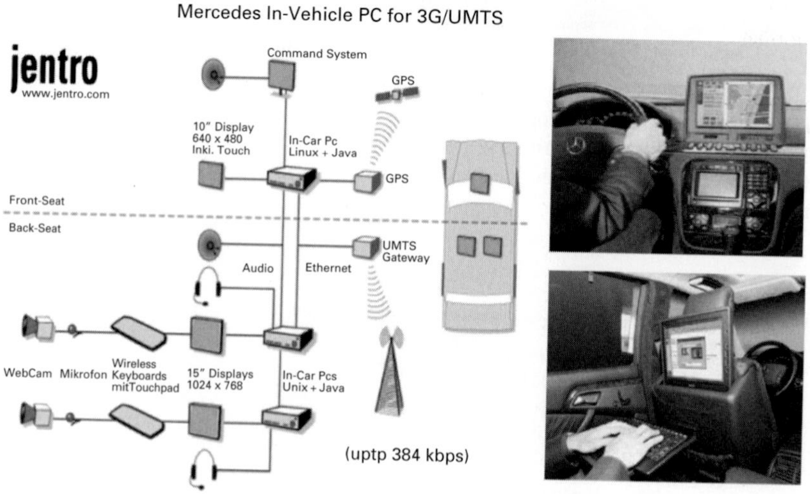

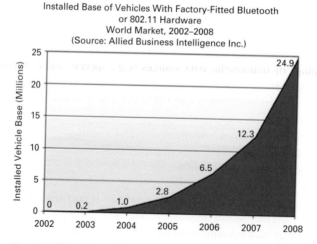

Figure 1.7 An example of new vehicles equipped with 3G mobile access provided by Jentro Technology (www.jentro.com).

Figure 1.8 The plot shows the exponentially growing number of WLAN and WPAN installations on vehicles (www.linuxdevices.com/news/NS2150004408.html).

Table 1.3 The bandwidth requirement of raw digital data without compression

Source	Bandwidth (Hz)	Sampling rate (Hz)	Bits per sample	Bitrate
Telephone voice	200–3400	8000 samples/s	12	96 kbps
Wideband speech	50–7000	16 000	14	224 kbps
Wideband audio (2 channels)	20–20 000	44 100 samples/s	16 per channel	1.412 Mbps (2 channels)
B/W documents		300 dpi (dots per inch)	1	90 kb per inch^2
Color image		512×512	24	6.3 Mb per image
CCIR-601 (NTSC)		720×576×25 (DVD)	24	248.8 Mbps
CCIR-601 (PAL)		720×576×25	24	248.8 Mbps
Source input format (SIF)		352×240×30 (VCD)	12	30 Mbps
Common intermediate format (CIF)		352×288×30	12	37 Mbps
Quarter CIF (QCIF)		176×144×7.5	12	2.3 Mbps
High definition DVD		1920×1080×30	24	1492 Mbps

vehicles in a local vicinity, e.g., inside a parking lot and moving with a slow speed yet still enjoying location-aware services.

1.3 Major components of multimedia networking

Multimedia is defined as information content that combines and interacts with multiple forms of media data, e.g., text, speech, audio, image, video, graphics, animation, and possibly various formats of documents. There are four major components that have to be carefully dealt with to allow the successful dissemination of multimedia data from one end to the other [1]. Such a large amount of multimedia data is being transmitted through Internet protocol (IP) networks that, even with today's broadband communication ability, the bandwidth is still not enough to accommodate the transmission of uncompressed data (see Table 1.3). The first major component of multimedia networking is the data compression (source encoding) of multimedia data sources (e.g., speech, audio, image, and video). For different end terminals to be able to decode a compressed bitstream, international standards for these data compression schemes have to be introduced for interoperability. Once the data are compressed, the bitstreams will be packetized and sent over the Internet, which is a public, best-effort, wide area network (as shown in Figure 1.9). This brings us to the second major component of multimedia networking, quality of service (QoS) issues [19] [20], which include packet delay, packet loss, jitter, etc. These issues can be dealt with either from the network infrastructure or from an application level.

Furthermore, wireless networks have been deployed widely as the most popular last-mile Internet access technology in homes, offices, and public areas in recent years. At the same time, mobile computing devices such as PDAs, smart phones, and laptops have been improved dramatically in not only their original functionalities but also their communication capabilities. This combination creates new services and an unstoppable trend of

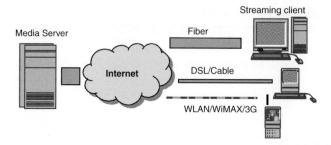

Figure 1.9 The compressed multimedia data are packetized and sent over the Internet, which is a public best-effort wide area network.

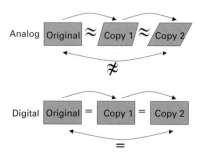

Figure 1.10 The proliferation of digital media makes illegal copying and falsification easy.

converting everything to wireless, for almost everything and everywhere [12]. In ensuring the effective dissemination of compressed multimedia data over IP-based wireless broadband networks, the main challenges result from the integration of wired and wireless heterogeneous networking systems; in the latter the QoS is further degraded by the dynamically changing end-to-end available bandwidth caused by the wireless fading or shadowing and link adaptation. This constitutes the third major component of today's multimedia networking. Moreover, the increased occurrence of wireless radio transmission errors also results in a higher bursty rate of packet loss than for wired IP networks. To overcome all these extra deficiencies due to wireless networks, several additional QoS mechanisms, spanning from physical, media access control (MAC), network and application layers, have to be incorporated.

There are numerous multimedia networking applications: digital broadcasting and IP streaming and meeting and/or messaging have been widely deployed. These applications will continue to be the main driving forces behind multimedia networking. The proliferation of digital media makes interoperability among the various terminals difficult and also makes illegal copying and falsification easy (see Figure 1.10); therefore, the fourth major component of multimedia networking consists of ensuring that the multimedia-networked content is fully interoperable, with ease of management and standardized multimedia content adapted for interoperable delivery, as well as intellectual property management and protection (i.e., digital rights management, DRM [21]), effectively incorporated in the system [22].

Providing an in-depth understanding of the four major components mentioned above, from both theoretical and practical perspectives, was the motivation for writing this book: it covers the fundamental background as well as the practical usage of these four components. To facilitate the learning of these subjects, specially designed multimedia coding and networking laboratory contents have been used in order to provide students with practical and hands-on experience in developing multimedia networking systems. The coverage and materials of this book are appropriate for a one-semester first-year graduate course.

1.4 Organization of the book

This book is organized as follows. Chapters 2–5 cover the first major component of multimedia networking, i.e., standardized multimedia data compression (encoding and decoding). More specifically, we discuss four types of medium, including speech, audio, image and video, each medium being covered in one chapter. The most popular compression standards related to these four media are introduced and compared from a tradeoff perspective. Thanks to the advances in standardized multimedia compression technologies, digital multimedia broadcasting is being deployed all over the world. In Chapter 6 we discuss several types of popular digital multimedia (video) broadcasting that are widely used internationally. Chapters 7 and 8 focus on QoS techniques for multimedia streaming over IP networks, ranging over the MAC, network, transport, and application layers of IP protocols. Several commercially available multimedia streaming systems are also covered in detail. In Chapters 9 and 10 we discuss specifically advances in wireless broadband technologies and the QoS challenges of multimedia over these wireless broadband infrastructures, again in terms of the layers of IP protocols. Chapter 11 deals with digital rights management (DRM) technologies for multimedia networking and the related standardization efforts. To provide readers with a hands-on learning experience of multimedia networking, many development software samples for multimedia data capturing, compression, streaming for PC devices, as well as GUI designs for multimedia applications, are provided in Chapter 12.

References
[1] Jerry D. Gibson, *Multimedia Communications: Directions and Innovations*, Communication, Networking and Multimedia Series, Academic Press, 2000.
[2] K. R. Rao, Z. S. Bojkovic, and D. A. Milovanovic, *Multimedia Communication Systems: Techniques, Standards, and Networks*, Prentice Hall, 2002.
[3] "Internet world stats: usage and population Statistics," http://www.internetworldstats.com/stats.htm.
[4] "Broadband internet access," http://en.wikipedia.org/wiki/Broadband_Internet_access.
[5] "IEEE 802.3 CSMA/CD (Ethernet)," http://grouper.ieee.org/groups/802/3/.
[6] "IEEE 802.11 wireless local area networks," http://grouper.ieee.org/groups/802/11/.
[7] "DLNA overview and vision whitepaper 2007," http://www.dlna.org/en/industry/pressroom/DLNA_white_paper.pdf.
[8] "Direct TV," http://www.directv.com/DTVAPP/index.jsp.
[9] "Digital video broadcasting: the global standard for digital television," http://www.dvb.org/.
[10] "The IPTV interoperability forum (IIF)," http://www.atis.org/iif/.

[11] S. Levi, "Designing encoders and decoders for mobile terrestrial broadcast digital television systems," in *Proc. TI Developer Conf.*, April 2006.

[12] A. Ganz, Z. Ganz, and K. Wongthavarawat, *Multimedia Wireless Networks: Technologies, Standards, and QoS*, Prentice Hall, 2003.

[13] "IEEE 802.15 Working Group for WPAN," http://www.ieee802.org/15/.

[14] "The 3rd Generation Partnership Project (3GPP)" http://www.3gpp.org/.

[15] "The 3rd Generation Partnership Project 2 (3GPP2)," http://www.3gpp2.org/.

[16] "The IEEE 802.16 Working Group on broadband wireless access standards," http://grouper.ieee.org/groups/802/16/.

[17] "The WiMAX forum," http://www.wimaxforum.org/home/.

[18] M. McMorrow, "Telematics – exploiting its potential," *IET Manufacturing Engineer*, 83(1): 46–48, February/March 2004.

[19] A. Tanenbaum, *Computer Networks*, Prentice Hall, 2002.

[20] M. A. El-Gendy, A. Bose, and K. G. Shin, "Evolution of the Internet QoS and support for soft real-time applications," *Proc. IEEE*, 91(7): 1086–1104, July 2003.

[21] S. R. Subramanya, Byung K. Yi, "Digital rights management," *IEEE Potential*, 25(2): 31–34, March/April 2006.

[22] W. Zeng, H. Yu, and C.-Y. Lin, *Multimedia Security Technologies for Digital Rights Management*, Elsevier, 2006.

[23] D. Niyato and E. Hossain, "Integration of WiMAX and WiFi: optimal pricing for bandwidth Sharing," *IEEE Commun. Mag.*, 45(5): 140–146, May 2007.

[24] Y.-Q. Zhang, "Advances in mobile computing," keynote speech in *IEEE Conf. on Multimedia Signal Processing*, Victoria BC, October 2006.

2 Digital speech coding

The human vocal and auditory organs form one of the most useful and complex communication systems in the animal kingdom. All speech (voice) sounds are formed by blowing air from the lungs through the vocal cords (also called the vocal fold), which act like a valve between the lung and vocal tract. After leaving the vocal cords, the blown air continues to be expelled through the vocal tract towards the oral cavity and eventually radiates out from the lips (see Figure 2.1). The vocal tract changes its shape with a relatively slow period (10 ms to 100 ms) in order to produce different sounds [1] [2].

In relation to the opening and closing vibrations of the vocal cords as air blows over them, speech signals can be roughly categorized into two types of signals: voiced speech and unvoiced speech. On the one hand, voiced speech, such as vowels, exhibit some kind of semi-periodic signal (with time-varying periods related to the pitch); this semi-periodic behavior is caused by the up–down valve movement of the vocal fold (see Figure 2.2(a)). As a voiced speech wave travels past, the vocal tract acts as a resonant cavity, whose resonance produces large peaks in the resulting speech spectrum. These peaks are known as formants (see Figure 2.2(b)).

On the other hand, the hiss-like fricative or explosive unvoiced speech, e.g., the sounds, such as s, f, and sh, are generated by constricting the vocal tract close to the lips (see Figure 2.3(a)). Unvoiced speech tends to have a nearly flat or high-pass spectrum (see Figure 2.3(b)). The energy in the signal is also much lower than that in voiced speech.

The speech sounds can be converted into electrical signals by a transducer, such as a microphone, which transforms the acoustic waves into an electrical current. Since most human speech contains signals below 4 kHz then, according to the sampling theorem [4] [5], the electrical current can be sampled (analog-to-digital converted) at 8 kHz as discrete data, with each sample typically represented by eight bits. This 8-bit representation, in fact, provides 14-bit resolution by the use of quantization step sizes which decrease logarithmically with signal level (the so-called A-law or μ-law [2]). Since human ears are less sensitive to changes in loud sounds than to quiet sounds, low-amplitude samples can be represented with greater accuracy than high-amplitude samples. This corresponds to an uncompressed rate of 64 kilobits per second (kbps).

In the past two to three decades, there have been great efforts towards further reductions in the bitrate of digital speech for communication and for computer storage [6] [7]. There are many practical applications of speech compression, for example, in digital cellular technology, where many users share the same frequency bandwidth and good compression allows more users to share the system than otherwise possible. Another example is in digital voice storage (e.g., answering machines). For a given memory size, compression [3] allows longer messages to be stored. Speech coding techniques can have the following attributes [2]:

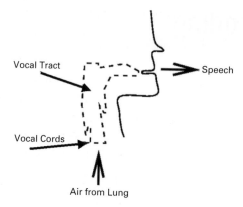

Figure 2.1 The human speech-production mechanism [3].

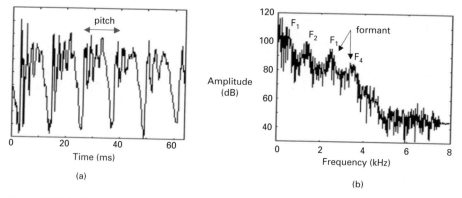

Figure 2.2 Voiced speech can be considered as a kind of semi-periodic signal.

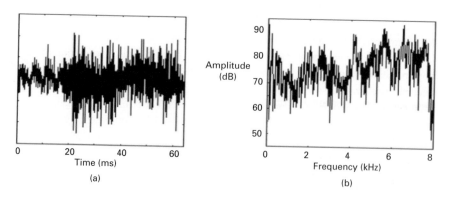

Figure 2.3 Hiss-like fricative or explosive unvoiced speech is generated by constricting the vocal tract close to the lips.

(1) *Bitrate*   This is 800 bps – 16 kbps, most 4.8 kbps or higher, normally the sample-based waveform coding (e.g., ADPCM-based G.726 [8]) has a relatively higher bitrate, while block-based parametric coding has a lower bitrate.

(2) *Delay*   The lower-bitrate parametric coding has a longer delay than waveform coding; the delay is about 3–4 times the block (frame) size.

(3) *Quality*   The conventional objective mean square error (MSE) is only applicable to waveform coding and cannot be used to measure block-based parametric coding, since the reconstructed (synthesized) speech waveform after decoding is quite different from the original waveform. The subjective mean opinion score (MOS) test [9], which uses 20–60 untrained listeners to rate what is heard on a scale from 1 (unacceptable) to 5 (excellent), is widely used for rating parametric coding techniques.

(4) *Complexity*   This used to be an important consideration for real-time processing but is less so now owing to the availability of much more powerful CPU capabilities.

2.1   LPC modeling and vocoder

With current speech compression techniques (all of which are lossy), it is possible to reduce the rate to around 8 kbps with almost no perceptible loss in quality. Further compression is possible at the cost of reduced quality. All current low-rate speech coders are based on the principle of *linear predictive coding (LPC)* [10] [11], which assumes that a speech signal s (n) can be approximated as an auto-regressive (AR) formulation

$$\hat{s}(n) = e(n) + \sum_{k=1}^{p} a_k s(n - k) \tag{2.1}$$

or as an all-pole vocal tract filter, $H(z)$:

$$H(z) = \frac{1}{A(z)} = \frac{1}{1 - \sum_{k=1}^{p} a_k z^{-k}}, \tag{2.2}$$

where the residue signal $e(n)$ is assumed to be white noise and the linear regression coefficients $\{a_k\}$ are called LPC coefficients. The LPC-based speech coding system is illustrated in Figure 2.4 [12]; note that a speech "codec" consists of an encoder and a decoder. This LPC modeling, which captures the formant structure of the short-term speech spectrum, is also called short-term prediction (STP).

Commonly the LPC analysis on synthesis filter has order p equal to 8 or 10 and the coefficients $\{a_k\}$ are derived on the basis of a 20–30 ms block of data (frame). More specifically, the LPC coefficients can be derived by solving a least squares solution assuming that $\{e(n)\}$ are estimation errors, i.e., solving the following normal (Yule–Walker) linear equation:

$$\begin{bmatrix} r_s(1) \\ r_s(2) \\ r_s(3) \\ \vdots \\ r_s(p) \end{bmatrix} = \begin{bmatrix} r_s(0) & r_s(1) & r_s(2) & \cdots & r_s(p-1) \\ r_s(1) & r_s(0) & r_s(1) & \cdots & r_s(p-2) \\ r_s(2) & r_s(1) & r_s(0) & \cdots & \cdots \\ \vdots & \vdots & \vdots & \ddots & \vdots \\ r_s(p-1) & r_s(p-2) & r_s(p-3) & \cdots & r_s(0) \end{bmatrix} \begin{bmatrix} a_1 \\ a_2 \\ a_3 \\ \vdots \\ a_p \end{bmatrix} \tag{2.3}$$

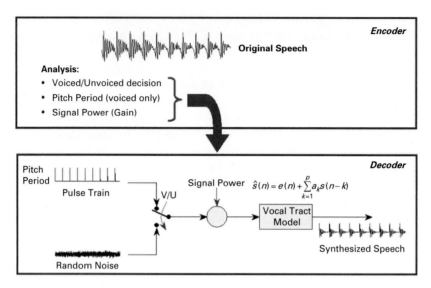

Figure 2.4 A typical example of a LPC-based speech codec.

where the autocorrelation $r_s(k)$ is defined as

$$r_s(k) = \sum_{n=0}^{N-k-1} s(n)s(n+k). \qquad (2.4)$$

Owing to the special Toeplitz matrix structure of the Yule–Walker linear equation, the LPC coefficients $\{a_k\}$ can be solved using the efficient Levinson–Durbin recursion algorithm [12]. A complete LPC-based voice coder (vocoder) consists of an analysis performed in the encoder (see the upper part of Figure 2.4), which determines the LPC coefficients $\{a_k\}$ and the gain parameter G (a side product of the Levinson–Durbin recursion in solving for the $\{a_k\}$ coefficients) and, for each frame, a voice/unvoiced decision with pitch period estimation. As shown in Figure 2.5, this is achieved through a simplified (ternary valued) autocorrelation (AC) calculation method. The pitch-period search is confined to $F_s/350$ and $F_s/80$ samples (i.e., 23–100 samples) or, equivalently, the pitch frequency is confined to between 80 to 350 Hz. Pitch period estimation is sometimes called long-term prediction (LTP), since it captures the long-term correlation, i.e., periodicity, of the speech signals. The autocorrelation function $R(k)$ is given by

$$R(k) = \sum_{m=0}^{N-k-1} x^c(m)x^c(m+k), \qquad (2.5)$$

where

$$x^c(n) = \begin{cases} +1 & \text{if } x(n) > C_L, \\ -1 & \text{if } x(n) < -C_L, \\ 0 & \text{otherwise}, \end{cases}$$

where C_L denotes the threshold, which is equal to 30% of the maximum of the absolute value of $\{x(n)\}$ within this frame. The 10 LPC coefficients $\{a_k\}$, together with the pitch period and gain parameters, are derived on the basis of 180 samples (22.5 ms) per frame and are encoded at 2.4

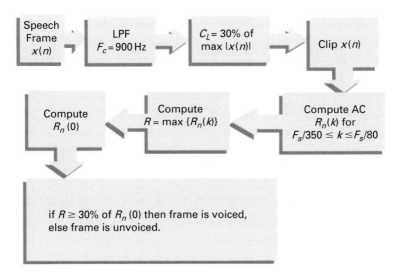

Figure 2.5 The voiced or unvoiced decision with pitch period estimation is achieved through a simplified autocorrelation calculation method (http://www.ee.ucla.edu/~ingrid/ee213a/speech/speech.html).

kbps for transmission or storage (according to the LPC-10 or FS-1015 standards) [13] [14]. The decoder is responsible for synthesizing the speech using the coefficients and parameters in the flow chart shown in the lower part of Figure 2.4. The 2.4 kbps FS-1015 was used in various low-bitrate and secure applications, such as in defense or underwater communications, until 1996, when the 2.4 kbps LPC-based standard was replaced with the new mixed-excitation linear prediction (MELP) coder [15][16] by the United States Department of Defense Voice Processing Consortium (DDVPC). The MELP coder is based on the LPC model with additional features that include mixed excitation, aperiodic pulses, adaptive spectral enhancement, pulse dispersion filtering, and Fourier magnitude modeling.

Even though the speech synthesized from the LPC vocoder is quite intelligible it does sound somewhat unnatural, with MOS values [9] ranging from 2.7 to 3.3. This unnatural speech quality results from the over-simplified representation (i.e., one impulse per pitch period) of the residue signal $e(n)$, which can be calculated from Eq. (2.5) after the LPC coefficients have been derived (see Figure 2.6). To improve speech quality, many other (hybrid) speech coding standards have been finalized, all having more sophisticated representations of the residue signal $e(n)$, as shown in Figure 2.6:

$$e(n) = s(n) - \sum_{k=1}^{p} a_k s(n - k). \tag{2.6}$$

To further improve the representation of the residue signal $e(n)$, long-term prediction (LTP) can be applied by first removing the periodic redundancy caused by the semi-periodic pitch movement. More specifically, each frame of speech (20 or 30 ms) is divided into four uniform subframes, each with N_{sf} samples, taking each of the subframe samples backwards to find the best-correlated counterpart (which has a time lag of p samples) having the necessary gain factor β. The LTP-filtered signal is called the excitation $u(n)$ and has an even smaller dynamic range; it can thus be encoded more effectively (see Figure 2.7). Different

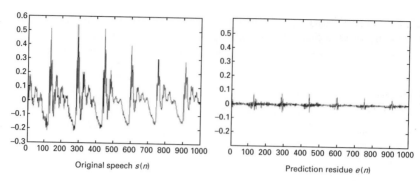

Original speech $s(n)$ Prediction residue $e(n)$

Figure 2.6 The prediction residue signal $e(n)$ of LPC can be calculated from Eq. (2.5) after the LPC coefficients have been derived.

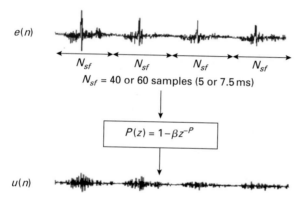

Figure 2.7 The LTP-filtered signal, the excitation $u(n)$, has an even smaller dynamic range than the unfiltered signal and can thus be encoded more effectively.

encoding of the excitation signals (with also some slight variations in STP analysis) leads to different speech coding standards (see Table 2.1), e.g.,

(1) *Regular pulse excitation (RPE)*   This is used mainly to encode the magnitude of selected (uniformly decimated) samples; e.g., GSM [17] [18] [19].
(2) *Code-excited linear prediction (CELP)*   This is used mainly to encode excitations based on pre-clustered codebook entries, i.e., magnitude and locations are both important; e.g., CELP [20], G.728 [21] [22], and VSELP [23].
(3) *Multiple pulse coding (MPC)*   This is used mainly to encode the locations of selected samples (pulses with sufficiently large magnitude); e.g., G.723.1 [24] and G.729 [25].

2.2   Regular pulse excitation with long-term prediction

The global system for mobile communications (GSM) [17] [18] [19] standard, the digital cellular phone protocol defined by the European Telecommunication Standards Institute (ETSI, http://www.etsi.org/), derives eight-order LPC coefficients from 20 ms frames and

Table 2.1 Various encodings of excitation signals (with also some slight variations in STP analysis) and the corresponding speech coding standards

Standards	Year	Bitrate (kbps)	MOS
PCM (PSTN)	1972	64	4.4
LPC-10 (FS-1015)	1976	2.4	2.7
	(1996)		(3.3)
G.726 (ADPCM, G.721)	1990	16, 24, 32, 40	4.1 (32 kbps)
GSM (RPE-LTP)	1987	13	3.7
CELP (FS-1016)	1991	4.8	3.2
G.728 (LD-CELP)	1992	16	4
VSELP (IS-54)	1992	8	3.5
G.723.1 (MPC-MLQ)	1995	6.3/5.3	3.98/3.7
G.729 (CS-ACELP)	1995	8	4.2

Table 2.2 There are 260 bits allocated for each GSM frame (20 ms), resulting in a total bitrate of 13 kbps

Parameters	Bits per subframe	Bits per frame
LPC coefficients	–	36
LTP lag	7	28
LTP gain	2	8
ORPE subsequence scaling factor	6	24
ORPE subsequence index	2	8
ORPE subsequence values	39	156
Total	56	260

uses a regular pulse excitation (RPE) encoder over the excitation signal $u(n)$ after redundancy removal with long-term prediction (LTP). More specifically, GSM sorts each subframe (5 ms, 40 samples) after LTP into four sequences:

(1) sequence 1: 0 3 6 9 ... 36
(2) sequence 2: 1 4 7 10 ... 37
(3) sequence 3: 2 5 8 11 ... 38
(4) sequence 4: 3 6 9 12 ... 39

Only one sequence, the highest-energy sequence among the four, per subframe is selected for encoding. Each sample of the selected sequence is quantized at three bits (instead of the original sampling of 13 bits). This is called an optimized RPE (ORPE) sequence. The bit allocation of each GSM frame (20 ms) is shown in Table 2.2 for the case when 260 bits are used per frame, resulting in a total bitrate of 13 kbps. The overall operations of a GSM encoder are shown in Figure 2.8 and those of a GSM decoder in Figure 2.9.

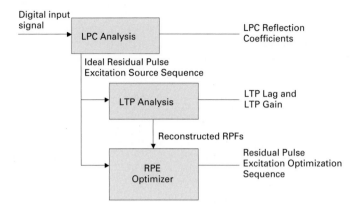

Figure 2.8 A GSM encoder.

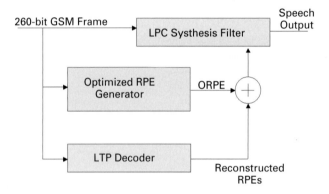

Figure 2.9 A GSM decoder.

2.3   Code-excited linear prediction (CELP)

The RPE uses a downsampled version of excitation signals to represent the complete excitation, while a code-excited linear prediction (CELP) coder uses a codebook entry from a vector quantized (VQ) codebook to represent the excitation; see Figure 2.10. In this figure, $P(z)$ is the LTP filter and $1/P(z)$ is used to compensate for the difference operation performed in the LTP filtering (i.e., recovering $u(n)$ back to $e(n)$); the $1/A(z)$ filter synthesizes the speech $\hat{s}(n)$ to be compared with the original speech $s(n)$. The objective of encoding the excitations is to choose the codebook entry (codeword) that minimizes the weighted error between the synthesized and original speech signals. This technique, referred to as *analysis by synthesis*, is widely used in CELP-based speech coding standards. The analysis by synthesis technique simulates the decoder in the encoder so that the encoder can choose the optimal configuration, or tune itself for the best parameters, to minimize the weighted error calculated from the *original* speech and the *reconstructed* speech (see Figure 2.11).

The perceptual weighting filter $A(Z)/A(Z/\gamma)$, $\gamma \approx 0.7$, is used to provide different weighting on the error signals by allowing for more error around the resonant formant

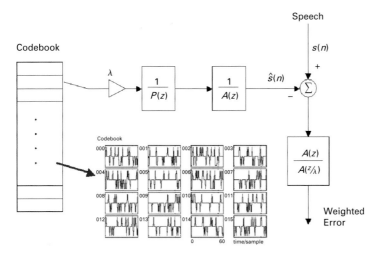

Figure 2.10 A CELP coder uses a codebook entry from a vector-quantized codebook to represent the excitation.

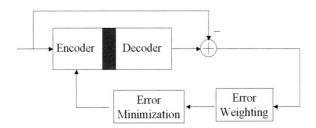

Figure 2.11 The analysis by synthesis technique simulates the decoder in the encoder so that the encoder can choose the optimal configuration to minimize the weighted error calculated from the original speech and the reconstructed speech.

frequencies (by widening the bandwidth of spectral resonance), since human ears are less sensitive to the error around those frequencies. A typical perceptual weighting filter frequency response, given the original LPC filter frequency response, is presented in Figure 2.12 for various values of γ.

The US Federal Standard FS-1016 [20] is based on CELP techniques with 4.8 kbps data rate. The index of the chosen codeword is encoded for transmission or storage. An effective algorithm for the codeword search was developed so that a CELP coder can be implemented in real time using digital signal processors. In FS-1016, 10 LPC coefficients were derived from each 30 ms frame and there are 512 codewords in the excitation codebook, each codeword having 7.5 ms (60 samples) of ternary valued (+1, 0, −1) excitation data. The FS-1016 is currently not widely used since its successor, MELP [15], provides better performance in all applications, even though CELP was used in some secure applications as well as adopted in MPEG-4 for encoding natural speech for 3G cellular phones.

Another CELP-based speech coder is the low-delay CELP G.728 (LD-CELP) [21] [22], which provides 16 kbps speech with a quality similar to that of the 32 kbps speech provided by the ADPCM waveform-based G.726 speech coding. The G.728 speech coder is widely used in

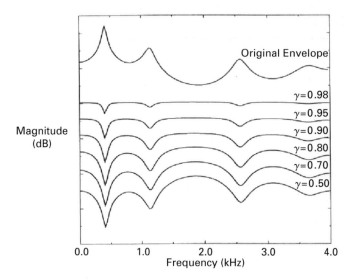

Original Envelope

$\gamma=0.98$
$\gamma=0.95$
$\gamma=0.90$
$\gamma=0.80$
$\gamma=0.70$
$\gamma=0.50$

Magnitude
(dB)

0.0   1.0   2.0   3.0   4.0
Frequency (kHz)

Figure 2.12 A typical perceptual weighting filter frequency response, given the original LPC filter frequency response, for various values of γ.

voice over cable or voice over IP (VoIP) teleconferencing applications through packet networks. For G.728, 50th-order LPC coefficients were derived recursively, on the basis of its immediate 16 past outputs (there was no need to transmit the LPC coefficients since they can be computed in the receiver side in a recursive backward-adaptive fashion) and there are 1024 codewords contained in the excitation codebook, each codeword having only five samples of excitation data.

The major drawback of CELP-based speech coding is the very large computational requirements. To overcome this requirement, the vector sum excited linear prediction (VSELP) speech coder [23], which also falls into the CELP class, utilizes a codebook with a structure that allows for a very efficient search procedure. This VSELP coder (see Figure 2.13), at 8 kbps with 20 ms/ frame, was selected by the Telecommunication Industry Association (TIA, http://www.tiaonline. org/) as the standard for use in North American TDMA IS-54 digital cellular telephone systems.

As shown in Figure 2.13, the VSELP codec contains two separate codebooks ($k=1$ or 2), each of which can contribute $2^M = 2^7 = 128$ codevectors (in fact there are only 64 distinct patterns since $u(n)$ and $-u(n)$ are regarded as the same signal pattern) if constructed as a linear combination of $M=7$ basis vectors $\{v_{k,m}(n), m=1, 2, \ldots, M\}$,

$$u_{k,i}(n) = \sum_{m=1}^{M} \theta_{i,k} v_{k,m}(n) \qquad \text{and} \qquad u(n) = v_i u_{1,i}(n) + v_2 u_{2,j}(n), \qquad (2.7)$$

where $u_{k,i}(n)$ denotes the ith codevector in the kth codebook; $v_{k,m}(n)$ denotes the mth basis vector of the kth codebook and $\theta_{i,m} = \pm 1$; $u(n)$ denotes the resulting combined excitation signal, which should be further compensated by inverse LTP filtering,

$$\frac{1}{p(z)} = \frac{1}{1 - \beta z^{-T}},$$

to recover $e(n)$ from $u(n)$. The pitch pre-filter and spectral post-filter are also used to further finetune the estimated parameters for a better synthesis.

The bit allocation of each VSELP frame (20 ms long) is shown in Table 2.3; 160 bits are used per frame, resulting in a total bitrate of 8 kbps.

Table 2.3 There are 160 bits allocated for each VSELP frame (20 ms), resulting in a total bitrate of 8 kbps

Parameters	Bits per subframe	Bits per frame
LPC coefficients	–	38
Energy – $R_q(0)$	–	5
Excitation codes (I, H)	7+7	56
Lag (L)	7	28
GS-P0-P1 code	8	32
\<unused\>	–	1
Total	**29**	**160**

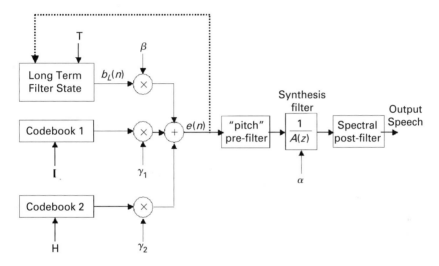

Figure 2.13 The VSELP codec contains two separate codebooks, each of which can contribute $2^7 = 128$ distinct code vectors.

2.4 Multiple-pulse-excitation coding

Another main category of speech coding is based on encoding only the locations of large enough pulses (i.e., the excitations $u(n)$ after LTP). The speech coder G.723.1 [24] [26] [27] provides near toll-phone quality transmitted speech signals. This type of speech coder has been standardized for internet speech transmission. The G.723.1 processes the speech using 30 ms frames, each 7.5 ms subframe containing 60 samples.

The 10th-order LPC analysis is based on a 60 sample subframe ($i = 0, 1, 2, 3$) but is only performed on the last subframe ($i = 3$) of each frame. The LPC coefficients are then converted into line spectral pairs (LSPs), which are defined to be the roots of $P(z)$ and $Q(z)$, where

$$P(z) = A(z) + z^{-P}A(z^{-1}),$$
$$Q(z) = A(z) - z^{-P}A(z^{-1}). \tag{2.8}$$

This ensures better stability during the quantization process.

The LSP vectors for the other three subframes of each frame can be derived using linear interpolation between the current frame's LSP vector, e.g., the $\{P_n\}$, and the previous frame's LSP vector, $\{P_{n-1}\}$, i.e.,

$$
p_{ni} = \begin{cases}
0.75p_{n-1} + 0.25p_n, & i = 0, \\
0.50p_{n-1} + 0.50p_n, & i = 1, \\
0.25p_{n-1} + 0.75p_n, & i = 2, \\
p_n, & i = 3.
\end{cases}
\tag{2.9}
$$

The speech coder G.723.1 also introduces a more efficient LTP technique to improve the accuracy of (pitch) redundancy removal based on open and closed loop analyses. This gives a lower encoding bitrate with better speech synthesis quality. A formant perceptual weighting filter, $W_i(z)$ (similar to the one used in the CELP analysis by synthesis technique), where

$$
W_i(z)\frac{A(z/r_1)}{A(z/r_2)} = \frac{1 - \sum_{j=1}^{10} a_{ij} z^{-j}(0.9)^j}{1 - \sum_{j=1}^{10} a_{ij} z^{-j}(0.5)^j}, \qquad 0 \le i \le 3,\ r_1 = 0.9,\ r_2 = 0.5
\tag{2.10}
$$

is constructed for every subframe, using the unquantized LPC coefficients $\{a_{ij}\}$ derived from the interpolated LSPs $\{P_n\}$ of each subframe, i.e., every subframe of input speech signal is first filtered to obtain perceptually weighted speech, $f(n)$;

$$
CR_{OL}(j) = \frac{\left(\sum_{n=0}^{119} f(n)f(n-j) \right)^2}{\sum_{n=0}^{119} f(n-j)f(n-j)}, \qquad 18 \le j \le 142.
\tag{2.11}
$$

The open-loop LTP estimates the pitch period for every two subframes (120 samples) and a cross-correlation criterion is used on perceptually weighted speech, $f(n)$ (see Eq. 2.11). The index j which maximizes CR_{OL} is selected and named $\hat{L}$. The open-loop LTP analysis is followed by a closed-loop LTP analysis, which estimates the pitch lag around the open-loop pitch lag $\hat{L}$ calculated earlier. More specifically, for subframes 0 and 2 the closed-loop pitch lag is selected in the range ± 1 of the open-loop pitch lag (coded with seven bits) and, for subframes 1 and 3, the lag differs from the subframe's open-loop pitch lag by -1, 0, $+1$ or $+2$ (coded with two bits).

Instead of directly determining the pulses from the excitation signal $u(n)$ resulting from the speech $s(n)$ passing through the STP and LTP operations, G.723.1 tries to determine a pure multiple-pulse signal $v(n)$, which can be filtered by a 20th-order FIR weighted synthesis filter $h(n)$ to produce $v'(n)$ so as to approximate $u(n)$, another effective use of the analysis by synthesis technique:

$$
v'(n) = \sum_{j=0}^{19} h(j)v(n-j), \qquad 0 \le n \le 59,
\tag{2.12}
$$

where $v(n)$ consists only of multiple pulses, i.e., six pulses for subframes 0 and 2, with magnitudes of either $+G$ or $-G$, the gain factor analyzed for this frame; and five pulses for subframes 1 and 3, also with magnitudes of either $+G$ or $-G$. This results in a 6.3 kbps multiple pulse coding (MPC) data rate (see Table 2.4), since

Table 2.4 The bit allocation for a 30 ms G.723.1 frame, which results in a 6.3 kbps multiple-pulse coding (MPC) data rate

Parameters	Subframe 0	Subframe 1	Subframe 2	Subframe 3	Total
LPC indices					24
Adaptive codebook lags	7	2	7	2	18
Combined gains	12	12	12	12	48
Pulse positions	20	18	20	18	73
Pulse signs	6	5	6	5	22
Grid index	1	1	1	1	4
Total					189

Table 2.5 The 5.3 kbps G.723.1 version locates at most four pulses from each subframe, and the four pulses have to be limited to one of four predefined groups

Sign	Positions
±1	0, 8, 16, 24, 32, 40, 48, 56
±1	2, 10, 18, 26, 34, 42, 50, 58
±1	4, 12, 20, 28, 36, 44, 52
±1	6, 14, 22, 30, 38, 46, 54

$$189 \frac{\text{bits}}{\text{frame}} \times 33 \frac{\text{frames}}{\text{seconds}} = 6.3 \text{ kbps.}$$

To reduce the bitrate further, G.723.1 also offers a 5.3 kbps version by locating at most four pulses from each subframe; the four pulses have to be limited to one of four predefined groups, as shown in Table 2.5.

Another multiple-pulse-coding-based speech coder is called the conjugate structure algebraic code-excited linear prediction (CS-ACELP) G.729 [25] [26] [27] and can achieve 32 kbps G.726 ADPCM toll-phone quality with only 8 kbps. It has been adopted in several Internet-based VoIP or session initiation protocol (SIP) phones. The G.729 uses 10 ms frames, with 5 ms (40 sample) subframes for excitation signal representation. Similarly to G.723.1, this speech codec allows four nonzero pulses (unit magnitude) and each pulse has to be chosen from a predefined group. This is called interleaved single-pulse permutation (ISPP), and the groups are as follows:

(1) pulse 1   0, 5, 10, 15, 20, 25, 30, 35 (3-bit encoding)
(2) pulse 2   1, 6, 11, 16, 21, 26, 31, 36 (3-bit encoding)
(3) pulse 3   2, 7, 12, 17, 22, 27, 32, 37 (3-bit encoding)
(4) pulse 4   3, 4, 8, 9, 13, 14, 18, 19, 23, 24, 28, 29, 33, 34, 38, 39 (4-bit encoding)

References

[1] J. D. Gibson, T. Berger, T. Lookabaugh, D. Lindbergh, and R. L. Baker, *Digital Compression for Multimedia: Principles and Standards*, Morgan Kauffman, 1998.

[2] W. B. Kleijn and K. K. Paliwal, *Speech Coding and Synthesis*, Elsevier Science, 1995.

[3] "Speech compression, by Data-Compression.com," http://www.data-compression.com/speech.html.

[4] J. Jerri, "The Shannon sampling theorem – its various extensions and applications: a tutorial review," *Proc. IEEE*, 65(11): 1565–1596, November 1977.

[5] A. V. Oppenheim, R. W. Schafer, and J. R. Buck, *Discrete-Time Signal Processing*, Second edition, Prentice Hall, 1999.

[6] A. Gersho, "Advances in speech and audio compression," *Proc. IEEE*, 82(6): 900–918, June 1994.

[7] A. S. Spanias, "Speech coding: a tutorial review," *Proc. IEEE*, 82(10): 1542–1582, October 1994.

[8] "40, 32, 24, 16 kbit/s adaptive differential pulse code modulation (ADPCM)," ITU G.726: December 1990, http://www.itu.int/rec/T-REC-G.726/e.

[9] "Methods for subjective determination of transmission quality," ITU Recommendation P.800; August 1996, http://www.itu.int/rec/T-REC-P.800-199608-I/en.

[10] S. Saito and F. Itakura, "The theoretical consideration of statistically optimum methods for speech spectral density," Report No. 3107, Electrical Communication Laboratory, NTT, Tokyo, December 1966.

[11] B. S. Atal, "The history of linear prediction," *IEEE Signal Process. Mag.*, 23(2): 154–161, March 2006.

[12] L. R. Rabiner and R. W. Schafer, *Digital Processing of Speech Signals*, Prentice Hall, 1978.

[13] T. E. Tremain, "The Government Standard linear predictive coding algorithm: LPC-10," *Speech Technol.*, 40–49, April 1982.

[14] J. P. Campbell Jr. and T.E. Tremain, "Voiced/unvoiced classification of speech with applications to the US government LPC-10E algorithm," in *Proc. Int Conf. on Acoustics, Speech, and Signal Processing (ICASSP)*, Vol. II, pp. 473–476, 1986.

[15] L. M. Supplee, R. P. Cohn, J. S. Collura, and A. V. McCree, "MELP: the new Federal Standard at 2400 bps," in *Proc. Int. Conf. on Acoustics, Speech, and Signal Processing (ICASSP)*, Vol. II: pp. 1591–1594, April 1997.

[16] M. A. Kohler, "A comparison of the new 2400 bps MELP Federal Standard with other standard coders," in *Proc. Int Conf. on Acoustics, Speech, and Signal Processing (ICASSP)*, Vol. II, pp. 1587–1590, April 1997.

[17] "Global system for mobile (GSM) world," http://www.gsmworld.com/index.shtml.

[18] P. Vary, K. Hellwig, R. Hofmann, R. J. Sluyter, C. Galand, and M. Rosso, "Speech codec for the European mobile radio system," in *Proc. Int. Conf. on Acoustics, Speech, and Signal Processing (ICASSP)*, Vol. I, pp. 227–230, April 1988.

[19] K. Hellwig, P. Vary, D. Massaloux, J.P. Petit, C. Galand, and M. Rosso, "Speech codec for the European mobile radio system," in *Proc. IEEE Global Communications Conf. (GLOBECOM)*, Vol. 2, pp. 1065–1069, November 1989.

[20] M. R. Schroeder and B. S. Atal, "Code-excited linear prediction (CELP): high-quality speech at very low bitrates," in *Proc. ICASSP'85*, pp. 937–940, March 1985.

[21] A. Kumar and A. Gersho, "LD-CELP speech coding with nonlinear prediction," *IEEE Signal Processing Letters*, 4(4): 89–91, April 1997.

[22] "Coding of speech at 16 kbit/s using low-delay code excited linear prediction," ITU-T Recommendation G.728, http://www.itu.int/rec/T-REC-G.728/e.

[23] I. A. Gerson, M. A. Jasiuk, "Vector sum excited linear prediction (VSELP) speech coding at 8 kbps," in *Proc. Int. Conf. on Acoustics, Speech, Signal Processing (ICASSP)*, pp. 461–464, April 1990.

[24] ITU-T Recommendation G.723.1, "Dual rate speech coder for multimedia communications transmitting at 5.3 and 6.3 kbps," http://www.itu.int/rec/T-REC-G.723.1/e.

[25] "Coding of speech at 8 kbit/s using conjugate-structure algebraic-code-excited linear prediction (CS-ACELP), ITU-T Recommendation G.729, http://www.itu.int/rec/T-REC-G.729/e.

[26] R. V. Cox, "Three new speech coders from the ITU cover a range of applications," *IEEE Commun. Mag.*, 35(9): 40–47, September 1997.

[27] R. V. Cox and P. Kroon, "Low bit-rate speech coders for multimedia communication," *IEEE Commun. Mag.*, 34(12): 34–42, December 1996.

3 Digital audio coding

In 1979, Philips and Sony set up a joint task force of engineers to design the new digital audio disk for recording audio music in digital format. After a year of experimentation and discussion, the task force produced the compact disk (CD) standard. The CD (see Figure 3.1(a)), made available on the market in late 1982, remains the standard physical medium for commercial audio recordings [1]. An audio CD consists of one or more stereo tracks stored using 16-bit pulse code modulation (PCM) coding at a sampling rate of 44.1 kHz (see Figure 3.1(b)). Standard compact disks have a diameter of 120 mm and can hold approximately 60–80 minutes of audio. From its origins as a music format, CD has grown to encompass other applications, e.g., the CD-ROM (read-only memory) and CD-R/W. Compact disks are now widely used as a data storage medium for computers and consumer electronics.

The original capacity requirement for recording digital audio is about 635 megabytes (MB) per hour or 1.411 Mbps, i.e.,

$$44100 \text{ (samples/s)} \times 2 \text{ (byte/sample)} \times 3600 \text{ (s/h)} \times 2 \text{ (channels)}$$
$$= 635 \text{ MB per hour of audio CD capacity.} \tag{3.1}$$

The famous computer-based waveform audio storage format, WAV (waveform audio format, with extension *.wav) [2], was originally developed by Microsoft and IBM. These audio files are usually uncompressed (i.e., lossless) but can act as a "wrapper" for various audio compression codecs, such as the lossy ADPCM scheme which achieves a compression ratio that is about 4 : 1 and drops the bitrate down to about 350 kbps. Unfortunately, this compression ratio is not satisfactory for many applications. To improve the data compression efficiency further we must resort to more advanced techniques.

In speech coding, we formulate the speech production mechanism using the LPC model and the LTP scheme along with the efficient representation of excitations, and this has successfully resulted in several low-bitrate speech coding standards. This sound production formulation approach cannot be applicable to digital music coding owing to the fact that music is produced by many potential sound generation mechanisms: specifically we have string (plucked and pulled), percussion, and wind. Therefore it is impractical to create one model for each sound generation mechanism. This fact leads us to investigate the sound reception mechanism, instead of sound generation, i.e., the human auditory system [3] [4]!

The human auditory system, consisting of our ears and brain, processes incoming audio signals in two different manners: physiologically and psychologically. Our ears consist of three fundamental physiological components [5], the inner, middle and outer ear (see Figure 3.2).

(1) The *outer ear* directs sounds through the ear canal towards the eardrum.
(2) The *middle ear* transforms sound pressure waves into mechanical movement on three small bones called "ossicles" (the hammer, anvil, and stirrup).

(a)

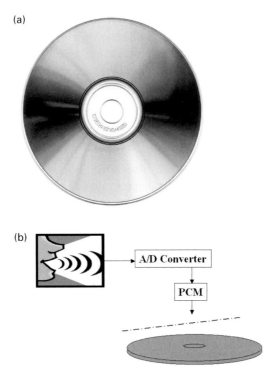

(b)

Figure 3.1 (a) The first compact disk (CD) was available on the market in late 1982. (b) An audio CD consists of one or more stereo tracks stored using 16-bit PCM coding at a sampling rate of 44.1 kHz.

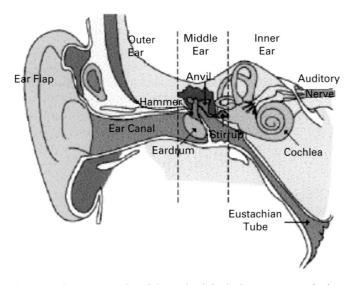

Figure 3.2 Our ears consist of three physiological components: the inner, middle, and outer ear [5].

(3) The *inner ear* houses the cochlea, a spiral-shaped structure for human hearing which sits in an extremely sensitive membrane called the basilar membrane. The cochlea converts the middle ear's mechanical movements to basilar membrane movement and eventually into the firing of auditory neurons, which, in turn, send electrical signals to the brain.

3.1   Human psychoacoustics

As mentioned above, human perception of sound involves some psychological effects. For example, the human ear is approximately logarithmic in its subjective response to increasing volume. Human hearing has a dynamic range of approximately 110 decibel (dB). Moreover, the ear's response to a fixed audio spectral distribution is subjectively different as the volume changes, e.g., the bass appears to become more pronounced as the volume increases and the subjective "richness" of tone increases over about a 60 dB loudness level. These facts lead us to the study of *human psychoacoustics* [6], as follows.

(1) *Hearing sensitivity* Put a person in a quiet room. Raise the level of a specific frequency tone until it is just barely audible. Vary the frequency and plot these auditory thresholds.

(2) *Frequency masking* Investigate how a loud tone affects neighboring frequency tones in human perception.

(3) *Temporal masking* Investigate how a loud tone affects subsequent (or preceding) tones in human perception.

We now consider each of these aspects in turn.

3.1.1   Hearing sensitivity

Our ears' cochleas do an approximate analysis of frequency in the range between 20 Hz to 20 kHz; this is similar to an imperfect Fourier transform with limited frequency resolution. Moreover, our ears' frequency resolution decreases with increasing frequency. Therefore we can define an absolute threshold of hearing (or the minimum audible field, MAF), which characterizes the amount of energy needed in a pure tone for it to be detected by a listener in a noiseless environment (see Figure 3.3).

More specifically, the absolute threshold of hearing, as defined using the sound pressure level (SPL), is illustrated in Figure 3.4 [6]. The SPL is defined mathematically as

$$SPL = 20 \log_{10} \frac{P}{P_0} \ (dB),$$

(3.2)

where P is the sound pressure of the stimulus in pascals and $P_0 = 20 \times 10^{-6}$ pascals.

By knowing the hearing sensitivity of our auditory system, it becomes obvious that we can take advantage of this during the quantization of audio samples. More specifically, in order to quantize the audio samples so as to compress the data, we have to choose enough bits per sample to reduce the resulting quantization noise to a satisfactory level. For example, if we uniformly quantize each audio sample with 12 bits (see Figure 3.5), the resulting quantization noise can be as low as –26 dB, which is far below the threshold of hearing. Since

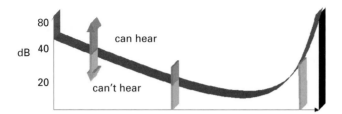

Figure 3.3 The absolute threshold of hearing characterizes the amount of energy needed in a pure tone for it to be detected by a listener in a noiseless environment.

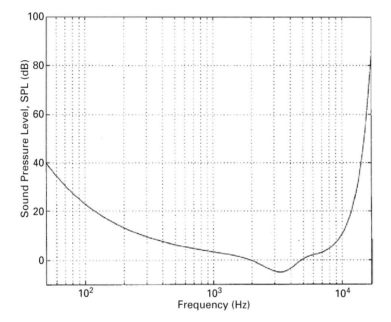

Figure 3.4 The absolute threshold of hearing as defined on the basis of the sound pressure level (SPL) (© IEEE 2000) [6].

our auditory system has non-uniform hearing sensitivity, one alternative is to divide the audible frequency range (20 Hz to 20 kHz) into several bands. Then the audio samples in different bands can be quantized with different number of bits to accommodate different tolerance levels of quantization noise (see Figure 3.6).

The next question to ask is, therefore, how to divide the audible frequency range into different bands. The human auditory system has a limited, frequency-dependent, resolution owing to the fact that the cochlea can be viewed as a bank of highly overlapping band-pass filters. A perceptually uniform measure of frequency can be expressed in terms of the width of *critical bands* [6]. The width of a critical band as a function of frequency is a good indicator of the human auditory system's frequency-dependent behavior. Many psychoacoustic effects are consistent with a critical-band frequency scaling. More specifically, the frequencies within a critical band are similar in terms of the ear's perception and can be

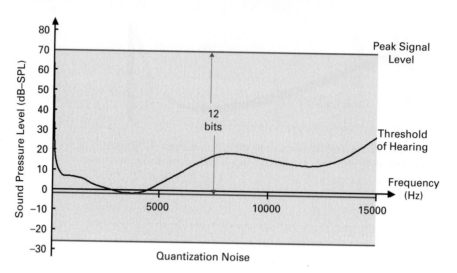

Figure 3.5 If we uniformly quantize each audio sample with 12 bits, the resulting quantization noise can be as low as −26 dB, which is far below the threshold of hearing.

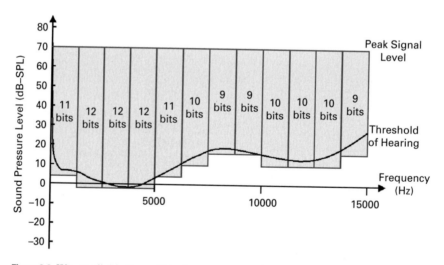

Figure 3.6 We can divide the audible frequency range (20 Hz to 20 kHz) into several bands, and the audio sample in different bands can be quantized with different numbers of bits to accommodate different tolerances of quantization noise.

processed separately from other critical bands. The bands have widths less than 100 Hz at the lowest audible frequencies and more than 4 kHz at the high end, i.e., human ears cannot distinguish between tones within 100 Hz at low frequencies and 4 kHz at high frequencies.

The audio frequency range can be partitioned into approximately 25 critical bands. Built upon the definition of critical bands, another way of defining the frequency unit, called a "bark," is introduced. A *bark* is defined to be the frequency unit for one critical band; thus for any frequency f (Hz), the corresponding bark number is given by

$$b = \begin{cases} f/100 & \text{for } f \le 500 \text{ Hz,} \\ 9 + 4 \log_2(f/1000) & \text{for } f > 500 \text{ Hz.} \end{cases} \tag{3.3}$$

3.1.2 Frequency masking

In the human ear's perception, different frequency channels can interfere with each other. More specifically, let us play a 250 Hz tone (the masker) at a fixed strong level (65 dB) and play a second test tone (e.g., 180 Hz) at a lower level simultaneously and raise its level until it is just distinguishable (the threshold of hearing). It can be observed that the louder sound (the masker) will distort the absolute threshold of hearing with respect to the logarithm of the frequency in kHz, see Figure 3.4, and make the quieter sound (the masked sound), which was originally audible, inaudible. This effect is called "frequency masking" (see Figure 3.7). In such frequency masking experiments hearing sensitivity thresholds can change drastically when multiple maskers are present, as shown in Figure 3.8 [7].

Advantage can be taken of this frequency masking in audio sample quantization for compression purposes. For example, as shown in Figure 3.9, we can reduce the number of bits required to quantize the audio samples in different frequency bands, since the (masking) hearing thresholds are greatly raised and the human tolerance of quantization noise is thus greatly increased.

3.1.3 Temporal masking

Similarly, a weak sound emitted soon after the end of a louder sound is masked by the louder sound. Even a weak sound just *before* a louder sound can be masked by

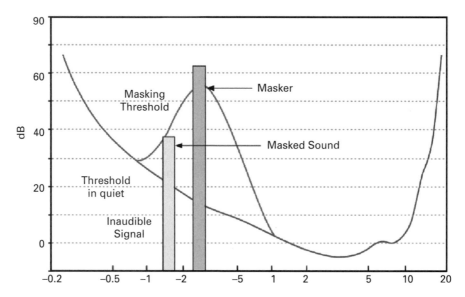

Figure 3.7 The louder sound (the masker) will distort the absolute threshold of hearing and make the quieter sound (the masked sound), which is originally audible, inaudible [7].

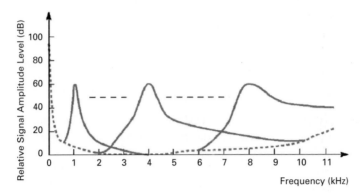

Figure 3.8 The frequency masking experiments can change hearing sensitivity thresholds (or the minimum audible field) drastically when multiple maskers are present.

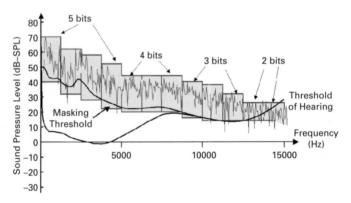

Figure 3.9 Owing to the effect of frequency masking, the number of bits needed to quantize the audio samples in different frequency bands can be further reduced.

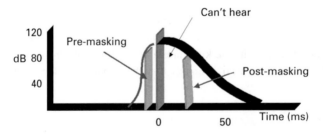

Figure 3.10 Forward and backward temporal masking.

the louder sound! These two effects are called post- and pre-temporal masking, respectively. The duration within which pre-masking applies is significantly less than one tenth of that of the post-masking, which is in the order of 50 to 200 ms (see Figure 3.10). The combined frequency and temporal masking effect is illustrated in Figure 3.11 [7].

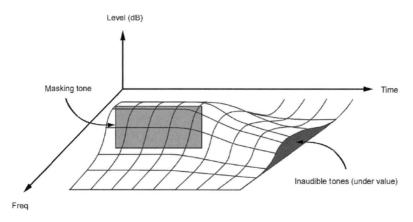

Figure 3.11 The combined frequency and temporal masking effect [7].

3.2 Subband signal processing and polyphase filter implementation

Owing to the frequency masking properties of human hearing, the best representation of audio is a frequency-domain representation, obtained by the use of a subband or transform filter bank. In order to analyze the psychoacoustics of the human ear's perception of the audio signal, the masking analysis and bitrate assignment must be prepared and processed in *frames* of a predetermined length of audio signal; this is similar to speech coding, where the STP is also processed in frames of a predefined length of human speech. Each audio coded frame contains mainly coded audio, but in addition:

(1) the peak level in each frequency subband;
(2) the masking level in each subband;
(3) the number of bits for each sample in each subband.

The encoder and decoder of a generic frame-based audio codec is shown in Figure 3.12. The encoder takes a frame of audio data and separates it into several non-overlapping subbands using a bank of filters that spans the whole audio frequency range (e.g., from 0 to 22 kHz). A discrete Fourier transform (or some kind of frequency transform such as discrete cosine transform, DCT) can be used to compute the overall frequency of this audio frame so that we can compare the signal energy with the masking thresholds for each subband on the basis of the psychoacoustics. The scale and quantize module uses the signal-to-mask ratios to decide how to apportion the total number of code bits available for the quantization of the subband signals so as to minimize the audibility of the quantization noise.

   The binary bits derived from the scaled and quantized samples can be compressed further using entropy analysis. Finally, the encoder takes the representation of the quantized subband samples and formats this data and side information into a coded bitstream. Ancillary data, not necessarily related to the audio stream, can also be inserted within the coded bitstream. The decoder basically reverses the operations of an encoder and restores the quantized and compressed bitstream back to a frame of audio samples.

   In an actual implementation of audio coding, such as MPEG audio, the codec divides each frame of the audio signal into 32 equal-width (uniform) frequency subbands rather than non-uniform critical bands, owing to the use of a relatively simple polyphase filter structure. The

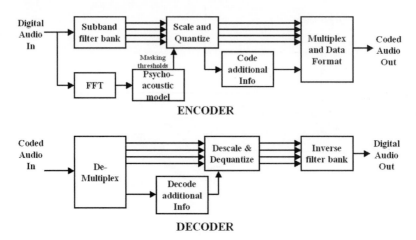

Figure 3.12 The encoder and decoder of a generic frame-based audio codec.

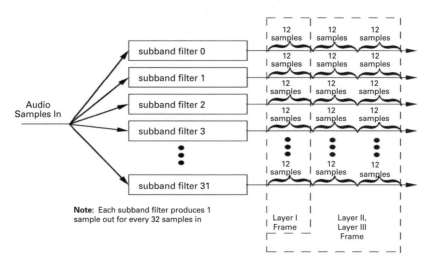

Figure 3.13 The MPEG audio codec divides each frame of audio signal into 32 equal-width (uniform) frequency subbands [8] (© IEEE 1995).

equal-width subband filtering used in MPEG audio coding is illustrated in Figure 3.13 [8]. In the subband filtering, the input signal $x(n)$, with sampling rate of F_s, is decomposed into several frequency subbands, using the uniform DFT filter bank approach (see Figure 3.14) [9] [10]. More specifically, $p_i(n)$, a version of $x(n)$ modulated by $W_N^{ni} = e^{-j\,2\pi ni/N}$, is equivalent to a shift in the ith subband frequency portion of $x(n)$ back to the origin, ready to be filtered by a prototype low-pass filter $H_N(z)$. According to multirate signal processing theory the modulated signals, after being filtered by the prototype low-pass filter $H_N(z)$, each only occupies an amount of bandwidth F_s/N. The band-limited filtered signal can thus be downsampled (decimated) to create $x_i(n)$ for the ith band without missing any information in the subband (presumably the decimation operation would create aliasing errors, while the low-pass filter $H_N(z)$ serves the purpose of anti-aliasing).

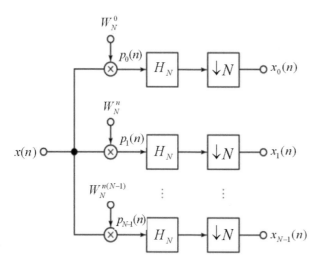

Figure 3.14 In subband filtering, the input signal $x(n)$ with sampling rate F_s is decomposed into several frequency subbands, using the uniform DFT filter bank approach.

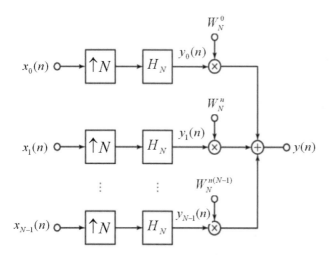

Figure 3.15 The N-band decimated signals $\{x_i(n)\}$ can be reconstructed via interpolation back to the original-rate version $\{y_i(n)\}$, with sampling rate F_s, using the uniform DFT filter bank approach.

Similarly, these N-band decimated signals, $\{x_i(n)\}$ can be reconstructed (interpolated) back to the original-rate version $\{y_i(n)\}$, with sampling rate F_s, using the uniform DFT filter bank approach (see Figure 3.15) [9] [10]. Note that the filter bank and its inverse do not provide lossless transformations. Even without quantization the inverse transformation cannot recover the original signal perfectly. However, by appropriate design of the prototype low-pass filter $H_N(z)$, the error introduced by the filter bank can be made small and inaudible. Moreover, adjacent filter bands have a major frequency overlap; therefore the signal at a single frequency can affect two adjacent filter bank outputs. To implement the decimator and interpolator effectively (see Figure 3.16), a polyphase filter structure can be adopted. More specifically,

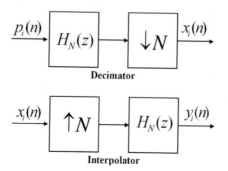

Figure 3.16 A typical representation of a decimator and an interpolator.

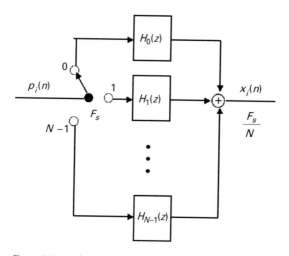

Figure 3.17 A decimator can be restructured by using N FIR subfilters, each of order $p-1$ (i.e., with filter length p).

the decimator consists of a prototype FIR low-pass filter $H_N(z)$, of order m and with impulse response $\{h(n)\}$, followed by a downsampler. The input $p_i(n)$ and the output $x_i(n)$ of a decimator can be mathematically formulated as in Eq. (3.4). It is assumed that the filter order $m+1$ is specifically selected so that $m+1=pN$, i.e., equal to a multiple of the conversion factor N (zero padding can be used). The decimator in Eq. (3.4) below can be restructured by using N FIR subfilters $\{H_k(z), k = 0, 1, 2, \ldots, N-1\}$, and each $(p-1)$th-order (i.e., filter-length p) subfilter in Eq. (3.5) is responsible for producing one decimated output in turn (see Figure 3.17):

$$x_i(n) = \sum_{r=0}^{m} h(r)p_i(Nn - r), \tag{3.4}$$

$$H_k(z): \quad h_k(n) \equiv h(Nn + k), \quad 0 \le k < N, \quad 0 \le n < p. \tag{3.5}$$

Similarly, to implement each interpolator, Figure 3.16, a polyphase filter structure can be adopted. More specifically, the interpolator consists of an up-sampler followed by a

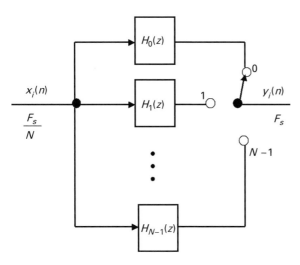

Figure 3.18 An interpolator can also be restructured by using N FIR subfilters, each of order $p-1$ (i.e., with filter length p).

prototype FIR low-pass filter $H_N(z)$ of order m, with the same impulse response $\{h(n)\}$. The input is the subband-processed $x_i (n)$, and the output $y_i(n)$ of an interpolator can be mathematically formulated as in Eq. (3.6).

$$y_i(n) = \sum_{r=0}^{m} h(r)\delta_N(n - r)x_i\left(\frac{n - r}{N}\right);$$ (3.6)

note that $x_i(n - r/N)$ is not defined except when $(n - r)/N$ is an integer. More specifically, $S_N(n-r)=1$ when $(n-r)/N$ is an integer; otherwise, it is defined to be 0. Again, it is assumed that the filter order is specifically selected so that $m + 1 = pN$, a multiple of the up-sampling factor N. The interpolator in Eq. (3.6) can be restructured by using N FIR subfilters $\{H_k(z), k = 0, 1, 2, \ldots, N - 1\}$, and each of $(p - 1)th$-order subfilters is responsible for producing one decimated output in turn (see Figure 3.18).

3.3　MPEG-1 audio layers

The Moving Picture Experts Group (MPEG-1) audio compression algorithm is the first international standard [11] for the digital compression of high-fidelity audio, jointly adopted by the International Organization for Standardization and the International Electrotechnical Commission (ISO/IEC) at the end of 1992. The MPEG-1 audio compression algorithm was originally proposed as one of three parts (audio, video, and system) in the compression standard, at a total bitrate of about 1.5 megabits per second (Mbps) [12].

The MPEG-1 audio compression accepts captured audio with sampling rates of 32, 44.1, or 48 kHz. The compressed bitstream can support one or two audio channels and have one of several predefined fixed bitrates ranging from 32 to 224 kbps per channel, equivalent to a compression ratio of 24 to 2.7. It is generally believed that with a 6 : 1 compression ratio (i.e., 16-bit stereo sampled at 48 kHz thus compressed to 256 kbps) and optimal listening conditions, expert listeners cannot distinguish between coded and original audio clips.

Moreover, using tradeoffs between coding complexity and compression ratio, MPEG-1 audio offers three independent layers of compression.

(1) *Layer 1* has the lowest complexity and highest bitrate, ranging from 32–224 kbps per channel with a target bitrate 192 kbps per channel. Only the hearing sensitivity and frequency masking psychoacoustic properties are used. Each frame contains 384 samples that are processed by 32 equal-width subbands (see Figure 3.13), each subband containing 12 samples. The layer 1 scheme has been adopted in Philips' Digital Compact Cassette (DCC) [13] at 192 kbps per channel.

(2) *Layer 2* has an intermediate complexity and is targeted for bitrates around 128 kbps per channel. Both frequency and temporal masking techniques are adopted. Each frame contains 1152 samples that are processed by 32 equal-width subbands, each subband using 36 samples (see Figure 3.13). Possible applications for this layer include the coding of audio for digital audio broadcasting (DAB) [14], the storage of synchronized video-and-audio sequences on CD-ROM, the full-motion extension of CD-interactive (CD-I) for multimedia gaming applications, video CDs (VCDs), and digital versatile disks (DVDs) for many movies and soap operas.

(3) *Layer 3* has the highest complexity and offers the best audio quality, particularly for bitrates around 64 kbps per channel. This layer's compressed data is also named the MP3 format. This format was developed in the late 1980s by the Fraunhofer Institute in conjunction with the University of Erlangen. The patent rights were granted by Thompson. A license is required to sell products that encode or decode MP3 as well as to broadcast commercial MP3 content. In addition to frequency and temporal masking, variable length lossless entropy (Huffman) coding, which takes into account the probabilities of the coded information, is also used to further reduce the redundancy. As in layer 2, each frame contains 1152 samples to be processed by 32 equal-width subbands (see Figure 3.13), each subband using 36 samples. This layer is well suited for audio transmission over the integrated services digital network (ISDN) and is now the most popular audio compression for Internet streaming.

3.3.1   Polyphase implementation of MPEG-1 filter bank

The polyphase implementation of an MPEG-1 filter bank is common to all three layers of MPEG-1 audio compression. The flow diagram from the ISO MPEG-1 audio standard for the MPEG-1 encoder filter bank is shown in Figure 3.19 [8].

The filtered and decimated output sample $S_t(i)$ for subband i at time t can be mathematically expressed as

$$S_t(i) = \sum_{k=0}^{63} \sum_{j=0}^{7} M(i,k) \left[C(k+64j) x(k+64j) \right], \qquad (3.7)$$

where $C(n)$ denotes one of the 512 coefficients of the analysis window defined in the standard and is derived from the prototype low-pass filter response $H_N(z) = \{h(n)\}$ for the polyphase filter bank, $x(n)$ is an audio input sample read from a 512-sample buffer, and $\{M(i, k)\}$ are the analysis matrix coefficients:

$$M(i,k) = \cos \left[\frac{(2i+1)(k-16)\pi}{64} \right]. \qquad (3.8)$$

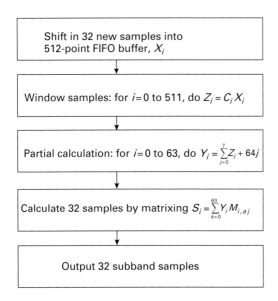

Shift in 32 new samples into
512-point FIFO buffer, X_i

Window samples: for $i = 0$ to 511, do $Z_i = C_i X_i$

Partial calculation: for $i = 0$ to 63, do $Y_i = \sum_{j=0}^{7} Z_i + 64j$

Calculate 32 samples by matrixing $S_j = \sum_{k=0}^{63} Y_i M_{i,aj}$

Output 32 subband samples

Figure 3.19 The flow diagram for the ISO MPEG-1 audio standard of the MPEG-1 encoder filter bank [8] (©IEEE 1995).

Note that for every 32 input samples the filter bank produces 32 output samples, i.e., each of the 32 subband filters downsamples its output by 32, to produce only one output sample in each subband for every 32 new audio samples.

3.3.2 MPEG-1 audio psychoacoustics

The audio psychoacoustics of MPEG-1 take advantage of the human auditory system's frequency-dependent characteristics, since the masking threshold (for frequency masking and/or temporal masking) at any given frequency is solely dependent on the signal energy within a limited-bandwidth neighborhood of that frequency, for the most efficient compression. More specifically, each band should be quantized with no more levels than is necessary to make the quantization noise inaudible. The encoder uses the frequency positions and the corresponding loudness to decide how best to represent the input audio signal with its limited number of code bits.

For this objective, a separate, independent, time-to-frequency mapping should be used instead of the filter bank because it needs finer frequency resolution for an accurate calculation of the masking thresholds. The MPEG-1 audio standard provides two example implementations of a psychoacoustic model based on a discrete Fourier transform (DFT) implemented by a fast Fourier transform (FFT) as shown in Figure 3.12. Model 1 is less complex than model 2 and makes more compromises to simplify the calculations. Either model works for any of the layers of compression. However, only model 2 includes specific modifications to accommodate layer 3. Model 1 uses a 512-sample FFT for layer 1 (for a 384-sample frame) and a 1024-sample FFT for layers 2 and 3 (for 1152-sample frame). However, model 2 uses a 1024-sample FFT for all layers (with the frame's 384 samples centered for layer 1). For layers 2 and 3, model 2 computes two 1024-sample psychoacoustic calculations for each frame. The first calculation centers the first half of the 1152 samples in the analysis window and the second calculation centers the second half. The model combines the results of the two calculations by using the higher of the two signal-to-mask ratios (SMRs) for each subband. The SMR is the ratio of the short-term signal power in each subband (or, for layer 3, group of bands) and the minimum masking threshold for that band. This in effect selects the lower of the two noise-masking thresholds for each subband. After the frequency mapping based on FFT is done, the following steps are required for a complete psychoacoustics analysis [4] [6].

(1) Group the spectral values according to critical bandwidths.
(2) Separate spectral values into tonal and non-tonal components on the basis of the local peaks of the audio power spectrum.
(3) On the basis of the separated tonal and non-tonal components, apply an empirically determined masking function to determine the masking threshold across the whole audio spectrum.
(4) Set the lower bound on the audibility of sound using the empirically determined absolute masking threshold.
(5) Find the masking threshold for each subband. Model 1 selects the minimum masking threshold within each subband, while model 2 selects the minimum of the masking thresholds covered by the subband only where the band is wide relative to the critical band in that frequency region. It uses the average of the masking thresholds covered by the subband when the band is narrow relative to the critical band.
(6) Calculate the signal-to-mask ratio (SMR) and pass this value to the bit (or noise) allocation section of the encoder.

3.3.3   Layer-3 audio bit allocations

The MP3 algorithm involves a much more sophisticated approach, which, however, is still based on the same filter bank as that used in layers 1 and 2. This algorithm compensates for some filter bank deficiencies by processing the subband filter outputs with a modified discrete cosine transform (MDCT) with window length 36. The MDCT further subdivides the subband outputs in frequency to provide better spectral resolution (now 576 bands) as shown in Figure 3.20 [15].

The polyphase filter used in MP3 is a cosine-modulated low-pass prototype filter with uniform-bandwidth parallel M-channel band-pass filters. This achieves *near perfect* reconstruction and has been called a pseudo quadrature mirror filter (P-QMF). The P-QMF is a linear phase finite impulse response (FIR) prototype filter with low-complexity

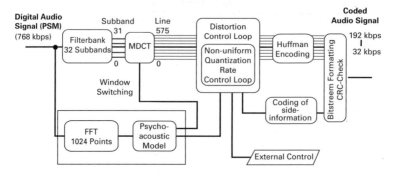

Figure 3.20 The MP3 algorithm compensates for some filter bank deficiencies by processing the subband filter outputs with a modified discrete cosine transform (MDCT) with window length 36 [15].

implementation (see Figures 3.17 and 3.18). The modified discrete cosine transform (MDCT) is defined as

$$X(k) = \sum_{n=0}^{2M-1} x(n) \cos\left[\frac{\pi}{M}\left(n + \frac{1}{2} + \frac{M}{2}\right)\left(k + \frac{1}{2}\right)\right] \quad k = 0, 1, \ldots, M-1, \quad (3.9)$$

where a window of two blocks ($2M = 36$) of data samples $\{x(n)\}$, is transformed into $M = 18$ frequency domain values $\{X(k)\}$. The MDCT is designed to be performed on consecutive blocks of a larger data set, where subsequent blocks are overlapped in such a way that the second half of one window coincides with the first half of the next window. This overlapping makes the MDCT especially attractive for signal compression applications, since it helps to avoid artifacts stemming from the block boundaries (see Figure 3.21(a)).

The inverse MDCT (IMDCT), owing to the different numbers of inputs and outputs, can still achieve perfect invertibility by *adding* the overlapped IMDCTs of subsequent overlapping windows, so that the original data can be reconstructed; this technique is known as time-domain aliasing cancellation (TDAC) and is shown in Figure 3.21(b):

$$y(n) = \frac{1}{N} \sum_{n=0}^{M-1} X(k) \cos\left[\frac{\pi}{M}\left(n + \frac{1}{2} + \frac{M}{2}\right)\left(k + \frac{1}{2}\right)\right] \quad n = 0, 1, \ldots 2M-1, \quad (3.10)$$

As shown in Figure 3.22, an MP3 encoder iteratively varies the quantizers using the selected scale factor bands, quantizes the MDCT values non-uniformly, counts the number of Huffman code bits required to code the audio data, and actually calculates the resulting distortion (noise). The scale factors are weights for ranges of frequency coefficients called scale factor bands (SFBs). If after quantization there are still scale factor bands with more than the allowed distortion, the encoder amplifies the MDCT values in those SFBs and effectively decreases the quantizer step size for those bands.

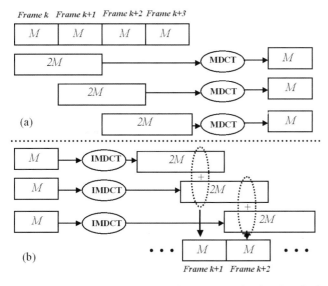

Figure 3.21 The MDCT is performed on consecutive blocks of a larger dataset, where subsequent blocks are overlapped in such a way that the second half of one window coincides with the first half of the next window [6] (© IEEE 1995).

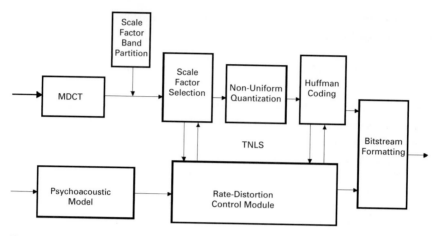

Figure 3.22 The two-nested-loop search (TNLS) procedure for MP3 rate control.

The quantized values are coded by Huffman coding, which will be discussed in more detail in Section 4.2.1. To adapt the coding process to the different local statistics of music signals, the optimum Huffman table is selected from a number of choices. To get an even better adaptation to the signal statistics, different Huffman code tables can be selected for different parts of the spectrum. More specifically, MP3 uses SFBs which cover several MDCT coefficients and have approximately critical bandwidths (unlike the system for layers 1 and 2, where there can be only one different scale factor for each subband). The SFBs in each frame are partitioned into a fixed number of band groups and each group shares the same 33 predesigned Huffman codebooks (HCBs), 31 for SFBs with large coefficients and two for SFBs with small coefficients: $+1$, 0, and -1.

In MP3, each scale factor starts with a minimum weight for an SFB. The number of SFBs depends on sampling rate and block size (e.g., there are 21 SFBs for a long block of 48 kHz input). For the starting set of scale factors, the encoder finds a satisfactory quantization step size in an inner quantization loop. In an outer quantization loop the encoder amplifies the scale factors until the distortion (quantization noise) in each SFB is less than the allowed distortion (minimum masking) threshold for that SFB, the encoder repeating the inner quantization loop for each adjusted set of scale factors. In special cases the encoder exits the outer quantization loop even if distortion exceeds the allowed distortion threshold for an SFB (e.g., if all scale factors have been amplified or if a scale factor has reached a maximum amplification). The MP3 encoder transmits the scale factors as side information using ad hoc differential coding and, potentially, entropy coding.

Before the quantization loops, the MP3 encoder can switch between long blocks of 576 frequency coefficients and short blocks of 192 frequency coefficients (sometimes called long windows or short windows). Instead of a long block, the encoder can use three short blocks for better time resolution. The number of scale factor bands is different for short blocks and long blocks. The MP3 encoder can use any of several different coding channel modes, including a single channel, two independent channels (left and right channels), or two jointly coded channels (sum and difference channels). If the encoder uses jointly coded channels, the encoder computes and transmits a set of scale factors for each of the sum and difference channels using the same techniques that are used for left and right channels. Or, if the encoder uses jointly coded channels then the encoder can instead use intensity stereo

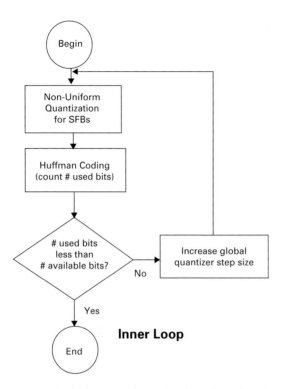

Figure 3.23 The inner (rate) iteration loop for MP3 rate control.

coding. Intensity stereo coding changes how the scale factors are determined for higher frequency SFBs and also changes how the sum and difference channels are reconstructed, but the encoder still computes and transmits two sets of scale factors for the two channels.

The MP3 quantizer raises its MDCT inputs to the power 3/4 before quantization to provide a more consistent signal-to-noise ratio over the range of quantizer values. This results in non-uniform quantization. The requantizer in an MP3 decoder relinearizes the values by raising its output to the power 4/3. The MP3 uses variable-length Huffman codes to encode the quantized samples in order to get better data compression. After quantization, the encoder arranges the 576 (32 subbands × 18 MDCT coefficients per subband) quantized MDCT coefficients in a predetermined order.

The optimum gain and scale factors for a given block, bitrate and output from the perceptual model are found by a two-nested-loop search (TNLS), as shown in Figure 3.22. Thus, through an iterative search of the best set of SFBs for both the inner rate loop and the outer noise control and distortion loop via an analysis-by-synthesis procedure, the MP3 can be encoded at an appropriate bitrate, with satisfactory distortion control.

The inner iteration loop is also called a rate loop (see Figure 3.23). In this loop, the Huffman code table assigns codewords to quantized values, commonly shorter codewords to (the more frequent) smaller quantized values. If the number of bits resulting from the coding operation exceeds the number of bits available to code a given block of data, this can be corrected by adjusting the global gain to result in a larger quantization step size, leading to smaller quantized values (shorter codewords). This operation is repeated with different quantization step sizes until the resulting bit demand for Huffman coding is sufficiently small.

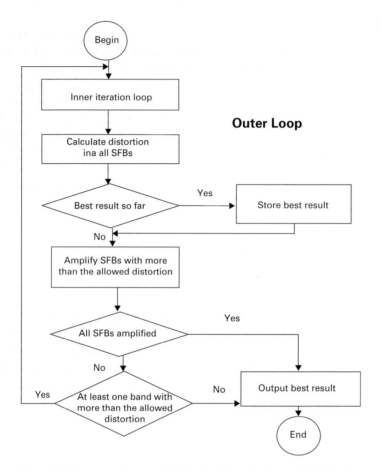

Figure 3.24 The outer (noise and distortion) iteration loop for MP3 rate control.

The outer iteration loop (see Figure 3.24), also called a noise control and distortion loop, is applied when the scale factors are allocated to each frequency band to shape the quantization noise according to the masking thresholds. If the quantization noise in a given band is found to exceed the masking threshold (the allowed noise) as supplied by the perceptual model, the scale factor for this band is adjusted to reduce the quantization noise. Since achieving a smaller quantization noise requires a larger number of quantization steps and thus a higher bitrate, the previous rate adjustment (inner) loop has to be repeated every time new scale factors are used. This outer loop is executed until the actual noise (computed from the difference between the original spectral values and the quantized spectral values) is below the masking threshold for every frequency band. The inner and outer loops are iterated alternately until any of following three conditions is met, then the iterations stop:

(1) None of the scale factor bands has more than the allowed distortion.
(2) The next iteration would cause the amplification for any of the bands to exceed the maximum allowed value.
(3) The next iteration would require all the scale factor bands to be amplified.

Real-time encoders also can include a time-limit exit condition for this process.

3.3.4 Joint stereo redundancy coding

The MPEG-1 audio compression algorithm supports two types of stereo redundancy coding: intensity stereo coding and middle/side (MS) stereo coding. All three layers support intensity stereo coding; MP3 also supports MS stereo coding. Both forms of redundancy coding exploit another perceptual property of the human auditory system. Psychoacoustic results show that above about 2 kHz and within each critical band, the human auditory system bases its perception of stereo imaging more on the temporal envelope of the audio signal than on its temporal fine structure.

In intensity stereo mode, the encoder codes some upper-frequency subband outputs with a single summed signal instead of sending independent left and right channel codes for each of the 32 subband outputs. The intensity stereo decoder reconstructs the left and right channels using only a single summed signal and independent left and right channel scale factors. With intensity stereo coding, the spectral shape of the left and right channels is the same within each intensity-coded subband but the magnitude is different.

The MS stereo mode encodes the left and right channel signals in certain frequency ranges as a middle channel (sum of left and right) and a side channel (difference of left and right). In this mode, the encoder uses specially tuned threshold values to compress the side channel signal further. Using joint stereo, the encoder analyzes each piece of the original audio file and chooses either *real stereo* (left channel and right channel) or *MS stereo* (audio is encoded using a middle channel that contains the similarities between the left and right channels and a side channel that contains the differences between the left and right channels). When there is too much difference between the left and right channels, real stereo will be used on that frame. Joint stereo is commonly used in MP3 with a bitrate of 128 kbps or lower to improve the quality.

3.4 Dolby AC3 audio codec

In order to provide a superior audio coding for multichannel surround sound, so that it can be used for High Definition Television (HDTV) while requiring a similar data rate as that used in the stereo sound system, AC3 was proposed and developed by Dolby Inc. for DVD, HDTV, home theater systems (HTSs), etc. [16] [17]. The AC3 audio codec, so-called Dolby Digital Surround audio, follows the recommendation made by the Society for Motion Picture and Television Engineers (SMPTE, http://www.smpte.org/home) that 5.1 channels (left, center, right, left surround, right surround, subwoofer, as shown in Figure 3.25) with a target bitrate of 320 kbps should be enough to provide the sound quality achieved by the 70 mm surround-sound format used in the cinema since 1979. The AC3 system also uses human psychoacoustic features to "mask" the inaudible audio signals. It delivers six totally separate channels of sound: the five main channels, except for the subwoofer channel, are full range (3–20 000 Hz) sound channels, while the sixth subwoofer channel, also called the low-frequency effects (LFE) channel, contains additional bass information to maximize the impact of scenes of explosions, crashes, etc. Because this channel has only a limited frequency response (3–120 Hz), it is sometimes referred to as the "0.1" channel, which results in the "5.1 channel" description of AC3 [18].

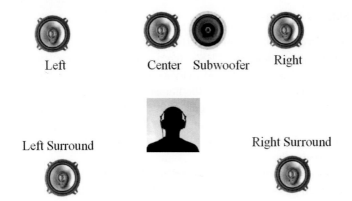

Figure 3.25 The 5.1 channels for AC-3 are: left, center, right, left surround, right surround, subwoofer (LFE) channels.

3.4.1   Bit allocation in AC3

As discussed in the section of MPEG-1 audio, owing to the frequency masking properties of human hearing the best representation of audio to use is a frequency-domain representation obtained by the use of a subband or transform filter bank.

Unlike MPEG-1 audio coding, which uses a forward adaptive method where the encoder calculates the bit allocation and explicitly codes the allocation into the coded bitstream, the AC3 uses a backward adaptive method. More specifically, the MPEG-1 forward adaptive encoder precisely calculates an optimum bit allocation within the limits of the psychoacoustic model employed. Even though the forward adaptive coding scheme can perform the adaptation dynamically without changing the installed decoder, since any modification of the psychoacoustic model resides only in the encoder, there is a cost in performance degradation owing to the need to deliver the explicit bit allocation to the decoder. For instance, the MPEG-1 layer-2 audio coder requires a data rate of approximately 4 kbps per channel to transmit the bit allocation information. While attractive in theory, forward adaptive bit allocation clearly does impose significant practical limitations on performance at very low bitrates.

However, a backward adaptive bit-allocation scheme generates the bit-allocation information from the coded audio data itself, without explicitly transmitted information from the encoder. The advantage of this approach is that none of the available data rate is used to deliver the allocation information to the decoder, and thus all the bits are available to be allocated to coding audio. The disadvantages of backward adaptive allocation come from the fact that the bit allocation must be computed in the decoder from information contained in the bitstream. The bit allocation is computed from information with limited accuracy and may contain small errors. Since the bit allocation is intended to be able to be performed in a low-cost decoder, the computation cannot be overly complex or else decoder costs would not be low. Since the bit-allocation algorithm available to the encoder is fixed, once decoders are deployed to the field the psychoacoustic model cannot be updated.

The AC3 coder makes use of hybrid backward and forward adaptive bit allocation, in which most of the disadvantages of backward adaptation are removed. This method involves a core backward adaptive bit allocator, which runs in both the encoder and the decoder. This psychoacoustic-model-based bit allocator is relatively simple but is quite accurate. The input to the core routine is the spectral envelope, which is part of the encoded audio data delivered to the decoder. The AC3 encoding (see Figure 3.26) can be summarized as the following six steps.

(1) Transform the windowed overlapping blocks of 512 samples into a sequence of frequency-coefficient blocks. Each individual frequency coefficient is represented as an exponent and a mantissa.
(2) The set of exponents is encoded into a coarse representation of the signal spectrum, the spectral envelope.
(3) This spectral envelope is used to determine how many bits should be used to encode each individual mantissa.
(4) The mantissa is then quantized according to the bit allocation.
(5) The spectral envelope and the coarsely quantized mantissas for six blocks ($256 \times 6 = 1536$ samples) are formatted into an AC3 frame.
(6) The AC3 bitstream (from 32 to 640 kbps) is a sequence of AC3 frames.

The AC3 decoder, as shown in Figure 3.27, basically reverses the encoding steps and converts the AC3 bitstream into the original time-domain waveform.

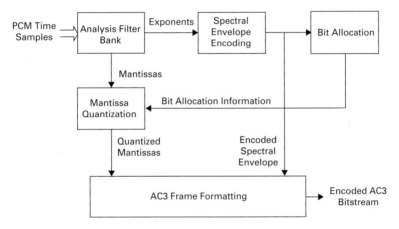

Figure 3.26 An AC3 encoding procedure [19] (© IEEE 2006).

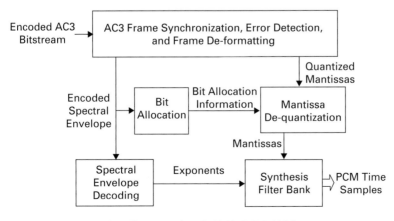

Figure 3.27 An AC3 decoding procedure [19] (© IEEE 2006).

3.4.2   Filter bank

The AC3 takes the overlapping blocks of 512 windowed samples (based on a proprietary 512-point Fielder window [16]) and transforms them into 256 frequency-domain points. Each transform block is formed from audio representing 10.66 ms (at a 48 kHz sample rate), although the transforms are performed every 5.33 ms due to the 50% overlap. During transient conditions where finer time resolution is needed, the block size is halved so that transforms occur every 2.67 ms. Bit allocation can occur in groups of transform coefficients sharing the same allocation or at the individual transform-coefficient level, neighboring coefficients receiving different allocations.

3.4.3   Spectral envelope

Each individual transform coefficient is coded into an exponent and a mantissa. The exponent allows for a wide dynamic range while the mantissa is coded with a limited precision, which results in quantizing noise. The set of coded exponents forms a representation of the overall signal spectrum and is referred to as the spectral envelope. The AC3 coder encodes the spectral envelope differentially in frequency. Since a change of at most ± 2 (a change of 1 represents a 6 dB level change) is required, each exponent can be coded as one of five changes from the previous (lower in frequency) exponent, $+2$, $+1$, 0, -1, -2. The first exponent (the DC term) is sent as an absolute value, and the rest of the exponents are sent as differentials. Groups of three differentials are coded into a 7-bit word. Each exponent thus requires approximately 2.33 bits to code. This method of transmitting the exponents is referred to as D15 and provides a very accurate spectral envelope.

To reduce the average number of bits used in encoding the spectral envelopes in D15, especially when the spectrum is relatively stable, the exponent differentials may be sent only occasionally. In a typical case, the spectral envelope is sent once every six audio blocks (32 ms), in which case the data rate required is < 0.39 bits per exponent. More specifically, since each individual frequency point has an exponent and since there is one frequency sample for each time sample, the D15 high-resolution spectral envelope requires < 0.39 bits per audio sample. When the spectrum of the signal is not stable, it is beneficial to send a spectral estimate more often. In order to keep the data-overhead for the spectral estimate from becoming excessive, the spectral estimate may be coded with a lower frequency resolution. For example the medium-frequency resolution method is D25, where a change is transmitted for every other frequency coefficient. This method requires half the data rate of the D15 method (or 1.16 bits per exponent) and has a frequency resolution that is smaller by a factor 2. The D25 method is typically used when the spectrum is relatively stable over two to three audio blocks and then undergoes a significant change [15].

Another compromise method is D45, where a change is transmitted for every four frequency coefficients. This method requires one quarter of the data rate of the D15 method (or 0.58 bits per exponent) and is typically used during transients for single audio blocks (5.3 ms). The transmitted spectral envelope thus has very fine frequency resolution for steady-state (or slowly changing) signals and fine time resolution for transient signals.

The AC3 encoder is responsible for selecting the exponent coding method to use for any given audio block. Each coded audio block contains a 2-bit field called an *exponent strategy.* The four possible strategies are D15, D25, D45, and REUSE. For most signal conditions a D15 coded exponent set is sent during the first audio block in a frame, and the following five audio blocks reuse the same exponent set. During transient conditions, exponents are sent more often. The encoder routine responsible for choosing the optimum exponent strategy may be improved or updated at any time. Since the exponent strategy is explicitly coded into the bitstream, all decoders will track any change in the encoder.

3.4.4 Coupling

Even though the coding techniques employed by AC3 are very powerful, when the coder is operated at very low bitrates there are signal conditions under which the coder would run out of bits. When this occurs, a coupling technique is invoked [16]. Coupling takes advantage of the fact that the ear is not able to independently detect the direction of two high-frequency signals which are very closely spaced in frequency. When the AC3 coder becomes starved of bits, channels may be selectively coupled at high frequencies. The frequency at which coupling begins is called the *coupling frequency.* Above the coupling frequency the channels to be coupled are combined into a *coupling* (or common) channel. Care must be taken with the phase of the signals to be combined in order to avoid signal cancellations. The encoder measures the original signal power of the individual input channels in narrow frequency bands, as well as the power in the coupled channel in the same frequency bands. The encoder generates *coupling coordinates* for each individual channel, which indicate the ratio of the original signal power within a band to the coupling channel power in the band. The coupling channel is encoded in the same manner as the individual channels; there is a spectral envelope of coded exponents and a set of quantized mantissas. The channels which are included in the coupling are sent discretely up to the coupling frequency. Above the coupling frequency, only the coupling coordinates are sent for the coupled channels. The decoder multiplies the individual channel-coupling coordinates by the channel coupling coefficients to regenerate the high-frequency coefficients of the coupled channels. The coupling process is audibly successful because the reproduced sound field is a close power match to the original.

The AC3 encoder is responsible for determining the coupling strategy. The encoder controls which of the audio channels are to be included in coupling and which will remain completely independent. The encoder also controls the frequency at which coupling begins, the coupling band structure (the bandwidths of the coupled bands), and the times at which new coupling coordinates are sent. The coupling strategy routine may be altered or improved at any time and, since the coupling strategy information is explicit in the encoded bitstream, all decoders will follow the changes. For example, early AC3 encoders often coupled channels with a coupling frequency as low as 3.5 kHz. As AC3 encoding techniques have improved, the typical coupling frequency has increased to 10 kHz [15].

3.5 MPEG-2 Advanced Audio Coding (AAC)

To advance audio coding technologies beyond MP3 and AC3, an effort was initiated to create a new audio coder which can produce indistinguishable quality at 64 kbps per mono channel. This effort led to the development of MPEG-2 Advanced Audio Coding (AAC) [20] [21] [22]. This combined research and development efforts from the world's leading audio coding laboratories, such as Fraunhofer Institute, Dolby, Sony, and AT&T. Technically, the AAC format can support up to 48 full-frequency sound channels and 16 low-frequency enhancement channels. It also supports sampling rates up to 96 kHz, twice the maximum afforded by MP3 and AC3 and is now the format used for songs downloaded from the popular iTunes music website (www.itune.com).

The AAC format, which was standardized in 1997, was built on a similar structure to MP3 and thus retains most of its design features. But unlike the previous MPEG layers, AAC uses a modular approach (see Figure 3.28), which allows new ideas to be developed and plugged into the basic structure and provides significantly more coding power [7]. This modular approach is summarized as follows.

(1) *Filter bank*   In contrast with MP3, which uses subbands and MDCT, AAC uses plain MDCT together with an increased window length ($8 \times 128 = 1024$ spectral lines per transform with 50% overlap), adapted window shape function, and transform block switching. The plain MDCT in AAC outperforms the filter banks of previous coding methods and provides better frequency selectivity for the filter bank. The long windows are nearly twice as long as those in MP3, providing better frequency resolution. In addition, AAC uses some short windows, which are smaller than those

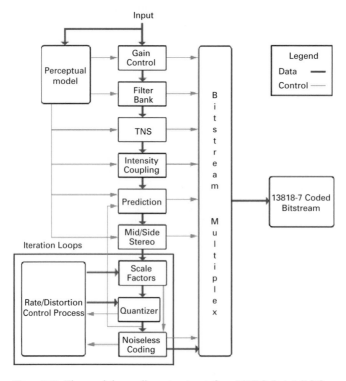

Figure 3.28 The modular coding structure of an MPEG-2 AAC [7].

in MP3, to provide better handling of transients and less pre-echo. Pre-echoes occur when a signal with a sharp attack begins near the end of a transform block immediately following a region of low energy (e.g., in audio recordings of percussive instruments) as shown in Figure 3.29(a) [6]. For a block-based algorithm like MP3 or AAC, when quantization and encoding are performed in order to satisfy the masking thresholds associated with the block-average spectral estimate, the

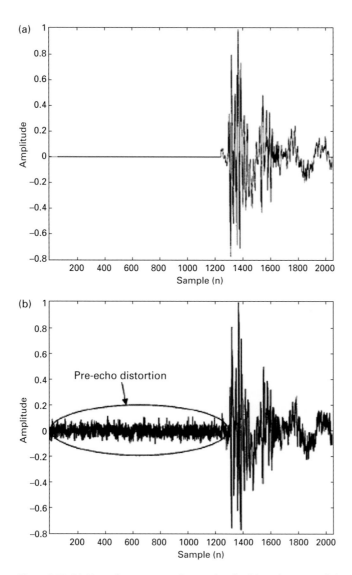

Figure 3.29 (a) Pre-echoes occur when a signal with a sharp attack begins near the end of a transform block immediately following a region of low energy. (b) When quantization and encoding are performed, the inverse transform will spread quantization distortion evenly in time throughout the reconstructed block [6] (© IEEE 2000).

inverse transform will spread quantization distortion evenly in time throughout the reconstructed block. This results in unmasked distortion throughout the low-energy region preceding in time the signal attack at the decoder (see Figure 3.29(b)) [6].

(2) *Temporal noise shaping (TNS)*   This is a frequency-domain technique that operates on the spectral coefficients generated by the analysis filter bank; it is applied only during input attacks susceptible to pre-echoes. The idea is to apply linear prediction (LP) across frequency (rather than time), since for an impulsive (transient) time signal it exhibits tonally in the frequency domain, i.e., it consists mainly of a few sinusoidal components in the frequency domain. Tonal sounds are easily predicted using an LP analysis. More specifically, parameters of a spectral LP "synthesis" filter are estimated via the application of standard minimum MSE estimation methods (e.g., Levinson–Durbin [26]) to the spectral coefficients. The resulting prediction residual is quantized and encoded using standard perceptual coding according to the original masking threshold. Prediction coefficients are transmitted to the receiver as side information to allow recovery of the original signal. The convolution operation associated with spectral-domain prediction is associated with multiplication in time. In a similar manner, the source-system (residual and vocal tract) separation is realized by LP analysis in the time domain for traditional speech codecs, as discussed in Section 2.1. Therefore, TNS effectively separates the time-domain waveform into an envelope and a temporally flat "excitation." Then, because quantization noise is added to the flattened residual, the corresponding time-domain multiplicative envelope shapes the quantization noise in such a way that it follows the original signal envelope. Temporal noise shaping clearly shapes the quantization noise to follow the input signal's energy envelope. It mitigates the effect of pre-echoes since the error energy is now concentrated in the time interval associated with the largest masking threshold, as shown in Figure 3.30 [6].

(3) *Prediction*   As discussed above, a signal which is transient in the time domain is tonal in the frequency domain, i.e., it consists mainly of a few sines, which can be predicted using an LP analysis in the spectral domain. The dual of this idea is that a tonal signal in the time domain has transient peaks in the frequency domain. A time-domain prediction module can then be used to enhance the compressibility of stationary audio by guiding the quantizer to very effective coding when there is a noticeable signal pattern, such as high tonality.

(4) *MS stereo*   This enables one to toggle middle or side stereo on a subband basis instead of on an entire-frame basis as in MP3. It also gives the ability to toggle the intensity stereo on a subband basis instead of using it only for a contiguous group of subbands.

(5) *Quantization*   By allowing finer control of quantization resolution, the given bitrate can be used more efficiently.

(6) *Huffman coding*   The Huffman code uses variable-length codewords to further reduce the redundancy of the scale factors and the quantized spectrum data. One scale factor Huffman codebook and 11 spectrum Huffman codebooks are used in MPEG-2 AAC.

(7) *Bitstream format*   Entropy coding is used to keep the redundancy as low as possible. The optimization of these coding methods together with a flexible bitstream structure has made further improvements in coding efficiency possible.

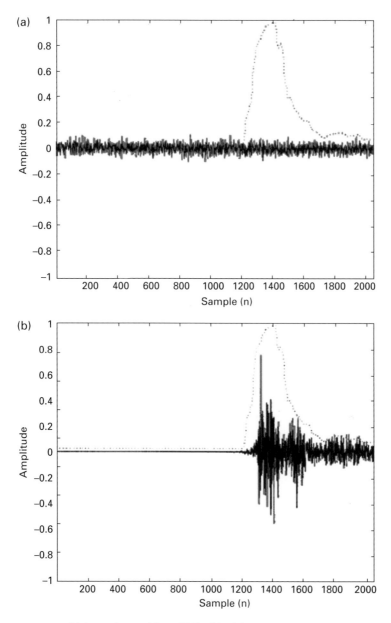

Figure 3.30 (a) Pre-echoes without TNS, (b) with TNS. It can be clearly seen that TNS mitigates the effect of pre-echoes: the error energy is now concentrated in the time interval associated with the largest masking threshold [6] (© IEEE 2000).

The MPEG-2 AAC system offers different tradeoffs between quality and complexity depending on the application. For this purpose three profiles have been defined: main profile, low-complexity (LC) profile, and scalable sample rate (SSR) profile [20] [21]. The main profile includes all available tools and delivers the best audio quality of the three profiles. The LC profile comes with a limited TNS tool and without prediction. This reduces audio quality for some special audio signals but saves a lot of computational power in both encoding and, especially, decoding. The SSR profile is a low-complexity profile with a different filter bank and a special gain control tool. In the scalable sampling rate profile of MPEG-2 AAC, the audio data is first split into four uniform subbands using a polyphase quadrature filter (PQF) [10]. For each of the four subbands an individual gain is transmitted as side information. The gain-controlled subband data is then transformed using an MDCT of length 256 (or 32 for transient conditions). The window used for the MDCT is either the Kaiser–Bessel derived (KBD) window [24] or the sine window, which has different spectral characteristics, suitable for different signals. For transient conditions, a shorter window is used for improved time resolution. The MDCT coefficients are predicted from the two preceding frames, using a separate least mean square (LMS) adaptive predictor [25] for every frequency band. This improves coding efficiency for stationary signals. Residuals after the prediction are non-uniformly quantized and coded using one of 12 different Huffman codes.

The AAC gives a performance superior to any known codec at bitrates greater than 64 kbps for stereo audio. More specifically, at 96 kbps for stereo audio it gives comparable quality to MPEG-1 Layer 2 at 192 kbps and to MP3 at 128 kbps. The Japanese Association of Radio Industries and Businesses (ARIB) selected MPEG-2 AAC as the only audio coding scheme for all Japan's digital broadcast systems.

3.6   MPEG-4 AAC (HE-AAC)

The MPEG-2 AAC system has been further enhanced and amended to become today's most efficient audio coding standard, the so-called High Efficiency AAC (HE-AAC or HE-AAC v1) and HE-AAC v2; the enhancement is intended mainly for delay-critical applications or for the scalable encoding of multimedia content. These two coding formats were standardized in the years 2003 and 2004 separately within MPEG-4 audio. The HE-AAC is the low-bitrate codec in the AAC family and is a combination of the MPEG-2 AAC LC (advanced audio coding low-complexity) audio coder and the spectral band replication (SBR) bandwidth expansion technique; HE-AAC is also known as AACPlus and can be used in multichannel operations. It is 30% more efficient than MPEG-2 AAC. Note that HE-AAC is not intended as a replacement for LC AAC but rather as its extension, and is proposed mainly for Internet, mobile, and broadcasting arenas. This encoder is targeted to medium-quality encoding at bitrates 24 kbps per channel and higher.

3.6.1   Spectral band replication (SBR)

When perceptual codecs are used at low bitrates, e.g., bitrates below 128 kbps, they can cause the perceived audio quality to degrade significantly. More specifically, the codecs either start to reduce the audio bandwidth or they introduce annoying coding artifacts, resulting from a shortage of bits, in their attempt to represent the complete audio bandwidth. In other words, low-bitrate audio coding can create audible artifacts that appear above the masking threshold, as shown in Figure 3.31. Both ways of modifying the perceived sound can be considered unacceptable above a certain level. For example, at 64 kbps stereo MP3 would either offer an audio bandwidth of only about 10 kHz or introduce a fair amount of coding artifacts. To overcome this difficulty, the spectral band replication (SBR) enhancement technique [26] [27] was introduced and is standardized in ISO/IEC 14496–3:2001/Amd.1:2003. The

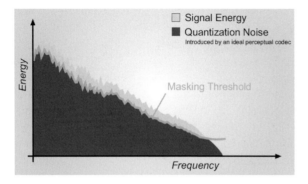

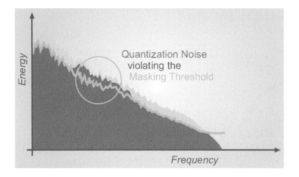

Figure 3.31 An example of bandwidth reduction due to low-bitrate audio coding [26] (© EBU 2006). (a) Inaudible quantization noise produced by an ideal perceptual coding process. (b) Low-bitrate audio coding going beyond its limits: audible artifacts appear above the masking threshold.

SBR offers the possibility of improving the performance of low-bitrate audio and speech codecs by either increasing the audio bandwidth at a given bitrate or improving coding efficiency at a given quality level.

Spectral band replication can be operated as follows: the codec itself transmits the lower frequencies of the spectrum while SBR synthesizes the associated higher frequency content using the lower frequencies and the transmitted side information. When applicable, it involves the reconstruction of a noise-like frequency spectrum by employing a noise generator with some statistical information (level, distribution, ranges), so the decoding result is not deterministic among multiple decoding processes of the same encoded data. The SBR idea is based on the principle that the human brain tends to consider high frequencies (high-band) to be either harmonic phenomena associated with lower frequencies (low-band) or noise and is thus less sensitive to the exact content of high frequencies in audio signals. Therefore a good approximation of the original input-signal high band can be achieved by a transposition from the low band, as shown in Figure 3.32. Besides pure transposition, the reconstruction of the high band (see Figure 3.33) is conducted by transmitting guiding information such as the spectral envelope of the original input signal or additional information to compensate for potentially missing high-frequency components [26]. This guiding information is referred to as SBR data. Also, efficient packaging of the SBR data is important to achieve a low data-rate overhead.

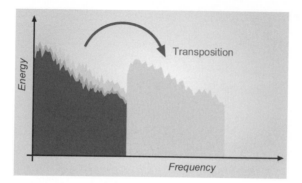

Figure 3.32 A good approximation of the original input signal high band can be achieved by a transposition from the low band [26] (© EBU 2006).

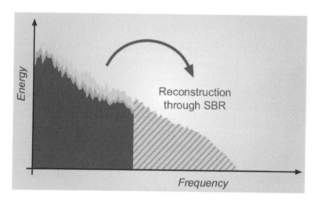

Figure 3.33 The reconstruction of the high band is achieved by transmitting SBR data, such as the spectral envelope of the original input signal or additional information, to compensate for potentially missing high-frequency components [26] (© EBU 2006).

On the encoder side, the original input signal is analyzed, the high-band spectral envelope and its characteristics in relation to the low band are encoded, and the resulting SBR data is multiplexed with the core coder bitstream. On the decoder side, the SBR data is first de-multiplexed before the core decoder is used on its own. Finally, the SBR decoder operates on its output signal, using the decoded SBR data to guide the SBR process. A full bandwidth output signal is obtained. Non-SBR decoders would still be able to decode the backward-compatible part of the core decoder, but the result is a band-limited output signal only.

The crossover frequency between the low band and the high band is chosen on the basis of different factors such as target bitrate and input sampling frequency. Generally the low band needs to cover the frequency range from DC up to around 4 to 12 kHz, depending on the target bitrate. The higher the crossover frequency between AAC and SBR, the higher the bitrate needed to fulfill the psychoacoustic masking threshold of the AAC encoder.

The limited frequency range that is covered by the AAC coder allows the use of a low sampling frequency of ≤ 24 kHz, which improves the coding efficiency significantly in comparison with the use of a higher sampling frequency of 48 or 44.1 kHz. Thus, AACPlus is designed as a dual-rate system, where AAC operates at half the sampling rate of SBR. Typical configurations are 16/32 kHz, 22.05/44.1 kHz, or 24/48 kHz sampling rates, while

Table 3.1 Typical examples of the crossover frequency between AAC and SBR [26]

Stereo bitrate (kbps)	AAC frequency range (Hz)	SBR frequency range (Hz)
20	0–4500	4500–15 400
32	0–6800	6800–16 900
48	0–8300	8300–16 900

8/16 or even 48/96 kHz are also possible. The resulting audio bandwidth can be configured flexibly and may also be dependent on the application or audio content type.

Table 3.1 shows typical examples of the crossover frequency between AAC and SBR as well as the audio bandwidth at a number of bitrates for stereo audio, using sample rates of 24/48 kHz in stereo, given a proper configuration of the HE-AAC encoder. The bitrate of the SBR data varies according to the encoder tuning but, in general, it is in the region of 1–3 kbps per audio channel. This is far lower than the bitrate that would be required to code the high band with any conventional waveform-coding algorithm [26]. Spectral band replication has not only been combined with AAC to create HE-AAC, it is also used with MP3 to create MP3PRO and with MPEG-2 for Musicam. In 2001, satellite-based XM Radio started a service using HE-AAC. The Digital Radio Mondiale (DRM) Consortium develops new digital services for the current analog services in long, medium, and shortwave bands. This consortium has selected HE-AAC as its audio coding scheme. Furthermore, HE-AAC is a mandatory format for second Session DVD-Audio and is part of the DVD Audio Recordable (DVD-AR) specification.

3.6.2   MPEG-4 HE-AAC v2

The MPEG-4 HE-AAC v2 standard [26] [27] incorporates the HE-AAC (AAC+) with a parametric stereo (PS) technique for extremely-low-bitrate audio, such as 32 kbps for stereo input. The PS technique transmits one combined mono channel plus 2–3 kbps of side information and achieves an efficiency nearly 50% better than AAC+ when used for Internet, mobile, broadcasting, and other domains with limited resources. The HE-AAC v2 codec is part of the 3GPP standard for the delivery of audio content to 3G devices.

Whereas SBR exploits the possibilities of a parameterized representation of the high band, the basic idea behind PS is to parameterize the stereo image of an audio signal as a "panorama," "ambience," or "time-phase difference" of the stereo channels, in order to enhance the coding efficiency of the codec. In the encoder, only a monaural downmix of the original stereo signal is coded after extraction of the PS data. Just as for SBR data, these parameters are then embedded as PS side information in the ancillary part of the bitstream. In the decoder, the monaural signal is decoded first. After that, the stereo signal is reconstructed, on the basis of the stereo parameters embedded by the encoder. Figure 3.34 shows the basic principle of the PS coding process [26]. Three types of parameter can be employed in a PS system to describe the stereo image, as follows.

(1) *Interchannel intensity difference (IID)*, describing the intensity difference between the channels.
(2) *Interchannel cross-correlation (ICC)*, describing the cross-correlation or coherence between the channels. The coherence is measured as the maximum of the cross-correlation as a function of time or phase.

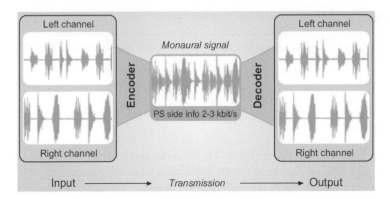

Figure 3.34 The basic principle of the parametric stereo coding process in HE-AAC v2 [26] (© EBU 2006).

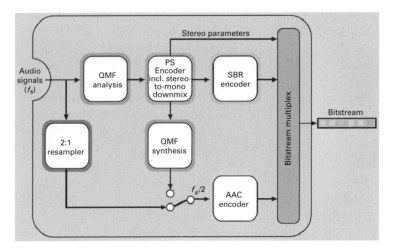

Figure 3.35 An MPEG-4 HE-AAC v2 encoder [26], showing the building blocks AAC, SBR, and PS (© EBU 2006).

(3) *Interchannel phase difference (IPD)*, describing the phase difference between the channels. This can be augmented by an additional *overall phase difference (OPD)* parameter, describing how the phase difference is distributed between the channels. The *Interchannel time difference (ITD)* can be considered as an alternative to the IPD.

As shown in Figure 3.35, AAC, SBR, and PS are basic building blocks of the MPEG-4 HE-AAC v2 profile. The AAC codec is used to encode the low band, SBR encodes the high band, and PS encodes the stereo image in a parameterized form. In a typical AAC + encoder implementation, the audio input signal at an input sampling rate of f_s is fed into a 64-band quadrature mirror filter (QMF) bank and transformed into the QMF domain. If the PS tool is used (i.e., for stereo encoding at bitrates below ~36 kbps) then the PS encoder extracts PS information based on the QMF samples. Furthermore, a stereo-to-mono downmix is applied. With a 32-band QMF synthesis, the mono QMF

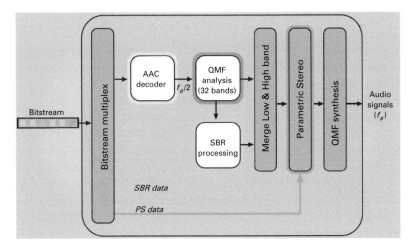

Figure 3.36 The building blocks of the MPEG-4 HE-AAC v2 decoder [26] (© EBU 2006).

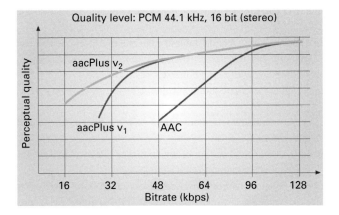

Figure 3.37 The anticipated audio quality vs. bitrate for the various codecs of the HE-AAC v2 family [26] (© EBU 2006).

representation is then transformed back into the time domain at half the sample rate of the audio signal, $f_s/2$. This signal is then fed into the AAC encoder. If the PS tool is not used then the audio signal is fed into a 2 : 1 resampler and again the downsampled audio signal is fed into the AAC encoder. The SBR encoder also works in the QMF domain; it extracts the spectral envelope and additional helper information to guide the replication process in the decoder. All encoded data is then multiplexed into a single bitstream for transmission or storage.

In the HE-AAC v2 decoder, the bitstream is first split into the AAC, SBR, and PS data portions. The AAC decoder outputs a time-domain low-band signal at a sample rate of $f_s/2$. The signal is then transformed into the QMF domain for further processing. The SBR processing results in a reconstructed high band in the QMF domain. The low and high bands are then merged into a full-band QMF representation. If the PS tool is used, it generates a stereo representation in the QMF domain. Finally, the signal is synthesized by a 64-band

QMF synthesis filter bank. The result is a time-domain output signal at the full f_s sampling rate (see Figure 3.36).

The audio quality of HE-AAC and of HE-AAC v2 has been evaluated in multiple double-blind listening tests conducted by independent entities such as the European Broadcasting Union (EBU), the Moving Pictures Expert Group (MPEG), the Third Generation Partnership Project (3GPP), and the Institut fur Rundfunktechnik (IRT).

Combining AAC with SBR and PS to produce the HE-AAC v2 results in a very efficient audio codec, providing high audio quality over a wide bitrate range, with only moderate gradual reduction of the perceived audio quality towards very low bitrates. Figure 3.37 allows a comparison of the anticipated audio quality versus bitrate for the various codecs of the HE-AAC v2 family. The diagram shows only a smooth degradation in audio quality of the HE-AAC v2 codecs towards low bitrates over a wide range down to 32 kbps for stereo audio. Even at bitrates as low as 24 kbps, HE-AAC v2 still produces a quality far higher than that of any other audio codec available. For multichannel 5.1 signals, HE-AAC provides a coding efficiency that is a factor 2 higher than Dolby AC3.

References

[1] "The inventor of the CD – Philips," http://www.research.philips.com/newscenter/dossier/optrec/index.html.

[2] "Wave PCM soundfile format," http://ccrma.stanford.edu/courses/422/projects/WaveFormat/.

[3] M. Bosi and R. E. Goldberg, *Introduction to Digital Audio Coding and Standards*, Kluwer Academic Publishers, 2002.

[4] P. Noll, "Digital audio coding for visual communications," *Proc. IEEE*, 83(6): 925–943, June 1995.

[5] "The human ear," NDT Resource Center, http://www.ndt-ed.org/EducationResources/HighSchool/Sound/humanear.htm.

[6] T. Painter and A. Spanias, "Perceptual coding of digital audio," *Proc. IEEE*, 88(4): 451–515, April 2000.

[7] Steve Church, "On beer and audio coding; why something called AAC is cooler than a fine Pilsner, and how it got to be that way," *Tech. Talk of Telos Systems*, http://www.telos-systems.com/techtalk/aacpaper_2/AAC_3.pdf.

[8] D. Pan, "A tutorial on MPEG/audio compression," *IEEE Multimedia*, 2(2): 60–74, Summer 1995.

[9] R. E. Crochiere and L. R. Rabiner, Chapter 7 in *"Multirate Digital Signal Processing,"* Prentice-Hall, 1983.

[10] J. H. Rothweiler, "Polyphase quadrature filters – a new subband coding technique," in *Proc. IEEE Int. Conf. on ASSP*, Boston, Vol. II, pp.1280–1283, 1983.

[11] P. Nolls, "MPEG digital audio coding," *IEEE Signal Proc. Mag.* 14(5): 59–81, September 1997.

[12] "Information technology – coding of moving pictures and associated audio for digital storage media at up to about 1.5 mbits/s – Part 3: Audio", ISO/IEC International Standard IS 11172–3, http://www.iso.ch/iso/en/CatalogueDetailPage.CatalogueDetail?CSNUMBER=22412&ICS1=35.

[13] G. C. Wirtz, "Digital compact cassette: audio coding technique," in *Proc. 91st AES Convention*, Preprint 3216, New York, 1991.

[14] "Digital audio broadcasting (DAB): Eureka 147," http://www.worlddab.org/.

[15] J. V. C. Preto Paulo, J. L. Bento Coelho, and C. R. Martins, "On the use of the MP3 format and m sequences applied to acoustic measurement," *in Proc. Forum Acusticum*, Seville, September 2002.

[16] C. C. Todd, G. A. Davidson, M. F. Davis *et al.,* "AC-3: Flexible perceptual coding for audio transmission and storage," in *Proc. 96th Convention of the Audio Engineering Society (AES),* Preprint 3796, February 26 – March 1, 1994.

[17] S. Vernon, "Design and implementation of AC-3 coders," *IEEE Trans. Consumer Electron.,* 41(30): 754–759, August 1995.

[18] "Dolby AC-3 audio coding, hosted by Joe Fedele," http://www.fedele.com/website/hdtv/dolbyac3.htm.

[19] G. A. Davidson, M. A. Isnardi, L. D. Fielder, M. S. Goldman, and C. C. Todd, "ATSC video and audio coding," *Proc IEEE,* 94(1): 60–76, January 2006.

[20] "IS 13818–7 (MPEG-2 advanced audio coding, AAC)," ISO/IEC JTC1/SC29/WG11 No. 1650, April 1997, http://www.iec.ch/cgi-bin/procgi.pl/www/iecwww.p?wwwlang=E &wwwprog=cat-det.p&progdb=db1&wartnum=035681.

[21] M. Bosi *et al.,* "ISO/IEC MPEG-2 Advanced Audio Coding," *J. Audio Eng. Soc.,* 45(10): 789–814, October 1997.

[22] "MPEG-2 advanced audio coding (AAC)," http://www.mpeg.org/MPEG/aac.html.

[23] N. Jayant and P. Noll, *Digital Coding of Waveforms: Principles and Applications to Speech and Video,* Prentice-Hall, 1984.

[24] J. F. Kaiser and R. W. Schafer, "On the use of the Io-sinh window for spectrum analysis," *IEEE Trans. Acoust., Speech, Signal Processing,* 28(2): 105–107, 1980.

[25] Simon Haykin, *Adaptive Filter Theory,* fourth edition, Prentice-Hall, 2001.

[26] S. Meltzer and G. Moser, "MPEG-4 HE-AAC v2: audio coding for today's digital media world," *EBU Technical Review,* January 2006.

[27] "MPEG-4 AAC (Information technology – coding of audiovisual objects – Part 3: Audio)," ISO/IEC IS 14496–3, 2001.

4 Digital image coding

Image compression is the application of data compression techniques to two-dimensional digital images $I(x, y)$, to reduce the redundancy of the image data for storage or transmission in an efficient form. Image compression can be classified into two categories: lossless or lossy. Lossless compression, which achieves smaller compression ratios than lossy compression, mainly takes advantage of the image contents containing a non-uniform probability distribution for a variable-length representation of the image pixels. Such images include technical drawings, icons or comics, and high-value contents such as medical imagery or image scans made for archival purposes. However, lossy compression methods, especially when they achieve a very high compression ratio, can introduce compression artifacts. Nevertheless, lossy compressions are especially suitable for natural images, such as photos, in applications where a minor (sometimes imperceptible) loss of fidelity is acceptable when it is desirable to achieve a substantial reduction in bitrate. Most of the state-of-the-art image compression standards use a combination of lossy and lossless algorithms to achieve the best performance.

The Joint Photographic Experts Group (JPEG) [1], a discrete cosine transform(DCT)-based technique, is the most widely used standardized lossy image compression mechanism; it was designed for compressing either full-color or gray-scale images of natural, real-world, scenes. It works well for photographs, naturalistic artwork and similar material if the compression ratio is about 20 : 1, which is much better than the 4 : 1 compression ratio provided by a lossless compression method such as the Graphics Interchange Format (GIF) [2]. However, it does not work so well for lettering, simple cartoons, or line drawings. The JPEG technique is lossy but it achieves much greater compression than is possible with lossless methods. The amount of "loss" can be varied by adjusting compression parameters that determine image quality and speed of decoding.

Despite the marketing success of JPEG, a growing number of new applications such as high-resolution imagery, digital libraries, high-fidelity color imaging, multimedia and Internet applications, wireless, medical imaging, etc. require additional and enhanced functionalities which cannot be offered by JPEG owing to some of its inherent shortcomings and design points, which were beyond the scope of JPEG when it was first developed. To overcome the deficiencies encountered in JPEG as well as to provide better efficiency in hierarchical JPEG coding, JPEG2000 [3] was proposed and standardized. This is a wavelet-based image compression standard, which was created with the intention of superseding the original DCT-based JPEG standard; it can achieve higher compression ratios without generating the characteristic "blocky and blurry" artifacts of the original JPEG standard. Owing to the use of a multi-resolution wavelet transform, JPEG2000 can offer effective scalable data representation in a natural way and thus allows more sophisticated hierarchical transmission.

4.1 Basics of information theory for image compression

Suppose that we are given a set of symbols (or alphabets), $A = \{a_1, a_2, \ldots, a_N\}$, which are generated (emitted) by a source where each symbol a_n is associated with an event or an observation that has occurrence probability p_n separately; $p_n \in \{p_1, p_2, \ldots, p_N\}$. It is usually convenient to ignore any particular feature of the event and only to observe whether it happened. Each event symbol can also be represented as a multiple m-tuple (m-dimensional) symbol. The *information* measure $I(a_n)$ of the symbol a_n is defined as

$$I(a_n) = -\log_b(p_n) = \log_b\left(\frac{1}{p_n}\right),$$

where the base b determines the resulting units of the information measurement. More specifically,

- $\log_2$ units are *bits* (from "binary units")
- $\log_3$ units are *trits* (from "trinary units")
- $\log_e$ or ln units are *nats* (from "natural logarithm units")
- $\log_{10}$ units are *hartleys* (after R. V. L. Hartley, who first proposed the use of a logarithmic measure of information)

If we also assume that the symbols are emitted independently, i.e., that successive symbols do not depend in any way on past symbols, then the average amount (expected value) of information we can get from each symbol emitted in the stream from the source is defined as the entropy $H(A)$ for the discrete set of probabilities $P = \{p_1, p_2, \ldots, p_N\}$:

$$H(A) = \bar{I}(A) = \sum_{n=1}^{N} p_n \log_b\left(\frac{1}{p_n}\right) \tag{4.1}$$

According to Shannon's information theory for communication [4], a source in a communication channel transmits a stream of input symbols selected from a discrete finite set of alphabets $A = \{a_1, a_2, \ldots, a_N\}$, which are to be encoded when sent through a channel that is potentially corrupted by noise (see Figure 4.1). At the other end of the channel, the receiver decodes the data stream and derives information from the sequence of output symbols $B = \{b_1, b_2, \ldots, b_M\}$. The channel can thus be characterized by a set of conditional probabilities $\{P(a_n|b_m)\}$. It should be noted that there are not necessarily the same numbers of symbols in the input and output sets, owing to the channel noise. Various related

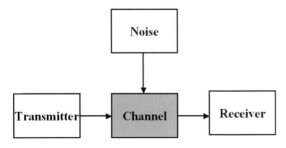

Figure 4.1 In a communication channel, a source transmits a stream of input symbols and the receiver decodes the data stream and derives information from the sequence of output symbols.

information and entropy measures due to the introduction of the channel can be defined. More specifically, the information (mutual information) associated with sending symbol a_n with probability $P(a_n)$ and observing a received symbol b_m (the known information) can be defined as

$$I(a_n; b_m) = \log\left(\frac{1}{P(a_n)}\right) - \log\left(\frac{1}{P(a_n|b_m)}\right) = \log\left(\frac{P(a_n|b_m)}{P(a_n)}\right). \tag{4.2}$$

We can then define the average mutual information over both sets of the symbols as

$$I(A; B) = H(A) + H(B) - H(A, B) \geq 0, \tag{4.3}$$

where $I(A;B) = 0$ if and only if A and B are independent, and

$$H(A) = \sum_{n=1}^{N} p(a_n) \log\left(\frac{1}{p(a_n)}\right),$$

$$H(B) = \sum_{m=1}^{M} p(b_m) \log\left(\frac{1}{p(b_m)}\right), \tag{4.4}$$

$$H(A, B) = \sum_{n=1}^{N} \sum_{m=1}^{M} p(a_n, b_m) \log\left(\frac{1}{p(a_n, b_m)}\right).$$

It can also be shown that

$$H(A, B) = H(A) + H(B|A) = H(B) + H(A|B), \tag{4.5}$$

where

$$H(A|B) = \sum_{n=1}^{N} \sum_{m=1}^{M} p(a_n|b_m) \log\left(\frac{1}{p(a_n|b_m)}\right),$$

$$H(B|A) = \sum_{n=1}^{N} \sum_{m=1}^{M} p(b_m|a_n) \log\left(\frac{1}{p(b_m|a_n)}\right), \tag{4.6}$$

According to Shannon's communication information theory [4], the maximum possible information that can be transmitted through a given channel is defined as the channel capacity C:

$$C = \max_{P(A)} I(A; B). \tag{4.7}$$

One important question is what mix of input symbols, $A = \{a_1, a_2, \ldots, a_N\}$, should be used in order to reach the channel capacity as nearly as possible? In other words, for any given channel, is there a way of encoding input symbols so that we can utilize the channel as near to capacity (as we wish) and, at the same time, have an error rate as close to zero as we wish? Shannon showed that it is possible using entropy coding to keep error rates low and still use the channel for information transmission at (or near) its capacity.

4.2   Entropy coding

An entropy encoding is a coding scheme that assigns codes to symbols so as to match code lengths with the probabilities of the symbols. Typically, entropy encoders are used to compress data by replacing symbols represented by equal-length codes with symbols represented by codes where the length of each codeword is proportional to the negative logarithm of the probability. Therefore, the most common symbols use the shortest codes. According to Shannon's source coding theorem, the optimal code length for a symbol is $\log_b(1/p)$, where b is the number of symbols used to create the output codes and p is the probability of the input symbol. Two of the most popular entropy encoding techniques are Huffman coding and arithmetic coding.

4.2.1 Huffman coding

Huffman coding, developed by David A. Huffman in 1952 [5], is an entropy encoding algorithm used for lossless data compression. It employs a variable-length binary code table for encoding a set of source symbols. The variable-length code table is based on the estimated probability of occurrence of each possible value of the source symbol. Huffman coding uses a specific method for choosing the representation for each symbol and results in a prefix-free code (also called an instantaneous code), i.e., the bit string representing some particular symbol is never a prefix of the bit string representing any other symbol. Huffman code expresses the most frequent symbols using shorter strings of bits than are used for less frequent source symbols. Huffman was able to design the most efficient compression method of this type, i.e., no other mapping of individual source symbols to unique strings of bits can produce a smaller average output size when the actual symbol *occurrence frequencies* agree with those used to create the code. Later, however, a method was found to do this in linear time if the input probabilities (also known as *weights*) are sorted in advance (see an example given in Figure 4.2). Huffman coding today is often used as a "back-end" to some other compression method. The PKZIP codec and some multimedia codecs such as JPEG and MP3 have front-end lossy compression schemes followed by the lossless Huffman coding.

Huffman code uses a probability-sorted binary tree from bottom up for building the variable-length look-up table, and the algorithm was proved to minimize the average codeword length (even though a Huffman code need not be unique). For example, given a set of source symbols $A = \{a_1, a_2, \ldots, a_6\}$ and the corresponding occurrence probabilities $P = \{0.12, 0.42, 0.04, 0.09, 0.02, 0.31\}$, the building of a Huffman binary tree based on probabilities (frequencies) in a bottom-up manner is illustrated in Figure 4.2. The resulting Huffman variable-length code assignment is shown in Figure 4.3.

Original Source		Source Reduction			
Symbol	*Probability*	*1*	*2*	*3*	*4*
a_2	0.42	0.42	0.42	0.42	0.58
a_6	0.31	0.31	0.31	0.31	0.42
a_1	0.12	0.12	0.15	0.27	
a_4	0.09	0.09	0.12		
a_3	0.04	0.06			
a_5	0.02				

Figure 4.2 A typical Huffman entropy code generation process [6].

Original Source			Source Reduction			
Symbol	Probability	Code	1	2	3	4
a_2	0.42	1	0.42 1	0.42 1	0.42 1	─0.58 0
a_6	0.31	00	0.31 00	0.31 00	0.31 00◄	0.42 1
a_1	0.12	011	0.12 011	0.15 010◄	0.27 01	
a_4	0.09	0100	0.09 0100◄	0.12 011◄		
a_3	0.04	01010	0.06 0101◄			
a_5	0.02	01011◄				

Figure 4.3 The variable-length Huffman code assignment [6].

Several desirable properties are associated with the Huffman code.

(1) *Block code*   Each source symbol is mapped into a fixed (variable-length) sequence of code symbols, without dynamically changing with the occurrence of combined symbols.

(2) *Instantaneous (prefix-free) code*   Decoding occurs without reference to the succeeding symbol's sequence.

(3) *Uniquely decodable*   Any string of code symbols can be decoded in only one way.

(4) *Close to entropy*   The average code length L is close to the expected amount of information, the entropy H, that can be retained through the Huffman code assignment:

$$H(A) = -\sum_{j=1}^{6} P(a_j) \log_2(P(a_j)) = 2.0278 \text{ bits} \qquad \text{(entropy)},$$

$$L = 0.42 \times 1 + 0.31 \times 2 + 0.12 \times 3 + 0.09 \times 4 + 0.04 \times 5 + 0.02 \times 5$$
$$= 2.06 \text{ bits per symbol.}$$

The frequencies or probabilities used can be generic for the application domain, based on average experience, or they can be the actual frequencies or probabilities found in the data being compressed. Huffman coding is optimal when the probability of each input symbol is a negative power of 2, e.g., $2^{-1} = 0.5$, $2^{-2} = 0.25$, $2^{-3} = 0.125$, ... Prefix-free codes tend to have a slight inefficiency for a small set of alphabets, where probabilities often fall between these optimal points. This inefficiency can be improved by the so-called "blocking" scheme, which expands the alphabet size by coalescing multiple symbols, especially when adjacent symbols are correlated, into "words" of fixed or variable length before the use of Huffman coding. The worst case for Huffman coding happens when the probability of a symbol exceeds $2^{-1} = 0.5$, making the upper limit of inefficiency unbounded. Such cases can sometimes be better encoded by either run-length coding or arithmetic coding.

To reduce the coding complexity when the size of the symbol set is too large, some variations of Huffman coding can be used [6]. Truncated Huffman coding or a Huffman shift can reduce the coding complexity by slightly sacrificing the coding efficiency, i.e., slightly increasing the average code length, as follows.

(1) *Truncated Huffman*   The probabilities for the lowest-probability symbols $\{a_{n+1}, a_{n+2}, \ldots, a_N\}$ (assuming all symbols have been ordered according to their probability) are aggregated together, i.e., one sets $P(a_J) = P(a_{n+1}) + P(a_{n+2}) + \cdots + P(a_N)$ and

the Huffman coding procedure is applied to the $n+1$ symbols, $\{a_1, a_2, \ldots, a_n, a_J\}$, to obtain the code assignment. The code assigned to the aggregated symbol a_J serves as the prefix code and is followed by a fixed-length code such as binary, for all the symbols belonging to this aggregated symbol.

(2) *Huffman shift* This variation divides all symbols into equal-size blocks and then uses the Huffman coding procedure to code all the symbols of the first block plus an aggregated symbol a_J that represents all the remaining symbols $\{a_{n+1}, a_{n+2}, \ldots, a_N\}$. The code assigned to the aggregated symbol a_J serves as the (repeated) prefix code for the rest of the blocks and is followed by the Huffman code for the first block.

Many other variations of Huffman coding exist, some of which use a Huffman-like algorithm; others find optimal prefix codes that for example put various restrictions on the output. Note that in the latter case the method need not be Huffman-like and in fact need not even be in polynomial time. An exhaustive list of papers on variations of Huffman coding is given in [7].

4.2.2 Arithmetic coding

Although Huffman coding is optimal for a symbol-by-symbol coding with a known input probability distribution, arithmetic coding can sometimes lead to better compression capability when the input probabilities are not precisely known. Arithmetic coding can be viewed as a generalization of Huffman coding. In practice arithmetic coding is often preceded by Huffman coding, as it is easier to find an arithmetic code for a binary input than for a non-binary input.

Arithmetic coding involves a non-block code without the one-to-one correspondence between source symbols and codewords of the Huffman code. An entire finite-length sequence of source symbols is assigned a single arithmetic codeword which defines an interval of real numbers between 0 and 1. As the number of symbols in the sequence increases, the interval used to represent it becomes smaller. The coding principle is based on the assumption that each symbol of a sequence can help to reduce the size of the interval in accordance with its probability [8]. Arithmetic coders produce near-optimal output for a given set of symbols and probabilities. Compression algorithms that use arithmetic coding start by determining a probabilistic model of the data, i.e., a prediction of the patterns that will be found in the symbols of the message. The more accurate this prediction is, the closer to optimality the output will be. In most cases, the probability model associated with the source symbols is regarded as a static model where the probabilities are fixed for each symbol regardless of where they appear in the sequence. More sophisticated models are also possible, e.g., a *higher-order* modeling changes its estimation of the current probability of a symbol on the basis of the symbols that precede it (the *context*). Models can even be *adaptive*, so that they continuously change their prediction of the data on the basis of what the stream actually contains. The decoder must have the same model as the encoder.

In the practical implementation of arithmetic coding, the encoder cannot calculate the fractions representing the endpoints of the interval with infinite precision and simply converts the fraction to its final form with finite precision at the end of encoding. Therefore, most arithmetic coders instead operate at a fixed limit of precision that the decoder will be able to match, and round up the calculated fractions to their nearest equivalents at that precision.

4.2.3   Context-adaptive binary arithmetic coding (CABAC)

The preceding discussion and analysis of arithmetic coding focused on coding a source set of multiple symbols, although in principle it applies to a binary symbol source as well. It is useful to distinguish the two cases since both the coder and the interface to the model are simpler for a binary source, producing bi-level (black and white, or 0 and 1) images. The context-adaptive binary arithmetic coder (CABAC) is an efficient technique for conditional entropy coding. It was previously applied to image coding for the compression of bi-level documents, e.g., by the Joint Bi-level Image Experts Group (JBIG) for black and white images or fax images.

Commonly, in CABAC an array of context variables, each with associated counts, is defined to represent different (geometric) probability distributions. More specifically, it uses reference (one- or two-dimensional) neighboring binary symbols, i.e., bits, to indicate which context index should be used. Thus it can efficiently exploit the correlations between the current binary symbol and its reference binary symbols. One example of context is shown in Figure 4.4, where the arithmetic encoding of any specific binary pixel of a black and white image is dependent on the corresponding probabilities for black and white pixels $\{p_b, p_w\}$, which are adaptively based on the values of seven context pixels shown in gray. The CABAC system utilizes a context-sensitive backward-adaptation mechanism for calculating the probabilities of the input binary symbols. Each bit is then coded using either an adaptive or a fixed probability model. Context values are used for appropriate adaptations of the probability models. There can be many contexts representing various different elements of the source data.

4.2.4   Run-length coding (RLC)

As discussed previously, a bad scenario for Huffman coding occurs when the probability of a symbol exceeds $2^{-1} = 0.5$. This will certainly happen when only two source symbols with

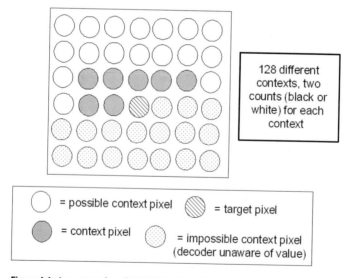

Figure 4.4 An example of CABAC, in which the arithmetic encoding of any specific binary pixel of a black and white image is dependent on the corresponding probabilities for black and white pixels, which is adaptively based on the values of seven context pixels.

unequal probabilities are encoded. Such situations can sometimes be better encoded by either arithmetic coding or run-length coding. Run-length coding (RLC) encodes sequences of binary symbols with *runs* of data (that is, sequences in which the same data value occurs in many consecutive data elements). Run-length coding can serve as a lossless coding method that records the run as a single data value and count rather than as the original run. This is most useful for data that contain many runs, such as the printed characters on a document image. For example, consider a single line of printed text in black ink (denoted as 1) on a white background (denoted as 0):

00000000111000000011000001111100011000.

The RLC algorithm for the above hypothetical scan line is

8(0)3(1)6(0)2(1)5(0)4(1)3(0)2(1)3(0).

Run-length encoding is often combined with other techniques, e.g., Huffman coding, to produce very efficient data compression techniques. More specifically, we can use Huffman coding to encode further these run numbers (e.g., 8, 3, 6, ...), which are treated as symbols with known probabilities. These combined techniques can be quite effective for compressing most faxed documents, using the JBIG standard, since such documents are mostly white space with occasional interruptions of black. The RLC method can also be extended to the two-dimensional case, when it is also known as relative address coding (RAC), and it has been used in JBIG2 [9] for compressing color fax documents [9].

Furthermore, RLC can be extended to multiple symbol sequences with one symbol having a higher probability than the rest of the symbols. As shown in the following numerical sequence, which contains many "0" symbols, this can be encoded using RLC specifically only for recording zero runs, denoted by #2, #1, #4, etc. Another way of representing this RLC is based on a sequence of two-tuple symbols, as also shown below; this sequence of two-tuple symbols can further be losslessly compressed by Huffman coding, as adopted for JPEG images:

$$0, 0, -3, 5, 1, 0, -2, 0, 0, 0, 0, 2, -4, 3, -2, 0, 0, 0, 1, 0, 0, -2$$
$$\Rightarrow \#2, -3, 5, 1, \#1, -2, \#4, 2, -4, 3, -2, \#3, 1, \# 2, -2$$
$$\Rightarrow (2, -3)(0, 5)(0, 1)(1, -2)(4, 2)(0, -4)(0, 3)(0, -2)(3, 1)(2, -2)$$

4.3   Lossy image compression

Lossless compression alone is not enough to compress most images to satisfactorily low bitrate. Therefore we have to resort to lossy compression approaches, which introduce distortion through the signal quantization process. The quantization can be performed either in the spatial or in the transform domain. Most spatial or transform domain samples of images can assume arbitrary values. For example, the integer *RGB* values $\{f(x, y)\}$ captured by a camera have to be converted to continuous (real) YC_bC_r values for further manipulation. Furthermore these values can also be transformed to other domains $\{F(u,v)\}$ for different types of processing, such as Fourier transform, discrete cosine transform (DCT), or wavelet transform. All these transformed values are inevitably continuously valued. However, in a digital implementation, continuous numbers have to be represented using a finite number of bits and these arbitrary real values have therefore to be represented as a digital sequence; this can be achieved via quantization.

Quantization is a non-linear and irreversible operation that maps a given continuous valued scalar sample x into a value $y = Q(x)$, whose level index can be represented by a finite number L of bits. There are two different types of quantization: uniform and non-uniform quantization. *Uniform quantization* defines the finite set of values from which to choose as uniformly spaced (see Figure 4.5). More specifically, we are allowed only L bits to represent the amplitude of a scalar sample. This means that, for each sample, we map its amplitude to only one of 2^L possible levels. Now if the input sample is known to have a dynamic range R ($= x_{\max} - x_{\min}$) then the same spacing Δ between these levels is equal to $\Delta = R/2^L$ for a uniform quantizer. Non-uniform quantization, however (see Figure 4.6), unevenly divides the quantization range into 2^L possible levels with different spacings $\{\Delta_i\}$. Let us define the decision boundaries $\{d_i, 0 \leq i \leq 2^L\}$ of these 2^L possible levels, where

$$
\begin{aligned}
&d_0 = x_{\min}, \quad d_{2^L} = x_{\max}, \\
&\Delta = d_i - d_{i-1}, \quad i = 1, 2, \ldots, 2^L, \text{ uniform quantization}, \\
&\Delta_i = d_i - d_{i-1}, \quad i = 1, 2, \ldots, 2^L, \text{ non-uniform quantization}, \quad (4.8) \\
&y = Q(x) = r_i, \quad \text{for} \quad d_{i-1} < x \leq d_i, \\
&x = y + e_q = r_i + e_q.
\end{aligned}
$$

Any scalar sample x that falls between the decision boundaries $(d_{i-1}, d_i]$ is quantized into (and is represented as) a finite number of L bits, which has a reconstruction magnitude r_i. Owing to this mapping from an arbitrary continuous scalar to a finite number 2^L of reconstruction levels, there is inevitably an error (the quantization noise e_q) created and this is the source of distortion in lossy image compression. Given the probability density function $P(x)$, of the scalar sample x, the optimal quantizer design, under the mean squared quantization error criterion (i.e., $\underset{\{d_i\},\{r_i\}}{\arg \min} E\{e_q^2\} = E\{(x - r_i)^2\}$) [10], is determined by the decision boundaries $\{d_i\}$ and reconstruction levels $\{r_i\}$ that minimize the mean squared quantization error $E\{e_q^2\}$.

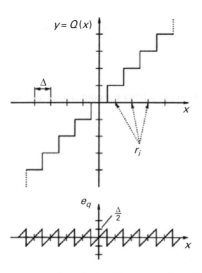

Figure 4.5 A typical uniform quantizer defines the finite set of values, which are uniformly spaced, to choose from.

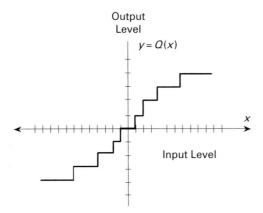

Figure 4.6 Non-uniform quantization unevenly divides the quantization range into 2^L possible levels with different spacings.

It has been shown that optimal quantizer (also called the Lloyd–Max quantizer) design implies that the optimal reconstruction level r_i should be equal to the centroid of the area under $P(x)$ for the specified decision interval $(d_{i-1}, d_i]$. The decision boundary d_i should be equal to the midpoint between two adjacent reconstruction levels, (r_i, r_{i+1}). In the very special case when a uniform quantizer is desired and $P(x)$ is also uniformly distributed in its dynamic range R, the quantizer is derived as

$$r_i = \frac{d_{i-1} + d_i}{2}, \quad \text{where} \quad d_i = d_{i-1} + \Delta = d_{i-1} + \frac{R}{2^L} \tag{4.9}$$

$$\text{and} \quad d_0 = x_{\min}, \quad d_{2^L} = x_{\max}.$$

4.4 Joint Photographic Experts Group (JPEG)

One of the most popular and comprehensive image coding standards is the JPEG standard. The name JPEG stands for Joint Photographic Experts Group, which is the international committee that created the lossy image compression standard ISO/IEC 10918-1/2/3 (now ITU-T Recommendations 81, 83, and 84) [11] [12] in 1994 for continuous tone still images, both grayscale and color. The compressed images, with extension *.jpg which employs JPEG compression, are commonly also called JPEG images. Owing to its adoption by the World Wide Web (WWW) and the digital still camera (DSC), JPEG is more or less regarded as the sole dominating format for storage and transmission of digital photographs.

There are many parameters associated with the JPEG compression process. By adjusting the parameters, one can trade off compressed image size against reconstructed image quality over a very wide range. For example, the image quality can be degraded with an obvious blocky effect at compression ratios higher than 100, or the image quality can also be almost indistinguishable from the original with compression ratios around 3. Usually the threshold of visible difference from the original image for the compression ratio is somewhere around 10 to 20, i.e., 1 to 2 bits per pixel for color images. Grayscale images normally cannot achieve as much compression ratio [13].

The objective image quality degradation is commonly defined on the basis of the peak signal-to-noise ratio (PSNR), which is the ratio of the maximum possible power of a signal and the power of the reconstruction errors, usually expressed in terms of the logarithmic decibel scale. The PSNR is most commonly used as a measure of the quality of reconstruction in image compression etc. More specifically, given the original image $\{f(x,y)\}$ of size $M \times N$ and its reconstructed (encoding followed by decoding) version $\{\hat{f}(x,y)\}$ from compression, the PSNR is defined as

$$PSNR = 10 \ \log_{10}\left(\frac{Max_I^2}{MSE}\right),$$

where

$$MSE = \frac{1}{MN}\sum_{x=0}^{M-1}\sum_{y=0}^{N-1}\left\| f(x,y) - \hat{f}(x,y)\right\|^2 \tag{4.10}$$

and Max_I is the maximum allowable pixel value of the image. When the pixels are represented using eight bits per sample, it equals 255. For color images with three RGB values per pixel, the definition of the PSNR is the same except that the MSE is the sum over all squared value differences divided by the image size and by 3. Typical values for the PSNR in image compression are between 30 and 40 dB.

A "baseline" lossy algorithm is defined by JPEG, with optional extensions for progressive and hierarchical coding. There is also a separate lossless JPEG compression mode, which typically gives about 2 : 1 compression, i.e., about 12 bits per color pixel. Most currently available JPEG hardwares and softwares handle only the baseline mode. The three different coding modes of JPEG are:

(1) *a lossy baseline mode*, often called a sequential baseline system, which is a DCT-based compression adequate for most applications. The image in this system is encoded in a single left-to-right, top-to-bottom scan;
(2) *an extended coding mode* for greater compression and higher-precision applications. This mode can also be used for progressive transmission, where the image is encoded in multiple scans for applications whose transmission time is long and the viewer prefers to watch the image building up in multiple coarse-to-clear passes;
(3) *a lossless encoding mode* in which the image is encoded to guarantee the exact recovery of every source image sample value, even though the compression performance is low compared with that of the lossy modes.

In mode (1), the sequential baseline system of JPEG, the DCT-based encoding and decoding compression procedures can be summarized in Figures 4.7 and 4.8 [14], where a block (8×8 pixels) of the single-component (e.g., grayscale) image with 8-bit precision (12-bit precision for some medical or special types of images) on both input and output data is performed by three sequential modules of operations: forward and inverse discrete cosine transform (DCT) computation, quantization and variable-length code assignment. For color image compression, an intercomponent transformation from RGB to YC_bC_r should be performed in advance before proceeding to the encoding process. For progressive mode-JPEG codecs, an image buffer exists prior to the entropy coding step, so that an image can be stored and then parceled out in multiple scans with successively improving quality. For the hierarchical mode of operation, the steps described are used as building blocks within a larger framework.

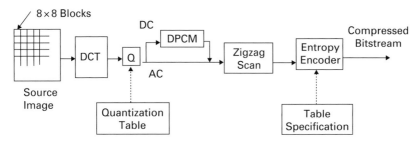

Figure 4.7 A DCT-based sequential baseline encoder of JPEG [14].

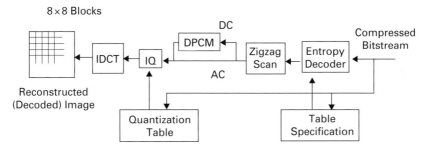

Figure 4.8 A DCT-based sequential baseline decoder of JPEG [14].

4.4.1　Source image data preparation: chroma subsampling

The source image data discussed in Figure 4.7 and the reconstructed image data in Figure 4.8 are mainly for single-component image data such as grayscale images. For color images we need to transform RGB into a luminance vs. chrominance color space (e.g., YC_bC_r) using the linear matrix operations shown below (see Eq. 4.11). The luminance component Y represents the grayscale intensity and the chrominance components C_bC_r are color information. The reason for doing this color transformation is that we can afford to lose more information in the chrominance components than in the luminance component, since human eyes are less sensitive to high-frequency chroma contents than to high-frequency luminance contents. After the transformation, the luminance component Y is left at full resolution (when the chroma components C_bC_r are also left at full resolution the format is called 4 : 4 : 4), while the chroma components C_bC_r are often reduced 2 : 1 or 4 : 1 horizontally and either 2 : 1 or 1 : 1 (no change) vertically. In JPEG, these alternatives are usually called the 4 : 1 : 1 and 4 : 2 : 2 chroma subsampling techniques (see Figure 4.9) before being sent to the single-component JPEG compression module shown in Figure 4.7. Of course, after the decoding process in Figure 4.9 an up-sampling (interpolation) is needed to recover the chrominance components to their original size, so that the transformed color images YC_bC_r can be transformed back to RGB for display or storage. The use of YC_bC_r components in JPEG encoding should enable us to reach a higher compression efficiency in coding the color images. Note that color space transformation is slightly lossy owing to the roundoff error introduced in the transformation

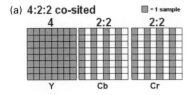

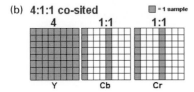

Figure 4.9 Chroma subsampling: (a): 4 : 2 : 2 format; (b) 4 : 1 : 1 format [15].

matrix coefficients, but the amount of error is much smaller than the quantization errors introduced later on. The transformations are as follows:

$$
\begin{bmatrix} Y \\ C_b \\ C_r \end{bmatrix} = \begin{bmatrix} 0.299 & 0.587 & 0.114 \\ -0.169 & -0.331 & 0.500 \\ 0.500 & -0.419 & -0.081 \end{bmatrix} \begin{bmatrix} R \\ G \\ B \end{bmatrix},
$$

$$
\begin{bmatrix} R \\ G \\ B \end{bmatrix} = \begin{bmatrix} 1.0 & 0.0 & 1.4021 \\ 1.0 & -0.3441 & -0.7142 \\ 1.0 & 1.7718 & 0.0 \end{bmatrix} \begin{bmatrix} Y \\ C_b \\ C_r \end{bmatrix}
$$

(4.11)

4.4.2   Block-based discrete cosine transform

For each component image, with or without chroma subsampling, we will group the pixel values $\{f(x,y)\}$ into 8×8 blocks for the discrete cosine transform (DCT), which creates a real valued (instead of complex valued, as produced by a Fourier transform) and energy-compact frequency map $\{F(u,v)\}$ of the two-dimensional data. The use of a DCT allows us to throw away the high-frequency information without affecting the low-frequency information. The DCT itself is reversible except for roundoff errors. The forward (FDCT) and inverse (IDCT) two-dimensional DCT for the 8×8 image block can be expressed as

$$
F(u, v) = \frac{1}{4} C(u)C(v) \sum_{x=0}^{7} \sum_{y=0}^{7} f(x,y) \cos\left[\frac{(2x+1)u\pi}{16}\right] \cos\left[\frac{(2y+1)v\pi}{16}\right],
$$

$$
f(x,y) = \frac{1}{4} \sum_{u=0}^{7} \sum_{v=0}^{7} C(u)C(v)F(u, v) \cos\left[\frac{(2x+1)u\pi}{16}\right] \cos\left[\frac{(2y+1)v\pi}{16}\right],
$$

where

$$
C(w) = \begin{cases} \dfrac{1}{\sqrt{2}}, & w = 0, \\ 1 & \text{otherwise.} \end{cases}
$$

(4.12)

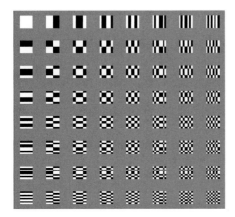

Figure 4.10 The two-dimensional DCT uses 64 orthogonal basis signals each containing one of the 64 unique two-dimensional spatial frequencies.

The two-dimensional DCT formulation in Eq. (4.12), owing to the separable property, can be broken up to become two identical one-dimentional equations. Hence the two-dimensional DCT can be calculated by first calculating the one-dimensional DCT of the rows, which can be represented as a matrix–matrix multiplication, and then calculating the one-dimensional DCT of the resulting columns, which involves another matrix–matrix multiplication. This procedure is called the row–column decomposition algorithm, and using it the number of calculations can be greatly reduced.

The FDCT can be viewed as a harmonic analyzer and the IDCT as a harmonic synthesizer. More specifically, the FDCT takes each 8×8 block of source image samples as its input and decomposes it into 64 orthogonal basis signals, each containing one of the 64 unique two-dimensional spatial frequencies (as shown in Figure 4.10), which comprise the input signal's spectrum. The output of the FDCT is a set of 64 basis-signal amplitudes or DCT coefficients (11–12 bits per coefficient), whose values are uniquely determined by the particular 64-point input signal. The coefficient with zero frequency in both dimensions is called the DC coefficient and the remaining 63 coefficients are called the AC coefficients. Because sample values typically vary slowly from point to point across an image, the FDCT processing step lays the foundation for achieving data compression by concentrating most of the signal in the lower spatial frequencies. For an 8×8 sample block from a typical source image, most spatial frequencies have zero or near-zero amplitude and need not be encoded.

At the decoder the IDCT reverses this processing step. It takes the 64 (quantized) DCT coefficients and reconstructs a 64-point output image signal by summing the basis signals. The FDCT introduces no loss to the source image samples if infinite computational precision can be used; it merely transforms them to a domain in which they can be more efficiently encoded (in reality this is not possible, therefore some additional errors are introduced). Many different algorithms by which the FDCT and IDCT may be approximately computed have been devised [16]. The FDCT or IDCT algorithms to be used in the proposed (JPEG) standard were not specified, except for some precision requirements of practical FDCT and IDCT implementations.

Table 4.1 (a) The default quantization table for luminance (Y component); (b) the default quantization table for chrominance (C_bC_r components)

16	11	10	16	24	40	51	61
12	12	14	19	26	58	60	55
14	13	16	24	40	57	69	56
14	17	22	29	51	87	80	62
18	22	37	56	68	109	103	77
24	35	55	64	81	104	113	92
49	64	78	87	103	121	120	101
72	92	95	98	112	100	103	99

(a)

17	18	24	47	99	99	99	99
18	21	26	66	99	99	99	99
24	26	56	99	99	99	99	99
47	66	99	99	99	99	99	99
99	99	99	99	99	99	99	99
99	99	99	99	99	99	99	99
99	99	99	99	99	99	99	99
99	99	99	99	99	99	99	99

(b)

4.4.3　Quantization of DCT coefficients

After output from the FDCT, each of the 64 DCT coefficients $F(u,v)$ is uniformly quantized using a prespecified 64-element quantization table (also called a normalization matrix), i.e., each coefficient is divided by a separate table entry $Q(u,v)$, where each entry specifies the step size of the quantizer for its corresponding DCT coefficient; this is followed by an integer-round operation resulting in the quantized FDCT coefficient $F^Q(u, v)$. Table 4.1(a) shows a default quantization table for luminance (the Y component) and Table 4.1(b) shows a default quantization table for chrominance (the C_bC_r components). These tables are in fact derived from psychovisual experiments by JPEG members and are included in the ISO standard. The quantization operation can be expressed as follows:

$$F^Q(u, v) = IntegerRound\left(\frac{F(u, v)}{Q(u, v)}\right). \tag{4.13}$$

The objective of this quantization step is to discard information which is not visually significant. Higher-frequency coefficients are always quantized less accurately (i.e., given larger quantization step sizes) than lower ones, since they are less visible to the eyes. Also, the luminance data is typically quantized more accurately than the chroma data, by using separate 64-element quantization tables. To obtain a different quality of JPEG encoded image, most existing encoders use a simple linear scaling of the default tables given in the JPEG standard, i.e., they employ a single user-specified "quality" setting to determine the scaling multiplier.

Dequantization is the inverse function and takes the quantized FDCT coefficients $F^Q(u, v)$ multiplied by the step size $Q(u, v)$ to return the result to a representation appropriate for input to the IDCT.

4.4.4 DC coding and zigzag sequence

After quantization, the DC coefficient, which commonly has a quite large value, is treated separately from the 63 AC coefficients. Because there is usually strong correlation between the DC coefficients of adjacent 8×8 blocks, the quantized DC coefficient is encoded as the difference from the DC term of the previous block in the encoding order (see Figure 4.12), as shown in Figure 4.11(a).

Finally, all the quantized coefficients are ordered into a "zigzag" sequence, as shown in Figure 4.11(b). This ordering helps to facilitate entropy coding by placing low-frequency coefficients (which are more likely to be nonzero) before high-frequency coefficients. Since

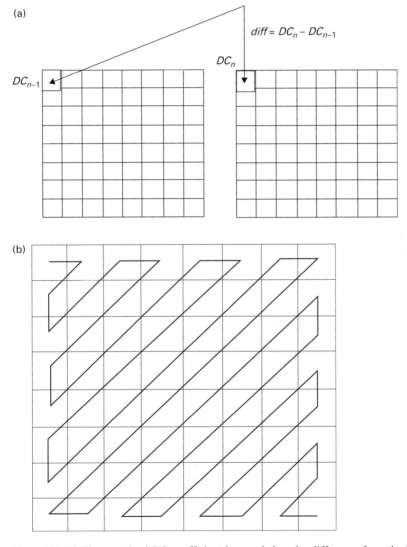

Figure 4.11 (a) The quantized DC coefficient is encoded as the difference from the DC term of the previous block in the encoding order. (b) The quantized coefficients are ordered into a "zigzag" sequence [14].

Table 4.2 (a) The FDCT coefficients $\{F(u,v)\}$; (b) the quantized FDCT coefficients $\{F^Q(u,v)\}$ derived from Eq. (4.13) [14]

235.6	−1.0	−12.1	−5.2	2.1	−1.7	−2.7	1.3
−22.6	−17.5	−6.2	−3.2	−2.9	−0.1	0.4	−1.2
−10.9	−9.3	−1.6	1.5	0.2	−0.9	−0.6	−0.1
−7.1	−1.9	0.2	1.5	0.9	−0.1	0.0	0.3
−0.6	−0.8	1.5	1.6	−0.1	−0.7	0.6	1.3
1.8	−0.2	1.6	−0.3	−0.8	1.5	1.0	−1.0
−1.3	−0.4	−0.3	−1.5	−0.5	1.7	1.1	−0.8
−2.6	1.6	−3.8	−1.8	1.9	1.2	−0.6	−0.4

(a)

15	0	−1	0	0	0	0	0
−2	−1	0	0	0	0	0	0
−1	−1	0	0	0	0	0	0
0	0	0	0	0	0	0	0
0	0	0	0	0	0	0	0
0	0	0	0	0	0	0	0
0	0	0	0	0	0	0	0
0	0	0	0	0	0	0	0

(b)

many of these quantized AC coefficients are zero valued, the multiple-symbol RLC lossless coding scheme discussed in Section 4.2.4 can be applied to represent the zigzag sequence. One such example is shown in Table 4.2 [14]: Table 4.2(a) shows the FDCT coefficients $\{F(u,v)\}$ and Table 4.2(b) shows the quantized FDCT coefficients $\{F^Q(u,v)\}$ derived from Eq. (4.13). The corresponding zigzag sequence with the leading differential DC value (assume that the quantized DC coefficient of the block to its immediate left is 12) and the corresponding multiple symbol RLC coding is as follows:

$$\{3, 0, -2, -1, -1, -1, 0, 0, -1, EOB\} \Rightarrow 3, (1, -2), (0, -1), (0, -1), (0, -1), (2, -1), EOB$$

where EOB denotes the end-of-block symbol.

4.4.5   Entropy coding

Entropy coding is then used to achieve additional compression losslessly by encoding the multiple-symbol RLC sequence of quantized DCT coefficients more compactly, using their statistical characteristics.

The JPEG proposal specifies two entropy coding methods, Huffman coding and arithmetic coding. The baseline sequential mode only allows Huffman coding, while arithmetic coding is used as an optional extension for other modes. The arithmetic coding option uses context-adaptive binary arithmetic coding (CABAC), which is identical to the coder used in JBIG, to achieve approximately 10% better compression:

$$3, (1, -2), (0, -1), (0, -1), (0, -1), (2, -1), EOB$$
$$\Rightarrow [2, 3], [(1, 2), -2], [(0,1), -1], [(0,1), -1], [(0,1), -1], [(2,1), -1], [(0,0)].$$

The DC coefficient and the 2-tuple RLC representation of the quantized AC coefficients are further converted into an intermediate representation, as shown above. More specifically, the intermediate representation $[S, V]$ for the quantized DC difference coefficient with value "$V=3$" is appended with a size category "$S=2$" on the basis of its range of magnitude. The rule of thumb for size category assignment follows a simple rule, as shown in Eq. (4.14) below; the 2-tuple RLC of quantized AC coefficients is also appended with a size category S, along with its run value R, in the form $[(R, S), V]$, where

$$-(2^S - 1) \leq V \leq -2^{S-1} \quad \text{or} \quad 2^{S-1} \leq V \leq 2^S - 1,$$
$$S = 0, 1, \ldots, 15 \ (DC), \qquad S = 0, 1, \ldots, 14 \ (AC) \tag{4.14}$$

and where the largest value limit (4-bit) for S is determined by the maximum allowable quantized DC or AC coefficients calculated on the 8-bit input image assumption.

The Huffman coding for quantized DC coefficients is derived from the frequencies (probabilities) of size category S. The resulting variable-length Huffman code is represented as $H(S) + B(V)$, where $H(S)$ is the Huffman code of numerical symbol S and $B(V)$ is the binary code for V (for negative values, complements are used). Similarly, the Huffman coding for quantized AC coefficients is also derived from the frequencies (probabilities) of 2-tuple symbols, the zero-run R, and the size category S. The resulting variable-length Huffman code is represented as $H(R, S) + B(V)$, where $H(R, S)$ is the Huffman code of the 2-tuple symbol (R, S) and $B(V)$ is the binary code for V (again for negative values complements are used).

Huffman coding requires that one or more sets of Huffman code tables be specified by the applications. The same tables that are used to compress an image are needed to decompress it. Huffman tables may be predefined and used within an application as defaults or computed specifically for a given image in an initial statistics-gathering pass prior to compression. In a baseline sequential JPEG, two sets of Huffman tables (each having one AC table and one DC table), one set for luminance and one for chrominance, are allowed. The JPEG proposal provides "typical" Huffman tables in the JPEG ITU-T81 standard, but it specifies no required Huffman tables and allows proprietary tables to be used.

4.4.6 Header information and JPEG decoding

A JPEG source image contains from one to 255 image components, sometimes called color or spectral bands or channels. Each component consists of a rectangular array of samples. A sample is defined to be an unsigned integer, with precision P bits, having any value in the range $[0, 2^P - 1]$, where P can be 8 or 12 and all samples of all components within the same source image must have the same precision.

The order in which the compressed 8×8 DCT blocks in the JPEG baseline mode are placed in the compressed data stream is a generalization of raster-scan order. Generally, the blocks are ordered from left to right and top to bottom (see Figure 4.12).

In a normal "interchange" JPEG file, all the compression parameters are included in the headers, so that the decoder can reverse the process. These parameters include the quantization tables and the Huffman coding tables. For specialized applications in closed systems, the specification permits those tables to be omitted from the file; this saves several hundred bytes of overhead but it means that the decoder must know a priori which tables were used by the compressor.

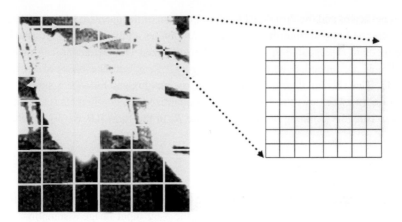

Figure 4.12 The order in which the compressed DCT blocks in JPEG baseline mode are placed in the compressed data stream is a generalization of raster-scan order [14].

The JPEG decoding first parses the codestream and then reverses the Huffman coding steps, using the Huffman tables, to obtain the quantized DCT coefficients. Then it multiplies the quantized DCT coefficients by the quantization table entries to produce approximate DCT coefficients. Since these are only approximate, owing to quantization errors the reconstructed pixel values are also approximate, but the errors are not highly visible. A high-quality decoder will typically add some smoothing steps (e.g., deblocking filtering) to reduce pixel-to-pixel discontinuities.

4.4.7 Trading off compression and picture quality

For color images with moderately complex scenes, all DCT-based modes of operation typically produce the levels of picture quality listed below for the indicated ranges of compression [14]; the unit "bits per pixel" means the total number of bits in the compressed image, including the chrominance components, divided by the number of samples in the luminance component.

(1) 0.25–0.5 bits/pixel   Moderate to good quality, sufficient for some applications
(2) 0.50–0.75 bits/pixel   Good to very good quality, sufficient for many applications
(3) 0.75–1.5 bits/pixel   Excellent quality, sufficient for most applications
(4) 1.5–2.0 bits/pixel   Usually indistinguishable from the original, sufficient for the most demanding applications.

4.4.8 JPEG progressive mode

The progressive mode of JPEG is intended to support the real-time transmission of images, where a low-quality preview can be sent very quickly followed by a refined version as time allows. This mode consists of the same FDCT and quantization steps as are used by baseline sequential mode. The key difference is that each image component is encoded in multiple scans rather than in a single scan. The first few scans encode a rough but recognizable version of the image which can be transmitted quickly in comparison to the total transmission time, and these are refined by succeeding scans until the level of picture quality established by the quantization tables is reached. The decoder must do essentially a full JPEG decode cycle for each scan: inverse DCT, up-sampling, and color conversion must all be done, not to mention any color quantization for 8-bit displays. So this scheme is useful only with fast decoders or slow transmission lines.

In order to achieve progressive transmission, an image-sized buffer memory is needed at the output of the quantizer, before the input to the entropy encoder. After each block of DCT coefficients is quantized, it is stored in the coefficient buffer memory, where the quantized DCT coefficient information can be viewed as a rectangle for which the axes are the DCT coefficients (in zigzag order) and their amplitudes. In a standard sequential baseline model of JPEG coding, each block of quantized DCT coefficients, after entropy coding, is sequentially transmitted. However, the buffered coefficients have to be partially encoded in each of the multiple scans in progressive transmission. There are two complementary methods by which a block of quantized DCT coefficients may be partially encoded. First, only a specified "band" of coefficients from the zigzag sequence is encoded within a given scan. This procedure is called "spectral selection," because each band typically contains coefficients which occupy a specific part of the spatial-frequency spectrum for that 8×8 block. Second, the coefficients within the current band need not be encoded to their full (quantized) accuracy in a single scan. Upon a coefficient's first encoding, the N most significant bits can be encoded first, where N is specifiable. In subsequent scans, the less significant bits can then be encoded. This procedure is called "successive approximation." Both procedures can be used separately, or mixed in flexible combinations, e.g., spectral selection slices the information in one dimension and successive approximation in the other.

4.4.9 JPEG hierarchical mode

The hierarchical mode is used to represent an image at multiple resolutions, e.g., to compress an image in terms of 256×256, 512×512, and 1024×1024 resolutions. Using the concept of spatial scaling, the higher-resolution images can be coded as differences from the next smaller image and thus will require many fewer bits than if they were stored independently at the higher resolution. However, the total number of bits will be greater than that needed to store just the highest-resolution frame in baseline form. The individual frames in a hierarchical sequence can be coded progressively if desired. The hierarchical mode of JPEG is not widely supported at present. It provides a "pyramidal" encoding of an image at multiple resolutions, each differing in resolution from its adjacent encoding by a factor 2 in either the horizontal or vertical dimension or both. The encoding procedure can be summarized as follows; see also Figure 4.13.

(1) Filter (low-pass) and downsample the original image by the desired number of multiples of 2 in each dimension.
(2) Encode this reduced-size image using one of the sequential DCT, progressive DCT, or lossless encoders described previously.
(3) Decode this reduced-size image and then interpolate and upsample it by a factor 2 horizontally and/or vertically, using the identical interpolation filter which the receiver must use.

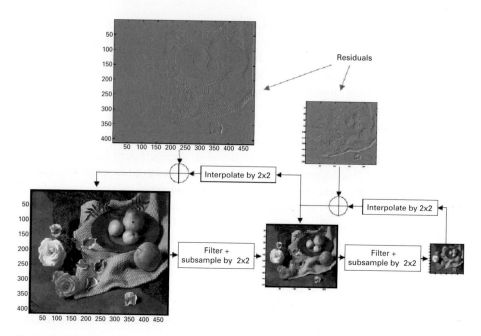

Figure 4.13 The hierarchical mode of JPEG encoding.

(4) Use this upsampled image as a prediction of the original at this resolution, and encode the difference image using one of the sequential DCT, progressive DCT, or lossless encoders described previously.

(5) Repeat steps (3) and (4) until the full resolution of the image has been encoded.

The encoding in steps (2) and (4) must be done using only DCT-based processes, lossless processes, or DCT-based processes with a final lossless process for each component. Hierarchical encoding is useful in applications in which a very-high-resolution image must be accessed by a lower-resolution display. An example is an image scanned and compressed at high resolution for a very-high-quality printer, where the image must also be displayed on a low-resolution PC video screen.

4.4.10   JPEG lossless mode

Owing to the roundoff errors introduced by quantization of the DCT coefficients, the lossless JPEG mode can no longer use DCT. For the same reason, one would not normally use color-space conversion or downsampling, although these are permitted by the standard. The lossless mode simply codes, using entropy coding, the difference between each pixel and the "predicted" value for the pixel where the encoders can use any source image precision from two to 16 bits per sample. A predictor combines the values of up to three neighboring samples (A, B, and C) to form a prediction of sample X. This prediction is then subtracted from the actual value of sample X, and the difference is encoded losslessly by either of the entropy coding methods, Huffman or arithmetic. Eight different predictor (combination) functions are permitted by JPEG. For the lossless mode of operation two different codecs are specified, one for each entropy coding method. Selections 1, 2, and 3 are one-dimensional predictors and selections 4, 5, 6, and 7 are two-dimensional predictors. A selection-value 0 can only be used for differential coding in the hierarchical mode of operation. The entropy coding is nearly identical to that used for the DC coefficient of baseline JPEG as described before.

The main reason for providing a lossless option is that it makes a good adjunct to the hierarchical mode: the final scan in a hierarchical sequence can be a lossless coding of the remaining differences, to achieve overall losslessness. Unfortunately, the lossless JPEG is not quite as useful as it may at first appear, because exact losslessness is not guaranteed unless the encoder and decoder have identical IDCT implementations (i.e., identical roundoff errors). Moreover, the use of downsampling or color space conversion is not allowed either, if you want true losslessness. Lossless codecs typically produce around 2 : 1 compression for color images with moderately complex scenes [8].

4.4.11 JPEG codestream

A typical JPEG codestream is illustrated in Figure 4.14. In such a codestream:

- a "Frame" is a picture, a "scan" is a pass through the pixels (e.g., to obtain the luminance component), a "segment" is a group of blocks, a "block" is an 8×8 group of pixels
- the Frame header includes
  - the sample precision
  - the width and height of the image
  - the number of components
  - the unique ID (for each component)
  - horizontal and vertical sampling factors (for each component)
  - the quantization table to use (for each component)

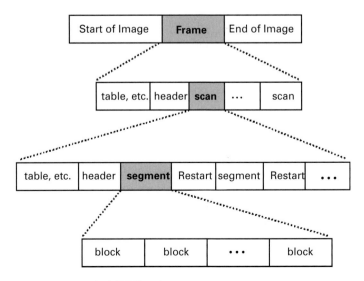

Figure 4.14 A typical JPEG codestream.

- the scan header includes
  - the number of components in the scan
  - the component ID (for each component)
  - the Huffman table (for each component)
- Misc. (can occur between headers) includes
  - quantization tables
  - Huffman tables
  - arithmetic coding tables
  - application data comments

4.5   JPEG2000

In spite of the growing popularity of JPEG image coding, especially on the World Wide Web and for digital still camera, there have been more and more observations of JPEG's shortcomings when applied to a growing number of new applications for which more advanced and enhanced capabilities and functionalities are demanded from image coding. More specifically, the shortcomings are as follows.

(1) *Distortion and artifacts*   When used in compressing large images, especially at high compression ratios, JPEG manifests well-known blocking or tiling artifacts, where each 8×8 block region develops well-defined edges and ringing artifacts, i.e., small waves appear next to sharp edges in the image.

(2) *Ineffective handling of high-quality images*   The limitations of JPEG's 64K-pixel sample size and the limitation to either 8-bit or 12-bit samples have proven to be too restrictive for many new imaging applications such as medical and high-resolution images.

(3) *Lack of effective color-space support*   This inefficiency severely hinders the adoption of JPEG in other graphic arts applications where consistent color information from image capture to editing, display, or printing has to be tightly managed.

(4) *Ineffective progressive and hierarchical modes*   The progressive JPEG mode trades gradually increasing resolution with time, owing to the limited bandwidth; this is no longer an issue, however, because of the growing popularity of broadband deployment. The hierarchical JPEG mode requires multiple JPEG encoding and decoding at different resolutions, in cases where better resolution is needed.

(5) *Poor lossless compression performance*   The lossless mode in JPEG is accomplished by a completely different (prediction) method than that for the lossy mode, and moving from one to the other requires complete decoding and recoding of the image.

To overcome these shortcomings, JPEG2000 [17] [18] was proposed and standardized. This standard was initiated at the Call for Proposal (CFP) in March 1997 with the intention of seeking to produce a new image coding standard to address the shortcomings discussed above and potentially accommodate new applications.

Based on the submission of more than 20 algorithms and their evaluation by the JPEG committee, a wavelet-based compression method was adopted as the backbone of the new standard. In March 1998, the first version of a reference test environment called the JPEG 2000 verification model (VM) was established. In December 2000, Part 1 (ISO/IEC 15444–1) of the JPEG2000 standard was finalized. The first six parts of JPEG2000 form the major portion of the development work by the JPEG committee during the period

1997–2001. In 2002, the committee introduced four new parts, Parts 8–11, to address new application areas. Part 12, which was formerly an amendment of Part 3, is a common activity with the MPEG committee working toward a common file format with MPEG-4. Part 13 is the most recent new part (March 2004), established with the aim of standardizing an entry-level JPEG2000 encoder [19].

As stated above, JPEG2000 is a wavelet-based image compression standard, which was created with the intention of superseding the original DCT-based JPEG standard and offering a long list of more advanced features to overcome the shortcomings of JPEG [20]:

(1) *Superior compression performance* At high bitrates, where artifacts just become imperceptible, JPEG2000 has a compression advantage over JPEG by roughly 20% on average. At lower bitrates, JPEG2000 has a much more significant advantage over certain modes of JPEG. The compression gains over JPEG are attributed to the use of discrete wavelet transform (DWT) and more sophisticated entropy encoding scheme.

(2) *Multiple resolution representation* JPEG2000 provides seamless compression of image components that are each from one to 16 bits per component sample. With tiling, it handles arbitrarily large image size in one single codestream.

(3) *Progressive transmission by pixel and resolution accuracy* JPEG2000 provides efficient and scalable codestream organizations which are progressive by pixel accuracy or by quality (SNR) and also by resolution or size.

(4) *Lossless and lossy compression* JPEG2000 provides both lossless and lossy compression from a single compression architecture with the use of a reversible (integer) wavelet transform.

(5) *Random codestream access and processing* JPEG2000 codestreams offer several mechanisms to support spatial random access or region-of-interest access at varying degrees of granularity.

(6) *Error resilience* JPEG2000 is robust to bit errors introduced by noisy communication channels such as wireless. This is accomplished by the inclusion of resynchronization markers, the coding of data in relatively small independent blocks, and the provision of mechanisms to detect and conceal errors within each block.

(7) *Sequential buildup capability* JPEG2000 allows for the encoding of an image from top to bottom in a sequential fashion without the need to buffer an entire image.

(8) *Flexible file format* The JP2 and JPX file formats allow for the handling of color-space information and metadata and for interactivity in networked applications as developed in JPEG Part 9, the JPIP protocol.

As of the end of 2007, however, JPEG2000 is not yet widely supported in web browsers and hence is not generally used on the World Wide Web or in consumer electronics.

4.5.1 Technical overview of JPEG2000

Figure 4.15 illustrates the basic building blocks of JPEG2000: a preprocessing step which typically consists of tiling, DC level shifting, and a multi-component transform (used for *RGB* color input), then a DWT, followed by a quantizer (for lossy compression only), an entropy coder (CABAC), and, finally, a bitstream organization step to prepare the final codestream of the compressed image [21] [24].

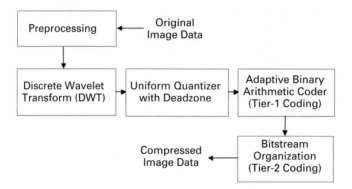

Figure 4.15 The basic building blocks of JPEG2000 [21].

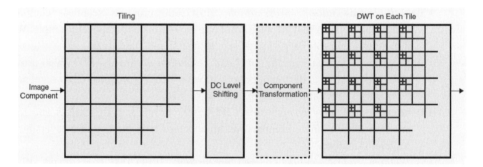

Figure 4.16 In the preprocessing stage, the input image is partitioned into rectangular and non-overlapping tiles of equal sizes that are compressed (wavelet-transformed) independently using their own set of specified compression parameters [26] (© IEEE 2001).

4.5.1.1 Preprocessing

In the preprocessing stage, the input image is partitioned into rectangular and non-overlapping tiles of equal sizes, except for tiles at the image borders, which are compressed (wavelet-transformed) independently using their own set of specified compression parameters (see Figure 4.16) [26]. The unsigned sample values in each component are level shifted (DC offset) by subtracting a fixed value from each sample to make its value symmetric around zero. The level shifted values can be subjected to a forward pointwise color intercomponent transformation to decorrelate the color data. A restriction is that the components must have identical bit-depths and dimensions. Two pointwise color intercomponent transforms are defined in JPEG2000. One is the irreversible color transform, which is the same as the JPEG color transform (i.e., from RGB to YC_bC_r, as shown in Eq. (4.11)). The other is the reversible color transform (RCT), which is an integer-to-integer transform introduced specifically for lossless coding. Forward RCT is specified by

$$Y = \left[\tfrac{1}{4}(R + 2G + B)\right], \qquad C_b = B - G, \qquad C_r = R - G \qquad (4.15)$$

and inverse RCT is specified by

$$G = Y - \left[\tfrac{1}{4}(C_b + C_r)\right], \qquad R = C_r + G, \qquad B = C_b + G.$$

4.5.1.2 Discrete wavelet transform and quantization

The discrete wavelet transform (DWT) is applied independently to the image components and decorrelates the image into different scale sizes, preserving much of its spatial correlation. A one-dimensional DWT consists of a low-pass filter (L) and a high-pass filter (H), which split a line of pixels into two subbands, each band containing half the original line size after downsampling. This is the so-called dyadic decomposition (see Figure 4.17). Application of the filters to two-dimensional images in the horizontal and vertical directions produces four subbands (LL, LH, HL, and HH). The LL subband is a lower resolution representation of the original image, and the missing details are filtered into the remaining subbands. The subbands contain the horizontal (LH), vertical (HL), and diagonal (HH) edges on the scale size defined by the wavelet (see Figure 4.18). Each subband can be decomposed further by subsequent wavelet transforms. Figures 4.19–21 illustrate the two-dimensional wavelet transform of an image up to three levels [21]. The JPEG2000 core coding system specifies a choice of two wavelet filters, the Daubechies 9/7 or the integer Daubechies 5/3. The 9/7 filter (given as floating point numbers) is primarily used for high-visual-quality compression (see Table 4.3). The integer 5/3 filter; which is shorter, can be implemented in integer arithmetic and the associated DWT is reversible, enabling lossless compression (see Table 4.4). To implement the DWT, there are two common filtering approaches. One is based on the standard convolution method for finite impulse response (FIR) filtering. The other is based on a lifting scheme that consists of a sequence of very simple filtering operations in which, alternately, the odd sample values of the signal are updated with a weighted sum of the even sample values and the even sample values are updated with a weighted sum of the odd sample values. The reason for introducing this lifting scheme implementation of DWT is that DWT requires significantly more memory than DCT. The lifting scheme, both for floating lossy and for integer lossless compression,

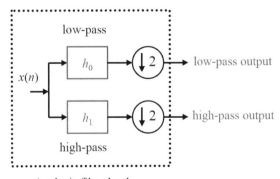

Analysis filter-bank

Figure 4.17 A one-dimensional DWT consists of a low-pass (L) and a high-pass (H) filter splitting a line of pixels into two subbands with each band containing half of the original line size after downsampling [21].

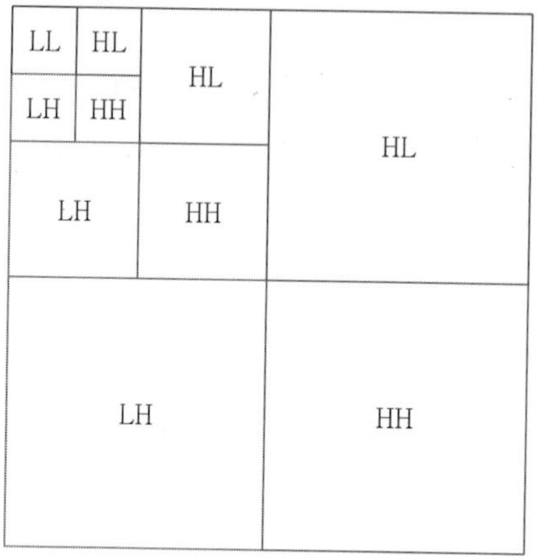

LL	HL		
LH	HH	HL	

(Figure represented by nested subband boxes: LL, HL, LH, HH in top-left; HL and LH, HH in mid sections; LH and HH in bottom sections; HL on the right.)

Figure 4.18 A three-level wavelet transform of an image tile [21].

Figure 4.19 Level 1 of a two-dimensional DWT [21].

is an alternative way to compute wavelet transform coefficients. More specifically, the three-step lifting algorithm can be summarized as follows [17]:

(1) *Split step*   The original data is split into even and odd subsequences.
(2) *Lifting step*   The odd and even subsequences are transformed using the prediction and update coefficients derived from the wavelet (L and H) filter coefficients.
(3) *Normalization step*   Normalization factors are applied to the transformed coefficients to get the resulting wavelet-transform coefficients.

Before proceeding with further manipulation of the subband information, the transform coefficients within each subband have to be quantized. The JPEG2000 core system adopts a simple uniform deadzone quantizer where the deadzone (see below) has exactly twice the step size of the remaining quantizing steps. This gives the optimal embedded structure for

Figure 4.20 Level 2 of a two-dimensional DWT [21].

Figure 4.21 Level 3 of a two-dimensional DWT [21].

achieving SNR scalability (see Figure 4.22). A deadzone is an interval, on the number line around zero, in which unquantized coefficients are quantized to zero. The use of a larger-size deadzone leads to substantial bit savings at low bitrates. Note that the quantization operation q of any subband DWT coefficient y can be denoted as,

$$q = \text{sgn}(y) \left\lfloor \frac{|y|}{\Delta_b} \right\rfloor,$$

$$\Delta_b = 2^{-\tau_b} \left(1 + \frac{\mu_b}{2^{11}}\right), \quad 0 \le \mu_b < 2^{11}, \quad 0 \le \tau_b < 2^5,$$

(4.16)

where $\lfloor x \rfloor$ denotes the floor function, which returns the largest integer not greater than x, and the step size parameter Δ_b for each subband b is specified in terms of an exponent τ_b and a mantissa μ_b according to the dynamic range of the subband. One quantization

Table 4.3 The Daubechies 9/7 filter (given in floating point numbers), primarily used for high-visual-quality compression. The quantities $h_0(n)$ and $h_1(n)$ are low- and high-pass filters

n	$h_0(n)$	n	$h_1(n)$
0	+0.60295	−1	+1.11509
±1	+0.26686	−2, 0	−0.391272
±2	−0.07822	−3, 1	−0.057543
±3	−0.01686	−4. 2	+0.09127
±4	+0.02675		

Table 4.4 The integer 5/3 filter, implemented in integer arithmetic; the associated DWT is reversible, enabling lossless compression. The quantities $h_0(n)$ and $h_1(n)$ are low- and high-pass filters

n	$h_0(n)$	n	$h_1(n)$
0	+6/8	−1	+1
±1	+2/8	−2, 0	−1/2
±2	−1/8		

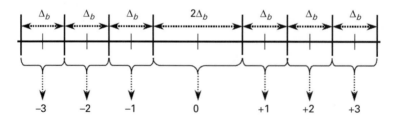

Figure 4.22 The wavelet transform coefficients within each subband are first quantized using a simple linear (uniform) deadzone quantizer [21].

step size is allowed per subband. These step sizes can be chosen so as to achieve a given level of "quality" (as in many implementations of JPEG) or perhaps in some iterative fashion, to achieve a fixed rate. The default is to quantize each coefficient rather finely and rely on subsequent truncation of the embedded codestreams to achieve precise rate control. The JPEG2000 standard places no requirement on the method used to select quantization step sizes. When the integer wavelet transform is employed, the quantization step size is essentially set to 1.0 (i.e., no quantization). In this case, precise rate control (or even fixed quality) is achieved through truncation of embedded codestreams.

4.5.1.3 Codeblock and precinct partition

After an L-level dyadic (pyramidal) DWT and quantization operations on any given YC_bC_r component of an image tile, the quantized wavelet coefficients are decomposed into several subbands, each containing wavelet coefficients that describe specific horizontal and vertical spatial-frequency resolutions (level) of the original image tile. This means that lower-frequency, less-detailed, information is contained in the lower transform levels while higher-frequency, more-detailed, information is contained in the higher transform levels. Each level is further partitioned into codeblocks and precincts, in a similar manner to the partitioning of the original image into tiles. This breakdown is necessary for coefficient modeling and coding and is done on a codeblock by codeblock basis.

Three spatially corresponding rectangles (one from each subband at each resolution level) comprise a precinct, and each precinct is further divided into non-overlapping rectangles called codeblocks (typically 64×64). The codeblock is the unit within which bitplane entropy coding is done. Precincts contain a number of codeblocks and are used to facilitate access to a specific area within an image, in order to process this area in a different way or to decode only a specific area of an image. The precinct size can be chosen independently by resolution; however, each side of a precinct must be a power of 2 in size.

Compressed data from a precinct are grouped together to form a *packet*. As we will see later, the JPEG2000 codestream is a collection of packets together with some header information describing the coding parameters and organization of the codestream. Codeblocks are the smallest geometric structures of JPEG2000. As for precincts, codeblocks are formed in the wavelet-transform domain. Unlike precincts, however, codeblocks are formed by partitioning each subband of the wavelet transform. Like precincts, each side of a codeblock is required to be a power of 2 in size. This forces the codeblock partition and the precinct partition to "line up." It should also be mentioned that while the precinct size can be resolution dependent, the codeblock size cannot. If the codeblock size exceeds the precinct size in any subband, the codeblocks are truncated to obey precinct boundaries. An example of the partitioning of wavelet subbands into precincts and codeblocks is illustrated in Figure 4.23.

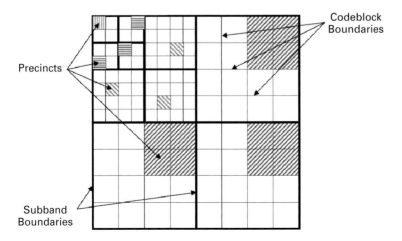

Figure 4.23 An example of the partitioning of the wavelet subbands into precincts and codeblocks.

Like other geometric structures, codeblocks help to produce the rich feature set of JPEG2000. Initial quantization and entropy coding are performed on codeblocks. Thus, codeblocks limit the amount of (uncompressed) wavelet data that must be buffered before it can be quantized and compressed. Furthermore, since each codeblock is entropy coded independently of others, codeblocks provide fine-grain random access to spatial regions and have important error-resilient compression applications [22].

4.5.1.4 Bitplane entropy coding

To create embedded bitstreams for the compressed image data, independent bitplane entropy encoding of the quantized wavelet coefficients of each subband is used. Since the codeblock in each subband is the unit within which bitplane entropy coding is done, each codeblock is indeed encoded independently so as to achieve effective rate control for the final codestream. This independent codeblock encoding approach, known as embedded block coding with optimized truncation (EBCOT) [23], offers many advantages including localized random access into the image, parallelization, improved cropping and rotation functionality, improved error resilience, efficient rate control, and maximum flexibility in arranging progression orders.

In essence, this coding is carried out as the context-adaptive binary arithmetic (CABAC) coding of bitplanes. Their bitstreams are coded with three coding passes per bitplane. This process, called context modeling, is used to assign information about the importance of each individual bit from a given coefficient. The codeblocks can then be grouped according to their significance. On the decoding side it is then possible to extract information according to its significance, allowing the most significant information to be seen first.

Regarding a quantized codeblock as an array of integers in sign-magnitude representation, consider a sequence of binary arrays with one bit from each coefficient. The first such array contains the most significant bit (MSB) of all the magnitudes. The second array contains the next MSB of all the magnitudes, and one continues in this fashion until the final array, which consists of the least significant bits of all the magnitudes. These binary arrays are referred to as bitplanes. Starting from the first bitplane having at least a single 1, each bitplane is encoded in three passes (referred to as sub-bitplanes). The scan pattern followed for the coding of bitplanes within each codeblock (in all subbands) is shown in Figure 4.24. This scan pattern is basically a column-wise raster within stripes of height 4. At the end of each stripe, scanning continues at the beginning (top left) of the next stripe, until an entire bitplane (of a codeblock) has been scanned.

Three coding passes The prescribed scan is followed in each of the three coding passes [23]. The decision about in which pass a given bit is coded is made on the basis of the "significance" of that bit's location (quantized coefficient) in relation to the "significance" of neighboring locations. A location is considered significant if a "1" has been coded for that location in the current or previous bitplanes. The first pass in a new bitplane is called the *significance propagation pass*. A bit is coded in this pass if its own location is not significant but the location of at least one of its eight connected neighbors, indicated as a "context window" in Figure 4.24b, is significant. If a bit is coded in this pass, and the value of that bit is 1, its location is marked as significant for the purpose of coding subsequent bits in the current and subsequent bitplanes. Also, the sign bit is coded immediately after the value-1 bit just coded. The second pass is called the *magnitude refinement pass*, in which all bits

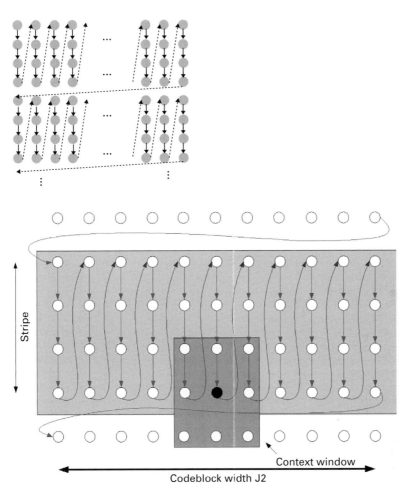

Figure 4.24 The scan pattern followed for the coding of bitplanes, within each codeblock (in all subbands) [21].

from locations that became significant in a *previous* bitplane are coded. The third and final pass is the *clean-up pass*, which takes care of any bits not coded in the first two passes.

All coding is done using context-dependent binary arithmetic coding. The arithmetic coder employed is the MQ-coder specified in the JBIG2 standard [9]. The coding for the first and third passes is usually identical, although run coding is sometimes employed in the third pass. Run coding occurs when all four locations in a column of the scan are insignificant and each has only insignificant neighbors. A single bit is then coded to indicate whether the column is identically zero or not. If not, the length of the zero run (0 to 3) is coded, reverting to the "normal" bit-by-bit coding for the location immediately following the 1 that terminated the zero run. The sign and magnitude refinement bits are also coded using contexts designed specifically for that purpose. Unlike JBIG or JBIG2, which use thousands of contexts, JPEG2000 uses no more than nine contexts to code any given type of bit (i.e., significance, refinement, etc.). This allows extremely rapid probability adaptation and decreases the cost of independently coded segments. There are some issues regarding

Figure 4.25 Four JPEG 2000 encoded pictures, based on three-pass bitplane coding, with different number of bitplanes used [21].

the arithmetic coding, as follows. The context models are always reinitialized at the beginning of each codeblock. Similarly, the arithmetic codeword is always terminated at the end of each codeblock (i.e., once, at the end of the last sub-bitplane). The best performance is obtained when these are the only reinitializations or terminations. It is allowable however, to reset (or terminate) at the beginning (or end) of every sub-bitplane within a codeblock. This frequent reset or termination, plus optionally restricting the context formation to include data from only the current and previous scan stripes is sufficient to enable parallel encoding of all sub-bitplanes within a codeblock (of course, parallel encoding of the code-blocks themselves is always a possibility).

Figure 4.25 shows four JPEG2000 encoded pictures based on three-pass bitplane coding, with different number of bitplanes used [21]. The top left figure has five bitplanes encoded and has a compression ratio 233 : 1. The top right figure has seven bitplanes encoded and has a compression ratio 47 : 1. The bottom left figure has nine bitplanes encoded and has a compression ratio 11 : 1. The bottom right figure has 11 bitplanes encoded and has a compression ratio 3 : 1.

4.5.1.4 Bitstream organization
As mentioned previously, three spatially corresponding rectangles (one from each subband at each resolution level) comprise a precinct, and each precinct is further divided into non-overlapping rectangles called codeblocks, which form the input to the entropy coder.

The individual bitplanes of the coefficients in a codeblock are coded within three coding passes. The compressed bitstreams from each codeblock in a precinct comprise the body of a packet. A collection of packets, one from each precinct of each resolution level, comprises a layer; each layer successively and monotonically improves the image quality at the given resolution (see Figure 4.26) [26]. According to this geometrical breakdown, the JPEG2000 stream starts with a main header containing information such as the uncompressed image size, tile size, number of components, bit depth of the components, coding style, transform levels, progression order, number of layers, codeblock size, wavelet filter type, quantization level, etc. The entire image data, grouped in codeblocks of LL, HL, LH, and HH subbands, follow the header rather than being mixed with the header information. Also, a table of contents can be stored on the encoder side, and this allows a decoder to call up a certain resolution on demand without first having to decode or download the entire JPEG2000 codestream. The JPEG2000 codec offers significant flexibility in the organization of the compressed bitstream, which enables such features as random access, region-of-interest coding, and scalability. The data representing a specific tile, layer, component, resolution, and precinct appear in the codestream in a continuous packet segment. More specifically, the data are divided up into packets by layer, precinct location, resolution level, and tile component. A packet contains the coding-pass data within one layer, for one precinct, from all subbands within one resolution level of one tile component. Contiguous sequences of packets from the same tile constitute tile parts.

In the JPEG2000 codestream shown in Figure 4.27 [25], the compressed codestreams associated with a certain number of sub-bitplanes from each codeblock in a precinct associated with a specific resolution level are collected together to form the body of a "packet." Therefore, a packet can be interpreted as one quality increment for one resolution

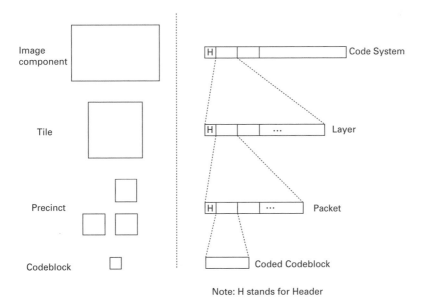

Note: H stands for Header

Figure 4.26 A collection of packets, one from each precinct of each resolution level, comprises a layer; each layer successively and monotonically improves the image quality at the given resolution [26] (© IEEE 2001).

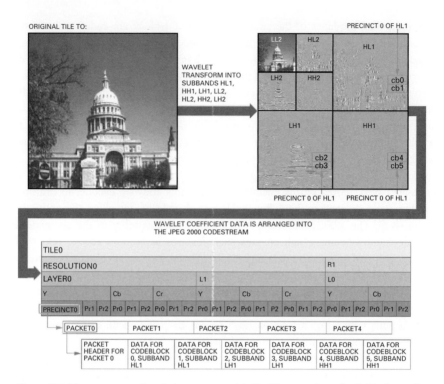

Figure 4.27 The compressed codestreams associated with a number of sub-bitplanes from each codeblock in a precinct associated with a specific resolution are collected together to form the body of a "packet" [25].

level at one specific spatial location (packet-partition locations correspond roughly to spatial locations). The body of a packet is preceded by a packet header. The packet header contains: block inclusion information for each codeblock in the packet (some codeblocks will have no coded data in any given packet); the number of completely zero bitplanes for each code-block; the number of sub-bitplanes included for each codeblock; and the number of bytes used to store the coded sub-bitplanes of each codeblock. It should be noted that the header information is itself coded in an efficient and embedded manner. The data contained in a packet header supplement the data obtained from previous packet headers within the same precinct location so as to just enable decoding of the current packet [23].

A "layer" is then a collection of some consecutive packets from all codeblocks in all subbands and (YC_bC_r) components. It can be interpreted as one quality increment for the entire image at full resolution. The number of layers per JPEG2000 image can range from 1 to 65 535 and is typically around 20. There is no restriction on the number of sub-bitplanes contributed by each codeblock to a given packet or a given layer. Thus, an encoder can format packets into layers for a variety of purposes. An example of such a layer decomposition is shown in Figure 4.28.

The JPEG2000 codec can contain a user-defined number of layers, each layer standing for the particular compression rate achieved by the quantization, rate-distortion, and context-modeling processes. Lower layers, for example, contain bitstreams from the lossy DWT transform that are heavily truncated, contain no coding passes, and thus provide the highest compression ratio and lowest quality. Higher layers can then contain bitstreams that are less

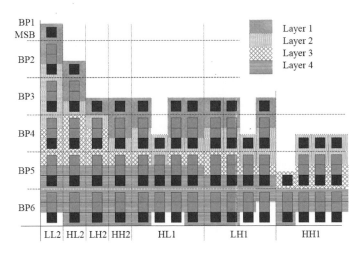

Figure 4.28 An example of JPEG2000 layer decomposition [21].

truncated and use a higher number of coding passes, thus providing low compression and high quality. Each packet is constructed to include all sub-bitplanes with (estimated) rate-distortion slope above a given threshold. This threshold is adjusted to achieve the desired size (bitrate) for the aggregate of all packets within the layer under construction. This provides very fine-grained quality (rate) progression at the expense of some additional overheads due to the large number of packet headers needed.

4.5.1.5 Bitstream progression

While JPEG provided different methods of generating progressive codestreams, with JPEG2000 the progression is simply a matter of the order in which the compressed bytes are stored in a file. JPEG2000 supports progression in four dimensions: in quality, in resolution, in spatial location, and in components. To achieve a different progression, the data need are merely reordered without any further decompression or recompression.

The first progression dimension is progression in quality (in terms of the signal-to-noise ratio, SNR). As more data are decoded sequentially from the beginning of a coded codestream, image quality is improved. It should be noted that the image quality does improve remarkably quickly with JPEG2000. An image is typically recognizable after only about 0.05 bits per pixel have been decoded. With 0.25 bits per pixel decoded, most major compression artifacts have disappeared. To achieve a quality corresponding to no visual distortion, between 0.5 and 2.0 bits per pixel are usually required, depending on the quality and resolution of the original imagery. Figure 4.29 shows a progression of three JPEG2000 encoded pictures with three levels of SNR [21].

The second progression dimension is progression in resolution. In this type of progression, the first few bytes are used to form a small "thumbnail" of the image. As more bytes are decoded, the resolution (or size) of the image increases by a factor 2 (on each side). Eventually, the full-size image is obtained (see Figure 4.30 for four different spatial resolution progressions).

The third progression dimension is progression in spatial location (as specified by precincts in different subband levels). With this type of progression, the imagery is decoded in a "sequential" fashion, from top to bottom. This type of progression is particularly useful for

Figure 4.29 Higher-quality and/or higher-resolution imagery requires more bits per pixel than lower-quality (noisy) and/or lower-resolution imagery [21]. The extent of the lighter-gray regions of the blocks shown to the left of each photograph indicates the percentage of bitplanes used to produce the quality of that photo: top photo, 30%; middle photo, 70%; bottom photo, 100%.

encoding and decoding that uses only moderate amounts of memory. Low-memory scanners can create spatially progressive compressed codestreams "on-the-fly," without buffering the entire image. Similarly, low-memory decoders for projectors or printers can render pixel data without buffering the entire compressed codestream.

The fourth and final progression dimension is progression in the number of components. JPEG2000 supports images with up to 16 384 components. Most images with more than four components are from scientific instruments (e.g., the multi-spectral images from

Figure 4.30 As more bytes are decoded, the resolution (or size) of the image increases by a factor 2 on each side. Eventually, the full size image is obtained [21].

LANDSAT). More typically, images are one-component (e.g., grayscale), three-component (e.g., RGB, YUV, XYZ, etc.), or four-component (CMYK). Overlay components containing text or graphics are also common. Component progression controls the order in which the data corresponding to different components are decoded. With progression by component, the grayscale version of an image would first be decoded, followed by the color information, followed by the overlaid annotations, text, etc.

It is worthwhile to mention that the progression type can be changed at various places within the codestream. For example, it is possible to progress by SNR at a given (reduced) resolution and then change to progression by SNR at a higher resolution. The packets included in the codestream will then be those needed for the higher-resolution subbands to "catch up" with the current layer of the lower-resolution image. This change in progression allows an icon to be displayed first, then a screen-resolution image, and finally if needed a print-resolution image. With a typical five-level transform, a 1024×1024 pixel print-resolution image can provide a 256×256 screen resolution image or a 32×32 icon.

4.5.2 JPEG2000 for Digital Cinema Initiatives

In the summer of 2004, Digital Cinema Initiatives (DCI), which is a joint venture of seven major Hollywood studios (Disney, Fox, Paramount, MGM, Sony Pictures Entertainment, Universal, and Warner Bros. Studios), selected JPEG2000 as the compression format to be used for the future digital distribution of motion pictures [27]. The final version of the DCI Digital Cinema Specification was published online in July 2005. According to this new specification, individual frames are to be compressed independently via JPEG2000 (also called Motion JPEG2000) for inclusion in essence tracks as currently being specified by the Society of Motion Picture and Television Engineers (SMPTE) DC28 committee.

A digital cinema system can be divided into four stages (see Figure 4.31): mastering, transport, storage and playback, and projection. At the mastering stage the movie is compressed, encrypted, and packaged for delivery to theaters. The data is then transported to the exhibition site, where it is decrypted, uncompressed, and played back.

The DCI venture was looking for a compression algorithm that was an open standard, so that many hardware manufacturers would be able to build digital cinema systems. The compression algorithm needed to support a high bit depth (e.g., 12 bits per color component) and a device-independent $X'Y'Z'$ color space without chroma subsampling. Significantly, the compression algorithm needed to support both 2K (up to 2160×1080 pixels) and 4K (up to 4096×2160 pixels) resolution digital cinema packages (DCPs) to be played by projectors installed in the theaters from the same file. The frame rate is set to be 24 frames per second. In addition, a rate of 48 frames per second is also allowed, for 2K content. In the case of a 2K DCP, the sites with 2K projectors can display the image without difficulty. However, the 4K playback system must upsample the 2K image for 4K

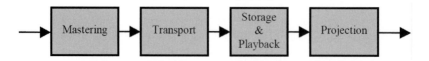

Figure 4.31 A digital cinema system can be divided into four stages: mastering, transport, storage and playback, and projection [27] (© IEEE 2006).

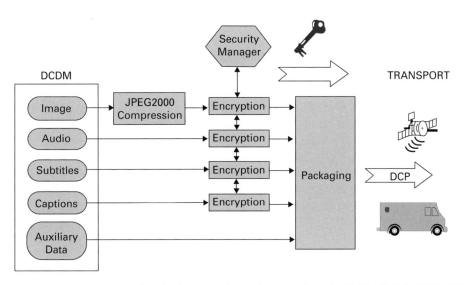

Figure 4.32 A typical procedure for the mastering and transporting of DCI files [27] (© IEEE 2006).

display. Similarly, when a 4K DCP is delivered to a site with a 2K projector, the playback system must downsample the content. The multi-resolution scalability properties of JPEG2000 can be utilized for this purpose. A typical procedure for the mastering and transporting of DCI files is shown in Figure 4.32 [27].

The DCI specification does not specify a particular mode of transport. It is envisioned that the transport can be via physical media or over a network. It is required that the content owners' encryption is not removed during transport. It is also required that all the data of the original files are held intact upon completion of the transport. Thus, no loss is allowed during transmission.

There are some special requirements imposed on JPEG2000 for use in DCI. For examples, image tiling is disallowed, i.e., the entire image should be encoded as a single tile. The maximum number of wavelet-transform levels is five for 2K content and six for 4K content. The codeblock sizes are restricted to be 32×32 and the precinct sizes at all resolutions are set to be 256×256, except for the lowest-frequency subband, where a precinct size of 128×128 is used. The progression order for a 2K distribution is required to be component–position–resolution–layer (CPRL). Progression-order-change (POC) marker segments are forbidden in 2K distributions. Only a single quality layer is allowed. There are constraints on the maximum sizes of codestream segments as well as on the total size of the codestream. These size constraints correspond to a maximum total rate of 250 Mbps. At these rates, the total size of a three-hour feature film is roughly 314 gigabytes.

4.5.3 New parts of JPEG2000

In 2002, the JPEG committee began investigating four application areas that were considered important for JPEG 2000, and it formally established four new parts of JPEG2000 to address these applications: JPIP, JPSEC, JP3D, and JPWL, respectively. These new parts are designed to address the standardization needs of specific application areas to which the rich set of technology from JPEG2000 applies [20].

4.5.3.1 JPIP – JPEG2000 Part 9: Interactive and progressive transmission

The JPIP application specifies a new protocol for interacting with the JPEG2000 contents in distributed applications in the Internet environment. The goal is to specify a network protocol (syntaxes and methods) that enables the interactive and progressive transmission of JPEG2000 coded data and files from a server to a client [29] [30]. JPEG2000 offers many desirable features in support of the interactive access of large images (resolution scalability, progressive refinement, spatial random access). However, JPEG2000 Part 1 describes only a codestream syntax, suitable for storing the compressed data in a file. One way to interact remotely with the image content is for an intelligent browsing client to access appropriate byte ranges from the file. Such an approach allows existing HTTP servers to be used for byte-range access [28]. However, a byte-range approach requires the inclusion of an index that such a client can read to determine the locations (byte ranges) of the relevant compressed data and header information. The JPIP protocol allows a client to request only the portions of an image (by region, quality, or resolution level) that are applicable to the client's need. The protocol also allows the client to access metadata or other content from the file. Although the terms client and server are typically used to refer to the image receiving and delivering applications, JPIP can be used, within both hierarchical and peer-to-peer networks, for bidirectional image data delivery on top of network transport protocols such as TCP, UDP, or HTTP [20].

As shown in Figure 4.33, the client of JPIP uses a view-window request to define the resolution, size, location, components, layers, and other parameters for the image and image-related data requested by the client. The server responds by delivering imagery and related data with precinct-based streams (JPP), tile-based streams (JPT), or whole images. The protocol allows for negotiation of the client and server capabilities and limitations. The client may request information about an image as defined in index tables from the server, which enables the client to refine its view-window request to image-specific parameters (e.g., byte-range requests). The JPIP protocol is designed to be transport neutral. A primary objective is that JPIP communication can be realized using HTTP as the underlying transport without interfering with the existing HTTP infrastructure. It can be used over several different types of transport, as shown in Figure 4.34.

4.5.3.2 JPSEC-JPEG2000 Part 8: Secure JPEG2000

The JPSEC protocol specifies tools and solutions to allow applications to generate, consume, and exchange secure JPEG2000 bitstreams [31]. Security issues that are targeted by

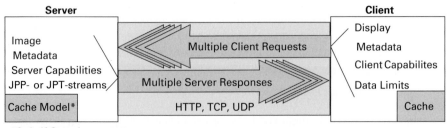

*Only if State is present

Figure 4.33 The JPIP protocol can be used within both hierarchical and peer-to-peer networks, for bidirection image data delivery, on top of network transport protocols such as TCP, UDP, and HTTP [20] (© IEEE 2005).

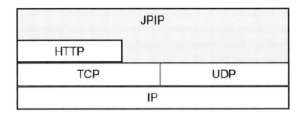

Figure 4.34 The JPIP protocol can be used over several different transports [20] (© IEEE 2005).

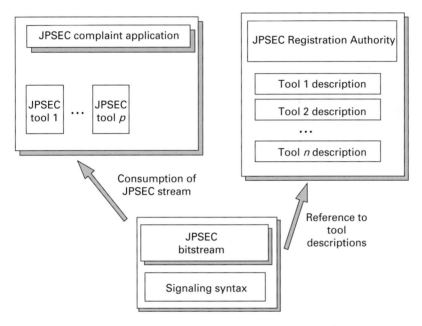

Figure 4.35 A number of JPSEC tools and tool descriptions have been preregistered [20] (© IEEE 2005).

this standard include, but are not limited to, authentication, data integrity, protection of contents and protection of technologies, privacy, conditional access, confidentiality, and transaction tracing. The JPSEC bitstreams can be created from an original image, from JPEG2000 encoded data, or from another JPSEC bitstream. The JPEG bitstream is associated with a number of JPSEC security services: confidentiality of the image data, integrity of the image data, authentication of the image data origin, etc. The signaling syntax specifies the security services associated with the image data, the JPSEC tools that are required to render the corresponding services, and the parts of the image data that are protected. A JPSEC-compliant application is one that is able to consume JPSEC bit-streams. In the JPSEC framework, all tools are registered by the JPSEC Registration Authority. This means that all JPSEC tools are associated with a unique identification number provided by the common registry. A number of JPSEC tools and tool descriptions have been preregistered. With this process, provision is made for future tools to be identified (see Figure 4.35) [20].

4.5.3.3 JP3D-JPEG2000 Part 10: Three-dimensional and floating point data

The JP3D application is a work item subdivision of JPEG2000 that will provide extensions of JPEG2000 for logically rectangular three-dimensional (3-D) data sets with no time component. JP3D is envisioned as a pure extension of JPEG2000 Parts 1 and 2. The potential markets for this work include applications that are inadequately served by the capabilities provided in JPEG2000 Part 2. Volumetric imagery, in particular, is handicapped by the fact that Part 2 does not treat 3-D data sources isotropically: all 3-D applications are potentially affected by the fact that Part 2 fails to enable a number of source coding features in the cross-component direction. The JPEG2000 work plan also envisioned a potential specification for coding a time series of JP3D images (e.g., a volumetric time series). Such sources are currently of interest in scientific visualization and medical imaging applications. Any such specification would be treated as an extension of Motion JPEG2000 for signaling sequences of JP3D frames and would be presented as an amendment or extension of Part 3.

4.5.3.4 JPWL-JPEG2000 Part 11: Wireless

The JPWL application defines a set of tools and methods to achieve the efficient transmission of JPEG2000 imagery over an error-prone wireless network [32]. Wireless networks are subject to the frequent occurrence of transmission errors, which, along with a low bandwidth, puts strong constraints on the transmission of digital images. Since JPEG2000 provides high compression efficiency, it is a good candidate for wireless multimedia applications. Moreover, owing to its high scalability, JPEG2000 enables a wide range of quality of service (QoS) strategies for network operators. However, to be suitable for wireless multimedia applications, JPEG2000 has to be robust to transmission errors. The main functionality of the JPWL system is to protect the codestream against transmission

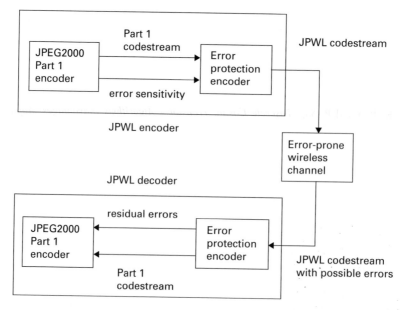

Figure 4.36 The encoder and decoder of a JPWL [20] (© IEEE 2005).

errors. Two other functionalities are a description of the degree of sensitivity of different parts of the codestream to transmission errors, and a description of the locations of residual errors in the codestream (see Figure 4.36) [20].

References

[1] ISO/IEC 10918–1, Joint Photographic Experts Group (JPEG), http://www.jpeg.org/jpeg/index.html.

[2] "Graphics interchange format, version 89a," CompuServe Incorporated, http://www.w3.org/Graphics/GIF/spec-gif89a.txt.

[3] "Information technology – JPEG2000 image coding system: Core coding system," ITU-T and ISO/IEC 15444–1, http://www.jpeg.org/jpeg2000/index.html.

[4] C. E. Shannon, "A mathematical theory of communication," *Bell Syst. Tech. J.*, 27: 379–423, 623–656, 1948.

[5] D. A. Huffman, "A method for the construction of minimum-redundancy codes," *Proc. IR E*: 1098–1102, September 1952.

[6] R. C. Gonzalez and R. E. Woods, *Digital Image Processing*, second edition, Prentice Hall, 2002.

[7] J. Abrahams, "Code and parse trees for lossless source encoding," in *Proc. conf. on Compression and Complexity of Sequences*, pp. 145–171, June 1997.

[8] P. G. Howard and J. S. Vitter, "New methods for lossless image compression using arithmetic coding," Brown University Department of Computer Science Technical Report No. CS9147, August 1991.

[9] P. G. Howard, F. Kossentini, B. Martins, S. Forchhammer, and W. J. Rucklidge, "The emerging JBIG2 standard," *IEEE Trans. Circuits Syst. Video Technol.*, 8: 838–848, November 1998.

[10] J. Max, "Quantization for minimum distortion," *IRE Trans. on Information Theory*, 6: 7–12, 1960.

[11] "Digital compression and coding of continuoustone still images, Part 1, Requirements and guidelines," ISO/IEC JTC1 Draft International Standard 109181, November 1991.

[12] "Digital compression and coding of continuoustone still images, Part 2, Compliance testing," ISO/IEC JTC1 Committee Draft 109182, December 1991.

[13] W. B. Pennebaker and J. L. Mitchell, *JPEG Still Image Data Compression Standard*, Van Nostrand Reinhold, 1993.

[14] Gregory K. Wallace, "The JPEG still picture compression standard," *Communi. of ACM*, 34(4): 30–44, April 1991.

[15] *Computer Desktop Encyclopedia*, The Computer Language Company, http://www.computerlanguage.com/.

[16] K. R. Rao and P. Yip, *Discrete Cosine Transform Algorithms, Advantages, Applications*, Academic Press, 1990.

[17] C. Christopoulos, A. Skodras, and T. Ebrahimi, "The JPEG 2000 still image coding system: an overview," *IEEE Trans. Consumer Electron.*, 46(4): 1103–1127, November 2000.

[18] "Information technology – JPEG 2000 image coding system – Part 1: Core coding system," ISO/IEC 15444–1:2000.

[19] D. S. Taubman and M. W. Marcellin, *JPEG2000: Image Compression Fundamentals, Standards and Practice*, Kluwer Academic Publishers, 2002.

[20] D. T. Lee, "JPEG 2000: retrospective and new developments," *Proc. IEEE*, 93(1): 32–41, January 2005.

[21] M. Rabbani, "JPEG2000: the basics", *tutorial talk in conf. on Visual Communication and Image Processing (VCIP)*, Lugano, 2003.

[22] M. W. Marcellin, M. J. Gormish, A. Bilgin, and M. P. Boliek, "An overview of JPEG-2000," in *Proc. Data Compression Conf. (DCC)*, pp. 523–541, Snowbird, 2000.

[23] D. Taubman, "High performance scalable image compression with EBCOT," *IEEE Trans. Image Process.*, 9(7): 1158–1170, July 2000.

[24] M. Rabbani and R. Joshi, "An overview of the JPEG 2000 still image compression standard," *Signal Process. Image Commun.*, 17: 3–48, January 2002.

[25] C. Bako, "JPEG2000 image compression," *Analog Dialogue* 38–09, September 2004, http://www.analog.com/analogdialogue.

[26] A. Skodras, C. Christopoulos, and T. Ebrahimi, "The JPEG 2000 still image compression standard," *IEEE Signal Process. Mag.*, 36–58, September 2001.

[27] Ali Bilgin and Michael W. Marcellin, "JPEG2000 for digital cinema," in *Proc. ISCAS 2006*, Kos Island, May 2006.

[28] S. Deshpande and W. Zeng, "Scalable streaming of JPEG 2000 images using hypertext transfer protocol," in *Proc. ACM Conf. on Multimedia Systems*, pp. 372–381, 2001.

[29] D. Taubman and R. Prandolini, "Architecture, philosophy, and performance of JPIP: Internet protocol standard for JPEG 2000," in *Proc. SPIE Conf. on Visual Communication and Image Processing*, Lugano, 2003.

[30] "JPEG 2000 image coding system – Part 9: Interactivity tools, API's and protocols," Int. Standards Org./Int. Electrotech. Comm. (ISO/IEC), ISO/IEC FCD2.0 15 444–9, December 2003.

[31] "JPEG 2000 image coding system – Part 8: Secure JPEG 2000," Int. Standards Org./Int. Electrotech. Comm. (ISO/IEC), ISO/IEC WD3.6 15 444–8, December 2003.

[32] "JPEG 2000 image coding system – Part 11: Wireless," Int. Standards Org./Int. Electrotech. Comm. (ISO/IEC), ISO/IEC WD2.0 15444–11, December 2003.

5 Digital video coding

Ever since the initial introduction of Sony's D-1 format, which digitally recorded an uncompressed standard definition (RGB) component video in 1983, the research and development efforts in digital video coding (compression) have been actively pursued. Apple released the first commercial version of QuickTime video for streaming and playback in 1991, and this set a fast pace for international standardization efforts for consumer digital video. Owing to the much larger amount of data (in bits per second, see Eq. (5.1)) generated by raw digital video, video coding (i.e., compression) becomes critically needed. More specifically, many types of video coding standard are now available to serve digital video playback, storage in CD or DVD, and broadcasting or streaming over the Internet. Digital videos are captured in two different forms: interlaced scan and progressive scan. Interlaced scan, which is the format used by analog broadcast TV systems, records the image in alternating sets of even and odd lines, each set of odd or even lines being referred to as a field, and a consecutive pairing of even and odd fields is called a *frame*: see Figure 5.1(a). A progressive scanning digital video records each frame as distinct, both fields being identical. Thus interlaced video captures twice as many fields per second as progressive video does when both operate at the same number of frames per second:

$$R = W \times H \times D \times F_s, \tag{5.1}$$

where W denotes the width (in pixels) of the video frame, H denotes the height of the video frame (in pixels), D denotes the depth (in bits per pixel), and F_s denotes the frame rate.

In comparison with digital image coding techniques, there are many additional technical issues and formats (see Table 5.1) involved in digital video coding, including the following.

(1) *Chroma subsampling*   Owing to the much larger amount of data required to be processed, chroma subsampling (see Section 4.4.1) becomes a critical step, and several formats are available. For example, in addition to the $4:4:4$ format for HDCAM-SR, the $4:2:2$ format for Betacam and DVCPRO50, and the $4:1:1$ format for DVCPRO and DVCAM, we can also find two versions of the $4:2:0$ format for DVD and some MPEG-1/MPEG-2 video (see Figure 5.1(b)) [1].

(2) *Intraframe compression*   This is a step very similar to the compression techniques used for coding an individual image frame, such as JPEG. It is used to remove spatial redundancy within a single frame, which can then be used as a reference for interframe compression.

(3) *Interframe compression*   This is one of the most distinct and critical steps for video coding. It is used to remove temporal redundancy between frames and is often referred to as motion estimation and compensation.

(4) *Entropy coding*   This is similar to the entropy coding used in image coding such as JPEG.

Table 5.1 Some formats in digital video

Image format	Resolution	Frame rate	YCbCr	Scan type
CCIR-601 (NTSC)	720×480	30	4 : 2 : 2	interlaced
				progressive
CCIR-601 (PAL)	720×576	25	4 : 2 : 2	interlaced
				progressive
SIF (NTSC)	352×240	30	4 : 2 : 0	progressive
SIF (PAL)	352×288	25	4 : 2 : 0	progressive
CIF	352×288	30	4 : 2 : 0	progressive
QCIF	176×144	30/15	4 : 2 : 0	progressive

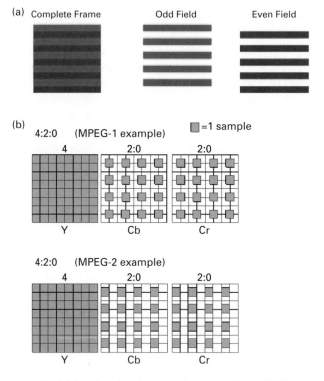

Figure 5.1 (a) Interlaced and progressive scan videos. (b) Two types of 4:2:0 format [1].

(5) *Rate control*   This is the determination of the quantization steps, in order that the overall bitrate can satisfy the specified data storage and transmission requirement.

5.1   Evolution of digital video coding

Digital video coding has gone through a long history of technological evolution (see Figure 5.2). The first practical digital video coding standard was H.261, proposed by the

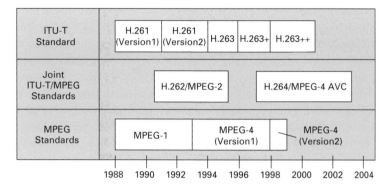

Figure 5.2 The evolution of digital video coding standards.

International Telecommunication Union (ITU-T) in 1990 [2], originally designed for transmission over ISDN lines on which data rates are multiples of 64 kbps. The coding algorithm was designed to be able to operate at a data rate between 40 kbps and 2 Mbps. The standard supports both CIF (352×288) and QCIF (176×144) video frames with the $4:2:0$ chroma subsampling. Owing to the pioneering effort of H.261 design and collaborative development, all subsequent international video coding standards (MPEG-1, MPEG-2/H.262, H.263, and even H.264) have been following a similar architectural design and standardization process. The H.261 standard is part of the H.320 group of standards for audio and visual communications. It forms the heart of many early-stage digital video applications in video conferencing and video communications systems, such as studio-based and desktop video conferencing, surveillance and monitoring, telemedicine, and computer-based training. In its later revision (the so-called version 2), H.261 allows a backward-compatible scheme for sending still picture graphics with 704×576 resolution.

As discussed in Chapter 3, MPEG-1 is a system-level standard which defines a group of audio and video (AV) coding and compression standards agreed upon by the MPEG committee. The MPEG-1 video coding standard [3] was originally designed with a targeted 1.5 Mbps data rate for 352×240 source input format (SIF) to replace the analog NTSC video and 352×288 SIF to replace the European analog PAL video resolution of progressive scanning video only. The MPEG-1 standard is used in the video CD (VCD) format, which has nearly the quality of a VHS tape. At present MPEG-1 is the most compatible format in the MPEG family; it is playable in almost all computers and VCD/DVD players.

The MPEG-2 video extends MPEG-1 in several respects, to make itself the generic coding of moving pictures and associated audio information. It has been widely used around the world to specify the format of the digital standard-definition (SD) or high-definition (HD) television signals broadcast by terrestrial ("over-the-air"), digital cable, and direct broadcast satellite TV systems. It also specifies the format of movies and other programs that are distributed on DVD and similar disks. More specifically, in its systems part (Part 1) MPEG-2 [4] defines two distinct (but related) container formats: the transport stream and the program stream. The transport stream is designed to carry digital video and audio over somewhat unreliable (networked) media, such as the DVB and ATSC digital broadcast applications. The program stream is designed for reasonably reliable (storage) media such as DVD and super VCD (SVCD) standards. The video part (Part 2) of MPEG-2, also known as ISO/IEC 13818–2 [4] [5] or ITU-T H.262, is similar to MPEG-1 but in addition provides support for interlaced video. MPEG-2 video is targeted at 3 Mbps and above for 720×480 (CCIR-601 for analog NTSC video) and 720×576 (CCIR-601 for analog PAL video) resolution. All standards-conforming MPEG-2 video decoders are fully capable of playing back MPEG-1 video streams. With some enhancements, MPEG-2 video and systems are also used in most HDTV transmission systems.

The H.263 video coding standard was developed as a low-bitrate video conferencing solution; it is an evolutionary improvement based on experience from the H.261, MPEG-1, and MPEG-2 standards. Its first version [6] was completed in 1995 and provided a suitable replacement for H.261 at all bitrates. It was further enhanced to a second version known as H.263v2 [7] (also known as H.263+ or H.263 1998) and to a third version known as H.263v3 (also known as H.263++ or H.263 2000). The H.263 ++ standard is the result of another new round of efforts to enhance video coding for real-time communication and related non-conventional services. Some new useful features were added in H.263++, but many of them have been incorporated in MPEG-4 H.264. Therefore, H.263++ never gained enough popularity for real-world product deployment. The added features include 4×4 block-size motion compensation with rate-distortion optimization, enhanced multi-frame reference picture selection, a new deblocking filter, a new intraspatial prediction, a new P-picture type in scalable enhancement layers, IDCT mismatch reduction with integer inverse transform, and error-resilient data partitioning, etc. [8] [9].

The H.263 codec was first designed to be utilized in H.324-based video conferencing and videotelephony systems over PSTN and other circuit-switched networks, but has since found use in H.323 for IP-based video conferencing, in H.320 for ISDN-based video conferencing, as well as in real-time-streaming protocol(RTSP)-based IP streaming and session-initiation-protocol(SIP)-based IP conferencing solutions. Most flash video contents, as used on sites such as YouTube (www.youtube.com/), Google Video (http://video.google.com/), MySpace (http://myspace.com/) etc. are encoded in H.263 format; H.263 video can be decoded with the free LGPL-licensed libavcodec library (part of the ffmpeg project), which is used by programs such as ffdshow, VLC media player (http://www.videolan.org/vlc/) and Mplayer (http://www.mplayerhq.hu/).

The next international standard was MPEG-4. This was undertaken as a major task in 1998 with intention of getting compatibility between many multimedia components: video, synthetic structures (graphics-like), audio, systems, reference software, test bitstreams, digital rights management, and so on. MPEG-4 is no longer a standard of video or audio coding algorithms, it is regarded more as a modular and extendable file type which is based on a combination of an open file specification and commercially licensed codecs, such as the Microsoft MPEG-4 v3 and Divx, etc. For example, an MPEG-4 file can contain, in addition to a video or audio stream, VRML (three-dimensional) objects, embedded interactivity, stage descriptions (placement of objects), and much more. The MPEG-4 video part alone (i.e., ISO 14496–2 MPEG-4 Part 2: Visual [10]) is quite large, with overcomplexity. There are many profiles and levels defined, the vast majority of which are unused by commercial applications. Most features included in MPEG-4 are left for individual developers to decide whether to implement them. This means that there are probably no complete implementations of the entire MPEG-4 set of standards. Moreover, some elements are not entirely clear and are open to various interpretations, and this is not to mention a somewhat annoying per-use and per-stream licensing policy. Therefore, even though the MPEG-4 standard was adopted by the International Streaming Media Alliance (ISMA) for the web-based streaming of multimedia data and also by the Third Generation Partnership Project (3GPP) for delivering multimedia data over 3G cell phones, it was held back and not widely distributed and deployed. Of course the more advanced video codecs, such as the H.264/AVC and Microsoft VC-1 (windows media video 9), have also aggressively moved on, which further slows down MPEG-4 distribution. Currently, Apple Quicktime 6 and RealMedia along with some consumer electronics can play back the MPEG-4 file format.

Owing to the increasing number of services and growing popularity of high-definition TV, there are needs for a higher efficiency of video coding than can be offered by MPEG-2. Moreover, other transmission media such as cable modem, xDSL, or 3G/UMTS offer much lower data rates than those of broadcast channels, and enhanced coding efficiency can enable the transmission of more video channels or higher-quality video representations within existing digital transmission capacities. This brings up the question of the major effort of the next enhanced video coding standard – the H.264 standard, also known as AVC and MPEG-4 Part 10 [11]. The H.264 standard was jointly developed by the ITU-T Video Coding Experts Group (VCEG) together with the ISO/IEC Moving Picture Experts Group (MPEG) as a collective partnership effort known as the Joint Video Team (JVT). This standard was designed to provide a "network friendly" video representation which can achieve a significant improvement in the rate-distortion efficiency, e.g., a factor 2 in bitrate savings, compared with MPEG-2 video; it has also achieved a significant coding gain over H.263, in the range of 25% to 50%, depending on the type of application. Most digital video broadcasting, video conferencing, video on demand, multimedia messenging, and video surveillance products now include H.264. The final drafting work on the first version of the H.264/AVC standard was completed in May 2003.

The H.264/AVC design covers a video coding layer (VCL), which efficiently represents the video content, and a network abstraction layer (NAL), which formats the VCL representation of the video and provides header information in a manner appropriate for conveyance by particular transport layers (such as the real-time transport protocol, RTP) or storage media. An NAL unit specifies a generic format for use in both packet-oriented and bitstream systems. The H.264/AVC video streams are commonly packetized for transportation over networks with the RTP protocol.

Windows Media 9 Series is the latest generation of digital media technologies developed by Microsoft, whose design objective is to enable the effective delivery of digital media through any network to any device. More specifically, in addition to Internet-based applications (e.g., subscription services, video on demand over IP, web broadcast, etc.), content compressed with Windows Media codecs is being consumed by a wide range of wired and wireless consumer electronic devices (e.g., mobile phones, DVD players, portable music players, car stereos, etc.). Windows Media content can also be delivered to consumers in physical formats, such as the secure digital (SD) memory card or CD or DVD. The Windows Media Video 9 (WMV-9) codec [12] is the latest video codec in this suite and is based on technology that can achieve state-of-the-art compressed video quality from very low bitrates (such as 160×120 at 10 kbps for modem applications) through to very high bitrates ($1280 \times 720 - 1920 \times 1080$ at 4–8 Mbps for high-definition video, and even higher bitrates for mastering).

To provide an alternative video codec to the H.264/AVC, the Society of Motion Picture and Television Engineers (SMPTE) built upon WMV-9 to form a new industry video coding standard – the SMPTE VC-1 (known formally as SMPTE 421M), which standardizes the decoder bitstream to facilitate independent implementation of interoperable encoders and decoders such as WMV-9. The crucial reason for creating the VC-1 standard is to ensure that VC-1 content is transport independent and container independent, allowing delivery over MPEG-2 and real-time transfer protocol (RTP) systems as well as advanced systems format (ASF). This enables device manufacturers and content services to create inter-operable solutions. The VC-1 standard contains coding tools for interlaced video sequences as well as progressive encoding. The main goal of VC-1 development and standardization is to support the compression of interlaced content without first converting it to progressive, making it more attractive to broadcast and video industry professionals. Both HD DVD and Blu-ray Disc have adopted VC-1 as a mandatory video standard (in addition to MPEG-2 and H.264/AVC), meaning that their video playback devices will be capable of decoding and playing video-content compressed using VC-1. Microsoft also designated VC-1 as the Xbox 360 video game console's official video codec, and game developers may use VC-1 for full-motion video included with games.

With the increasing deployment of heterogeneous networks, especially the rapidly growing wireless networks, where the end-to-end available bandwidth is dynamically changing, the need for scalable video coding techniques is getting continually greater. The MPEG com-mittee started a new work item on scalable video coding (SVC) with the title "Ad hoc group on exploration of interframe wavelet technology in video coding" in December 2001. It was then widely believed that so-called three-dimensional wavelet coding would be the key technique for achieving a wide range of scalability in combination with high coding effi-ciency. In response to the "Call for proposals on scalable video coding technology" issued by MPEG in October 2003, 14 technical proposals were submitted and evaluated in March 2004; among them were 12 wavelet-based proposals and two proposals were extensions of H.264/AVC, one of which was contributed by the image communication group of the Fraunhofer's Heinrich Hertz Institute (HHI) in Germany. The subjective tests showed that the HHI proposal outperformed all other proposals in terms of coding efficiency for one tested scalability scenario. After further subjective tests for different scalability scenarios in June 2004 and October 2004 that verified the superior coding efficiency of the HHI solution in comparison to all other submitted proposals, MPEG chose the HHI proposal as a starting point of its SVC standardization project in October 2004. In January 2005, MPEG and the Video Coding Experts Group (VCEG) of the ITU-T agreed to finalize the SVC project jointly as an amendment of their H.264/AVC standard, and the scalable coding scheme developed by the image communication group of the HHI was selected as the first working draft (WD-1) and was consented to in July 2007 as Version 8 of H.264/AVC [13] [14].

5.2   Compression techniques for digital video coding

As with image coding, the objectives of video coding are bitrate reduction in storage and transmission by the use of both statistical and subjective redundancies, and to take advantage of probabilistic information using entropy coding techniques to reduce the symbol encoding length losslessly. There exists a practical tradeoff between coding gain (compression ratio) and video quality, as well as between the compression ratio and implementation complexity.

High video compression is achieved by degrading the video quality, as normally measured by the frame-by-frame (or temporal average of) peak signal-to-noise ratio (PSNR), Eq. (4.10), which gives the objective quality of the decoded image frames in relation to the original image frames prior to encoding. It should be noted that the degree of video quality degradation (due to both objective degradation and visible artifacts)

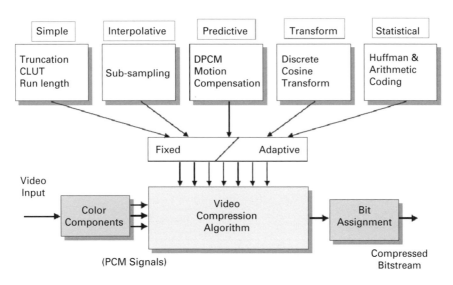

Figure 5.3 The techniques used in most video coding algorithms [15].

depends on the complexity of the video scene as much as on the sophistication of the compression technique, i.e., for simple textures in image frames and low video activity a good video reconstruction with no visible artifacts may be achieved even with simple video compression techniques. In general, the compression techniques used in most video coding algorithms can be summarized as in Figure 5.3 [15].

5.2.1 Simple techniques

Most simple techniques are used mainly for very-low-complexity operations or for lossless coding purposes. For example, the truncation technique can be adopted to reduce the number of bits used in representing either the spatial domain (RGB or YC_bC_r) or transform domain (DCT or wavelet) data values. The quantization of DCT or wavelet coefficients can be regarded as one type of truncation.

A color look-up table (CLUT) can be used to convert the logical color numbers stored in each pixel of video memory into physical colors, normally represented as RGB triplets, that can be displayed on a computer monitor. The number of logical color entries in the palette is the total number of colors which can appear on screen simultaneously, e.g., a common compression example would use a palette of $2^8 = 256$ colors (i.e., each addressed by an 8-bit pixel value), where each color can be chosen from a total of $2^{24} = 16.7$ million colors (i.e., an 8-bit output for each of red, green and blue).

As discussed in Chapter 4 run-length coding (RLC) is another useful technique for the lossless compression of finite sets of source symbols which have uneven frequencies of occurrence. When combined with entropy coding, Huffman coding, or arithmetic coding, RLC can be a very effective tool prior to the entropy or other coding in converting the data from one symbol representation to another symbol representation more adequate for entropy coding; an example is the conversion from quantized DCT coefficients to the intermediate representation used in JPEG (see Section 4.4.5.).

5.2.2   Subsampling and interpolation

As discussed previously, almost all video coding techniques make extensive use of chroma subsampling, such as $4:2:2, 4:2:0, 4:1:1$, etc. The fundamental principle behind chroma subsampling is to reduce the spatial dimensions of the chrominance components of input video (the horizontal dimension and/or the vertical dimension) so that the number of pixels to be coded prior to the encoding process can be reduced. Chroma subsampling for reducing the spatial dimensions can also be applied in the temporal direction to reduce the frame rate prior to coding. At the receiver the decoded images are interpolated for display. This technique makes use of specific physiological characteristics of the human eyes and thus removes the subjective redundancy contained in the video data.

5.2.3   Entropy coding

It is obvious that, as mentioned earlier, the average number of bits per symbol can be reduced if symbols with lower probability are assigned longer codewords and symbols having higher probability are assigned shorter codewords. Variable-length coding or entropy coding constitutes one of the most basic elements of today's multimedia coding standards, especially in combination with transform domain or predictive coding techniques. The two main techniques have been well discussed in Chapter 4: one is Huffman coding and the other is arithmetic coding. Entropy coding can be preceded by a run-length coding procedure to achieve further data reduction.

5.2.4   Predictive coding and motion estimation

When the correlation between spatially or temporally adjacent pixels or transform-domain coefficients is strong, prediction coding can be effectively applied. In basic predictive coding systems, an approximate prediction of the pixel or transform coefficient to be coded is made from previously coded information that has been transmitted. The difference between the actual pixel or transform coefficient and the prediction values (the prediction error) is usually quantized and entropy coded. This is the well-known differential pulse-code modulation (DPCM) technique. Predictive methods can also be combined with a run-length coding procedure. One such example was discussed in connection with coding the quantized DC coefficient of each block using that of the neighboring block, in the context of JPEG coding. For interframe (temporal) predictive coding, motion-compensated prediction (so-called motion estimation) has found wide application in standard video coding schemes. Motion-compensated transform coding, a two-dimensional extension of DPCM techniques, aims to exploit temporal and spatial redundancies by using some form of motion compensation followed by a transform coding. The key step in removing temporal redundancy is motion estimation, in which a motion vector is predicted between the current frame and the reference frame. Following the motion estimation, a motion compensation stage is applied to obtain the residual image, i.e., the pixel differences between the current frame and a reference frame. This residual is then compressed using transform coding or a combination of transform and entropy coding. Most video compression standards employ block motion estimation techniques: one motion vector is assumed to be representative for the motion of a "block" (commonly called a macroblock) of adjacent pixels when the temporal correlation between two adjacent image frames is being used. The main advantages of using fixed-size block motion estimation are the simplicity of the algorithm and the fact that no segmentation information needs to be transmitted.

Figure 5.4 shows a (fixed-size) block-matching-based approach for motion compensation, where one *motion vector mv* is estimated for each block in the current frame to be encoded. The image frame is first divided into equal size (16×16) macroblocks, as shown in Figure 5.4(a). The motion vector for each macroblock points to a reference block of the same size in a previously coded frame (say frame $n-1$ or an even earlier frame). This reference block is regarded as the most similar spatially displaced counterpart of this macroblock. The amount of spatial displacement is called the *motion vector*. The motion-compensated prediction error is calculated by subtracting each pixel in a block from its motion shifted counterpart in the reference block of the previous frame. Usually both the prediction error and the motion vector of this macroblock are transmitted to the receiver. The motion-compensated DPCM technique (used in combination with the subsequent DCT-based

(a)

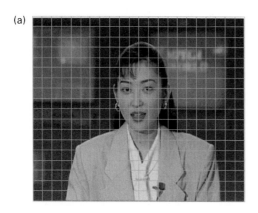

☐ 16×16 Macroblock

(b)

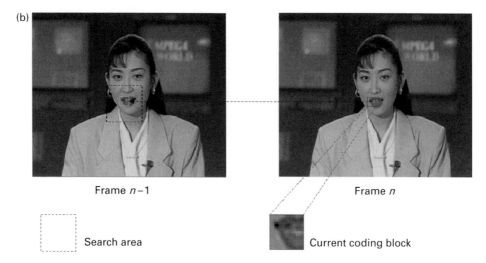

Figure 5.4 (a) A video frame is partitioned into equal size (16×16) macroblocks for motion estimation and compensation. (b) The search area for performing the similarity measure.

transform coding) has proven to be highly efficient and robust for video data compression and has become a key element in the success of today's state-of-the-art coding standards.

In order to determine the motion vector for each macroblock, it is general practice to define a search area (commonly centered around the same location as that of the macroblock in the current frame) in the reference frame for similarity matching (see Figure 5.4 (b)). The size of the search area is determined by assuming how fast the motion is between the current and reference frames. Within the search area, the macroblock $\{a(i, j)\}$ of size $N \times N$ will find its best motion match with motion vector (u^*, v^*) (see Figure 5.5) according to the following criterion, involving minimization of the sum of the absolute differences (SAD):

$$(u^*, v^*) = \arg\min_{(u,v)} SAD(u, v) = \arg\min_{(u,v)} \sum_{i=1}^{N} \sum_{j=1}^{N} |a(i,j) - b(i+u, j+v)| \qquad (5.2)$$

where $\{b(i, j)\}$ is the searched block within the search area from the reference frame. The output of the motion-estimation algorithm comprises the motion vector for each block and the pixel value differences between the blocks in the current frame and the "matched" blocks in the reference frame. We call this difference signal the motion compensation error or simply the block error.

For a fixed motion-estimation block size of 16×16 pixels, the maximum displacement can be as large as ± 64 pixels from the block's original position. A traditional motion-estimation algorithm uses an exhaustive search (ES) (also called a full search, see Figure 5.6), in which every possible integer displacement within a presumed square search region (in Figure 5.6, $u \in [-7, +7]$, $v \in [-7, +7]$) is searched, and this requires very high computational costs. Since, in all the video coding standards, the methods and the reliability and accuracy of motion estimation are not defined this leaves room for some innovation in the implementation. Thus, many fast motion-estimation techniques have been proposed for video compression [16]. Since in most cases motion estimation constitutes a high percentage of the computational load

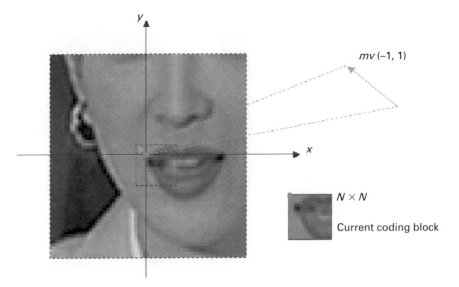

Figure 5.5 Within the search area, the macroblock finds its best motion match with the motion vector (u^*, v^*).

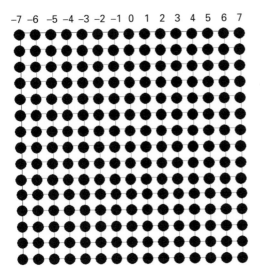

−7 −6 −5 −4 −3 −2 −1 0 1 2 3 4 5 6 7

Figure 5.6 An exhaustive search within the motion estimation search area.

on the video encoder, there is a need for a fast, simple and efficient motion-estimation algorithm. Several fast search strategies are called for, normally using some kind of sampling mechanism instead of an exhaustive search, which is the most straightforward while computationally expensive method.

5.2.4.1 Rectangular-pattern fast block search

Many fast algorithms have been proposed to pursue low computational complexity. The most popular example is three-step search (3SS) [17], which employs rectangular search patterns (a central point plus eight neighboring points) and searches for the best motion vectors in terms of the minimum sum of absolute differences, SAD in a coarse-to-fine search pattern. The 3SS has pretty robust and near optimal performance. Figure 5.7 shows the displacements (indicated as 1, 2, 3) visited by the 3SS for the square search region in Figure 5.6 ($u \in [-7, +7]$, $v \in [-7, +7]$).

The 3SS uses a uniformly allocated point checking pattern in its first step, but this becomes inefficient for the estimation of small motions. Moreover, experimental results show that the block motion field of a real-world image sequence is usually gentle, smooth, and slowly varying. Use of the 3SS results in a center-biased global minimum motion vector distribution instead of a uniform distribution. To overcome the deficiencies introduced by 3SS, a four-step search (4SS) algorithm was proposed [18]. As shown in Figure 5.8, for the maximum motion displacements of ±7, the 4SS algorithm utilizes a center-biased search pattern with nine checking points on a 5×5 window in the first step instead of the 9×9 window in the 3SS. The center of the search window is then shifted to the point with minimum SAD. The search window sizes of the next two steps are dependent on the location of the minimum SAD points. If the minimum SAD point is found at the center of the search window, the search will go to the final step (step 4) with 3×3 search window. Otherwise, the search window size is maintained in 5×5 for step 2 or step 3, depending on whether the minimum SAD point is on a horizontal or vertical neighbor or on a diagonal neighbor, and in the final step, the search window is reduced to 3×3; the search stops at this small search window.

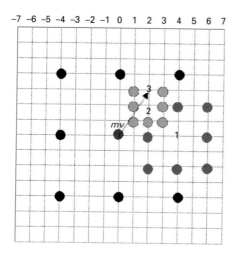

Figure 5.7 The three-step search (3SS) for motion estimation.

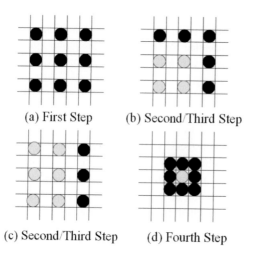

(a) First Step   (b) Second/Third Step

(c) Second/Third Step   (d) Fourth Step

Figure 5.8 The four-step search (4SS) algorithm for motion estimation [18] (© IEEE 1996).

From the algorithm, we find that the intermediate steps of the 4SS may be skipped, and a jump to the final step with a 3×3 window can be made, if at any time the minimum BDM point is located at the center of the search window. Using this 4SS pattern, the whole 15×15 displacement window can be covered even with small, 5×5 and 3×3, search windows. The computational complexity of the 4SS is in general less than that of the 3SS, while the performance in terms of quality is as good. The 4SS is also more robust than the 3SS and it maintains its performance for image sequences with complex movements such as camera zooming and fast motion. Hence it is a very attractive strategy for motion estimation. Figure 5.9 shows two search path examples based on 4SS for the ± 7 search window size.

To take advantage of the smooth block-motion field of a real-world image sequence and a center-biased search pattern, a block-based gradient descent search (BBGDS)

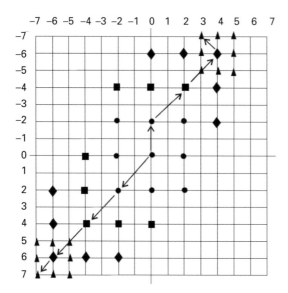

Figure 5.9 Two search path examples based on 4SS for the ±7 search window size [18] (© IEEE 1996).

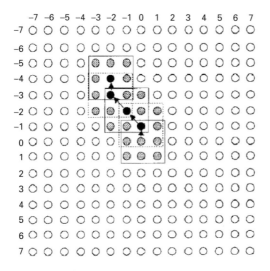

Figure 5.10 The block-based gradient descent search (BBGDS) motion-estimation method [19] (© IEEE 1996).

motion-estimation method was proposed [19]. The BBGDS utilizes the statistical nature of the motion by assuming that the global minimum of SAD has a monotonic distortion in its immediate neighborhood. More specifically, BBGDS only searches around the center point. If the optimum is found at the center then the procedure stops, which is the case more than 80% of the time. Otherwise, BBGDS proceeds to search around the point where the minimum was found. The procedure continues until the winning point is a center point of the checking block or the checking block hits the boundary of the predefined search range. The search procedure of the BBGDS algorithm is illustrated in Figure 5.10, where the BBGDS starts by initializing the checking block so that its center pixel is at the origin; then the procedure is as follows.

(1) The SAD values are evaluated for all nine points in the checking block.
(2) If the minimum occurs at the center, the procedure stops; the motion vector points to the center. Otherwise, the checking block is reset so that its center is the winning pixel (the minimum SAD point), and step 1 is repeated.

Note that except for the first iteration, most of the pixels in the checking block will have been checked already in a previous pass. The search procedure of the BBGDS always moves the search in the direction of optimal gradient descent. This is the direction where one expects the SAD value to approach its minimum. The procedure is illustrated in Figure 5.10, where the motion vector $(-2, -4)$ is found.

5.2.4.2 Non-rectangular-pattern fast block search

It is estimated that 52.76% to 98.70% of the motion vectors are enclosed in a circular support with a radius of two pixels and centered on the position of zero motion. Moreover, the block displacement of real-world video sequences could be in any direction but is mainly in the horizontal and vertical directions (e.g., camera panning). Using these two crucial observations, a *diamond search* (DS) algorithm was developed [20]. The proposed DS algorithm employs two search patterns, the *large diamond search pattern* (LDSP) and the *small diamond search pattern* (SDSP). The LDSP comprises nine checking points, eight of which surround a central point, composing a diamond shape, and the SDSP consisting of five checking points that form a smaller diamond shape, as shown in Figure 5.11.

In the searching procedure of the DS algorithm, LDSP is repeatedly used until a step in which the minimum BDM occurs at the center point. The search pattern is then switched from LDSP to SDSP as this final search stage is reached. Among the five checking points in SDSP, the position yielding the minimum BDM provides the motion vector of the best-matching block. The DS algorithm can be summarized as follows.

(1) The initial LDSP is centered at the origin of the search window, and the nine checking points of LDSP are tested. If the minimum SAD point calculated is located at the center position, go to step 3; otherwise, go to step 2.
(2) The minimum SAD point found in the previous search step is re-positioned as the center point, to form a new LDSP. If the new minimum SAD point obtained is located at the center position, go to step 3; otherwise, recursively repeat this step.

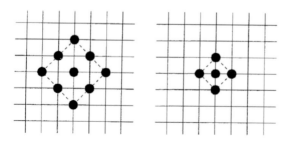

Figure 5.11 The diamond search algorithm employs two search patterns, the large diamond search pattern and the small diamond search pattern [20] (© IEEE 1997).

(3) Switch the search pattern from LDSP to SDSP. The minimum SAD point found in this step is the final solution for the motion vector and points to the best-matching block.

In the example shown in Figure 5.12, the DS search path leads to the motion vector $(4, -2)$ in five search steps, i.e., four steps of LDSP and finially one step of SDSP. There are 24 search points in total with nine, five, three, three, and four search points being taken at each step, sequentially.

5.2.5 Transform domain coding

As in still image coding, transform coding, such as the discrete cosine transform (DCT), has been also used extensively for video coding. The purpose of transform coding is to decorrelate the image frame content and to encode transform coefficients rather than the original pixels of the video frames. Such DCT-based implementations are used in most image and video coding standards owing to their high decorrelation performance and the availability of fast DCT algorithms suitable for real-time software and hardware implementations.

The DCT is mainly used in two different ways in video coding: in the first it is mainly used for reducing the spatial redundancy of a single video frame (as in JPEG image coding) and in the second it is combined with the motion compensation technique to first reduce the temporal redundancy, using the motion estimation, this is followed by a DCT to reduce the spatial redundancy further, i.e., the DCT is applied to the motion compensation error $\{e(i, j)\}$ of each block, where

$$e(i,j) = a(i,j) - b(i + u^*, j + v^*). \tag{5.3}$$

Since a human viewer is sensitive to reconstruction errors related to low spatial frequencies rather than to high frequencies, a frequency adaptive weighting (quantization) of the DCT coefficients according to human visual perception (perceptual quantization) is desirable to improve the visual quality of the decoded images for a given bitrate. As discussed in Chapter 4,

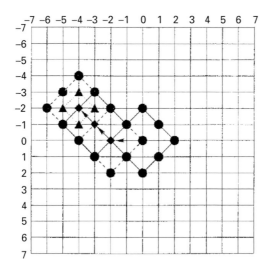

Figure 5.12 The DS search path leads to the motion vector $(4, -2)$ in five search steps [20] (© IEEE 1997).

the DCT coefficients are quantized prior to coding. The quantization operation will result in a significant number of zero valued coefficients; advantage is taken of this by the use of run-length coding followed by entropy coding. When DCT is combined with motion estimation to become a hybrid coding scheme, both temporal and spatial redundancies can be reduced. Such a hybrid DCT–DPCM coding has become the core method for all recent video coding standards.

5.2.6   Rate control in video coding

Rate control is an essential part of most video encoders. Rate control in video coding basically determines which frames are coded and the number of bits or quality level of the encoded frames. At lower bitrates, the perceived quality of coded video can be determined by tradeoffs between the frame rate and the quantized frame quality. However, at higher bit rates it is essential to maintain the full frame rate, and the difficulty lies in maintaining an appearance with uniform (high) quality as well. Most video standards do not specify how to control the bitrate, so that the exact rate control of the encoder is generally left open to user specifications. Ideally, the encoder should balance the quality of the decoded images against the available channel bandwidth. This complexity of the rate control problem is compounded by the fact that video sequences contain widely varying contents and motions. More specifically, if we set the same quantization step size (so-called constant-quality video coding) then, owing to varying video contents and motion amount, the resulting bitrate will be dynamically changing. However, if we want to achieve constant bitrate coding then a dynamically changing quantization step size is needed. This results in two different types of rate controls, i.e., constant bitrate (CBR) and variable bitrate (VBR) video coding. For CBR video coding, rate control designers mainly focus on improving the matching accuracy between the target bitrate and actual bitrate and satisfying the low-latency and buffer constraints. In CBR applications, the fluctuation of video quality cannot be avoided owing to the varying content in natural scenes. By taking the advantage of the encoder buffer, smoothing the video quality is possible as the buffer can tolerate limited bitrate fluctuation provided that it does not either overflow or underflow. However, VBR video coding is employed in some applications in which natural video frames need to be presented in a stabler way. Variable bitrate video can also be incorporated in a VBR transmission networking infrastructure. In cases where the delay or rate constraint is not as strict as in real-time video coding, VBR rate control is expected to present video at constant quality for the end users or to be adaptive to both the source and channel conditions [21].

The rate control of video coding can be formulated as follows. Given the desired bitrate and video sequences with a certain complexity, how can we efficiently encode the video sequences to achieve the highest quality (objective or subjective) of the encoded video? The problem of optimal bit allocation has been researched extensively in the literature. It still remains as a very active area of research in visual communication owing to the emergence of new video codecs and an increasing number of heterogeneous networks [22].

Generally speaking, the rate control algorithm consists of two main parts: the first explores the rate–distortion (R–D) relationship of the video source; the second involves the design of optimal or suboptimal control techniques that achieve optimization aims while satisfying the constraints of the applications [21]. According to rate–distortion (R–D) theory [23] [24], the distortion D is a decreasing function of the bitrate R (see Figure 5.13). A decreasing distortion leads to an increasing rate and vice versa. So the fundamental problem in rate control can be stated as follows:

$$\text{minimize } D \text{ subject to } R \leq R_{\max}, \tag{5.4}$$

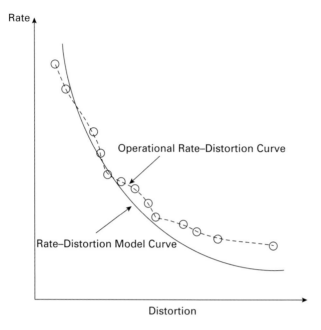

Figure 5.13 According to rate–distortion (R–D) theory, the distortion D is a decreasing function of the bitrate R [21].

where R_{max} denotes the maximum bitrate. In other words, rate control consists of achieving the maximum picture quality (minimum distortion) without exceeding the maximum permitted bitrate where quality is typically represented by the peak signal-to-noise ratio (PSNR).

In R–D theory, the R–D function is defined to describe the lower bound for the rate at a given distortion. However, there is no guarantee that this lower bound can be achieved in practical video coding schemes. In comparison, operational rate–distortion (ORD) theory [25] is more applicable for video compression. This theory applies to lossy data compression with a finite number of possible R–D pairs $\{R(q_i), D(q_i)\}$, the plotting of which generates the ORD function, with

$$R(q_i) \geq R(q_j) \Rightarrow D(q_i) < D(q_j),$$

where the set of all admissible quantizers is defined as $Q = \{q_0, q_1, \ldots, q_{M-1}\}$, $R(q_i)$ denotes the rate, and $D(q_i)$ denotes the distortion for a particular quantizer q_i, respectively. The ORD function represents the convex curve of the specific compression scheme in such a way that the optimal solution for rate control, i.e., the optimal quantizer achieving minimum distortion at a given bitrate, can be obtained. However, ORD-function-based rate control schemes do not work efficiently in many practical video coding applications, especially for real-time video coding, since the generating of the ORD curve by actual R–D pairs has high computational complexity and introduces intolerable delays. So, in many video coding systems, model-based rate control schemes are adopted. The R–D relationships are approximated by R–D models. Some models are derived using the statistical properties of video signal and R–D theory and others are developed empirically and benefit from various regression techniques [21]. But these R–D models may suffer from relatively large esti-mation errors as shown in Figure 5.13, where the circles correspond to R–D pairs of the admissible quantizers and the solid line shows an approximation by an R–D model. Owing to the strict delay or complexity requirements of video coding applications, model-based R–D functions have been widely used in practical video systems. The model-based approach assumes various input distributions and quantizer characteristics such as are found in the quadratic model, the exponential model, the normalized rate distortion model, or the spline approximation model [21]. Using the model-based approach, we can get closed-form solutions. The drawback of this approach is that if the statistical distribution of the input video sequence does not fit the model (which is the case most of time) then the quality of the encoded video will not be good. The ORD-based approach has been used in practical coding environments, where the inputs are completely arbitrary and the quantizers are from a finite set of admissible quantizers.

There are two different types of ORD-based approach: feedforward bitrate control and feedbackward bitrate control (see Figure 5.14) [22]. In feedforward bitrate control, the exact relation between the encoder settings, e.g., quantizer step size, bitrate, and quality, is assumed to be available. Using this relation, the optimum setting of the encoder can be computed exactly. The drawback of feedforward rate control approaches is that the input images have been coded independently, so that the problem of selecting the best encoding strategy for a frame is not considered at the macroblock level. The only way that the optimal solution is feasibly found is if the numbers of quantization choices are relatively small in the feedforward control. Thus, the application of feedforward bitrate control is more useful for non-real-time applications. Feedbackward rate control, however, takes account of the real-time transmission status of the out-going channel, with the feedback of the output buffer, to determine the best encoding rate. The formulation of feedbackward is normally much more involved, with higher computational complexity. It is possible to employ a combination of feedforward and feedbackward ORD frameworks to formalize the problem of optimizing the encoder rate control on a macroblock-by-macroblock basis within each frame of a video sequence [22].

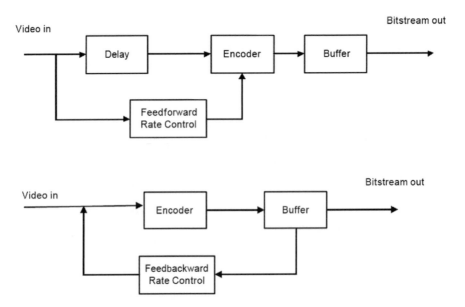

Figure 5.14 The two different ORD-based approaches, feedforward rate control and feedbackward rate control [22].

5.3   H.263 and H.263+ video coding

In order to achieve better compression performance and channel error robustness levels than can be achieved by H.261, Version 1 of the international standard ITU-T H.263, entitled "Video coding for low bitrate communications" [6], provides better picture quality at low bitrates with little additional complexity. It also includes four optional modes aimed at improving compression performance. The H.263 standard has been adopted in several videophone-terminal standards, notably ITU-T H.324 (PSTN), H.320 (ISDN), and H.310 (B-ISDN). The H.263 Version 2 standard, also known as H.263+ in the standards community, was officially approved as a standard in January 1998 [7]. It is an extension of H.263 providing 12 new negotiable modes and additional features. These modes and features improve compression performance, allow the use of scalable bitstreams, enhance performance over packet-switched networks, support custom picture size and clock frequency, and provide supplemental display and external usage capabilities.

5.3.1   The ITU-T H.263 standard

Figure 5.15 shows a block diagram of an H.263 baseline encoder, in which motion-compensated prediction is used first, to reduce temporal redundancies. Discrete-cosine-transform (DCT)-based algorithms are then used for encoding the motion-compensated prediction difference frames. The quantized DCT coefficients, motion vectors, and side information are entropy coded using variable-length codes (VLCs). The two switches represent intra or inter mode selection, which is not specified in the standard; this can be made at the macroblock level. The performance of the motion-estimation process, usually measured in terms of the associated SAD values, can be used to select the coding mode (intra or inter). If a macroblock does not change significantly with respect to the reference picture then an encoder can choose not to encode it, and the decoder will simply repeat the macroblock located at the subject macroblock's spatial location in the reference picture.

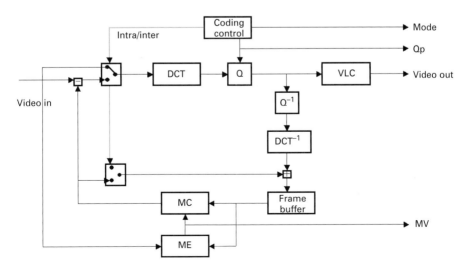

Figure 5.15 A block diagram of an H.263 baseline encoder.

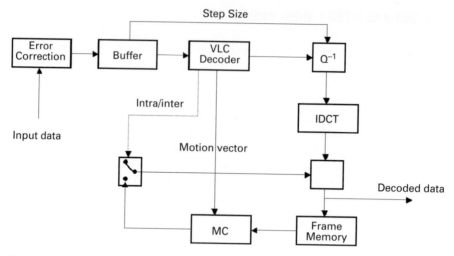

Figure 5.16 A block diagram of an H.263 baseline decoder.

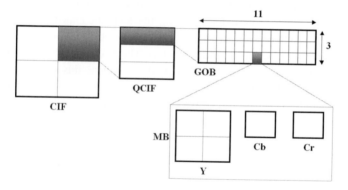

Figure 5.17 The picture structure of a CIF resolution.

Figure 5.16 illustrates an H.263 video decoder, which in fact can be regarded as a subset of video encoder that requires to decode the video for motion-compensation purposes during the encoding process.

5.3.1.1 Video frame structure

The H.263 standard supports five standardized picture formats: sub-QCIF, QCIF, CIF, 4CIF, and 16CIF. The luminance component of the picture is sampled at these resolutions, while the chrominance components are downsampled by a factor 2 in both the horizontal and vertical directions. The picture structure is shown in Figure 5.17 for the CIF resolution. Each CIF picture, which is four times a QCIF picture, in the input video sequence is divided into macroblocks (MBs) consisting of four luminance blocks, each in turn consisting of 8×8 pixels. A group of blocks (GOB) is defined as an integer number of macroblock rows, a number that is dependent on picture resolution. For example, a GOB could consist of a single macroblock row at QCIF resolution.

Interpicture temporal prediction based on motion estimation and compensation is supported by H.263. The coding mode in which temporal prediction is used is called inter mode, otherwise it is called intra mode.

5.3.1.2 Motion estimation and compensation

In baseline H.263, each 16×16 macroblock is predicted from the previous frame. One motion vector, a two-dimensional displacement vector (u^*, v^*), is estimated for each macroblock, which implies an assumption that each pixel within the macroblock undergoes the same amount of translational motion. Both the horizontal and vertical components of the motion vectors may be of half-pixel accuracy, but their values may lie only in the [16, 15.5] range, limiting the search window used in motion estimation. A positive value of the horizontal or vertical component of the motion vector represents a macroblock spatially to the right or below the macroblock being predicted, respectively. Half-pixel accuracy can be achieved after the best integer-motion vector, which corresponds to the minimum SAD values calculated within the search window, is found, and then bilinear interpolation is used to calculate half-pixel values for SAD comparisons around the optimal integer motion vector with the reference macroblock (see Figure 5.18). For the coding of inter coded P-pictures, the previously encoded-then-decoded I- or P-picture has to be stored in a frame buffer in both the encoder and decoder, so that motion estimation and compensation can be performed on a macroblock basis. Only one motion vector is estimated between this frame and the reference (I or P) frame for each macroblock to be encoded. These motion vectors are coded and transmitted to the receiver. The motion-compensated prediction error is calculated by subtracting each pixel in a macroblock from its motion-shifted counterpart in the previous frame. An 8×8 DCT is then applied to each of the 8×8 blocks contained in the macroblock, and this is followed by quantization of the DCT coefficients with subsequent run-length coding and entropy coding (VLC).

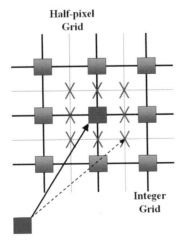

Figure 5.18 Half-pixel accuracy for motion estimation can be determined after the best integer-motion vector is found. A bilinear interpolation is used to calculate half-pixel values for SAD comparisons around the optimal integer motion vector with the reference macroblock.

5.3.1.3 DCT transform

The 8×8 DCT specified by H.263 is used to decorrelate the 8×8 blocks of original pixels or motion-compensated difference pixels, and to compact their energy into as few coefficients as possible. The most common algorithm for implementing the 8×8 DCT is that which consists of eight-point DCT transformations of the rows and the columns, respectively. The original 8×8 block of pixels can be recovered using an 8×8 inverse DCT (IDCT). Although exact reconstruction can be theoretically achieved, it is often not possible using finite-precision arithmetic. While forward DCT errors can be tolerated, inverse DCT errors must meet the H.263 standard if compliance is to be achieved.

5.3.1.4 Quantization

Unlike the use of different step sizes in the quantization table (see Table 4.1 for JPEG image coding), in H.263 quantization is performed using the same step size within a macroblock (i.e., using a uniform 8×8 quantization table for all four DCT blocks within the same macroblock). The uniform quantization levels can be set in the range from 2 to 62, except for the first coefficient (the DC coefficient) of an intra block, which is uniformly quantized using a step size 8. The quantizers consist of equally spaced reconstruction levels with a deadzone centered at zero. After the quantization process, the reconstructed picture is stored so that it can be used later for prediction of the future picture.

5.3.1.5 Entropy coding

Entropy coding is first performed by means of variable-length codes (VLCs). Motion vectors are first predicted by setting their components' values to the median values of neighboring motion vectors already transmitted, i.e., the motion vectors of the macroblocks to the left, above, and above right of the current macroblock. The difference motion vectors between the current macroblock and the predicted (median) motion vector are then VLC coded (see Figure 5.19). More specifically,

$$MV_d = MV - MV_p \ (MV_p \text{ is the predictor}), \tag{5.5}$$

$$MV_p = \text{median}(MV1, MV2, MV3). \tag{5.6}$$

As in JPEG coding, prior to entropy coding the quantized DCT coefficients are arranged in a one-dimensional array by scanning them in zigzag order (see Figure 4.11(b)). The

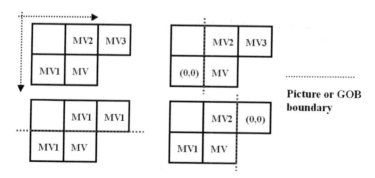

Figure 5.19 The difference motion vectors between the current macroblock and the predicted (median) motion vector are VLC coded.

rearranged array is coded using a three-dimensional run-length VLC table representing the 3-tuple $H(R, S) + B(V)$, as discussed in Section 4.4.4 for JPEG. This coding method produces a compact representation of the 8×8 DCT coefficients, as a large number of the coefficients are normally quantized to zero and the reordering results (ideally) in the grouping of long runs of consecutive zero values. Other information such as prediction types and quantizer indication is also entropy coded by means of VLCs.

5.3.1.6 Optional modes

In addition to the core encoding and decoding algorithms described above, H.263 includes four negotiable advanced coding modes: unrestricted motion vectors, advanced prediction, *PB* frames, and syntax-based arithmetic coding. The first two modes are used to improve inter-picture prediction. For example, when the unrestricted motion vector mode is used, motion vectors can take on values in the range [31.5, 31.5] instead of [16, 15.5], which allows the motion vectors to point outside the picture boundaries. The advanced prediction mode allows for the use of four motion vectors per macroblock, one for each of the four 8×8 luminance blocks with median-value prediction from the motion vectors of three neighboring luminance blocks. The *PB*-frame mode improves temporal resolution with little bitrate increase. When the syntax-based arithmetic coding mode is enabled, arithmetic coding replaces the default (Huffman) VLC coding. These optional modes allow developers to trade compression performance and complexity. More detailed description of such modes can be found in [26].

5.3.2 The ITU-T H.263+ standard

The H.263+ or H.263 Version 2, which is backward compatible with H.263, was developed with the intention of broadening the range of applications and to improve compression efficiency. This standard offers many improvements over H.263. More specifically, it allows the use of a wide range of custom source formats; in H.263 only five video source formats, defining picture size, picture shape, and clock frequency, can be used. This added flexibility opens H.263+ to a broader range of video scenes and applications, such as wide format pictures, resizeable computer windows, and higher refresh rates. Moreover, picture size, aspect ratio, and clock frequency can be specified as part of the H.263+ bitstream. H.263+ also introduces a larger motion search range based on the video frame sizes, e.g., [−32, 31.5] for frame sizes up to CIF and [−64, 63.5] for 4CIF, and it can be as large as [−256, 255.5] for video frame sizes up to 2048×1152. Another major improvement of H.263+ over H.263 is scalability, which can improve the delivery of video information in error-prone, packet-lossy, or heterogeneous environments by allowing multiple display rates, bitrates, and resolutions to be available at the decoder. Furthermore, picture segment dependences may be limited, which is likely to reduce error propagation.

In addition to an enhanced unrestricted motion vector mode added in H.263+, 12 new optional coding modes were introduced in the H.263+ video coding standard. The new unrestricted motion vector mode in H.263+, which is different from that of H.263, adopts new reversible VLCs (RVLCs) to encode the difference between an actual motion vector and the corresponding predicted motion vector. The idea behind RVLCs is that decoding can be performed by processing the received motion vector part of the bitstream in the forward and reverse directions. If an error is detected while decoding in the forward direction then the motion vector data is not completely lost, as the decoder can proceed in the reverse direction. This improves the error resilience of the bitstream. Furthermore, the motion vector range is extended to up to ±256, depending on the picture size. This is very useful given the wide range of new picture formats available in H.263+. Among the 12 new optional coding modes, some are intended to improve coding efficiency, some for overall picture quality enhancement, some for improved error robustness of mobile video or video over an unreliable transport environment, and some address scalability issues [26].

5.3.3   Rate control in H.263+

An important motivation for the development of the emerging H.263+ coding standards was enhancement of the quality of highly compressed video for two-way real-time communi-cations. In these applications, the delay produced by bits accumulated in the encoder buffer must be very small, typically below 100 ms, and a rate control strategy encodes the video with high quality and maintains a low buffer delay. A rate control technique for H.263+ was proposed in [27] that achieved these two objectives by intelligent selection of the quant-ization parameter values in typical DCT-based video coders. Using Lagrange optimization, the distortion is minimized subject to the target bit constraint of choosing optimal quant-ization parameters Q^*. This method was adopted as a rate control tool in the test model TMN8 of H.263+. The idea is based on the principle that more bits should be allocated to macroblocks with more "perceptual" activity, which can be measured based on the variance σ^2 of the DCT coefficients in that macroblock. It was shown in [27] the average or expected number of bits B_i produced by the ith macroblock in a frame is

$$B_i = A\left(K\frac{\sigma_i}{Q_i^2} + C\right),$$ (5.7)

where σ_i is the macroblock's empirical standard deviation, C is referred to as the overhead rate, A is the number of pixels in a macroblock, and K is a constant with typical value $e/\ln 2$. In the distortion model, the typical distortion measure D for the encoded macroblocks can be given as

$$D = \frac{1}{N}\sum_{i=1}^{N} a_i^2 \frac{Q_i^2}{12},$$ (5.8)

$$Q_1^*, \ldots, Q_N^* = \underset{Q_1, \ldots, Q_N}{\arg\min} \frac{1}{N}\sum_{i=1}^{N} a_i^2 \frac{Q_i^2}{12},$$ (5.9)

$$B = \sum_{i=1}^{N} B_i,$$

where N is the number of macroblocks in a frame, B is the target number of bits for a frame, and a_i is the distortion weighting of the ith macroblock. The TMN8 rate control mechanism tries to find the optimal quantization step sizes $\{Q_i^*\}$ of all macroblocks by minimizing the total distortion. By substituting Eq. (5.7) into Eq. (5.9) and setting partial derivatives to zero in Eq. (5.8) we can obtain the optimal quantization values:

$$Q_i^* = \sqrt{\frac{AK}{(\beta_i - AN_iC)}\frac{\sigma_i}{a_i}\sum_{k=1}^{N} a_k\sigma_k}, \quad i = 1, \ldots, N,$$ (5.10)

where N_i is the number of macroblocks that remain to be encoded in the frame and β_i denotes the number of bits left for encoding the frame, with $\beta_i = B$ at the initialization stage. Equation (5.10) can be treated as the key solution for the TMN8 rate control technique.

5.4 MPEG-1 and MPEG-2 video coding

The Moving Picture Experts Group (MPEG) was started in 1988 as a working group within ISO/IEC with the aim of defining standards for the digital compression of audio-visual signals. MPEG-1 was the first standard, finalized by MPEG in 1993 as ISO/IEC 11172. The video part, Part 2, of the MPEG-1 standard was developed to operate principally from storage media offering a continuous transfer rate of about 1.5 Mbps [3]. The MPEG-1 video coding is a generic standard which supports a very wide range of applications profiles with a large diversity of input parameters to be specified by the users. The MPEG-2 video coding, Part 2 of ISO/IEC 13818 standard [5], further extends the capabilities of MPEG-1 video coding and is capable of coding standard-definition television (SDTV) at bitrates from about 3–15 Mbps and high-definition television (HDTV) at 15–30 Mbps. As for MPEG-1 video, which is adopted as the video CD storage format, MPEG-2 video also specifies the format of movies and other programs that are distributed on DVDs and similar disks. Moreover, MPEG-2 video also specifies the format of the digital television signals that are broadcast by terrestrial (over-the-air), cable, and direct broadcast satellite TV systems.

5.4.1 MPEG-1 video coding

The MPEG-1 video algorithm was developed in parallel with JPEG and H.261 standardization activities, therefore the basic MPEG-1 (as well as the MPEG-2) video coding technique is very similar to the H.261 hybrid DCT/DPCM block-based scheme, which includes macroblock structure, motion compensation, and conditional replenishment.

5.4.1.1 Picture organization in MPEG-1 video

As shown in Figure 5.20 [28], there are six hierarchical layers in an MPEG-1 video bitstream: the video sequence, group of pictures (GOP), picture, slice, macroblock, and block layers. More specifically, the MPEG-1 coding algorithm partitions each video sequence into many non-overlapping groups of pictures, each always beginning with an intra coded frame (I-picture) and followed by several forward predictive frames (P-pictures), in each case at a distance of some frames. In the remaining gaps are bidirectional (i.e., bidirectional predicted and bidirectional interpolated) frames (B-pictures), which were introduced into MPEG-1 to support important functionalities for accessing video from storage media, such as random access and fast forward (FF) and fast reverse (FR) playback, and to improve further the coding efficiency gained from motion estimation (see Figure 5.21). For examples, for MPEG-1 coding based on motion-compensated temporal redundancy, each I-picture requires about 300K bits, each P-picture requires about 65K–100K bits (slow–fast motion), and each B-picture requires only about 7K–18K bits (slow–fast motion).

As shown in Figure 5.21, with the next I-frame a new GOP begins. Typically, every 15 frames (0.5 second of a 30 frames per second video sequence) with the frame organization ($I_1 B_2 B_3 P_4 B_5 B_6 P_7 B_8 B_9 P_{10} B_{11} B_{12} P_{13} B_{14} B_{15}$) is made into a GOP. However, the standard is flexible about this and allows users to specify the GOP structure (e.g., how many pictures in a

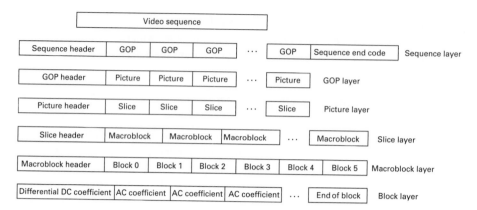

Figure 5.20 There are six hierarchical layers in an MPEG-1 video bitstream: the video sequence, group of pictures, picture, slice, macroblock, and block layers [28] (© Elsevier 2005).

I	B	B	P	B	B	P	B	B	P	B	B	P	B	B

Figure 5.21 A 15-frame GOP structure with I-, P- and B-pictures.

GOP and how many B-pictures in between a reference picture pair, either P-to-P, I-to-P, or P-to-I). Note that the encoding and decoding as well as the transmission ordering of video frames in a GOP should be rearranged, i.e., should become (I_1 P_4 B_2 B_3 P_7 B_5 $B_6 P_{10}$ B_8 B_9 $\ldots$), to allow the B-pictures to be encoded and decoded with all the reference frames ready for motion estimation. This inevitably introduces additional coding delay as well as buffer memory, which may not be tolerable for applications such as videotelephony or video conferencing.

Each video frame, commonly of size 352×240 pixels (the source input format, SIF), is further partitioned into non-overlapping 16×16 "macroblocks"; each macroblock contains four blocks of data from both the luminance (Y) and the corresponding chrominance ($C_b C_r$) bands with $4 : 1 : 1$ subsampling, each with a size of 8×8 pixels. Moreover, a slice structure is also employed in the MPEG-1 for the synchronization of coding parameter (mode) changes within a single frame. A slice consists of a contiguous sequence of macroblocks in a raster scan order (from left to right and from top to bottom). In an MPEG coded bitstream, each slice starts with a slice-header which is a clear-codeword (a clear-codeword is a unique bit-pattern which can be identified without decoding the variable-length codes in the bitstream). Owing to the clear-codeword slice-header, slices are the lowest level of units which can be accessed in an MPEG coded bitstream without decoding the variable-length codes. Slices are important in handling the channel errors. If a bitstream contains a bit error then the error may cause error propagation, owing to the variable-length coding. The decoder can regain synchronization at the start of the next slice. Having more slices in a bitstream allows better error termination, but the overhead will increase.

5.4.1.2 Coding of MPEG-1 I-P-B frames

A block diagram for an MPEG-1 video encoder is shown in Figure 5.22(a) and the corresponding decoder is shown in Figure 5.22(b); they are similar to those for an H.263 codec (see Figures 5.15 and 5.16). In the case of an intra coded frame without reference to any past

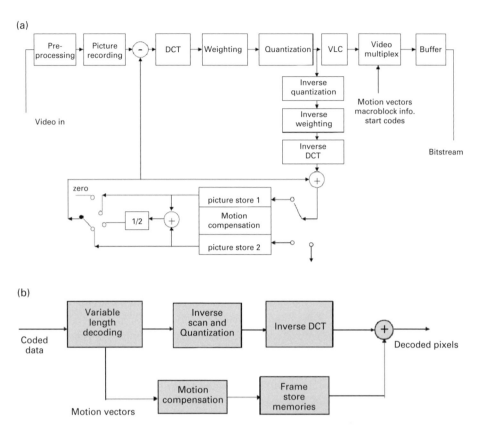

Figure 5.22 (a) The block diagrams of an MPEG-1 video encoder. (b) The corresponding MPEG-1 video decoder.

or future frames, the DCT is applied to each 8×8 luminance and chrominance block and, after output of the DCT, each of the 64 DCT coefficients is uniformly quantized (Q) using a predefined quantization matrix (similar to Table 4.1). Just as for JPEG encoding, the quantized DC coefficient is encoded as the difference between the quantized DC value of the previous block and the quantized DC value of the current block. The nonzero quantized AC values of the remaining DCT coefficients and their locations are then zigzag scanned and run-length entropy coded using variable-length code (VLC) tables.

As shown in Figure 5.22(a), the coding of inter coded P-pictures in MPEG-1 is very similar to the coding of P-pictures in H.263, where the previously encoded-then-decoded I- or P-picture has to be stored in a frame buffer in both the encoder and decoder, so that motion estimation and compensation can be performed on a macroblock basis with half-pixel accuracy (see Figure 5.23). A video buffer is needed to ensure that a constant target bitrate output is produced by the encoder. The quantization step size in MPEG-1 can be adjusted for each macroblock in a frame to achieve a given target bitrate and to avoid buffer overflow and underflow.

The coding of B-pictures is similar to that of P-pictures except that the motion vectors can reference the past reference picture, the future reference picture, or both. In B-pictures, each bidirectional motion-compensated macroblock, based on the SAD values, can be chosen to

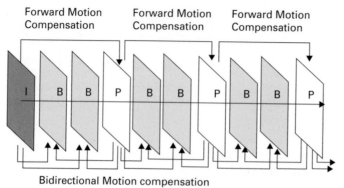

Bidirectional Motion compensation

Figure 5.23 The coding of inter coded P-pictures in MPEG-1: the previously encoded-then-decoded I- or P-picture has to be stored in a frame buffer in both the encoder and decoder, so that motion estimation and compensation can be performed on a macroblock basis with half pixel accuracy.

have one of two motion vectors, a forward motion vector which references to a best matching block in the past reference I- or P-pictures or a backward motion vector which references to a best matching block in the future reference I- or P-pictures. Motion-compensated prediction can also be formed by the average of the two referenced motion-compensated blocks from the forward and backward motion vectors. By averaging the past and the future reference blocks, the SAD value can sometimes be further reduced. In this case, both motion vectors will be sent and the prediction errors will be computed on the basis of the current macroblock and the average of the best matching macroblocks. B-pictures normally provide better compression than I- or P-pictures. Since B-pictures are not used as reference pictures, they do not propagate prediction errors.

The MPEG-1 decoder, as shown in Figure 5.22(b), uses the reverse process to reproduce a macroblock of a frame at the receiver. After decoding the variable-length words contained in the video buffer the pixel values of the prediction error are reconstructed. The motion-compensated pixels from the reference frame (the past reference, the future reference or both) contained in the video buffer are added to the prediction error to recover the particular macroblock of the current frame. Every MPEG-1 compatible decoder must be able to support at least video-source parameters up to TV size, i.e., a minimum number of 720 pixels per line, a minimum number of 576 lines per picture, a minimum frame rate of 30 frames per second, and a minimum bitrate of 1.86 Mbps [15]. In addition to the six-layer MPEG-1 bitstream syntax shown in Figure 5.20, above the video sequence layer there is also a system layer in which the video sequence is packetized. The video and audio bitstreams are then multiplexed into an integrated data stream defined in the system part.

5.4.1.3 Conditional replenishment in MPEG-1 video

The key for efficient coding of MPEG-1 video sequences is the selection of appropriate prediction modes of macroblocks in the motion-compensation process to achieve conditional replenishment. The MPEG-1 standard categorizes three macroblock (MB) coding types, Skipped_MB, Inter_MB, and Intra_MB [15], as follows.

(1) *Skipped_MB*   Its motion vector is zero and all the quantized DCT coefficients are zero. No information about the macroblock is coded or transmitted to the receiver.

(2) *Inter_MB* Motion-compensated prediction from the past and future reference frame is used. The MB type, the MB address, and, if required, the motion vector(s), the DCT coefficients, and quantization step size are transmitted. The *Inter_MB* can further be divided into forward-predicted, backward-predicted, and averaged macroblocks.

(3) *Intra_MB* No prediction is used from the previous frame (intraframe prediction only). Only the MB type, the MB address, and the DCT coefficients and quantization step size are transmitted to the receiver.

As a rule of thumb, the macroblocks in P-pictures and B-pictures can assume any of the above three types but Skipped_MB is not allowed in I-pictures. The first and last macroblocks in a slice must always be coded. Non-intra-coded macroblocks in P- and B-pictures can be skipped. For a skipped macroblock, the decoder just copies the macroblock from the previous picture.

5.4.1.4 Rate control in MPEG-1 video

By adjusting the quantizer step size to quantize the DCT coefficients, the MEPG-1 encoding algorithms is able to tailor the bitrate (and thus the quality of the reconstructed video) to specific application requirements. Coarse quantization of the DCT coefficients enables the storage or transmission of video with high compression ratios but, depending on the level of quantization, may result in significant coding artifacts. The MPEG-1 standard allows the encoder to select different quantizer values for each coded macroblock, and this enables a high degree of flexibility in allocating bits in images where they are needed to improve image quality. Furthermore it allows the generation of both constant and variable bitrates for storage or real-time transmission of the compressed video.

The MPEG-1 rate control works by allocating the total number of bits among the various types of pictures (I-, P- and B-frames). The MPEG video committee suggests that P-pictures be allotted about 2–5 times as many bits as B-pictures and I-pictures up to three times as many bits as P-pictures to give good results for typical natural scenes. Under normal circumstances, the encoder monitors the buffer status and adjusts the quantizer step size to avoid both buffer overflow and underflow problems. In MPEG-1, the adaptive quantization algorithm [29] is recommended. Each MB is classified using class $cl(r, c)$, which is based on the coding complexity; r and c denote the row and column coordinates of the MB. The quantizer step size for each class $Q(r, c)$, with an overall minimum step size Q_{low} is assigned according to

$$Q(r, c) = Q_{low} + \Delta Q \, cl(r, c) \qquad (5.11)$$

where ΔQ is typically 1 or 2.

5.4.2 MPEG-2 video coding

The MPEG committee continued its standardization efforts, with a second phase (MPEG-2) in 1991, to provide a video coding solution for emerging applications such as digital cable TV distribution; digital storage media (such as DVD), networked database services via various networks, as well as satellite and terrestrial digital broadcasting were seen to benefit from the

Table 5.2 MPEG-2 profiles, defined as specific subsets of the MPEG-2 bitstream syntax and functionality to support a class of applications

Abbr.	Name	Frames	YC$_b$C$_r$	Streams	Comment
SP	Simple profile	P, I	4 : 2 : 0	1	no interlacing
MP	Main profile	P, I, B	4 : 2 : 0	1	
422P	4 : 2 : 2 profile	P, I, B	4 : 2 : 2	1	
SNR	SNR profile	P, I, B	4 : 2 : 0	1–2	SNR: signal-to-noise ratio
SPA	Spatial profile	P, I, B	4 : 2 : 0	1–3 ⎫	low-, normal-, and high-
HP	High profile	P, I, B	4 : 2 : 2	1–3 ⎭	quality decoding

Table 5.3 MPEG-2 levels, defined to support applications which have differing quality requirements

Abbr.	Name	Pixel/line	Lines	Frame rate (Hz)	Bitrate (Mbps)
LL	Low level	352	288	30	4
ML	Main level	720	576	30	15
H-14	High 1440	1440	1152	30	60
HL	High level	1920	1152	30	80

increased quality expected to result from the new MPEG-2 standardization. The specification of the standard is intended to be generic, and the standard aims to facilitate bitstream interchange between different applications and transmission and storage media. Basically, MPEG-2 can be seen as a superset of the MPEG-1 coding standard and was designed to be backward compatible to MPEG-l, i.e., every MPEG-2 compatible decoder can decode a valid MPEG-1 bitstream.

To achieve better functionality and quality, additional prediction modes were developed in MPEG-2 to support efficient coding of *interlaced video*. In addition, *scalable video* coding extensions were introduced to provide extra functionality, such as the embedded coding of digital TV and HDTV, and graceful quality degradation in the presence of transmission errors. The MPEG-2 standard was designed to cover a wide range of applications, however, implementation of the full syntax may not be practical for most applications; MPEG-2 thus introduced the concept of "profiles" and "levels" to stipulate conformance between types of equipment not supporting the full implementation [5]. Profiles and levels specify conformance points that provide interoperability between encoder and decoder implementations within applications of the standard and between various applications that have similar functional requirements. A profile, as seen in Table 5.2, is defined as a specific subset of the MPEG-2 bitstream syntax and functionality that supports a class of applications (e.g., low-delay video conferencing applications, or storage media applications). Within each profile, several levels (see Table 5.3) are defined to support applications which have different quality requirements (e.g., different resolutions). Levels are specified as a set of restrictions on some of the parameters (or their combination) such as the sampling rates, frame dimensions, and bitrates in a profile. As

Table 5.4 Applications that can be implemented in the allowed range of values of a particular profile at a particular level of MPEG-2

Profile and level	Resolution (px)	Frame rate max. (Hz)	Sampling	Bitrate (Mbps)	Examples of applications
SP@LL	176×144	15	4 : 2 : 0	0.096	Wireless handsets
SP@ML	352×288 320×240	15 24	4 : 2 : 0	0.384	PDAs
MP@LL	352×288	30	4 : 2 : 0	4	Set-top boxes
MP@ML	720×480 720×576	30 25	4 : 2 : 0	15 (DVD: 9.8)	DVD, SD-DVB
MP@H-14	1440×1080 1280×720	30 30	4 : 2 : 0	60 (HDV: 25)	HDV
MP@HL	1920×1080 1280×720	30 60	4 : 2 : 0	80	ATSC 1080i, 720p60, HD-DVB (HDTV)
422P@ML	720×480 720×576	30 25	4 : 2 : 2	50	Sony IMX using I-frame only, Broadcast "contribution" video (I&P only)
422P@H-14	1440×1080 1280×720	30 60	4 : 2 : 2	80	Potential future MPEG-2-based HD products from Sony and Panasonic
422P@HL	1920×1080 1280×720	30 60	4 : 2 : 2	300	Potential future MPEG-2-based HD products from Panasonic

seen in Table 5.4, applications are implemented in the allowed range of values of a particular profile at a particular level.

The MPEG-2 core algorithm at the main profile specifies the non-scalable coding of both progressive and interlaced video sources. It is expected that most MPEG-2 implementations will at least conform to the main profile at main level, which supports the non-scalable coding of digital video with approximately digital TV parameters, i.e., a maximum sample density of 720 samples per line and 576 lines per frame, a maximum frame rate of 30 frames per second, and a maximum bitrate of 15 Mbps. The algorithm defined with the MPEG-2 simple profile is basically identical to that in the main profile, except that no B-picture prediction modes are allowed at the encoder. Thus, the additional implementation complexity and the additional frame stores necessary for the decoding of B-pictures are not required for MPEG-2 decoders that only conform to the simple profile.

5.4.2.1 MPEG-2 non-scalable coding modes

The MPEG-2 algorithm defined in the main profile is a straightforward extension of the MPEG-1 coding scheme to accommodate the coding of interlaced video, while retaining the full range of functionality provided by MPEG-1. The MPEG-2 coding algorithm, which is identical to the MPEG-1 standard, is also based on the general hybrid DCT/DPCM coding scheme (see Section 5.2), incorporating a macroblock structure, motion compensation, and coding modes for the conditional replenishment of macroblocks. MPEG-2 supports $4:2:0$, $4:2:2$ and $4:4:4$ chrominance subsampling but uses a slightly different $4:2:0$ format from MPEG-1, as shown in Figure 5.1. The concept of I-pictures, P-pictures, and B-pictures as introduced in MPEG-1 is fully retained in MPEG-2 to achieve efficient motion prediction and to assist random access functionality. The slices defined in MPEG-2 have to begin and end in the same horizontal row of macroblocks, which is different from the slice definition in MPEG-1, where a slice can cross macroblock row boundaries.

Field and frame pictures   MPEG-2 introduced the concept of frame pictures and field pictures along with particular frame prediction and field prediction modes to accommodate the coding of progressive and interlaced videos (see Figure 5.1(a)). Normally, progressive video requires a higher picture rate than the frame rate of an interlaced video to avoid a flickering display. The main disadvantage of the interlaced format is that when there are object movements the moving object may appear distorted when two fields are merged into a frame. Interlaced video also tends to cause horizontal picture details to dither and thus introduces more high-frequency noise. For interlaced sequences it is assumed that the coder input consists of a series of odd and even fields that are separated in time by a field period. Two fields of a frame may be coded separately. In this case each field is separated into adjacent non-overlapping macroblocks and the DCT is applied on a field basis. Alternatively, two fields may be coded together as a frame as in the conventional coding of progressive video sequences. Here, consecutive lines of the top and bottom fields are simply merged to form a frame. Notice that both frame pictures and field pictures can be used in a single video sequence.

Field and frame motion prediction   In MPEG-2, an interlaced picture can be encoded as a frame picture or as field pictures separately. Accordingly, MPEG-2 defines two different motion-compensated prediction types: frame-based and field-based motion-compensated prediction. Frame-based prediction forms a prediction based on the reference frames. Field-based prediction is based on reference fields. For the Simple profile, where bidirectional prediction cannot be used, MPEG-2 introduced dual-prime motion-compensated prediction to explore the temporal redundancies between fields efficiently. Figure 5.24 shows three types of motion-compensated prediction [28]. Note that all motion vectors in MPEG-2 are specified with a half-pixel resolution. As shown in Figure 5.24(a), in frame-based prediction for frame pictures the whole interlaced frame is considered as a single picture. It uses the same motion-compensated predictive coding method as in MPEG-1, i.e., each 16×16 macroblock can have only one motion vector for each forward or backward prediction. Two motion vectors are allowed in the case of bi-directional prediction.

Field-based prediction for frame pictures considers each frame picture as two separate field pictures. Separate predictions are formed for each 16×8 block of the macroblock, as shown in Figure 5.24(b). Thus, field-based prediction in a frame picture needs two sets of motion vectors. A total of four motion vectors are allowed in the case of bidirectional prediction. Each field prediction may select either field 1 or field 2 of the reference frame.

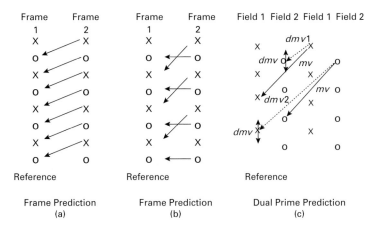

Figure 5.24 Three types of motion-compensated prediction for interlaced video [28] (© Elsevier 2005).

In field-based prediction for field pictures, the prediction is formed from the two most recently decoded fields. The predictions are made from reference fields, independently for each field, being each field considered as an independent picture. The block size for prediction is 16×16; however, it should be noted that a 16×16 block in the field picture corresponds to a 16×32 pixel area in a frame picture. Field-based prediction for field picture needs only one motion vector for each forward or backward prediction. Two motion vectors are allowed in the case of bi-directional prediction.

There is another mechanism called 16×8 prediction in field pictures, in which two motion vectors are used for each macroblock. The first motion vector is applied to the 16×8 block in field 1 and the second motion vector is applied to the 16×8 block in field 2. A total of four motion vectors is allowed in the case of bi-directional prediction.

Dual-prime motion-compensated prediction can be used only for P-pictures. Once the motion vector *mv* for a macroblock in a field of given parity (field 1 or field 2) is known relative to a reference field of the same parity, it is extrapolated or interpolated to obtain a prediction of the motion vector for the opposite-parity reference field. In addition, a small correction is also made to the vertical component of the motion vectors to reflect the vertical shift between lines of field 1 and field 2. These derived motion vectors are denoted by *dmv*1 and *dmv*2 (the dotted line) in Figure 5.24(c). Next, a small refinement *differential motion vector dmv* is added. The choice of *dmv* values $(-1, 0, +1)$ is determined by the encoder. The motion vector *mv* and its corresponding *dmv* value are included in the bitstream so that the decoder can also derive *dmv*1 and *dmv*2. In calculating the pixel values of the prediction, the motion-compensated predictions from the two reference fields are averaged, which tends to reduce the noise in the data.

Dual-prime prediction is used mainly for low-delay coding applications such as videophone and video conferencing. For low-delay coding using the simple profile, B-pictures should not be used. Without using bidirectional prediction, dual-prime prediction has been developed for P-pictures to provide better prediction than forward prediction.

Field and frame DCT MPEG-2 has two DCT modes: frame-based and field-based. In frame-based DCT, a 16×16-pixel macroblock is divided into four 8×8 DCT blocks. This mode is suitable for the blocks in the background or in a still image that have little motion, because these blocks have a high correlation between pixel values from adjacent scan lines. In field-based DCT, a macroblock is divided into four DCT blocks in which pixels from the same field are grouped together into one block. This mode is suitable for blocks that have motion because, as explained, motion causes distortion and may introduce high-frequency noise into the interlaced frame.

Alternate scan   MPEG-2 also defines two different zigzag scanning orders: "zigzag" and "alternate." The "zigzag" scan used in MPEG-1 is suitable for progressive images where the frequency components have equal importance in each horizontal and vertical direction. In MPEG-2, the "alternate" scan is introduced. This is based on the fact that interlaced images tend to have higher-frequency components in the vertical direction. Thus, the scanning order has more weight for the higher vertical frequencies than for the same horizontal frequencies. In MPEG-2, the selection between these two zigzag scan orders can be made on a picture basis.

5.4.2.2 MPEG-2 scalable coding modes

In order to accommodate networking environments that are increasingly heterogeneous, it is highly desirable for MPEG-2 to be equipped with scalability. It was found impossible to develop one generic MPEG-2 scalable coding scheme capable of satisfying all the diverse application requirements envisaged, so MPEG-2 has standardized three scalable coding schemes: signal-to-noise ratio (quality) scalability, spatial scalability, and temporal scalability. Each scalability scheme is targeted to assist applications with particular requirements [15]. The scalable coding modes provide algorithmic extensions to the non-scalable scheme defined in the MAIN profile. It is possible to combine different scalability modes into a hybrid coding scheme; i.e., interoperability between services with different spatial resolutions and frame rates can be supported by combining the spatial scalability and temporal scalability modes into a hybrid-layered coding scheme. Interoperability between HDTV and SDTV services can be provided also, with a certain resilience to channel errors, by combining the spatial scalability extensions with the SNR scalability mode. The MPEG-2 syntax supports up to three different scalable layers.

SNR scalability   This mode was primarily developed to provide graceful degradation of quality, on the basis of frequency (DCT domain) scalability, of the video in the prioritized transmission media. Both layers in Figure 5.25(a) encode the video signal at the same spatial resolution [28]. At the base layer the DCT coefficients are coarsely quantized to achieve moderate image quality at a reduced bitrate. The enhancement layer encodes the *difference* between the non-quantized DCT coefficients and the quantized coefficients from the base layer with a finer quantization step size. By doing this, moderate video quality can be achieved by decoding only the lower-layer bitstreams while the higher video quality can be achieved by decoding both layers, as shown in Figure 5.25(b). The mode is implemented as a simple and straightforward extension to the MAIN Profile MPEG-2 coder and achieves excellent coding efficiency.

Spatial scalability Spatial scalability was developed to support displays with different spatial resolutions at the receiver. For example, by encoding standard-definition TV (SDTV) in the base layer, with the enhancement layer the overall bitstream can provide HDTV resolution. The input to the base layer is usually created by downsampling the original video to create a low-resolution video for providing the basic spatial resolution, as shown in Figure 5.26(a) [28]. To generate a prediction for the enhancement-layer video

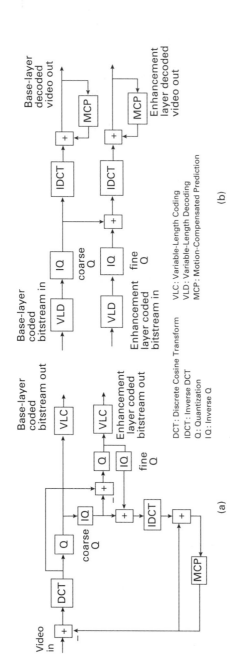

Figure 5.25 MPEG-2 SNR scalability. (a) Both layers encode the video signal at the same spatial resolution with different quantization parameters. (b) Moderate video quality can be achieved by decoding only the base-layer bitstreams while the higher video quality can be achieved by decoding both layers [28] (© Elsevier 2005).

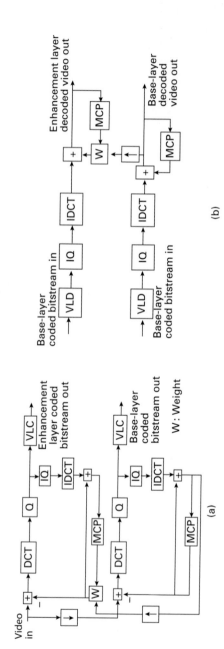

Figure 5.26 MPEG-2 spatial scalability. (a) The downsampled video is used to create a low-resolution base layer. (b) The lower-resolution video is interpolated and weighted before being added to the motion-compensated prediction from the enhancement layer [28] (© Elsevier 2005).

signal input, the decoded lower-layer video signal is upsampled by spatial interpolation and is weighted and combined with the motion-compensated prediction from the enhancement layer. The selection of weights is done on a macroblock basis and the selection information is sent as a part of the enhancement-layer bitstream. The base- and enhancement-layer coded bitstreams are then transmitted over the channel. As seen in Figure 5.26(b) [28], at the decoder the lower-layer bitstreams are decoded to obtain the lower-resolution video. The lower-resolution video is interpolated and weighted before being added to the motion-compensated prediction from the enhancement layer. In the MPEG-2 video standard, the spatial interpolator is defined as a linear interpolation or a simple averaging for missing samples. Spatial scalability can flexibly support a wide range of spatial resolutions but adds considerable implementation complexity to the MAIN Profile coding scheme.

Temporal scalability The MPEG-2 temporal scalability tool was developed with an aim similar to that of spatial scalability: to accommodate different temporal resolutions or frame rates. In temporal scalable coding, the base layer is coded at a lower frame rate by dropping some bidirectional frames which are used as enhancement-layer data for providing better temporal resolution, as seen in Figure 5.27(a). The target applications of this scalability include video over wireless channels, where the video frame rate may need to drop when the channel condition is poor. The decoded base-layer pictures provide motion-compensated predictions for encoding the enhancement layer. Another useful application of MPEG-2 temporal scalable coding is to support stereoscopic video, where layering is achieved by providing a prediction of one of the images in stereoscopic video (i.e., the left view) in the enhancement layer using coded images from the opposite view transmitted in the base layer. As seen in Figure 5.27(b), in this application not only the base-layer but also the enhancement-layer data are used as a reference for decoding the enhancement-layer data.

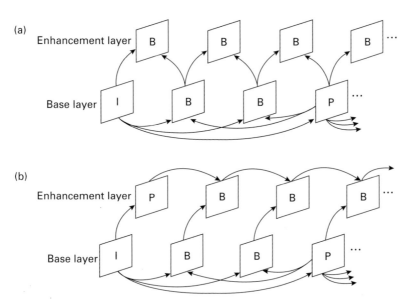

Figure 5.27 MPEG-2 temporal scalability: (a) The base layer is coded at a lower frame rate by dropping some bidirectional frames which are used as the enhancement-layer data for providing better temporal resolution. (b) Temporal scalable coding is used to support stereoscopic video.

5.5   MPEG-4 video coding and H.264/AVC

The MPEG-4 standard absorbs many features of the MPEG-1, MPEG-2 and other related standards and was introduced primarily to compress audio and visual (AV) digital data, under the formal standard ISO/IEC 14496 [10]. The MPEG-4 project started as a standard for video compression at very low bitrates. After working on this project for two years, the committee members, realizing that the rapid development of multimedia applications and services will require ever more compression standards, revised their approach. Instead of a compression standard, they decided to develop a set of tools (a "toolbox") to deal with AV objects. All the video coding techniques (e.g., H.263, MPEG-1/2, etc.) discussed so far have been based on pixels. Each video frame is a rectangular set of pixels and the algorithm looks for correlations between the pixels in a frame and between frames. The compression paradigm adopted for MPEG-4, however, is based on objects; that is why MPEG-4 was also named the "coding of audio-visual objects." Thus MPEG-4 relies on an object-based representation of the video data. It was standardized with the following features for handling audio and visual media objects (see Figure 5.28) [30] [31].

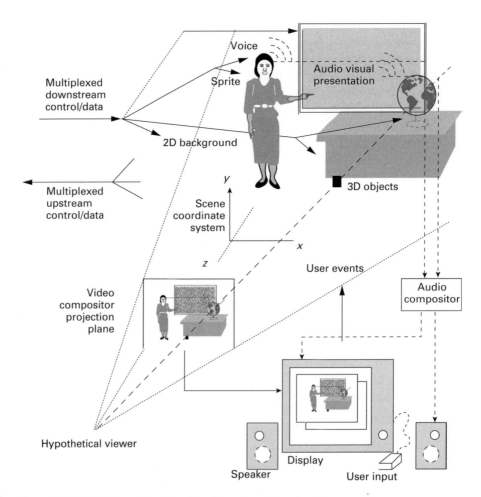

Figure 5.28 MPEG-4, an object-based multimedia coding standard.

(1) *Media objects coding* allows the data compression of media objects which include audio, visual, or audio-visual content. These two-dimensional or three-dimensional media objects can be natural or synthetic signals, meaning they could have been recorded with a camera or microphone or could be graphics or animation generated by a computer.

(2) *Media object composition* enables the composition of these media objects to create compound media objects that form audio-visual scenes.

(3) *Media object multiplex* involves the multiplexing and synchronizing of data associated with media objects for transport over network channels providing a QoS appropriate for the nature of the specific media objects.

(4) *Media object interaction* allows interaction with the audio-visual scene at the receiver's end or, via a back channel, at the transmitter's end.

Thanks to object-based representation, MPEG-4 can enjoy the most important feature of multimedia, i.e., interactivity. The passage from a pixel-based representation to an object-based representation can be performed using manual, semi-automatic, or automatic segmentation techniques. The inverse operation is achieved by rendering, blending, or composition. At this point it is important to note that, since the input information is pixel-based, all tools used in MPEG-4 still operate in a pixel-based approach where care has been taken to extend their operation to arbitrarily shaped objects.

Uses for the MPEG-4 standard are web-based streaming media and CD distribution, mobile and cellular media dissemination, videophone, and broadcast television. MPEG-4 is still a developing standard and is divided into a number of parts. The key parts related to video coding are MPEG-4 Part 2 [10] (MPEG-4 SP/ASP, used by codecs such as DivX, XviD, and 3ivx and by Apple Quicktime 6) and MPEG-4 Part 10 (MPEG-4 H.264/AVC, used by Quicktime 7, and by next-generation DVD formats like HD DVD and Blu-ray Disk and many digital video broadcasting standards).

Most of the features included in MPEG-4 are left to individual developers to implement, if they so desire. To deal with this, the standard also includes the concept of "profiles" and "levels," similarly to MPEG-2, allowing a specific set of capabilities to be defined in a manner appropriate for a subset of applications (see Table 5.5).

5.5.1   Object-based video coding

As mentioned above, the central concept in MPEG-4 video is that of the *video object (VO)*. Each VO is characterized by intrinsic properties such as shape, texture, and motion. MPEG-4 considers a scene to be composed of several VOs: this representation is more amenable to interactivity with the scene content than pixel-based or block-based representations, though the standard does not specify the method for creating VOs. Since the main objective of MPEG-4 visual is for coding purposes, MPEG-4 simply provides a standard convention for describing VOs, such that all compliant decoders will be able to extract VOs of any shape from the encoded bitstream, as necessary. The decoded VOs may then be subjected to further manipulation as appropriate for the application at hand.

A general block diagram of an MPEG-4 video encoder is shown in Figure 5.29, The video is first represented by VOs as required by the application. The coding control unit decides, possibly on the basis of the requirements of the user or the capabilities of the decoder, which VOs are to be transmitted, the number of layers, and the level of scalability suited

Table 5.5 Profiles and levels in MPEG-4; these allow a specific set of capabilities to be defined in a manner appropriate for a subset of applications

Profile	Level	Maximum size	Maximum objects	Maximum kbps
Simple profile	0	QCIF	1	64
	1	QCIF	4	64
	2	CIF	4	128
	3	CIF	4	384
Advanced simple profile	0	QCIF	1	128
	1	QCIF	4	128
	2	CIF	4	384
	3	CIF	4	768
	4	2CIF	4	3000
	5	4CIF	4	8000
Simple scalable profile	1	CIF	4	128
	2	CIF	4	256
Fine grain scalable profile	0	QCIF	1	128
	1	QCIF	4	128
	2	CIF	4	384
	3	CIF	4	768
	4	2CIF	4	3000
	5	4CIF	4	8000
Core profile	1	QCIF	4	384
	2	CIF	16	2000
Core scalable profile	1	QCIF	4	768
	2	CIF	8	1500
	3	4CIF	16	4000
Main profile	2	CIF	16	768
	3	2CIF	32	1500

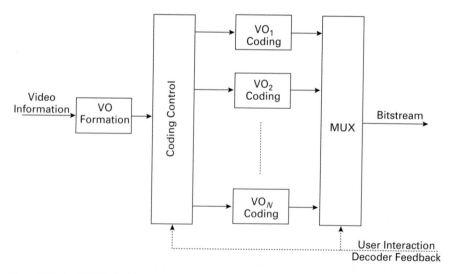

Figure 5.29 An MPEG-4 video encoder splits the video into VOs as required by the application [32].

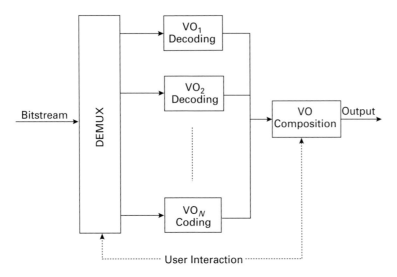

Figure 5.30 An MPEG-4 decoder decomposes an incoming bitstream into its individual VO bitstreams. Each VO is then decoded, and the result is composed [32].

to the current video session. Each VO is encoded independently of the others. The multiplexer then merges the bitstreams representing the different VOs into a video bitstream. Figure 5.30 shows a block diagram of an MPEG-4 decoder, where an incoming bitstream is first decomposed into its individual VO bitstreams. Each VO is then decoded, and the result is composed [32]. The composition handles the way in which the information is presented to the user. For a natural video, composition is simply the layering of 2D VOs in the scene. The VO-based structure has certain specific characteristics. In order to be able to process data available in a pixel-based digital representation, the texture information for a VO (in the uncompressed form) is represented in *YUV* color coordinates. Up to 12 bits may be used to represent a pixel component value. Additional information regarding the shape of the VO is also available. Both shape and texture information are assumed to be available for specific snapshots of VOs called *video object planes (VOPs)* [32].

MPEG-4 video also supports sprite-based coding in the case of natural video, where a large composite image results from the blending of pixels belonging to various temporal instances of the video object. A VOP may be thought of as just the portion of the sprite that is visible at a given instant of time. If we can encode the entire information about a sprite then VOPs may be derived from this encoded representation as necessary. Sprite-based encoding is particularly well suited for representing synthetically generated scenes. As far as video coding is concerned, a sprite captures spatio-temporal information in a very compact way and therefore achieves a high coding efficiency. Figure 5.31 shows an example of how a sprite is generated for the background of a test sequence called "Stefan," using the first 200 frames of the sequence. Note that the sequence has been previously segmented in order to exclude foreground objects. Because of camera motion, the sprite results in an extended view of the background. The foreground player, the VO, can thus be composed (by combining a portion of the static background with the VOP sequence) in terms of a sequence of VOPs on the decoder side.

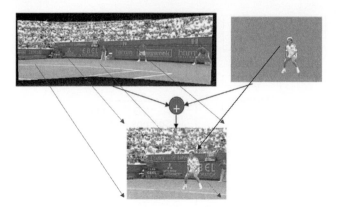

Figure 5.31 An example of a sprite generated for the background of a test sequence called "Stefan" using the first 200 frames [32].

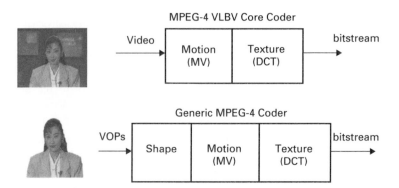

Figure 5.32 Generic MPEG-4 visual coding, where the input is an extracted VOP and shape coding, motion estimation, and texture coding are consecutively applied to create the MPEG-4 video bitstream.

5.5.2   Coding of VOPs

A natural video object consists of a sequence of 2D VOPs. The efficient coding of VOPs exploits both temporal and spatial redundancies. Thus a coded representation of a VOP includes its shape, its motion, and its texture. Figure 5.32 shows in the lower flow chart generic MPEG-4 visual coding, where the input is an extracted VOP and shape coding, motion estimation, and texture coding are consecutively applied to create the MPEG-4 video bitstream. Of course, in applications where shape information is not explicitly required, such as when each VOP is a rectangular frame, the shape coding scheme may be disabled, which results in the MPEG-4 very-low-bitrate video (VLBV) coder, which is similar to the H.263 or MPEG-1 video coder, as shown in the upper flow chart of Figure 5.32. More specifically, Figure 5.33 shows the block diagram of an MPEG-4 VOP encoder. The VOs are compressed by coding their corresponding VOPs using a hybrid coding scheme somewhat similar to those of previous MPEG standards. Details of the VOP coding technique are shown in the figure. This technique is implemented in terms of *macroblocks* (blocks of

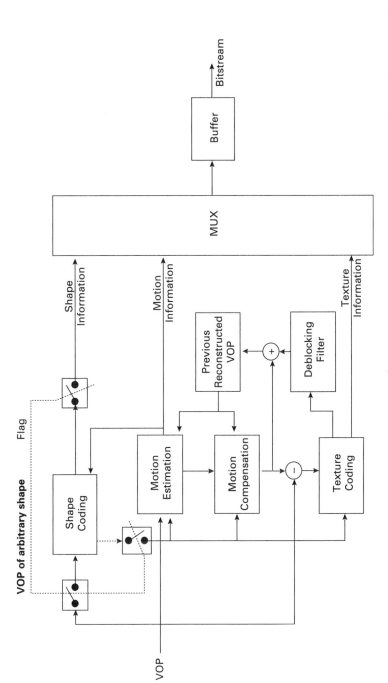

Figure 5.33 An MPEG-4 VOP encoder, where the VOs are compressed by coding their corresponding VOPs in a hybrid coding scheme [32].

16×16 pixels) [32]. Grouping the encoded information in macroblocks can facilitate resynchronization if there are transmission errors. A VOP has two basic types of information associated with it: shape information and texture information. The shape information, also referred to as the alpha plane, needs to be specified explicitly since in general VOPs are expected to have arbitrary shapes. Thus, the VOP encoder essentially consists of two encoding schemes: one for shape, and one for texture. The same coding scheme is used for all VOPs in a given VO [32].

5.5.3   VOP shape coding

When encoding the shape information in arbitrarily shaped VOPs, beside the shape information available for the corresponding VOP the shape coding scheme also relies on motion estimation to compress the shape information further. The shape information for a VOP is specified based on two components; the binary shape and the grayscale shape. The binary shape uses a simple array of binary labels (called a bitmap), arranged in a rectangular window with sides that are multiples of 16 pixels in the horizontal and vertical directions, corresponding to the bounding box of the VOP, to specify whether an input pixel belongs to the VOP. However, the grayscale shape uses a transparency value, ranging from 0 (completely transparent) to 255 (opaque), to denote each pixel of the VOP.

For encoding the binary shape, the rectangular bounding box is then partitioned into blocks of 16×16 samples (hereafter referred to as shape blocks) and the encoding and decoding processes are performed block by block [32]. Moreover, grayscale shape information is encoded using a block-based motion-compensated DCT similar to that used for texture coding, allowing lossy coding only.

5.5.4   VOP texture coding

In the case of intra VOPs (I-VOPs), the term "texture" refers to the information present in the gray or chroma values of the pixels forming the VOP. In the case of predicted VOPs (B-VOPs and P-VOPs), the residual error after motion compensation is considered as the texture information. The MPEG-4 video standard also uses an adapted version of the block-based discrete cosine transform (DCT) method to code the texture information of an arbitrarily shaped VOP. The VOP texture is split into macroblocks of size 16×16. Of course, this implies that the blocks along the boundary of the VOP may not fall completely on the VOP, that is, some pixels in a boundary block may not belong to the VOP. Such boundary blocks are treated differently from the non-boundary blocks.

The blocks that lie completely within the VOP are encoded using a standard 2D 8×8 block DCT. The luminance and chrominance blocks are treated separately. Thus, six blocks of DCT coefficients are generated for each macroblock. The DCT coefficients are quantized to compress the information. The DC coefficient is quantized using a step size of 8. The MPEG-4 video algorithm offers two alternatives for determining the quantization step to be used for the AC coefficients. One alternative follows an approach similar to that used in H.263, where a quantization parameter determines how the coefficients are quantized. The same quantization step size is applied to all coefficients in a macroblock but may change from one macroblock to another, depending on the desired image quality or target bitrate. The other alternative is a quantization scheme similar to that used in MPEG-2, where the quantization step may vary depending on the position of the coefficient. After appropriate quantization, the DCT coefficients in a block are scanned in zigzag fashion in order to create a string of coefficients from the 2D block. The string is compressed using run-length coding and entropy coding.

Macroblocks that straddle the VOP boundary are encoded using one of two techniques: *repetitive padding* followed by conventional DCT or *shape-adaptive DCT* (SA-DCT); the latter was considered only in Version 2 of the MPEG-4 visual standard. Repetitive padding consists in assigning a value to the pixels of the macroblock that lie outside the VOP. The padding is applied to 8×8 blocks of the macroblock in question. Only the blocks straddling the VOP boundary are processed by the padding procedure. When the texture data is the residual error after motion compensation, the blocks are padded with zero values. For intra coded blocks, the padding is performed in a two-step procedure called *low-pass extrapolation (LPE)*. This procedure is as follows [32].

(1) The mean of the pixels in the block that belongs to the VOP is computed. This mean value is then used as the padding value, that is,

$$f_{r,c|(r,c)\notin VOP} = \frac{1}{N} \sum_{(x,y)\in VOP} f_{x,y}, \qquad (5.12)$$

where N is the number of pixels of the macroblock in the VOP. This is also known as *mean-repetition* DCT.

(2) The averaging operation given in Eq. (5.12) is used for each pixel $f_{r,c}$, where r and c are the row and column position of each pixel in the macroblock outside the VOP boundary. Starting from the top left corner $f_{0,0}$ of the macroblock, the padding proceeds row by row to the bottom right pixel:

$$f_{r,c|(r,c)\notin VOP} = \frac{f_{r,c-1} + f_{r-1,c} + f_{r,c+1} + f_{r+1,c}}{4}. \qquad (5.13)$$

The pixels appearing on the right-hand side of Eq. (5.13) must lie within the VOP, otherwise they are not considered and the denominator is adjusted accordingly. Once the block has been padded, it is coded in a similar fashion to an internal block.

In the SA-DCT-based scheme [33], the number of coefficients generated is proportional to the number of pixels of the block belonging to the VOP. The SA-DCT is computed as a separable two-dimensional DCT. For example, transformation of the block shown in Figure 5.34(a) is performed as follows. First, the active pixels of each column are adjusted to the top of the block (see Figure 5.34(b)). Then, for each column, the one-dimensional DCT is computed for only the active pixels in the column, with the DC coefficients at the top (see Figure 5.34(c)). This results in a possibly different number of coefficients for each column. As shown in Figure 5.34(d), the rows of coefficients generated in the column DCT are then adjusted to the left before computing the row DCT. The two-dimensional SA-DCT coefficients are laid out as shown in Figure 5.34(e), with the DC coefficient at the top left corner. The binary mask of the shape and the DCT coefficients are both required in order to decode the block correctly. The coefficients of the SA-DCT are then quantized and entropy coded in a similar way to that explained in the previous section.

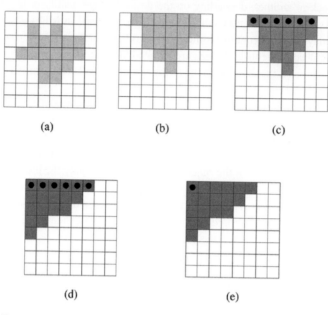

Figure 5.34 Steps in computing shape-adaptive DCT for an arbitrarily shaped two-dimensional region. (a) the region to be transformed, with shaded blocks representing the active pixels; (b) top-adjusted columns; (c) column DCT coefficients with DC coefficients marked with black spots; (d) left-adjusted rows of coefficients; (e) the final two-dimensional SAD-CT with the DC coefficient marked as a black spot [33] (© Elsevier 1995).

5.5.5 Motion-compensated coding

Motion estimation for the texture and gray scale shape is performed using the luminance values. The algorithm consists of the following three steps [32].

(1) *Padding of the reference VOP*   This step is applied only to arbitrarily shaped VOPs.
(2) *Full-search polygon matching with single-pixel accuracy*   The search range in this step is $-2^{R+3} \le MV_x, MV_y \le 2^{R+3}$, where MV_x and MV_y denote the horizontal and vertical components of the motion vector in single-pixel units, and $1 \le R \le 7$. The value of R is defined independently for each VOP. The error measure, $SAD\,(MV_x, MV_y)$, is defined as

$$SAD(MV_x, MV_y) = \sum_{i=0}^{15} \sum_{j=0}^{15} |I(x_0 + i, y_0 + j)$$

$$- L(x_0 + i + MV_x, y_0 + j + MV_y)|$$
$$\times BA(x_0 + i, y_0 + j) - (NB/2 + 1)\delta(MV_x, MV_y), \quad (5.14)$$

where
- (x_0, y_0) denotes the top left coordinates of the macroblock,
- $I(x, y)$ denotes the luminance sample value at (x, y) in the input VOP,
- $L(x, y)$ denotes the luminance sample value at (x, y) in the reference VOP,
- $BA(x, y)$ is 0 when the pixel at (x, y) is transparent and 1 when the pixel is opaque,
- NB denotes the number of non-transparent pixels in the macroblock,

- $\delta(MV_x, MV_y)$ is 1 when $(MV_x, MV_y) = (0, 0)$ and 0 otherwise, and
- the solidus "/" here denotes integer division with truncation towards zero.

(3) *Polygon matching with half-pel accuracy* Starting from the motion vector estimated in step (2), half-sample search in a $(\pm 0.5) \times (\pm 0.5)$ pixel (pel) window is performed using $SAD(MV_x, MV_y)$ as the error measure. The estimated motion vector (x, y) must stay within the range $-2^{R+3} \leq MV_x, MV_y \leq 2^{R+3}$.

When applied to binary shape macroblocks, however, the motion estimation algorithm is different from the algorithm applied to the texture planes; more details about this algorithm are available in [32].

The MPEG-4 standard has not been widely deployed and used, owing to the fact that it is a major task to render all MPEG-4 parts compatible, ranging from video, synthetic structures (graphics-like), audio, systems, reference software, test bitstreams, digital rights management, and so on. The MPEG-4 video part alone (i.e., ISO 14496–2 MPEG-4 Part 2 Visual) is quite large, with overcomplexity, i.e., there are many profiles and levels, the vast majority of which are unused by commercial applications. Moreover some elements are not entirely clear and are open to interpretation; it also takes a long time to complete the license arrangements. Another major factor in hindering its deployment is newer and better codecs that have moved on, such as the H.264/MPEG-4 AVC and VC-1 evolved from Microsoft Windows Media Technology.

5.6   H.264/MPEG-4 AVC

The H.264/MPEG-4 AVC standard, in short H.264/AVC, is the latest video coding standard of the ITU-T Video Coding Experts Group (VCEG) and the ISO/IEC Moving Picture Experts Group (MPEG). These two groups formed a joint video team (JVT) in 2000 to develop "advanced video coding" (AVC) under MPEG-4 Part 10 [11], and the final draft was approved in Spring 2003. H.264/AVC has enhanced visual quality at very low bitrates (low-delay end-to-end), particularly at rates below 24 kbps for mobile video applications. It has become the most widely accepted video coding standard since the deployment of MPEG-2. It covers all common video applications ranging from digital broadcasting, mobile services, and video conferencing to IPTV, HDTV, and HD video storage. Since the completion of the first version of the H.264/AVC standard, the JVT experts group has done further work to extend the capabilities of H.264/AVC with important new enhancements, known as the fidelity range extensions (FRExt) [47] [48], which further broaden the application domain of the new standard toward areas such as professional contribution, distribution, or studio post-production. Another set of extensions of H.264/AVC for scalable video coding (SVC) aims at creating a functionality that allows the reconstruction of video signals with various spatio-temporal resolutions from parts of the coded video representation (i.e., from partial bitstreams). Also, standardization of the multi-view video coding (MVC) extension is also under development.

The video coding of H.264/AVC is quite similar to that of MPEG-2 video coding, which consists of a hybrid of block-based temporal and spatial prediction in conjunction with block-based transform coding [34]. Figure 5.35 shows a block diagram for such an encoder and decoder. As for most existing video codecs, the first picture of a sequence or a random access point is typically coded in *intra* (intra-picture) mode. An intra-picture prediction and compensation is introduced in AVC to reduce the spatial redundancy within the same

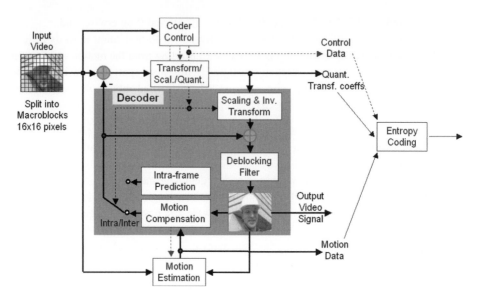

Figure 5.35 An H.264/AVC video encoder; the shaded area denotes the corresponding decoder structure [34] (© IEEE 2003).

picture. For all remaining pictures of a sequence or between random access points, *inter* (inter-picture) coding is typically utilized. Inter coding employs inter-picture temporal prediction (i.e., motion compensation) using other previously decoded pictures. The *residual* of the prediction (either intra or inter), which is the difference between the original input samples and the predicted samples for the block, is transformed. The transform coefficients are then scaled and approximated using scalar quantization. The quantized transform coefficients are entropy coded and transmitted together with the entropy-coded prediction information for either intra- or inter-frame prediction. As for other video codecs, the encoder contains a model of the decoding process (see the large shaded area in Figure 5.35) so that it can compute the same prediction values as those computed in the decoder, for the prediction of subsequent blocks in the current picture or subsequent coded pictures. The decoder inverts the entropy coding processes, performs the prediction process as indicated by the encoder using the prediction-type information and motion data. It also inverse-scales and inverse-transforms the quantized transform coefficients to form the approximated residual and adds this to the prediction. The result of that addition is then fed into a deblocking filter, which provides the decoded video as its output.

Each picture, which can be an entire frame or a single field, of an AVC video sequence is partitioned into fixed-size *macroblocks* that each cover a rectangular picture area of 16×16 samples of the luma component and, in the case of video in $4:2:0$ chroma sampling format, 8×8 samples of each of the two chroma components. All luma and chroma samples of a macroblock are either spatially or temporally predicted, and the resulting prediction residual is represented using transform coding. Each color component of the prediction residual is subdivided into blocks. Each block is transformed using an integer transform, and the transform coefficients are quantized and entropy coded. The macroblocks are organized in *slices* (see Figure 5.36), which represent regions of a given picture that can be decoded independently of each other. The H.264/AVC standard

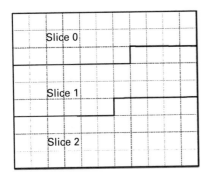

Figure 5.36 The macroblocks in H.264/AVC are organized in *slices*, which represent regions of a given picture that can be decoded independently of each other [34] (© IEEE 2003).

supports five slice-coding types. The simplest is the intra slice (I-slice), where all macroblocks are coded without reference to any other pictures in the video sequence. Prior-coded images can be used to form a prediction signal for macroblocks of the predictive slice (P-slice) and the bipredictive slice (B-slice). The remaining two slice types are SP (switching P) and SI (switching I) slices, which are specified for efficient switching between bitstreams coded at various bitrates [11].

5.6.1 Innovative video coding features of H.264/AVC

There are several innovative technologies behind the new H.264/AVC standard. The key objective of the new innovative features added in AVC is the achievement of a substantially higher degree of diversification, sophistication, and adaptability than most prior video coding standards [34] [35].

5.6.1.1 Spatial intra-picture prediction

Each macroblock can be transmitted in one of several coding types, depending on the slice-coding type. In all slice-coding types, at least two intra macroblock coding types are supported. All intra coding types in H.264/AVC rely on prediction of samples in a given block conducted in the spatial domain. The types are distinguished by their underlying luma prediction block sizes of 4×4, 8×8 (FRExt only), and 16×16, whereas the intra prediction process for chroma samples operates in an analogous fashion but always with a prediction block size equal to the block size of the entire macroblock's chroma arrays. In each of those intra coding types, and for both luma and chroma, spatially neighboring samples of a given block that have already been transmitted and decoded are used as a reference for spatial prediction of the given block's samples. The number of encoder-selectable prediction modes in each intra coding type is either four (for chroma and 16×16 luma blocks) or nine (for 4×4 and 8×8 luma blocks). As illustrated in Figure 5.37(a) [48] for the case of 8×8 spatial luma prediction, luma values of each sample in a given 8×8 block are predicted from the values of neighboring decoded samples. In addition, as a distinguishing feature of the 8×8 intra coding type, the reference samples are smoothed by applying a low-pass filter prior to performance of the actual prediction step. Eight different prediction directions plus an additional averaging (so-called DC) prediction mode (corresponding to mode 2 and not shown in Figure 5.37(a)) can be selected by the encoder. The 4×4 and 16×16 intra

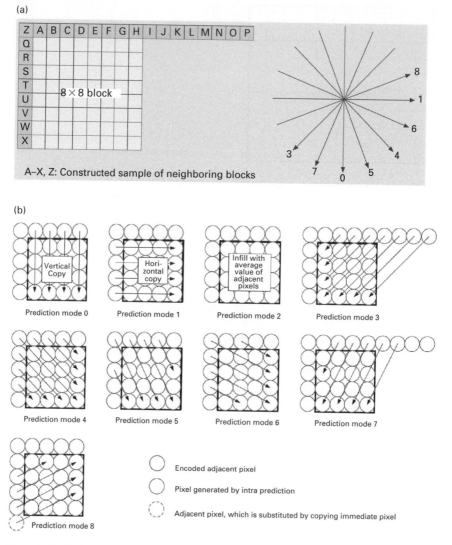

Figure 5.37 The intra prediction mode. (a) Eight different prediction directions plus an additional averaging prediction mode for 8×8 blocks can be selected by the encoder for spatial intra-picture prediction [48] (© IEEE 2006). (b) The situation for 4×4 blocks.

prediction types operate in a conceptually similar fashion except that they use different block sizes and do not include the smoothing filter. Figure 5.37(b) shows in detail the action of the nine intra 4×4 prediction modes. For example, for mode 0 (vertical prediction), the samples above the 4×4 block are used as predictions for the samples of the same column within the block, as indicated by the arrows. Mode 1 (horizontal prediction) operates in a similar manner except that the samples to the left of the 4×4 block are referenced. For mode 2 (DC prediction), the adjacent samples are averaged as indicated in the figure. The remaining six modes are diagonal prediction modes, referred to as diagonal-down-left, diagonal-down-right, vertical-right, horizontal-down, vertical-left, and horizontal-up prediction. As their names indicate, they are suited to predict textures with structures in the direction specified.

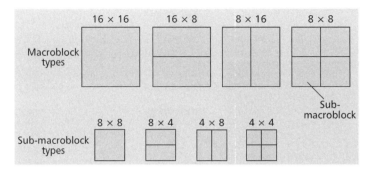

Figure 5.38 The H.264/AVC standard supports variable block sizes for motion estimation.

5.6.1.2 Variable-block-size motion compensation and rate–distortion optimization

In addition to the intra macroblock coding types, various predictive or motion-compensated coding types are allowed in P-slices. Each P-type macroblock is partitioned into fixed-size blocks, e.g., 16×16, 16×8, 8×16, and 8×8 samples are supported in luma blocks. When the macroblock is partitioned into four so-called sub-macroblocks each of size 8×8 luma samples, one additional syntax element is transmitted for each 8×8 sub-macroblock to specify whether the corresponding sub-macroblock was further coded, using motion-compensated prediction, with luma block sizes of 8×8, 8×4, 4×8, or 4×4 samples. Figure 5.38 illustrates the partitioning. The motion estimation for each predictive-coded $M \times N$ luma block is obtained by displacing a corresponding area of a previously decoded reference picture, where the displacement is specified by a translational motion vector and a picture reference index. Thus, if the macroblock is coded using four 8×8 sub-macroblocks, and each sub-macroblock is coded using four 4×4 luma blocks, a maximum of 16 motion vectors may be transmitted for a single P-slice macroblock. The motion vector precision is at the granularity of one quarter-pixel between luma samples. If the motion vector points to an integer-sample position then the prediction signal is formed by the corresponding samples of the reference picture. For finer resolution of motion estimation, the prediction signal is obtained using interpolation between integer-sample positions. The prediction values at half-sample positions are obtained by separable application of a one-dimensional six-tap finite impulse response (FIR) filter, and prediction values at quarter-sample positions are generated by averaging the samples at integer- and half-sample positions. The prediction values for the chroma components are obtained by bilinear interpolation. The advantage of using this variable-block-size motion-compensated coding can be seen in Figure 5.39 where the rapid motion of a bicycle wheel can be isolated to improve the coding efficiency.

The H.264/AVC standard adopts a rate-distortion optimization (RDO) algorithm to find the best motion vector, the best reference frame, and the best intra prediction mode and to select the best macroblock mode. This standard supports not only multiple inter modes (16×16, 16×8, 8×16, 8×8, 8×4, 4×8, 4×4) with different block types, but also skip mode and intra modes. The RDO framework as used in H.264 assigns a specific motion-prediction mode (various block sizes) so as to minimize $J_{MB} = D_{MB} + \lambda R_{MB}$:

$$\min_{mode}(J_{MB}) = \min_{mode}(D_{MB} + \lambda R_{MB}), \tag{5.15}$$

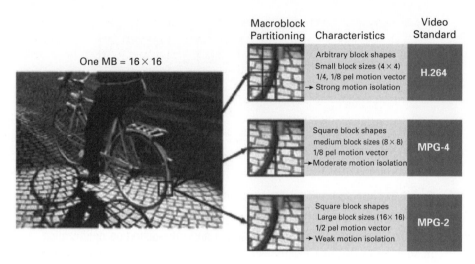

Macroblock Partitioning	Characteristics	Video Standard
	Arbitrary block shapes Small block sizes (4 × 4) 1/4, 1/8 pel motion vector → Strong motion isolation	H.264
	Square block shapes medium block sizes (8 × 8) 1/8 pel motion vector → Moderate motion isolation	MPG-4
	Square block shapes Large block sizes (16 × 16) 1/2 pel motion vector → Weak motion isolation	MPG-2

One MB = 16 × 16

Figure 5.39 The rapid motion of a bicycle wheel can be isolated to improve the coding efficiency in H.264/AVC.

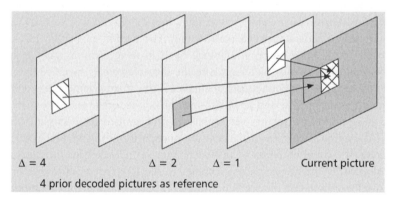

Δ = 4   Δ = 2   Δ = 1   Current picture

4 prior decoded pictures as reference

Figure 5.40 Multi-picture motion-compensated prediction [48] (© IEEE 2006).

where the distortion in the macroblock, D_{MB}, is the sum of the SAD distortion contributions of the individual pixels,

$$D_{MB} = \sum_{i \in MB} D_n(i). \qquad (5.16)$$

Note that R_{MB} denotes the bitrate, and λ is a Lagrange multiplier. The optimal encoding mode for each macroblock can thus be determined.

5.6.1.3 Multi-frame motion compensation

The H.264/AVC standard supports multi-picture motion-compensated prediction in a manner similar to that known as enhanced reference picture selection in H.263 v.3 [8] [9]. That is, more than one prior-coded picture can be used as a reference for motion-compensated prediction. Figure 5.40 illustrates this concept, which can be also extended to B-pictures as described below. For multi-frame motion-compensated prediction, the encoder stores decoded reference pictures in a multi-picture buffer. The reference index parameter, giving the location of the reference picture, is transmitted for each motion-compensated 16×16, 16×8, or 8×16 macroblock partition or 8×8 sub-macroblock. In addition to the macroblock modes described above, a P-slice macroblock can also be coded in the so-called skip mode. For this mode, quantized prediction error signals, motion vectors, or reference index parameters are not transmitted. The reconstructed signal is computed in a manner similar to the prediction of a macroblock with partition size 16×16 and fixed reference picture index equal to 0. In contrast with previous video coding standards, the motion vector used for reconstructing a skipped macroblock is inferred from the motion properties of neighboring macroblocks rather than being inferred as zero (i.e., no motion).

5.6.1.4 Motion-compensated prediction of B-slices

In comparison with prior video coding standards, the use of B-slices in H.264/AVC is much more versatile, e.g., B-pictures themselves can be used as reference pictures for motion-compensated prediction. Thus, the only substantial difference between B-and P-slices in H.264/AVC is that in the coding of B-slices, some macroblocks or blocks may use a weighted average of two distinct motion-compensated prediction values for building the prediction signal. In fact, the *generalized B-pictures* in H.264/AVC allow any arbitrary pair of reference pictures (called list 0 and list 1) to be utilized for the prediction of each region (as exemplified in Figure 5.40). This flexibility has been shown to have greater importance in its use as a fundamental part of the new scalable video coding (SVC) [13] and multiple view coding (MVC) extensions [36] [37]. Depending on which reference picture list is used for forming the prediction signal, three different types of inter-picture prediction are distinguished in B-slices: list 0, list 1, and bipredictive, where the last uses a superposition of list 0 and list 1 prediction signals and is the key feature provided by B-slices. With a partitioning similar to that specified for P-slices, the three different inter-picture prediction types in B-slices can be chosen separately for each macroblock partition or sub-macroblock partition. Additionally, B-slice macroblocks or sub-macroblocks can also be coded in so-called direct mode without the need to transmit any additional motion information [48]. If no prediction residual data are transmitted for a direct-coded macroblock, it is also referred to as skipped, and skipped macroblocks can be indicated very efficiently, as for the skip mode in P-slices.

5.6.2 Transform, scaling, and quantization

In contrast with prior video coding standards, such as MPEG-2 or H.263, which use a 2D DCT of size 8×8 as the block transform, H.264/AVC specifies a set of integer transforms of different block sizes. In all Version-1-related profiles, a 4×4 (separable) integer transform is applied to both the luma and chroma components of the prediction residual signal. The transformation matrix H for one-dimensional transformation after row-column decomposition (see Section 4.4.2) of a 2D integer transform is given as

$$H = \begin{bmatrix} 1 & 1 & 1 & 1 \\ 2 & 1 & -1 & -2 \\ 1 & -1 & -1 & 1 \\ 1 & -2 & 2 & -1 \end{bmatrix}. \tag{5.17}$$

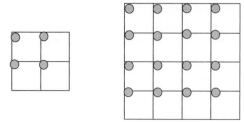

Figure 5.41 Four-tap Hadamard transforms and two-tap Haar–Hadamard transforms are applied to all the resulting DC coefficients. The small circles denote the left upper corner value of each square.

Note that the inverse transform is defined by an exact integer transform; inverse-transform mismatches are thus avoided.

An additional $M \times N$ transform stage is further applied to all resulting DC coefficients in cases where the luma component of a macroblock is coded using the 16×16 intra coding type (with $N = M = 4$) as well as in cases where both chroma components (with $N = 2$, $M = 4$) depend on the chroma format. For these additional transform stages, separable combinations of the four-tap Hadamard transform and two-tap Haar–Hadamard transform are applied (see Figure 5.41). Besides the important property of low computational complexity, the use of these small-block-size transforms in H.264/AVC has the advantage of significantly reducing ringing artifacts.

For the quantization of transform coefficients, H.264/AVC uses uniform-reconstruction quantizers similarly to H.263. One of 52 quantizer step size scaling factors is selected for each macroblock by a quantization parameter (QP). The scaling operations are arranged so that there is a doubling in quantization step size for each increment of 6 in the value of the QP. The quantized transform coefficients of a block are generally scanned in a zigzag fashion for the entropy coding methods described below. In addition to the basic step-size control, the FRExt amendment also supports encoder-specified scaling matrices for perceptual-tuned frequency-dependent quantization step sizes; these are similar to the normalization matrix used in JPEG and MPEG-2 video.

5.6.3   In-loop deblocking filter

Owing to coarse quantization at low bitrates, block-based coding typically results in visually noticeable discontinuities along the block boundaries (so-called blocking artifacts), which potentially can diffuse further into the interior of blocks owing to the motion-compensated prediction process. The removal of such blocking artifacts can provide a substantial improvement in the accuracy of motion estimation and in perceptual quality. To overcome such artifacts, H.264/AVC introduces a deblocking filter that operates within the predictive coding loop, and thus constitutes a required component of the decoding process. Using this filtering process, the blockiness is reduced with little effect on the sharpness of the content. Consequently, the subjective quality is significantly improved. At the same time, the filter reduces the bitrate by typically 5–10 percent while producing the same objective quality as the non-filtered video.

5.6.4   Entropy coding

In H.264/AVC, two entropy coding configurations are supported for encoding many syntax elements: a lower-complexity configuration based on context-adaptive variable-length coding (CAVLC) and a higher-complexity configuration based on context-based adaptive binary arithmetic coding (CABAC). More specifically, the lower-complexity configuration (e.g., for software-based applications) uses the exponential-Golomb code for nearly all syntax elements except those of quantized transform coefficients, for which the more sophisticated CAVLC method is employed [48]. When using CAVLC, the encoder switches between different VLC tables for various syntax elements, depending on the values of the previously transmitted syntax elements in the same slice. The higher-complexity config-uration further improves the entropy coding performance by using CABAC. As depicted in Figure 5.42, the CABAC design is based on three components: binarization, context modeling, and binary arithmetic coding [38]. Binarization enables efficient binary arith-metic coding by mapping non-binary syntax elements to sequences of bits referred to as bin strings. The bins of a bin string can each be processed in either an arithmetic coding mode or a bypass mode. The latter is a simplified coding mode and is chosen for selected bins such as sign-information or lesser-significance bins in order to speed up the overall decoding (and encoding) processes. The arithmetic coding mode provides the largest compression benefit; using it a bin may be context-modeled and subsequently arithmetically encoded. In com-parison with CAVLC, CABAC can typically provide reductions in bitrate of 10–20 percent for the same objective video quality when coding SDTV/HDTV signals.

5.6.5   Profiles and levels

The H.264/AVC standard also defines several profiles and levels for various application requirements. The profile defines a set of syntax features for use in generating conforming bitstreams, whereas the level places constraints on certain key parameters of the bitstream such as the maximum bitrate and maximum picture size. All decoders conforming to a specific profile and level must support all features included in that profile when constrained as specified for the level. Encoders are not required to make effective use of any particular set of features supported in a profile and level but must not violate the syntax feature set and associated constraints. This implies in particular that conformance to any specific profile and level, although it ensures interoperability with decoders, does not provide any guarantee of end-to-end reproduction quality.

Table 5.6 lists several key profiles of H.264/AVC and their corresponding main features. The application requirements for these profiles are shown in Table 5.7. The picture sizes and frame rates supported in different levels of H.264/AVC are given in Table 5.8. Comparative simulation results, targeting at video streaming applications, have been reported for H.264/ AVC (see Figure 5.43) [35]. The measure of fidelity is based on PSNR on luma, which is the most widely used such objective video quality measure. The four codecs compared use bitstreams conforming to the profiles of the following standards:

- MPEG-2 Visual, main profile (MPEG-2);
- H.263, high latency profile (HLP);
- MPEG-4 Visual, advanced simple profile (ASP);
- H.264/AVC, main profile (MP).

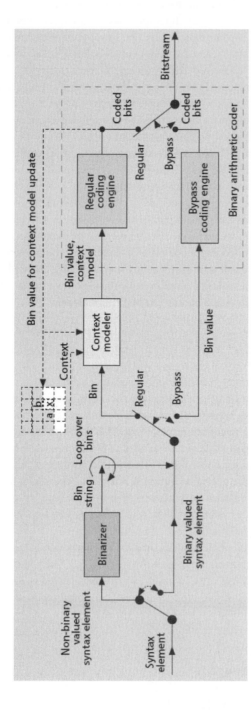

Figure 5.42 The CABAC design in H.264/AVC is based on three components: binarization, context modeling, and binary arithmetic coding [48] (© IEEE 2006).

Table 5.6 The current key profiles of H.264/MPEG-4 AVC and their corresponding main features

	Baseline	Main	Extended	High
I- and P-slices	X	X	X	X
Deblocking filter	X	X	X	X
Quarter-pixel motion compensation	X	X	X	X
Variable block size (16×16, 4×4)	X	X	X	X
CAVLC/UVLC (default)	X	X	X	X
Error resilience: flexible MB order, ASO, red. slices	X		X	
SP/SI slices			X	X
B-slices		X	X	X
Interlaced coding		X	X	X
CABAC		X		
Data partitioning			X	

Table 5.7 Application requirements for different H.264/AVC profiles

Application	Requirements	H.264 profiles	MPEG-4 profiles
Broadcast TV	Coding efficiency, reliability (over a controlled distribution channel), interlace, low-complexity decoder	main	ASP[a]
Streaming video	Coding efficiency, reliability (over a controlled packet-based network channel), scalability	extended	ARTS[b] or FGS[c]
Video storage and playback	Coding efficiency, interlace, low-complexity encoder & decoder	main	ASP
Video conferencing	Coding efficiency, reliability, low latency, low-complexity encoder & decoder	baseline	SP[d]
Mobile video	Coding efficiency, reliability, low latency, low-complexity encoder & decoder, low power consumption	baseline	SP[d]
Studio distribution	Lossless or near-lossless, interlace, efficient transcoding	main, high	Studio profile

[a] Advanced simple
[b] Advanced real-time simple
[c] Fine grain scalable
[d] Simple

These profiles generally support low to medium bitrates and picture resolutions, QCIF resolution at 10–256 kbps and CIF resolution at 128–1024 kbps being common. As shown in the PSNR versus bitrate curves in Figure 5.43, for the sequence "Tempete" the H.264/AVC significantly outperforms the other codecs. A similar superior performance is achieved by H.264 for other video sequences [39].

Table 5.8 The picture sizes and frame rates supported in different levels of H.264-AVC

Level number	Picture type & frame rate
1	QCIF @ 15 fps
1.1	QCIF @ 30 fps
1.2	CIF @ 15 fps
1.3	CIF @ 30 fps
2	CIF @ 30 fps
2.1	HRR @ 15 or 30 fps
2.2	SDTV @ 15fps
3	SDTV: $720 \times 480 \times 30i$, $720 \times 576 \times 25i$, 10 Mbps (max.)
3.1	$1280 \times 720 \times 30p$
3.2	$1280 \times 720 \times 60p$
4	HDTV: $1920 \times 1080 \times 30i$, $1280 \times 720 \times 60p$, $2K \times 1K \times 30p$, 20 Mbps (max.)
4.1	HDTV: $1920 \times 1080 \times 30i$, $1280 \times 720 \times 60p$, $2K \times 1K \times 30p$, 50 Mbps (max.)
4.2	HDTV: $1920 \times 1080 \times 60i$, $2K \times 1K \times 60p$
5	SDTV/D-Cinema: $2.5K \times 2K \times 30p$
5.1	SDTV/D-Cinema: $4K \times 2K \times 30p$

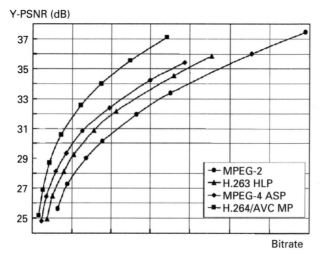

Figure 5.43 Comparative simulation results, targeting video streaming applications, for H.264-AVC [35] (© IEEE 2005).

5.7 Windows Media Video 9 (WMV-9)

The Windows Media 9 Series includes a variety of audio and video codecs, which are key components supporting the authoring and playback of digital media in Microsoft products. The Windows Media Video 9 (WMV-9) codec is the latest video codec in this suite and is based on technology that can achieve state-of-the-art compressed video quality from very low bitrates (such as 160×120 at 10 kbps for modem applications) through very high bitrates (1280×720 or 1920×1080 at 4–8 Mbps for high-definition video and even higher bitrates for mastering) [12].

5.7.1 Special Features of WMV-9

The internal color format for WMV-9 is 8-bit $4:2:0$. As seen in Figure 5.44, the WMV-9 codec, like most other standardized video codecs, uses a block-based motion-compensation and spatial transform scheme. The motion-compensated frame difference, or residual error, is transformed using a linear energy-compacting transform and then quantized and entropy coded. On the decoder side, the quantized transform coefficients are entropy decoded, dequantized, and inverse-transformed to produce an approximation to the residual error, which is then added to the motion-compensated prediction to generate the reconstruction.

As for most other standardized video codecs, WMV-9 has intra (I), predicted (P), and bidirectionally predicted (B) frames for both progressive and interlaced sequences. The B-frames in WMV-9 are not used as a reference for subsequent frames. This allows some shortcuts to be taken during the decoding of B-frames without the introduction of drift or long-term visual artifacts. Some innovations behind WMV-9 are [12]:

 (1) an adaptive block-size transform;
 (2) a limited-precision transform;
 (3) motion compensation;
 (4) quantization and dequantization;
 (5) advanced entropy coding;
 (6) an in-loop deblocking filter;
 (7) advanced B-frame coding;
 (8) interlace coding;
 (9) overlap smoothing;
(10) low-rate tools; and
(11) fading compensation.

In addition, WMV-9 uses many techniques and approaches for encoding side information, including motion vectors, B-frame-related quantities, fading parameters, etc.

5.7.1.1 Adaptive block-size transform

The H.264/AVC standard makes the first attempt to use the 4×4 integer transform and claims to have achieved the advantage of reducing ringing artifacts at edges and discontinuities. It is known that smaller transforms are better in areas with discontinuities because they produce fewer ringing artifacts. However, trends and textures, especially periodic textures, are better preserved when the transforms have a larger support. The WMV-9 codec takes the approach of allowing an 8×8 block to be transformed on the basis of the

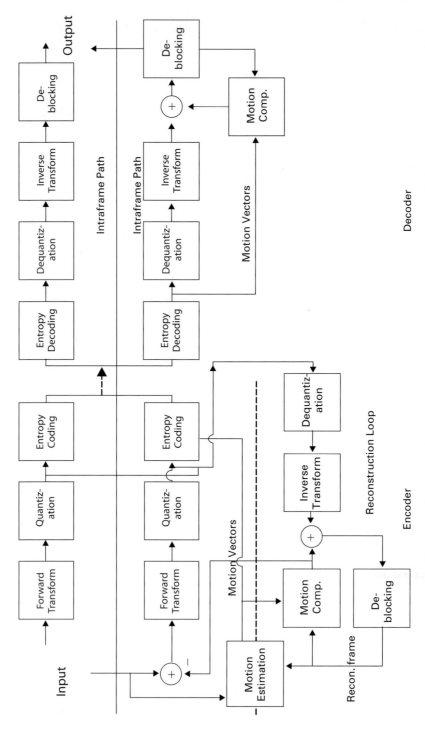

Figure 5.44 The WMV-9 codec, like most other standardized video codecs, uses a block-based motion-compensation and spatial transform scheme [12] (© Elsevier 2004).

transforms of one 8×8 block, or two 8×4 blocks, or two 4×8 blocks, or four 4×4 blocks. This allows WMV-9 to use the adaptive transform size and shape that is best suited for the underlying data. The specific transform configuration used must be signaled as part of the bitstream, which is possible at the frame, macroblock, or block level and allows for coarse and fine level specification of the transform type, especially when the data is non-stationary.

5.7.1.2 Limited-precision transforms

To reduce the computational complexity involved in the decoding of the four transform sizes used by the adaptive block transform described above, an inverse transform that can be implemented in 16-bit fixed point arithmetic is used to significantly reduce the computational complexity of the decoder in comparison with a transform requiring 32-bit or floating point arithmetic. The WMV-9 transforms are designed to be separable: both the one-dimensional 4- and 8-point transforms (approximating to a DCT transform) involve small integers and can be operated in 16-bit. The transform matrices for a one-dimensional 8-point inverse transformation and a one-dimensional 4-point inverse transformation are presented in Eq. (5.18) [12]:

$$
H_8^{-1} = \begin{bmatrix}
12 & 12 & 12 & 12 & 12 & 12 & 12 & 12 \\
16 & 15 & 9 & 4 & -4 & -9 & -15 & -16 \\
16 & 6 & -6 & -16 & -16 & -6 & 6 & 16 \\
15 & -4 & -16 & -9 & 9 & 16 & 4 & -15 \\
12 & -12 & -12 & 12 & 12 & -12 & -12 & 12 \\
9 & -16 & 4 & 15 & -15 & -4 & 16 & -9 \\
6 & -16 & 16 & -6 & -6 & 16 & -16 & 6 \\
4 & -9 & 15 & -16 & 16 & -15 & 9 & -4
\end{bmatrix},
$$

$$
H_4^{-1} = \begin{bmatrix}
17 & 17 & 17 & 17 \\
22 & 10 & -10 & -22 \\
17 & -17 & -17 & 17 \\
10 & -22 & 22 & -10
\end{bmatrix}.
$$

(5.18)

On the basis of the one-dimensional matrices of the 4- and 8-point inverse transforms in Eq. (5.19), additional restrictions are imposed on 2D implementations. First, the rows of the dequantized transform coefficients are inverse-transformed. This is followed by a rounding operation and in turn by inverse transformation of the columns and another rounding operation.

5.7.1.3 Motion compensation

Like H.264/AVC, WMV-9 also allows a resolution of one quarter-pixel in the motion prediction. H.264/AVC permits motion vectors to reference areas as small as 4×4, which imposes higher computational overheads on the decoder side since motion compensation is usually the most significant computational component when there are many filtering steps on a per-pixel basis. By taking a compromised alternative, WMV-9 also allows half-pixel resolution. The 16×16 motion vectors are default size, but 8×8 motion vectors are also permitted for frames which are signaled as containing mixed motion vector resolutions.

Moreover, special attention is paid on the interpolation filters used for generating sub-pixel predictors since they are critical for the quality of the motion compensation. More specifically, shorter filters are computationally simpler but have poor frequency response and are adversely influenced by noise. Longer filters, referencing more pixels, are computationally more difficult to implement. Moreover, images have strong local and transient characteristics that tend to get blurred with long filters. WMV-9 trades off these considerations by using two sets of filters for motion compensation. The first is an approximately bicubic filter with four taps, and the second is a bilinear filter with two taps. From the 3-tuplet information (the MV resolution, size of predicted area, and filter type) related to the motion compensation, WMV-9 defines four motion vector modes:

(1) mixed block size (16×16 and 8×8), quarter-pixel, bicubic;
(2) 16×16, quarter-pixel, bicubic;
(3) 16×16, half-pixel, bicubic;
(4) 16×16, half-pixel, bilinear.

The combined motion vector mode is signaled at frame level. In general, higher bitrates tend to use the modes at the top of the list and lower bitrates tend to use modes at the bottom of the list. The consolidation of the three criteria into one leads to more compact decoder implementation with no significant performance loss.

5.7.1.4 Quantization and dequantization

The WMV-9 codec uses multiple transform sizes, but the same quantization rules apply to all coefficients, i.e., the quantization is scalar, so that each transform coefficient is independently quantized and coded. In most video coding standards, all quantization intervals except the deadzone are of the same size, the deadzone typically being larger. The use of a larger deadzone leads to substantial bit savings at low bitrates. The quantization intervals and dequantization points (i.e., reconstruction levels) are shown in Figure 5.45(a). In practice, a deadzone may be used on the encoder side, but this would not be revealed by the reconstruction set. Therefore, a regular uniform dequantization, where the reconstruction levels are all equally spaced, is used (see Figure 5.45(b)). This type of quantization performs very well at high bitrates.

The WMV-9 codec allows both deadzone and regular uniform quantization, i.e., both variations shown in Figure 5.45 are possible. The specific type of quantization is signaled at the frame level and the appropriate dequantization rule is applied to all coefficients within the frame by the decoder. On the encoder side, a quantization-parameter-based rule allows for an

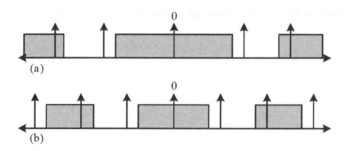

Figure 5.45 The WMV-9 codec's quantization and dequantization rules, showing (a) deadzone and (b) regular uniform quantization. The arrows show reconstruction levels, and the gray boxes show the recommended quantization bins (for alternate intervals) [12] (© Elsevier 2004).

automatic switch from regular uniform quantization to deadzone uniform quantization as the parameter increases. Although this rule works well across a variety of sequences, other factors such as the noise within a sequence and the rate control parameters may be used in more sophisticated encoders to fine tune this changeover. Allowing for a rule for switching between deadzone and regular uniform quantization is a key factor in the superior performance of WMV-9 at both low and high bitrates.

Subsequent to quantization, the two-dimensional transform coefficients are reordered into a linear array. This is done by means of a (zigzag) scanning process. Multiple scan arrays are used for this re-indexing process. These arrays are tuned, using their edge orientations, to cluster nonzero coefficients efficiently in macroblocks of the frame. The scan array used may be inferred from causal data.

5.7.1.5 Advanced entropy coding

The WMV-9 codec uses multiple tools to provide efficient entropy coding which lossly encodes symbols into the bitstream. Although WMV-9 uses simple variable-length codes for encoding quantized transform coefficients, a high degree of efficiency is achieved by allowing the use of multiple code tables for encoding each alphabet. One out of a possible set of code tables, determined by the frame-level quantization parameters, is chosen for a particular alphabet and this is signaled in the bitstream.

The motion vectors and coded block patterns are themselves entropy coded using one of the code tables. Some information such as the motion vector resolution, the skip macroblock, and the frame or field switch is represented as a bitplane where each bit signifies the corresponding value for a macroblock. An efficient method of encoding such bitplanes is also included in the syntax of WMV-9. These and other innovations are used for efficient entropy encoding in WMV-9.

5.7.1.6 In-loop deblocking filter

In-loop deblocking filtering is introduced in H.264/AVC to be performed on the reconstructed frame prior to its use as a reference frame for the subsequent frame(s). The WMV-9 scheme also uses an in-loop deblocking filter to remove block-boundary discontinuities. Therefore, the encoder and decoder must perform the same filtering operation. Since the intention of loop filtering is to smooth out the discontinuities at block boundaries, the filtering process mainly operates on the pixels that border neighboring blocks [12].

5.7.1.7 Interlace coding

Two types of interlace picture coding mode are supported by WMV-9, i.e., field picture coding, and frame picture coding. In field picture coding, the two fields that make up a frame are coded separately. A field is divided into macroblocks that can be either intra or inter coded. An intra coded macroblock in a field is coded in the same manner as progressive-picture intra macroblock coding. Inter coded macroblocks may contain either one (16×16) or four (8×8) motion vectors. As shown in Figure 5.46, when one motion vector is used the entire macroblock is compensated from one previous field. However, when four motion vectors are used, each of the four luminance blocks may be compensated from different fields.

In frame picture coding, both fields in an interlaced frame are coded jointly. Each macroblock in an interlaced frame contains samples contributed from two fields, i.e., eight lines of the top field and eight lines of the bottom field [12]. For an intra coded macroblock,

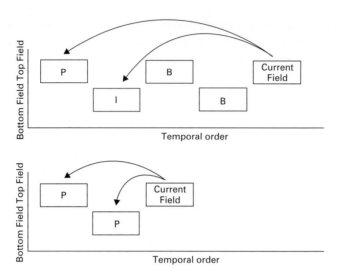

Figure 5.46 Each motion vector can refer to either of two previously encoded fields for motion compensation in WMV-9 [12] (© Elsevier 2004).

the encoder has the option of re-ordering the luminance portion according to fields, with the aim of increasing the spatial correlation prior to transform coding. An inter coded macro-block, however, may be motion compensated in one of two modes. In the frame motion-compensation mode, each macroblock is motion compensated without regard to the field structure. This is similar to progressive coding, where the macroblock is compensated using either one motion vector (16×16) or four motion vectors (8×8). In the field motion-compensation mode, the fields inside a macroblock are compensated separately using either two field motion vectors or four field motion vectors. When two field motion vectors are used, the top field lines are compensated using one motion vector and the bottom field lines are compensated using the other motion vector. When there are four field motion vectors, each field inside the macroblock is compensated using two motion vectors. More specif-ically, each field is further subdivided into two 8×8 blocks, each block having its own motion vectors. After motion compensation, the residual can be re-ordered, in the same way as in intra macroblock coding, prior to transform, and the re-ordering process is independent of the type of motion compensation used.

5.7.1.8 Advanced B-frame coding

The operation of B-frames in WMV-9 is broadly similar to that in MPEG, in the sense that it includes multiple prediction modes. The key innovations that differentiate WMV-9 are motion vector prediction for higher efficiency, the explicit inclusion of timing for the scaling of motion vectors, and the prediction of B-fields in interlaced video coding. The WMV-9 codec employs some new algorithms to make B-frames more efficient in reducing the bitrate by means of the following:

(1) explicit coding of the B-frame's temporal position relative to its two reference frames;
(2) intra coded B-frames, commonly known as B/I frames, which typically occur at scene changes and can be used when it is more economical to code the data as intra rather than P or B.

(3) improved motion vector (MV) coding efficiency, due to MV prediction based on a "forward predicts forward, backward predicts backward" rule which involves the separate buffering of the forward and backward components of each MV, including those corresponding to the direct mode;

(4) allowing bottom B-fields in interlace coding to refer to top fields from the same picture.

5.7.1.9 Overlap smoothing

Overlap smoothing is a post-processing technique used to reduce blocking artifacts in intra data, i.e., both intra frames as well as intra areas in predicted (P- and B-) frames. As discussed previously, the in-loop deblocking filter can be used to reduce blocking artifacts during the block-based intra coding, while it can also smoothen block-aligned true edges.

In order to further minimize blocking artifacts while retaining block-aligned true edges, cross-block correlations can be exploited by means of a lapped transform in the decoder [40]; this is a transform whose input spans, besides the data elements in the current block, a few adjacent elements in neighboring blocks. The "overlap" refers to the use of data points in neighboring blocks for computing the current transform. On the reconstruction side, the inverse transform influences all data points in the current block as well as a few data points in neighboring blocks. In two dimensions, the lapped transform is a function of the current block together with select elements of blocks to the left, top, right, bottom, and possibly top-left, top-right, bottom-left, and bottom-right.

5.7.1.10 Low-rate tools

The WMV-9 video codec also provides some tools and algorithms specifically geared to accommodate low-bit-rate (LBR) scenarios, e.g., sub-100 kbps. One example is the spatial scalability, offered with the aim of scaling down the dimensions of each coded frame by arbitrary resizing factors while keeping the coded frame size of I-pictures intact. The decoder is informed that these frames have been scaled down, and it will upscale the decoded image appropriately before displaying them. By operating at a downscaled level, the range of quantization is extended beyond the normal range (usually 1–31) by a factor $\sqrt{2}$ each time any dimension is scaled down by a factor 2; this allows the WMV-9 encoder to operate at low bitrates.

5.7.1.11 Fading compensation

To combat the challenging motion-compensation issues faced on video sequences that include fading, WMV-9 proposes a fading compensation that improves the performance by the effective detection of fading prior to motion compensation. The proposed fading compensation not only improves the overall compression efficiency of WMV-9 on sequences with fading but also sequences with other global illumination changes. By computing an error measure for the current video image relative to the original reference video image, fading can thus be detected by comparing the error measure with a threshold. If fading is detected, the encoder computes the fading parameters, which specify a pixel-wise linear first-order transform of the reference image. The fading parameters are quantized and signaled to the decoder. The encoder and decoder use the quantized fading parameters to transform the original reference frame into a new reference frame, which is used for motion compensation. This process uses motion compensation to find better predictors for each block, and thus code more blocks as inter blocks.

5.7.2   WMV performance

Overall, WMV-9 achieves 15% to 50% compression improvements (in terms of PSNR) over its predecessor WMV-8, and Microsoft's ISO MPEG-4 video codec (simple profile). These figures are based on experiments conducted with 13 typical MPEG clips [12], where a fixed quantization step size is set for all codecs and used with the same mode selection strategy, as is usually the case in MPEG and ITU standard tests. Higher gains are achieved by WMV-9 than by MPEG-2, i.e., WMV-9 commonly requires less than 50% of the bits used in MPEG-2 to achieve the same PSNR.

Figure 5.47(a), (b) shows comparable PSNR performances of WMV-9 and H.264/AVC for the video sequences "Glasglow" and "Stefan," respectively. The H.264/AVC codec used was the baseline without use of multiple reference frames; accordingly, the WMV-9 Main profile is used without B-frames. The PSNRs of WMV-9 and H.264/AVC would be somewhat higher, of course, if their more complex features were used. The results suggest that the PSNR performance of both these codecs is data dependent and relatively similar. The technical features offered by WMV-9 also provide favoring reasons for use in high definition content. Moreover, it was shown that the H.264/AVC decoding complexity is likely to be twice that of the WMV-9 codec, or at least that there is a significant computational benefit of WMV-9 over H.264/AVC on the decoder side [12].

5.8   Scalable extension of H.264/AVC by HHI

Thanks to the wider and wider range of varying connection qualities and receiving terminal devices, Internet access networks are getting more and more heterogeneous. Varying connection quality is the result of the best-effort nature of the distributed sharing of these networks, with time-varying data-throughput requirements of a varying number of users and the dynamically changing signal strengths in wireless access. The variety of terminal devices with different capabilities ranges from cell phones with small screens and restricted processing power to high-end PCs with high-definition displays. These facts call for scalable video coding (SVC), which provides the graceful degradation functionality for using parts of the complete encoded bitstream while achieving an R–D performance at any supported spatial, temporal, or SNR (fidelity) resolution. This serves the various needs or preferences of end users and allows varying terminal capabilities or network conditions.

As mentioned earlier, in January 2005, MPEG and the Video Coding Experts Group (VCEG) of the ITU-T agreed to finalize jointly the SVC project as an Amendment of their H.264/AVC standard, and the scalable coding scheme developed by the image communication group of the Heinrich Hertz Institute (HHI) was selected as the first Working Draft (WD-1), and H.264/AVC ISO/IEC 14496–10 Version 8 (including SVC extension) obtained consent in July 2007 [13] [14]. A remarkable feature of this proposal was that most components of H.264/AVC are used as specified in the standard (e.g. motion-compensated prediction, intra prediction, transform coding, entropy coding, deblocking filter), while only a few components have been added or modified. The key features of the scalable extension of H.264/AVC (in short, H.264/SVC) are [13]:

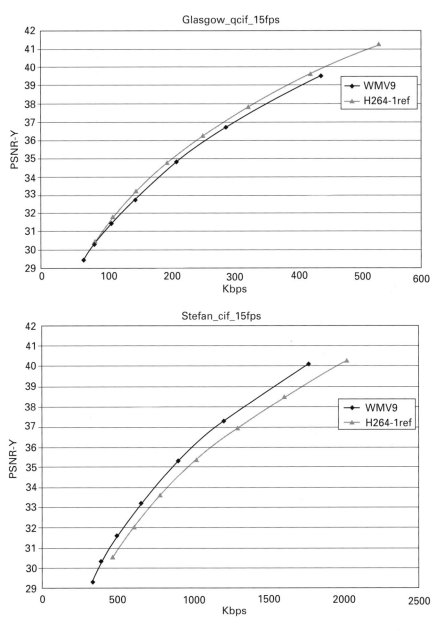

Figure 5.47 Comparable PSNR performances of WMV-9 and H.264/AVC for the video sequences (a) "Glasglow" and (b) "Stefan" [12] (© Elsevier 2004).

(1) an hierarchical prediction structure;
(2) a layered coding scheme with switchable interlayer prediction mechanisms;
(3) base-layer compatibility with H.264/AVC;
(4) fine granular quality scalability using progressive refinement slices;
(5) usage and extension of the NAL unit concept of H.264/AVC.

5.8.1   Coding structure of H.264/SVC

The layered coding scheme for achieving a wide range of spatial, temporal, and SNR scalability of H.264/SVC is illustrated in Figure 5.48 [13], which shows an example of an encoder structure with two spatial layers and combined scalability.

In each (spatial) layer, an independent hierarchical motion-compensated prediction structure with layer-specific motion parameters is employed. This hierarchical structure provides a temporal scalable representation of a sequence of input pictures that is also suitable for efficiently incorporating spatial and quality scalability. The redundancy between the different layers is exploited by various interlayer prediction concepts that include prediction mechanisms for motion parameters as well as texture data. A base-layer coding of the input pictures of the lower layer (layer 0) is obtained by transform coding similar to that of H.264/AVC; the corresponding NAL units contain motion information and texture data. These base-layer coding representations determine the minimum bitrate at which a spatio-temporal resolution can be decoded. The NAL units of the base-layer representation of the lower layer are compatible with standard H.264/AVC. For base-layer representations of the enhancement layers, additional syntax elements have been included in the macroblock layer syntax in order to signal the usage of the interlayer prediction mechanisms. The reconstruction quality of a layer can be improved by an additional SNR scalable coding based on so-called progressive-refinement slices, which represent refinements of the texture data (the intra and residual data). Since the coding symbols of the progressive refinement slices are ordered similarly to a coarse-to-fine description, these NAL units of SNR scalable coding can be arbitrarily truncated to support fine granular quality scalability (FGS) or flexible bitrate adaptation. Bitstreams for reduced spatial and/or temporal resolution can be obtained simply by discarding from the global bitstream the NAL units (or network packets) that are not needed for decoding the spatio-temporal target resolution. The NAL units that correspond to progressive refinement slices can also be arbitrarily truncated to further reduce the bitrate and the associated reconstruction quality.

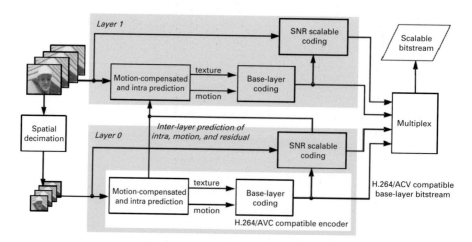

Figure 5.48 Basic coder structure for the H.264/SCV [13] (© IEEE 2004).

5.8.2   Hierarchical prediction structure

One key element of the scalable video codec is a hierarchical motion-compensated prediction structure for achieving temporal scalability. As shown in Figure 5.49, an example of
the hierarchical prediction structure for a group of eight pictures with dyadic temporal
scalability is depicted. The first picture of a video sequence group is the key picture, which
is intra coded as an instantaneous decoder refresh (IDR) picture and is coded commonly in
regular intervals. A key picture and all pictures that are temporally located between the key
picture and the previous key picture form a group of pictures (GOP). The key pictures are
either intra coded (to produce an I-picture) (e.g., in order to enable random access) or inter
coded (to produce a P-picture) by using previous key pictures as references for motion-
compensated prediction. The sequence of key pictures is independent of any other pictures
of the video sequence, and in general it represents the minimal temporal resolution that
can be decoded. Furthermore, the key pictures can be considered as resynchronization
points between the encoder and decoder. Hierarchical prediction refers to motion pre-
diction in multiple stages, which creates layers of B-slices ($I \rightarrow P \rightarrow B_0 \rightarrow B_1 \rightarrow B_2$) in an
hierarchical manner.

The remaining pictures of a GOP are hierarchically predicted. For the $N=8$ example in
Figure 5.49, the picture in the middle (B_0) is predicted by using the surrounding key pictures
(I and P) as references. It depends only on the key pictures and represents the next higher
temporal resolution together with the key pictures. The pictures of the next temporal level

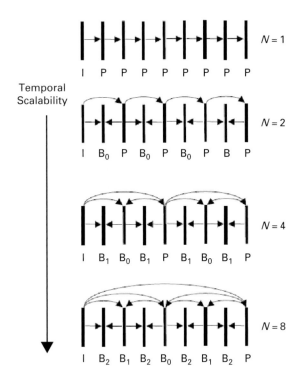

Figure 5.49 An example of the hierarchical prediction structure for a group of eight pictures
with dyadic temporal scalability [13].

(B$_1$) are predicted by using only the pictures of the lower temporal resolution as references, etc. It is obvious that this hierarchical prediction structure inherently provides temporal scalability; but it has turned out that it also offers the possibility of efficiently integrating quality and spatial scalability.

5.8.3   Interlayer prediction

Interlayer prediction is designed to enable the usage of as much lower-layer information as possible, in order to improve the rate–distortion efficiency of higher layers in the prediction of motion vectors and residual signals. Usually the interlayer predictor has to compete with the temporal predictor and, especially for sequences with slow motion and high spatial detail, the temporal prediction signal mostly represents a better approximation of the original signal than an up-sampled lower-layer reconstruction [42].

5.8.3.1   Interlayer motion prediction
In order to take advantage of the lower base-layer (layer 0) motion data in the spatial enhancement layer coding (see Figure 5.48), additional macroblock modes have been introduced in the spatial enhancement layer. In dyadic spatial scalability, a 16×16 macroblock in a spatial enhancement layer corresponds to an 8×8 (co-located) sub-macroblock in its corresponding base layer. The macroblock partitioning is obtained by upsampling the partitioning of the corresponding 8×8 block in the lower-resolution layer. The reference picture indices are copied from the corresponding base-layer blocks, and the estimated associated motion vectors are scaled by a factor 2. These scaled motion vectors are either used unmodified or refined by an additional quarter-sample motion vector refinement. In this case, the partitioning data of this macroblock together with the associated reference indices and motion vectors are derived from the corresponding data of the co-located 8×8 block in the base layer; this is called *interlayer motion prediction*.

5.8.3.2   Interlayer residual prediction
The idea of predicting motion vectors for the higher enhancement layers using the lower base layers can also apply to the prediction of motion-compensated residuals; this is called *interlayer residual prediction*. More specifically, the residual signal of the corresponding 8×8 base-layer sub-macroblock is block-wise upsampled using a bilinear filter and then used to predict the residual signal of the corresponding enhancement-layer macroblock, so that only the corresponding difference signal needs to be coded in the enhancement layer. The upsampling of the base-layer residual is done on a transform block basis in order to ensure that no filtering is applied across transform block boundaries, by which disturbing signal components could be generated [43].

5.8.3.3   Interlayer intra prediction
Furthermore, when the co-located 8×8 sub-macroblock in its base layer is intra coded, the prediction signal of the enhancement-layer macroblock is obtained by interlayer intra prediction, for which the corresponding reconstructed intra signal of the base layer is upsampled. For this prediction it is generally required that the lower layer is completely decoded, including the computationally complex operations of motion-compensated prediction and deblocking. However, this problem can be circumvented when the interlayer

intra prediction is restricted to those parts of the lower-layer picture that are intra coded. With this restriction, each supported target layer can be decoded with a single motion-compensation loop.

For upsampling the luma component, one-dimensional four-tap FIR filters are applied horizontally and vertically. The chroma components are upsampled by using a simple bilinear filter. Filtering is always performed across sub-macroblock boundaries using samples of neighboring intra blocks. When the neighboring blocks are not intra coded, the required samples are generated by specific border extension algorithms. In this way, the reconstruction of inter coded macroblocks in the base layer is avoided, thus providing so-called single-loop decoding [44]. To prevent disturbing signal components in the prediction signal, the H.264/AVC deblocking filter is applied to the reconstructed intra signal of the base layer before upsampling.

5.8.4   Fidelity quality scalability

Fidelity quality scalability can be considered as a special case of spatial scalability for which the picture sizes of the base and enhancement layers are identical; it is also referred to as coarse-grain fidelity scalable (CGS) coding (or coarse-grain SNR scalable coding). The same interlayer prediction mechanisms as for spatial scalable coding are employed, but without use of the corresponding upsampling operations, and the residual prediction is directly done in the transform domain. When utilizing interlayer prediction for coarse-grain fidelity quality scalability in SVC, a refinement of the texture information is typically achieved by requantizing the residual texture signal in the enhancement layer with a smaller quantization step size than that used for the preceding CGS layer. However, this multi-layer concept for fidelity quality scalable coding only allows a few preselected bitrates to be supported in a scalable bitstream once the quantization step sizes are chosen, i.e., the number of supported bitrates is identical to the number of layers and switching between different CGS layers can be done only at defined points in the bitstreams. This is why it is called coarse-grain fidelity scalable coding. One issue related to motion-compensated prediction for fidelity quality scalable coding is the *drift* [45], the effect whereby the motion-compensated prediction loops at the encoder and decoder are not synchronized, e.g., because fidelity quality refinement packets have been discarded from a bitstream.

To improve the growing inefficiency caused by CGS when the rate difference between successive CGS layers gets smaller, i.e., the number of layers increases, another technique which supports fine-granular SNR scalability, so-called progressive refinement (PR) slices, was introduced [42]. Each PR slice represents a refinement of the residual signal that corresponds to a bisection of the quantization step size (a QP increase by a factor 6). These signals are represented in a way such that only a single inverse transform has to be performed for each transform block at the decoder side. The ordering of transform coefficient levels in PR slices allows the corresponding PR NAL units to be truncated at any arbitrary byte-aligned point, so that the quality of the SNR base layer can be refined in a fine-grained way.

One final issue regarding SNR scalability is the effective extraction of partial bitstreams from the scalable bitstreams. There are many possible ways of extracting a substream with a specific average bitrate from a given fidelity quality scalable bitstream. The same average bitrate can be adjusted by discarding different fidelity quality refinement packets. Thus, the obtained average reconstruction error that corresponds to the given target bitrate may depend on the extraction method used. A very simple method may consist of randomly discarding refinement packets until the requested bitrate is reached. Alternatively, in a more sophisticated method, a priority identifier is assigned to each packet by an encoder. During bitstream extraction, first all packets with the lowest priority value are discarded and if the target bitrate is not reached then packets with the next priority value are discarded, etc. The priority identifiers can either be fixed by the encoder on the basis of the coder structure that is being employed they can be determined by a rate–distortion analysis.

References

[1] *Computer Desktop Encyclopedia*, The Computer Language Company, http://www. computerlanguage.com/.

[2] "Line transmission of non-telephone signals. Video codec for audio visual services at $p \times 64$ kbits/s," ITU-T Recommendation H.261, March 1993, http://www.h261.com/doc/h.261.pdf.

[3] "Coding of moving pictures and associated audio for digital storage media at up to about 1.5 Mbit/s – Part 2: Video," Int. Standards Org./Int. Electrotech. Comm. (ISO/IEC) JTC 1, ISO/IEC 11 172–2 (MPEG-1), Mar. 1993, http://www.chiariglione.org/mpeg/standards/mpeg-1/mpeg-1.htm.

[4] "Information technology – generic coding of moving pictures and associated audio information," ISO/IEC JTC1 CD 13818, 1994.

[5] "Generic coding of moving pictures and associated audio information – Part 2: Video," Int. Telecommun. Union-Telecommun. (ITU-T) and Int. Standards Org./Int. Electrotech. Comm. (ISO/IEC) JTC 1, Recommendation H.262 and ISO/IEC 13 818–2 (MPEG-2 Video), November 1994.

[6] "Video coding for low bitrate communication," ITU-T Recommendation H.263, ITU Telecom. Standardization Sector of ITU, March 1996.

[7] "Video coding for low bitrate communication," Draft ITU-T Recommendation H.263 Version 2, ITU Telecom. Standardization Sector of ITU, September 1997.

[8] "Recommended simulation conditions for H.263v3," in *Proc. ITU-T Study Group 16, Video Coding Experts Group* (Question 15), Doc. Q15D62, Tampere, April 1998.

[9] "Report of the ad hoc committee on H.263++ development," in *Proc ITU-T Study Group 16, Video Coding Experts Group* (Question 15), Doc. Q15F09, Seoul, November 1998.

[10] "Coding of audio-visual objects – Part 2: Visual," Int. Standards Org./Int. Electrotech. Comm. (ISO/IEC) JTC 1, ISO/IEC 14 496–2 (MPEG-4 Visual version 1), 1999–2003.

[11] "Advanced video coding for generic audiovisual services," Int. Telecommun. Union-Telecommun. (ITU-T) and Int. Standards Org./Int. Electrotech. Comm. (ISO/IEC) JTC 1, Recommendation H.264 and ISO/IEC 14 496–10 (MPEG-4) AVC, 2003.

[12] S. Srinivasan, P. Hsu, T. Holcomb *et al.*, *Windows Media Video 9: Overview and Applications*, Signal Process. Image Commun., special issue on technologies enabling movies on Internet HD DVD and Dcinema, 19(9): 851–875, 2004.

[13] H. Schwarz, D. Marpe, and T. Wiegand, "Overview of the scalable video coding extension of the H.264/AVC standard", *IEEE Trans. Circuits Syst. Video Technol.*, 17(9): 1103–1120, September 2007.

[14] "Advanced video coding for generic audiovisual services," ITU-T Rec. H.264 and ISO/IEC 14496–10 (MPEG-4 AVC), ITU-T and ISO/IEC JTC 1, Version 8 (including SVC extension): consent given in July 2007.

[15] R. Schafer and T. Sikora, "Digital video coding standards and their role in video communications," *Proc. IEEE*, 83(6): 907–924, June 1995.

[16] M. Manikandan, P. Vijayakumar, and N. Ramadass, "Motion estimation method for video compression: an overview," in *Proc. IFIP International Conf. on Wireless and Optical Communications Networks*, April 2006.

[17] R. Li, B. Zeng, and M. L. Liou, "A new three-step search algorithm for block motion estimation," *IEEE Trans. Circuits Syst. Video Technol.*, 4: 438–442, 1994.

[18] L.-M. Po, W.-C. Ma, "A novel four step search algorithm for fast block motion estimation," *IEEE Trans. Circuits Syst. Video Technol.*, 6(3): 313–317, June 1996.

[19] L. K. Liu and E. Feig, "A block-based gradient descent search algorithm for block motion estimation in video coding," *IEEE Trans. Circuits Syst. Video Technol.*, 6(4): 419–423, August 1996.

[20] S. Zhu and K. K. Ma, "A new diamond search algorithm for fast block-matching motion estimation," in *Proc. 1997 Int. Conf. on Information, Communications and Signal Processing (ICICS)*, Vol. 1, pp. 292–296, September 1997.

[21] Z. Chen and K. N. Ngan, "Recent advances in rate control for video coding," *Signal Process. Image Commun.*, 22(1): 19–38, 2007.

[22] A. G. Nguyen and J.-N. Hwang, "A novel hybrid HVPC/mathematical model rate control for low-bit-rate streaming video," *Signal Process. Image Commun.*, 17(5): 423–440, May 2002.

[23] T. Berger, *Rate Distortion Theory*, Prentice-Hall, 1971.

[24] T. Berger and J. D. Gibson, "Lossy source coding," *IEEE Trans. Information. Theory*, 44: 2693–2723, 1998.

[25] G. M. Schuster and A. K. Katsaggelos, *Rate Distortion Based Video Compression*, Kluwer, 1997.

[26] G. Cote, B. Erol, M. Gallant and F. Kossentini, "H.263 + : video coding at low bitrate," *IEEE Trans. Circuits Syst. Video Technol.*, 8(7): 849–866, November 1998.

[27] J. Ribas-Corbera and S. Lei, "Rate control in DCT video coding for low-delay communications," *IEEE Trans. Circuits Syst. Video Technol.*, 9(2): 172–185, February 1999.

[28] S. Aramvith and M. T. Sun, *MPEG-1 and MPEG-2 Video Standards, Image and Video Processing Handbook*, second edition, Academic Press, 2005.

[29] E. Viscito and C. Gonzales, "A video compression algorithm with adaptive bit allocation and quantization," in *Proc. Visual Communications and Image Processing Conf. (VCIP'91)*, Vol. 1605, Boston, MA, pp. 58–72, SPIE 1991.

[30] S. Battista, F. Casalino, and C. Lande, "MPEG-4: a multimedia standard for the third millennium, *Part 1*," *IEEE Multimedia*, 6(4): 74–83, October–December 1999.

[31] S. Battista, F. Casalino, and C. Lande, "MPEG-4: a multimedia standard for the third millennium, *Part 2*," *IEEE Multimedia*, 7(1): 76–84, January–March 2000.

[32] T. Rbrahim, F. Dufaux, and Y. Nakaya, "*MPEG-4: natural video coding – Part II*," Chapter 9 in *Multimedia Systems, Standards, and Networks*, A. Puri and T. Chen (eds.), Marcel Dekker, 2000.

[33] T. Sikora, "Low complexity shape-adaptive DCT for coding of arbitrarily shaped image segments," *Signal Proc. Image Commun.*, 7: 381–395, November 1995.

[34] T. Wiegand, G.J. Sullivan, G. Bjntegaard, and A. Luthra, "Overview of the H.264/AVC video coding standard," *IEEE Trans. Circuits Syst. Video Technol.*, 13(7): 560–576, July 2003.

[35] G. J. Sullivan and T. Wiegand, "Video compression – from concepts to H.264/AVC standard," *Proc. IEEE*, 93(1): 18–31, January 2005.

[36] M. Flierl, A. Mavlankar, and B. Girod, "Motion and disparity compensated coding for multi-view video," *IEEE Trans. Circuits Syst. Video Technol.*, 17(11): 1474–1484, November 2007.

[37] Joint Video Team (JVT), *JMVM Software Manual and JMVM 3.0*, 2007.

[38] D. Marpe, H. Schwarz, and T. Wiegand, "Context-adaptive binary arithmetic coding in the H.264/AVC video compression standard," *IEEE Trans. Circuits Syst. Video Technol.*, 13(7): 620–636, July 2003.

[39] T. Wiegand, H. Schwarz, A. Joch, F. Kossentini, and G. J. Sullivan, "Rate-constrained coder control and comparison of video coding standards," *IEEE Trans. Circuits Syst. Video Technol.*, 13(7): 688–703, July 2003.

[40] T. D. Tran, J. Liang, and C. Tu, "Lapped transform via time domain pre- and post-filtering," *IEEE Trans. Signal Process.*, 51(6): 1557–1571, June 2003.

[41] Joint Video Team (JVT) of ISO/IEC MPEG & ITU-T VCEG (ISO/IEC JTC1/SC29/WG11 and ITU-T SG16 Q.6). *Proc. 23rd Meeting*, San Jose CA, April 2007, http://ftp3.itu.ch/av-arch/jvt-site/2007_04_SanJose/JVT-W132.zip.

[42] H. Schwarz, D. Marpe, and T. Wiegand, "SVC core experiment 2.1: inter-layer prediction of motion and residual data," ISO/IEC JTC 1/WG11, doc. M11043, Redmond WA, July 2004.

[43] H. Schwarz, D. Marpe, and T. Wiegand, "Further results for the HHI proposal on combined scalability," ISO/IEC JTC 1/SC29/WG11, doc. M11399, Palma de Mallorca, October 2004.

[44] H. Schwarz, D. Marpe, and T. Wiegand, "Constrained inter-layer prediction for single-loop decoding in spatial scalability," in *Proc. ICIP'05*, Genoa, September 2005.

[45] J.-R. Ohm, "Advances in scalable video coding," *Proc. IEEE*, 93(1): 42–56, January 2005.

[46] H. Schwarz, D. Marpe, and T. Wiegand, "Comparison of MCTF and closed-loop hierarchical B pictures," in *Proc. Joint Video Team Conf.*, doc. JVT-P059, Poznan, July 2005.

[47] G. J. Sullivan *et al.*, "Draft text of H.264/AVC fidelity range extensions amendment," ISO/IEC MPEG and ITUT VCEG, JVT-L047, Redmond WA, July 2004.

[48] D. Marpe, T. Wiegand, and G. Sullivan, "The H.264/MPEG advanced video coding standard and its applications," *IEEE Commun. Mag.*, 134–143, August 2006.

6 Digital multimedia broadcasting

Building upon the fast growing technological advance of video compression in the 1980s, along with the availability of affordable fast computing processors and digital memories in the early 1990s, the evolution in use of digital multimedia broadcasting proceeded rapidly (see Table 6.1). The arrival of digital broadcasting was significant; what was happening was not just a simple move from an analog system to a digital system. Rather, digital broadcasting permits a level of quality and flexibility unattainable with analog broadcasting and provides a wide range of convenient services, thanks to its high picture and sound quality, interactivity, and storage capability. European broadcasters initiated the first attempt to implement a complete direct-to-home satellite digital television program delivery infrastructure having a capacity in excess of 100 channels from a single satellite. This was the digital video broadcasting (DVB) project in 1993, and the main standardization work for satellite (DVB-S) and cable (DVB-C) delivery systems was completed in 1994 [1] [2]. The fixed terrestrial version (DVB-T) was soon added to the DVB family to offer one-to-many broadband wireless data broadcasting based on roof-top antenna and the use of IP packets.

All these DVB sub-standards basically differ only in the specifications to the physical representation, modulation, transmission, and reception of the signal. Digital video broadcasting is, however, much more than a simple replacement for existing analog television transmission. More specifically, DVB provides superior picture quality with the opportunity to view pictures in standard format or wide screen (16 : 9) format, along with mono, stereo, or surround sound. It also allows a range of new features and services including subtitling, electronic program guides (EPGs), multiple audio tracks, interactive content, multimedia content, etc. Moreover, these programs may be linked to World Wide Web material to be disseminated through the Internet.

Even though DVB was a European initiative, it is rapidly becoming the worldwide standard for digital TV. At the time DVB was being developed in Europe, a parallel program of standards and equipment development was also going on in the US under the aegis of the Advanced Television Systems Committee (ATSC). Instead of using the technical foundations of DVB, the ATSC committee adopted different standards such as the ATSC digital television (DTV) standard A/53 [3] and the digital audio compression (AC3) standard A/52 [4]. These ATSC standards, established in 1995, were the world's first standards for DTV, and they established the precedent for system quality and flexibility that separates DTV from all the existing analog television systems. The nature of the ATSC DTV system is to provide new features that build upon the infrastructure within the broadcast plant and the receiver.

Early DTV receivers were deployed in December 1998. In 1999, a formal process began to get the four network affiliates (ABC, CBS, NBC, Fox) in the top 30 markets on the air with DTV. The early versions had implementation issues, with poor RF performance owing

Table 6.1 Various digital broadcasting standards [11]; CMMB denotes China Mobile Multimedia Broadcasting

Region	Fixed reception standards	Mobile reception standards
Europe, India, Australia, Southeast Asia	DVB-T	DVB-H
North America	ATSC	DVB-H
Japan	ISDB-T	ISDB-T one-segment
Korea	ATSC	T-DMB
China	DVB-T/T-DMB/CMMB	

to problems with tuner overload and multipath equalization. In 2001, the fourth-generation (4G) receivers had removed any doubts regarding the viability of outdoor DTV reception and, in 2005, the fifth-generation (5G) performance convinced many critics that successful indoor reception equivalent to NTSC could be achieved. These advances have come primarily in tuner design, carrier and clock synchronization, and multipath cancellation.

By the encapsulation of Internet protocol (IP) based services in DVB data streams, DVB networks are further opened up for transmission of graphics, photos, and data [5]. The advent of the multimedia home platform (MHP, http://www.mhp.org/) [6], which is a set of Java-based interactive TV middleware specifications developed by the DVB Project, was another milestone since with MHP software applications can be run on all sorts of terminal devices. By taking advantage of the MHP, the receiver manufacturers of DVB can thus target multiple markets rather than developing products to the specification of a particular broadcaster. The first MHP services were launched on the DVB-T platform in Finland in 2002. The core of MHP has also been adopted in a compatible manner by non-DVB systems such as ATSC and Blu-ray Disk Association, through the development of the globally executable MHP (GEM, http://www.mhp.org/mhp_technology/gem/) specification.

The DVB-T platform [7] is a flexible system which can provide services for both fixed roof-top antenna and portable or mobile receptions, with some tradeoff between bitrate and signal robustness. The cell size can be up to 100 km and the channel numbers up to 54 with 5–32 Mbps per channel. To accommodate this flexibility, DVB-T specifications allow hierarchical modulation, where two separate data streams are modulated onto a single DVB-T stream. One stream, the high-priority (HP) stream, is embedded within a low-priority (LP) stream. Receivers with "good" reception conditions can receive both streams while those with poorer reception conditions may only receive the high-priority stream. Broadcasters can target two different types of DVB-T receiver with two completely different services. Typically, the LP stream is of higher bitrate but lower robustness than the HP stream. For example, a broadcaster could choose to deliver HDTV in the LP stream and regular DTV in the HP stream.

In spite of the flexible ability of DVB-T to offer a roof-top antenna service to fixed receivers and a mobile service to moving receivers, the DVB community was asked to offer TV-type services to a small handheld device such as a mobile phone. This introduces more challenges to the DVB transmission system serving such devices. More specifically, the transmission system of these battery powered mobile devices must be equipped with capabilities to minimize the power consumption and so increase the battery usage duration, to seamlessly maintain access to the services when receivers leave a given transmission cell and enter a new one, to reliably receive the TV programs delivered in an environment suffering severe mobile multipath channels and high levels of manmade noise, etc. Moreover, these mobile devices should be capable of handling a number of reception scenarios: indoor, outdoor, pedestrian, and the inside of a vehicle moving with various speeds. To overcome these technical challenges, a new sub-standard called digital video broadcasting – handheld (DVB-H) based on IP-based design services was proposed and was published as the European Telecommunications Standards Institute (ETSI) Standard EN 302 304 in November 2004 [8] [9] [10] [11].

Another, competing, digital video broadcasting standard for handheld devices was developed in Korea, terrestrial digital multimedia broadcasting (T-DMB) standard. The T-DMB standard, which was built upon the European digital audio broadcasting standard (DAB, also known as Eureka-147 [12]), was commercialized for the first time in the world in December 2005 [13] [14] to offer diverse multimedia broadcasting services in the high-speed mobile environment. Some T-DMB services are also deployed in Europe, e.g., by the German leading mobile telecommunication firm Debitel AG (using Samsung's SGH-P900 T-DMB phones) and in China. The T-DMB standard uses the stream mode data service to carry multimedia data. The MPEG-4 binary format for scenes (BIFS) is also available in T-DMB for interactive services. Another application of T-DMB is visual radio, which is a radio service that can send two video frames per second [15].

The digital broadcasting offered in Japan is based on a highly versatile system called integrated services digital broadcasting (ISDB) [16] [17]. In Japan, digital-satellite-based ISDB was launched in 2000, and terrestrial-based ISDB (ISDB-T) began in December 2003. The ISDB-T was also chosen by Brazil for their DTV broadcasting systems, in 2006. Another characteristic is that most TV broadcasters of ISDB are providing data broadcasting services and TV services to cellular phones and HDTV services to moving vehicles. The ISDB system was designed to provide integrated digital information services consisting of audio, video, and data via satellite, terrestrial, or cable TV network transmission channels.

The ISDB-T system is designed to provide reliable high-quality video, sound, and data broadcasting not only for fixed receivers but also for mobile receivers. The system is also designed to provide flexibility, expandability, and commonality or interoperability for multimedia broadcasting. Like DVB's hierarchical modulation, the ISDB-T system supports hierarchical transmissions of up to three layers (Layers A, B, and C). The transmission parameters can be changed in each layer. In particular, the center segment of this hierarchical transmission can be received by one-segment handheld receivers. Owing to the common structure of the physical-layer modulation which divides the fullband spectrum into multiple segments, a one-segment receiver can "partially" receive a program transmitted on the center segment of a fullband ISDB-T signal (partial reception is the name given to the means by which a receiver picks out only part of the transmission bandwidth).

6.1   Moving from DVB-T to DVB-H

The two most important aspects of the DVB-T system are the use of the MPEG-2 transport stream (TS) and the use of the coded orthogonal frequency-division multiplex (COFDM) format [19]. More specifically, once the video and audio data in DVB-T are compressed according to MPEG-2 compression standards, they are packed by the packetized elementary stream (PES) layer and then each PES packet is divided into fixed-length packets of 188 bytes.

The multiplexing of different streams and program specific information (PSI) tables are accomplished at the MPEG-2 TS layer (see Figure 6.1).

The MPEG-2 TS is a communications protocol for audio, video, and data which is specified in MPEG-2 Part 1, Systems (ISO/IEC standard 13818-1 [20] [21]). Its design goal is to allow the multiplexing of digital video and audio and to synchronize the output. The transport stream protocol offers features to enable error correction for transportation over unreliable media, and is used in broadcast applications such as DVB, T-DMB, and ATSC. In MPEG-2 TS a packet is the basic unit of data in a transport stream. A transport stream consists of a sequence of fixed-size transport packets of 188 bytes: each packet comprises 184 bytes of payload and a 4-byte header. One of the key items in this 4-byte header is the 13-bit packet identifier (PID), which plays a key role in the operation of the transport stream. Additional optional transport fields, as signaled in the optional adaptation field, may follow to occupy some bytes of the 184-byte payload. The communication medium may add some error-correction bytes on top of the 188 bytes to the packet, e.g., ATSC transmission adds 20 bytes of Reed–Solomon forward error correction to create a packet that is 208 bytes long. The 188-byte packet size was originally chosen to allow MPEG-2 TS to be transported over ATM systems.

The broadcasting network interface receives the transport streams, channel encodes, and modulates them before sending them to the transmission medium. Depending on the broadcasting medium, different modulations are used based on the multicarrier modulation technique, coded orthogonal frequency-division multiplexing (COFDM). The OFDM, also referred to as multicarrier or discrete multitone modulation, is a technique which divides a high-speed serial information signal into multiple lower-speed sub-signals, so that the system can more effectively transmit these sub-signals simultaneously at different frequencies in parallel (more details about OFDM will be discussed in Section 9.1.3). To cope with the frequency-dependent multipath interference potentially encountered in OFDM, forward error correction (FEC) coding is incorporated with soft-decision decoding (thus called coded

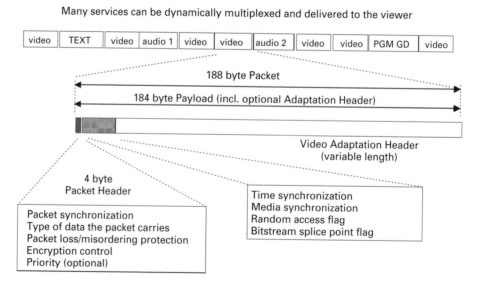

Figure 6.1 A typical representation of the MPEG-2 TS data format [20].

OFDM); the coding and decoding is integrated in a way which is specially tailored to frequency-dependent channels and brings much better performance than that of an uncoded OFDM. Various modulations used in COFDM for DVB are as follows.

- Satellite (DVB-S): QPSK
- Cable (DVB-C): 16-QAM, 32-QAM, 64-QAM
- Terrestrial (DVB-T): QPSK, 16-QAM, 64-QAM, non-uniform 16-QAM, non-uniform 64-QAM

The DVB-H [9] [10] is specifically designed for broadcasting TV contents to battery-powered handheld mobile devices, along with a related set of specifications for IP data-cast. For the delivery of mass media content, DVB-H broadcast networks should be preferred over point-to-point digital cellular networks such as 3G UMTS (Universal Mobile Telecommunications System) mobile systems. Largely based on the successful DVB-T specification for digital terrestrial television [6] [7], DVB-H further adds a number of features to take account of the limited battery life of small handheld devices and the particular environments in which such receivers must operate. The use of a technique called time slicing, where bursts of data are received periodically, allows the receiver to power off when it is inactive, leading to significant power savings. The DVB-H system also employs additional forward error correction (FEC) to improve further the already excellent mobile performance of DVB-T. The protocol stack of DVB-H is shown in Figure 6.2. As seen in the figure, DVB-H adopts the IP Datacast specifications, which are essential to the convergence of broadcast networks and wireless telecommunications networks. With IP Datacast, all content is delivered in the form of IP data packets (i.e., the IP network-layer header, UDP transport header, and RTP/RTCP QoS status monitoring header, etc.), which are mainly used for distributing digital contents over the Internet

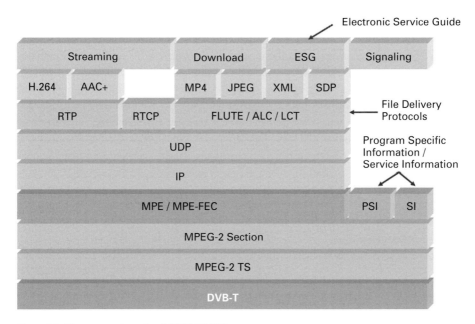

Figure 6.2 The protocol stack of DVB-H [22].

(more information about IP-based multimedia streaming will be given in Chapter 7). This provides the additional advantage that customers can enjoy all IP-based multimedia content with their mobile handsets or wireless PDAs, in addition to the traditional broadcast TV content. Moreover, DVB-H can be further enhanced to overcome the challenges presented by the use of battery powered handheld devices with a small, possibly built-in, antenna and a small screen. More specifically, the enhanced elements in the link layer are time slicing (see Section 6.1.1) [8] and multiprotocol encapsulated forward error correction (MPE-FEC) coding (see Section 6.1.2) [23]. The mandatory use of time-slicing technology in DVB-H can enable longer battery power for handheld devices, which receive the contents in bursts that are buffered and played back. Between the bursts the receiver front-end can be shut down to save the battery power considerably (up to about 90%–95%) and also enable smooth and seamless frequency handover when the user leaves one service area and enters a new cell. Furthermore, the optional use of MPE-FEC in DVB-H gives an improvement in noise immunity performance and Doppler performance in mobile channels for fast-moving vehicles and, moreover, also improves tolerance to impulse interference. It should be emphasized that neither time slicing nor MPE-FEC technology in the link layer can touch the DVB-T physical layer in any way. This means that the existing receivers for DVB-T are not disturbed by the DVB-H signal, i.e., DVB-H is totally backward compatible to DVB-T.

To accommodate the more restricted data rates suggested for individual DVB-H services and the small displays of typical handheld terminals, the traditional MPEG-2-based audio and video coding schemes used in DVB-T do not suit DVB-H well. It was therefore suggested that MPEG-2 video coding should be replaced by the more efficient MPEG-4 H.264/AVC [24] video coding standards and that MPEG-2 audio coding should be replaced by the more efficient MPEG-4 HE AAC [25]. In addition, the physical layer in DVB-H should have four extensions to the existing DVB-T physical layer [8]:

(1) upgrading of the transmitter parameter signaling (TPS) to indicate the presence of DVB-H services and the possible use of MPE-FEC to enhance and speed up the service discovery;
(2) adoption of a new (4K) mode, i.e., 4K subcarriers, of orthogonal frequency-division multiplexing (OFDM) for trading mobility and single-frequency network (SFN) cell size, allowing single-antenna reception in medium SFNs at very high speeds (all the modulation formats, QPSK, 16QAM and 64QAM with non-hierarchical or hierarchical modes, are also possible for use in DVB-H). Note that in an SFN, all transmitters radiate the same signal on the same frequency and a receiver may receive signals from several transmitters. As long as the range of delays of the received multipath signals does not exceed the designed tolerance of the system, all the received-signal components contribute usefully to better reception quality;
(3) definition of a new way of using the symbol interleaver of DVB-T in order to bring a basic tolerance to impulse noise and also to improve robustness in a mobile environment;
(4) use of the 5 MHz channel bandwidth in non-broadcast bands (e.g., in the USA).

As shown in Figure. 6.3, a DVB-H receiver consists of a DVB-T demodulator, a mandatory time-slicing module, an optional MPE-FEC module, and a DVB-H terminal [8]. The DVB-T demodulator recovers the MPEG-2 transport stream (TS) packets from the received RF signal of DVB-T. It offers three different transmission modes (with 2K, 4K, and 8K OFDM

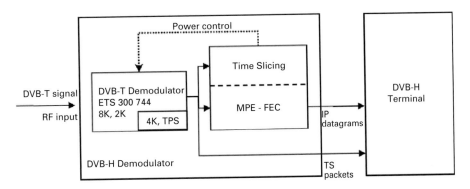

Figure 6.3 A DVB-H receiver [8] (© IEEE 2006).

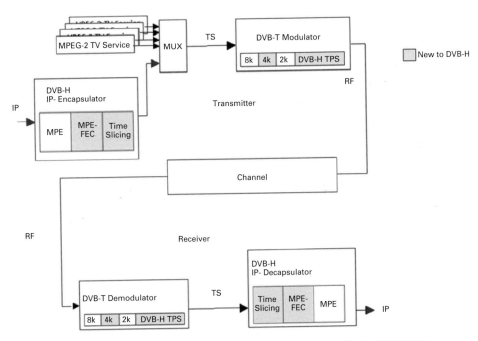

Figure 6.4 A conceptual description of the use of the DVB-H service [8] (© IEEE 2006).

subcarriers respectively) with the corresponding signaling. The time-slicing module controls the receiver to decode the wanted service and shut off during the other service bits. It aims to reduce receiver power consumption while also enable a smooth and seamless frequency handover. The MPE-FEC module, provided by DVB-H, offers in addition to error correction in the physical-layer transmission a complementary FEC function that allows the receiver to cope with particularly difficult reception situations. An example of the use of DVB-H for the transmission of IP services is given in Figure 6.4, where both traditional MPEG-2 services and time-sliced "DVB-H services" are carried over the same multiplex. The handheld terminal decodes and uses IP services only. Note that the 4K mode and the in-depth interleavers are not available, for compatibility reasons, in cases where the multiplex is shared between services intended for fixed DVB-T receivers and services for DVB-H devices.

6.1.1   Time slicing in DVB-H

Since IP datagrams are used to packetize the DVB-H content, the standard DVB way of carrying IP datagrams in an MPEG-2 TS is to use multiprotocol encapsulation (MPE), in which each IP datagram is encapsulated into one MPE section. A stream of MPE sections are then put into an elementary stream (ES), i.e., a stream of MPEG-2 TS packets with a particular program identifier (PID). Each MPE section has a 12-byte header, a 4-byte cyclic redundancy check (CRC-32) tail and a payload length (identical to the length of the IP datagram). A typical situation for future handheld DVB-H devices may be to receive audio and video services transmitted over IP on ESs having a fairly low bitrate, probably in the order of 250 kbps. The MPEG-2 TS may, however, have a bitrate of e.g., 10 Mbps. The particular ES of interest thus occupies only a fraction (in this example, 2.5%) of the total MPEG-2 TS bitrate. In order to reduce the power consumption drastically the time-slicing technique is introduced, so that the receiver demodulates and decodes only the 2.5% portion of interest and not the full MPEG-2 TS. More specifically, the MPE sections of a particular ES are sent in high-bitrate bursts instead of with a constant low bitrate. During the time between the bursts, i.e., the off-times, no sections of the particular ES are transmitted. This allows the receiver to power off completely during off-times as shown in Figure 6.5. The receiver will, however, have to know when to power on again to receive the next burst. In a particular burst the start time of the following burst in the same ES is signaled via a parameter in the header of all sections of the burst, which makes the signaling very robust against transmission errors. During the off-time, bursts from other time-sliced ESs are typically transmitted. The peak bitrate of the bursts is potentially the full MPEG-2 TS bitrate, but it could also be any lower peak value allocated for the ES. If the value is lower than the peak bitrate, the MPEG-2 TS packets of a particular burst may be interleaved with MPEG-2 TS packets belonging to other ESs.

6.1.2   MPE-FEC in DVB-H

With MPE-FEC [23], the IP datagrams of each time-sliced burst are protected by Reed–Solomon (RS) forward error correction (FEC) codes, which are calculated from the IP datagrams of the burst. The RS(n, k) code is one kind of erasure code where a message of size k symbols (each symbol represents a concatenation of several bits, e.g., a byte) can be used to create $n-k$ additional redundant (parity) symbols. On the basis of the original k transmitted symbols plus $n-k$ redundant symbols, any k of the n transmission symbols can

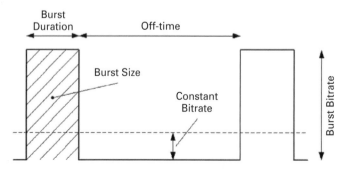

Figure 6.5 Principle of time slicing [8] (© IEEE 2006).

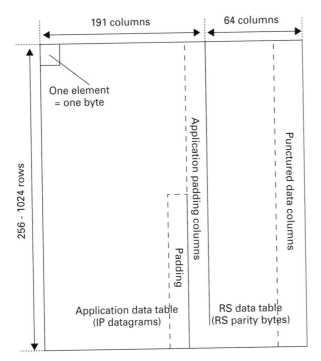

Figure 6.6 An MPE-FEC data frame [8] (© IEEE 2006).

be used to decode the original message. For more details of RS and erasure codes in general, refer to Section 10.1.2. The RS data are encapsulated into MPE-FEC sections, which are also part of the burst and are sent immediately after the last MPE section of the burst, in the same ES but with a table_id different from the MPE sections, which enables the receiver to discriminate between the two types of section in the ES. For the calculation of the RS data an MPE-FEC frame is used. The MPE-FEC frame consists of an application data table (ADT), which hosts the IP datagrams (and possible padding), and an RS data table, which hosts the RS data, as shown in Figure 6.6.

The number of rows in the MPE-FEC frame is signaled in the service information (SI) and may take any of the values 256, 512, 768, or 1024. The number of columns is 191 for the ADT and 64 for the RS data table. The IP datagrams of a particular burst are introduced vertically, column by column, in the ADT, starting in the upper left corner. If an IP datagram does not end exactly at the bottom of a column, the remaining bytes continue from the top of the next column. If the IP datagrams do not exactly fill the ADT, the remaining byte positions are padded with zeros. On each row the 64 parity bytes of the RS data table are then calculated from the 191 IP datagram bytes (and padding bytes, if applicable) of the same row, using the Reed–Solomon code RS(255,191). This provides a large virtual time interleaving, since all RS data bytes are calculated from IP datagrams distributed all over the burst.

6.2 T-DMB multimedia broadcasting for portable devices

Terrestrial digital multimedia broadcasting (T-DMB), which was developed in Korea on the basis of the European Digital Audio Broadcasting standard (DAB, Eureka-147), has some similarities with the competing mobile TV standard DVB-H for handheld devices in terms of the use of multimedia data compression techniques. More specifically, T-DMB uses MPEG-4 H.264/AVC (baseline profile at level 1.3) for video coding and MPEG-4 bit-sliced arithmetic coding (BSAC) [28] or MPEG-4 HE AAC [25] (see Section 3.6) for audio coding. The audio and video compressed data are encapsulated in an MPEG-2 transport stream (TS) and an MPEG-4 synchronization layer (SL) for streaming. Terrestrial digital multimedia broadcasting also defines the interactive data service functionality in order to provide convergent services in broadcasting and telecommunication. It adopts the MPEG-4 binary format for scenes (BIFS) as an option to enable interactive broadcasting services; MPEG-4 BIFS [26] is the part of the MPEG-4 system used for MPEG-4 scene description. In addition, it can overlay two-dimensional graphics, such as the virtual reality modeling language (VRML), with the video. Interaction between the various graphic objects is also possible. For example, a button can be made to be displayed in the corner of the broadcast program. If the button is clicked, the object linked to the button can be displayed to the users. A link to a website for a more sophisticated level of interactive service is also possible if the return channel is secured.

Another application of T-DMB is Visual Radio, which is a radio service that is able to send two video frames per second. The concept of Visual Radio is very similar to the Slideshow offered by the multimedia object transfer (MOT) transport protocol for the transmission of multimedia content in DAB data channels to various receivers with multimedia capabilities [27]. However, the technology behind Visual Radio is totally different. Visual Radio also uses MPEG-4 BSAC for digital audio broadcasting with a sequence of H.264/AVC image frames. In essence, it is a variation of T-DMB. Currently, widely varying content is broadcast through Visual Radio, e.g., image clips from a music video are sent during a radio music show along with information relevant to the music being played, pictures and graphics relevant to the news are broadcast to aid understanding, and game images are also broadcast during a live sports event.

The Transport Protocol Experts Group (TPEG), which was developed by the European Broadcasting Union (EBU) as an end-user-oriented application for the delivery of road traffic messages, can be organized as a self-contained stream of data to be broadcast over T-DMB using the transparent data channel (TDC) specification [27]. The TPEG data stream contains all messages as well as provisions for synchronization and error detection. This means that the TPEG data stream is carried in a completely transparent way over a virtually stream-oriented channel, allowing "in-order" reception of data bytes, i.e., bytes are received in the order in which they are transmitted. As an alternative transport mechanism, the transportation of TPEG data within MOT is also possible. In Korea, both mechanisms are used for TPEG applications.

6.2.1   MPEG-4 BSAC audio coding in T-DMB

Bit-sliced arithmetic coding (BSAC) is an MPEG-4 general audio coding tool based on the perceptual coding approach, as used in the MPEG-2/4 advanced audio coding (AAC) scheme. The compression methods of BSAC are similar to those of AAC except in the lossless coding algorithms, ensuring that the coding efficiency of BSAC is almost the same as that of AAC. The MPEG-4 BSAC-based audio service in T-DMB supports the standardized stereo audio broadcasting at sampling rates 24, 44.1, or 48 kHz. Bit-sliced arithmetic coding allows adaptive bitrate control, a smaller initial buffer, and a seamless play of digital audio. The audio service of T-DMB provides CD-quality audio for audio-only broadcasting and a radio quality better than analog FM for the audio accompanying video.

Bit-sliced arithmetic coding is mainly based on MPEG-4 AAC and uses most of its tools. It improves upon MPEG-4 AAC, however, by offering a fine-grain bitstream scalability and error resilience. Its compression rate is comparable with the AAC main profile. The scalable coder achieves audio coding at different bitrates and qualities by processing the bitstream in an ordered set of layers. The base layer is the smallest sub-set of the bitstream that can be decoded independently to generate the audio output. The remaining part of the bitstream is organized into a number of enhancement layers where each enhancement layer improves the audio quality. Bit-sliced arithmetic coding supports a wide range of bitrates from the low bitrate stream of 16 kbps per channel (kbps/ch) at the base layer to the higher bitrate stream of 64 kbps/ch at the top layer. It offers a fine-grain bitstream scalability at 1 kbps/ch in the full range of 16–64 kbps/ch. This fine-grain scalability of BSAC comes from the bit-sliced coding technique. In bit-sliced coding, the quantized spectral values are first grouped by the frequency bands. Then the bits in each group are processed in slices, i.e., "bit-sliced," in order, starting with the most significant bits (MSBs) and finishing with the least significant bits (LSBs). As shown in Figure 6.7, the most significant bits across the groups form the first bit-slice. This bit-slice is fed into the lossless coding part and then transmitted in the base layer. The next most significant bits form the second bit-slice, which is processed and transmitted in the first enhancement layer. This process continues until the least significant bits are processed and transmitted in the top enhancement layer. Each enhancement layer adds 1 kbps/ch of bitstream and thus provides fine-grain scalability in BSAC.

The bit-slices are then fed into the lossless coding part, which is based on an entropy coding process. In BSAC, the entropy coding is carried out by an arithmetic coder, rather than the Huffman coder used in AAC. The arithmetic coder in BSAC enhances the coding efficiency of the bit-sliced audio streams. The error resilience feature of BSAC is implemented by segmented binary arithmetic (SBA) coding, where multiple layers of audio streams are grouped again into segments. Therefore, with SBA coding, any error propagation can be constrained to a single segment in BSAC by re-initializing the arithmetic coder after several enhancement layers.

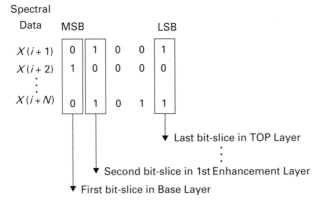

Figure 6.7 Bit slicing in MPEG-4 BSAC [14] (© IEEE 2004).

6.2.2   System specification of T-DMB

Figure 6.8 shows the specification for a T-DMB system for the transmission of audio, video, and data services which is designed for mobile reception and supports about 2 Mbps of useful data rate in a 1.536 MHz channel. The video and audio used in T-DMB video and Visual Radio are encoded using the MPEG-4 H.264/AVC video codec and the MPEG-4 error resilient BSAC or MPEG-4 HE AAC, respectively. The encoded elementary stream (ES) is multiplexed together with the MPEG-4 BIFS data used for bidirectional transport services. MPEG-4 Systems specifies the standard for packetization, synchronization, and multiplexing; it also specifies essential information, such as a scene description, object descriptions, and the synchronization among various types of media streams, for describing and communicating the coded representation of the interactive multimedia contents. The scene description, based on MPEG-4 BIFS, addresses the organization of audio-visual objects in a scene, in terms of both spatial and temporal attributes. This information allows the proper composition and rendering of individual objects. The object description identifies and describes the media streams and associates them appropriately to a scene description. Terrestrial digital multimedia broadcasting adopts MPEG-4 Core profile/level 1 for its scene description and graphics in interactive broadcasting.

To use the MPEG-2 transport infrastructure in T-DMB, whose interactive contents are mainly described by MPEG-4 in terms of a variable number of media (audio, video) streams, another multiplexing mechanism used to carry MPEG-4 contents over MPEG-2 systems, called the "MPEG-4 over MPEG-2 standard," is adopted in T-DMB. In order to carry MPEG-4 presentations over MPEG-2 systems, ESs are packetized into MPEG-4 sync. layer (SL) packet streams, which contain timing information such as composition and decoding time stamps. When the SL packetized ESs are multiplexed into MPEG-2 TS streams, two types of encapsulation method are used, according to the ES type. One is 14496-section packetization and the other is packetized ES (PES). Since BIFS scene description streams and object description streams should be transmitted periodically, 14496 section is used for

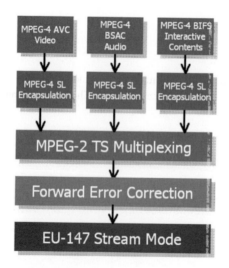

Figure 6.8 Multimedia specification architecture of a T-DMB system [37] (© IEEE 2007).

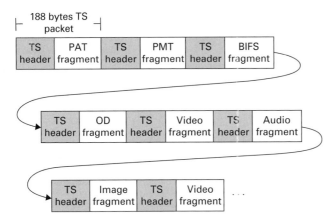

Figure 6.9 An example of an MPEG-2 time slice for T-DMB [37] (© IEEE 2007).

these streams; audio-visual streams are packetized into PES. Then those 14496-section and PES streams are multiplexed into a TS stream.

The MPEG-2 TS stream for carrying MPEG-4 streams contains program specific information (PSI) such as a program association table (PAT) and a program map table (PMT). The program association table conveys the list of programs and program ID (PID) for each single program. The program map table conveys the PID of ESs included in the program and the initial object descriptor (IOD) of the current presentation. The IOD includes descriptions and identifiers on a scene description stream and an object descriptor (OD) stream, respectively. Also, the IOD includes information about the profiles on elementary streams required by terminals. Object descriptors have elementary stream descriptors which include information on media objects such as ES IDs, media universal resource locators (urls), and synch. layer (SL) configuration descriptions, as well as the relationship between object streams. An example of such an MPEG-2 TS is shown in Figure 6.9 [37].

6.3   ATSC for North America terrestrial video broadcasting

The Advanced Television Systems Committee (ATSC) digital television (DTV, A/53) standard describes a system designed to transmit high-quality video and audio, and ancillary data, within a single 6 MHz terrestrial television broadcast channel. The design emphasis on quality resulted in the advent of digital HDTV and multichannel surround sound. The system can deliver about 19 Mbps in a 6 MHz terrestrial broadcasting channel and about 38 Mbps in a 6 MHz cable television channel. The ATSC standard can be better described as a layered architecture that separates picture formats, compression coding, data transport, and transmission, as illustrated in Figure 6.10 [29]. In ATSC, the audio is compressed on the basis of the Dolby AC3 (ATSC A/52B) [30] coding standard and the video is compressed using the ISO/IEC MPEG-2 (ATSC A/53) coding standard [31]. The video may be high definition (HDTV) or standard definition (SDTV) [32]. The former has a resolution approximately twice that of conventional television and a picture aspect ratio 16 : 9 (horizontal : vertical). The audio may be monophonic, stereo, or multichannel [33] [34] [35]. Various forms of data may supplement the main video and audio program (e.g., closed captioning, descriptive text, or

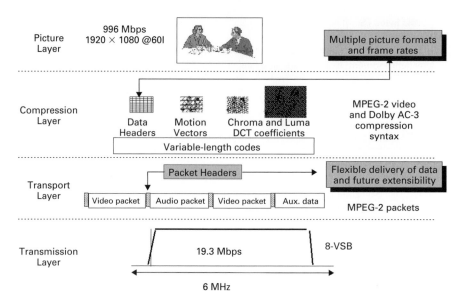

Figure 6.10 Layered architecture of the ATSC DTV system [29] (© IEEE 2006).

commentary) or perhaps one or more stand-alone services (e.g., a stock or news ticker). On the receiving end, after an ATSC receiver demodulates the RF signal and before a complete program can be assembled and presented, the receiver first processes the digital bitstream extracted from the received signal to yield a collection of program elements (video, audio, and/or data) that match the service(s) that the consumer selected. This selection is made using system and service information, transmitted as part of the digital signal.

6.3.1   Video and audio subsystems in ATSC

As shown in Figure 6.11, the video and audio subsystems provide the data compression techniques appropriate for application to the video, audio, and ancillary digital data streams. The ATSC standard employs the MPEG-2 video stream syntax (Main Profile at High Level) for the coding of video and the AC3 for the coding of audio. The ATSC DTV standard defines six video formats for HDTV (1920×1080 and 1280×720) and 12 video formats for SDTV (704×480 and 640×480), as shown in Table 6.2 and Figure 6.12; ATSC consumer receivers are designed to decode all HDTV and SDTV streams, providing program service providers with maximum flexibility. The ancillary data defined in the A/53 standard is a broad term that includes control data and supplementary data. By delivering the ancillary data as a separate payload, it can provide independent services as well as data elements related to an audio- or video-based service.

6.3.2   Service multiplex and transport in ATSC

The service multiplex and transport subsystem, as shown in Figure 6.11, refers to the means of dividing each bitstream into "packets" of information, the means of uniquely identifying each packet including packet type, and the appropriate methods of interleaving or multiplexing video bitstream packets, audio bitstream packets, and data bitstream packets into a single

Table 6.2 The 18 formats comprised by ATSC HDTV and SDTV [29] (© IEEE 2006)

Vertical lines	Pixels	Aspect ratio	Picture rate
1080	1920	16 : 9	60 interlaced
			30 progressive
			24 progressive
720	1280	16 : 9	60 progressive
			30 progressive
			24 progressive
480	704	16 : 9 and 4 : 3	60 progressive
			60 interlaced
			30 progressive
			24 progressive
480	640	4 : 3	60 progressive
			60 interlaced
			30 progressive
			24 progressive

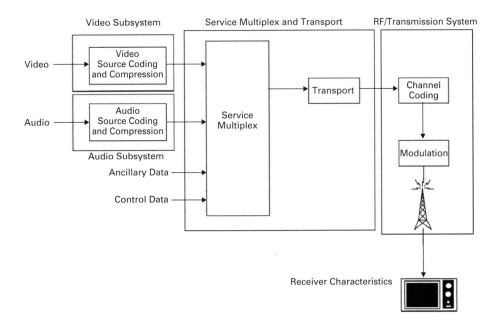

Figure 6.11 Video and audio subsystems for ATSC [29] (© IEEE 2006).

transport mechanism. The structure and relationships of these essence bitstreams are carried in service information bitstreams, also multiplexed in the single transport mechanism.

Similarly to the transport of DVB-T and DVB-H and T-DMB, the ATSC system employs the MPEG-2 transport stream (TS) syntax (see Figure 6.1) for the packetization and multi-plexing of video, audio, and data signals for digital broadcasting systems. The MPEG-2 TS syntax was developed for applications where channel bandwidth or recording media capacity is

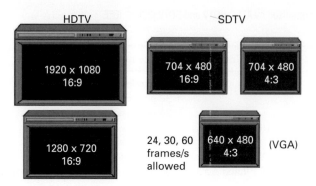

Figure 6.12 There are 18 formats altogether for ATSC HDTV and SDTV [29].

limited and the requirement for an efficient transport mechanism is paramount. It also provides the critical timing information for the receiver to perform video and audio synchronization.

6.3.3   RF transmission subsystem in ATSC

The RF transmission system, as shown in Figure 6.11, refers to channel coding and modulation. The channel coder takes the packetized digital bitstream, reformats it, and adds additional information that assists the receiver in extracting the original data from the received signal, which, owing to transmission impairments, may contain errors. In order to provide protection against both burst and random errors, the packet data is interleaved before transmission and Reed–Solomon FEC codes are added. The modulation (or physical layer) uses the digital bitstream information to modulate a carrier for the transmitted signal. Eight-vestigial-sideband(VSB) modulation is used for terrestrial broadcasting and was designed for spectral efficiency, maximizing the data throughput with a low receiver SNR threshold requirement, high immunity to both co-channel and adjacent channel interference, and a high robustness to transmission errors [29]. The recently developed enhanced-VSB (E-VSB) mode involves the transmission of a backward compatible signal within the standard 8-VSB symbol stream that can be received at a lower SNR than that of conventional 8-VSB. The E-VSB mode allows broadcasters to trade off some of their data capacity for additional robustness.

6.3.4   Data services and interactivity in ATSC

The ATSC standard consists of a suite of data broadcast standards that enable a wide variety of data services, ranging from streaming audio, video, or text services to private data delivery services. ATSC receivers may include personal computers, televisions, set-top boxes, or other devices. Generally speaking, ATSC data broadcast applications targeted to consumers can be classified by the degree of *coupling* to the main video programming, as follows [29].

(1) The viewers of *tightly coupled* data services can tune to TV programs and simultaneously receive data enhancement.
(2) A *loosely coupled* TV program might send supplementary data, which are related to the program but are not closely synchronized with it in time.

(3) *Non-coupled* data are typically contained in separate "data-only" virtual channels, which may be intended for real-time viewing, such as a 24-hour news headline or stock ticker service.

The advanced common application platform (ACAP, A/101) standard is a platform adopted by ATSC for interactive television (iTV) services. Under the ACAP, the interactive programming content runs on a single platform, the so-called common receiver, which contains a well-defined architecture, execution model, syntax, and semantics. As a "middleware" specification for interactive applications, ACAP provides the interoperability of various content or application programs or data, which are received and run uniformly on all brands and models of receivers. *Interactive television* (iTV) includes a broad array of applications, including:

- customized news, weather, and traffic;
- stock market data, including personal investment portfolio performance in real time;
- enhanced sports scores and statistics on a user-selective basis;
- games associated with a program;
- online real-time purchase of everything from groceries to software without leaving home;
- video on demand (VOD).

With the rapid adoption of digital video technology in the cable, satellite, and terrestrial broadcasting industries, the stage is set for the creation of an iTV segment that introduces to a mass consumer market a whole new range of possibilities. One typical scenario of iTV for ATSC DTV systems is shown in Figure 6.13. The ACAP standard is intended to offer consumers advanced interactive services while giving content providers, broadcasters, cable

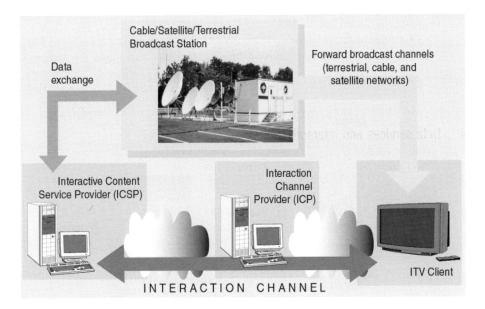

Figure 6.13 A typical iTV scenario for ATSC DTV systems.

and satellite operators, and consumer electronics manufacturers the technical details necessary to develop interoperable services and products.

6.4   ISDB digital broadcasting in Japan

Digital broadcasting in Japan is based on a highly versatile system called integrated services digital broadcasting (ISDB) [17] [18]. The ISDB system was designed to provide integrated digital information services consisting of audio, video and data via satellite, terrestrial, or cable TV network transmission channels. The ISDB was designed to allow the system easily to accommodate new services such as high-quality audio and video and data broadcasting, through the digitization of valuable broadcasting transmission channels.

In the year 2000, Japan launched the world's first HDTV satellite broadcasting, called BS, which consists of 1080 effective scanning lines (interlaced) as the video input format for MPEG-2 (high profile and main level, see Tables 5.3, 5.4) video source coding and up to 5.1 channel audio signals for MPEG-2 AAC audio coding. The BS digital broadcasting system can provide a data transmission rate of more than 52 Mbps within a bandwidth of 34.5 MHz. The transmission scheme became Recommendation ITU-R BO.1408 under the name of ISDB-S and has since become an international standard system. The major broadcasters in Japan have been broadcasting seven HDTV programs via direct BS channels since December 2000.

However, the Japanese Digital Terrestrial Broadcasting System Development Group had been set up by the Association of Radio Industries and Businesses (ARIB) in 1996, and ISDB-T services deployed by NHK began in December 2003 in the Tokyo, Osaka, and Nagoya areas; the services were later expanded to the entire nation. Furthermore, since the channel bandwidth of terrestrial broadcasting is the same as that of the cable television system, digital broadcasting services have been implemented by a number of cable television operators (ISDB-C) on the basis of the retransmission of ISDB-T and ISDB-S broadcasting (see Figure 6.14).

To coexist with the current analog TV service, the 6 MHz transmission bandwidth for ISDB-T digital terrestrial television broadcasting is divided into 14 segments (using a

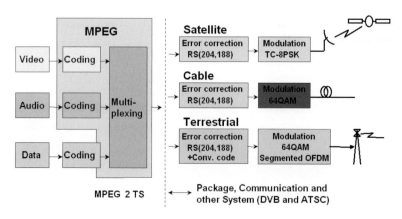

Figure 6.14 The system architecture of the ISDB system [36].

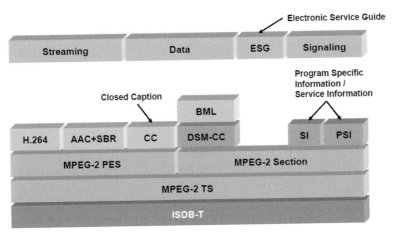

Figure 6.15 The protocol stack of ISDB-T One-Segment for mobile reception of TV contents [22].

scheme called band-segmented-transmission orthogonal-frequency-division multiplexing, BST-OFDM), of which 13 segments are for signal transmission and the remaining segment is for a guard band between channels. Among these 13 segments, one SDTV channel requires 4–5 segments, one HDTV requires 12 segments, and mobile handheld devices will access one segment. Therefore ISDB-T One-Segment, whose protocol stack is shown in Figure 6.15 [22], is thus reserved as the name of the ISDB-T service for reception on cell phones, laptop computers, and vehicles. The broadcaster can select the combination of segments to use, and so this choice of segment structure allows for flexibility of service. For example, ISDB-T can transmit mobile TV and HDTV using one TV channel or change to 3 SDTV using the same bandwidth, a switch that can be performed anytime.

References

[1] "Digital video broadcasting (DVB); DVB specification for data broadcasting," ETSI EN 301 192 V1.4.1 (2004-11), European Telecommunications Standards Institute.
[2] "Digital video broadcasting (DVB); specification for service information (SI) in DVB systems," ETSI EN 300 468 V1.6.1 (2004-11), European Telecommunications Standards Institute.
[3] "ATSC digital television standard," ATSC: A/53D, Advanced Television Systems Committee, Washington DC, July 2005.
[4] "Digital audio compression (AC-3) standard, ATSC: A/52B," Advanced Television Systems Committee, Washington, DC, August 2001.
[5] U. Reimers, "DVB – the family of international standards for digital video broadcasting," *Proc. IEEE*, 94(1): 173–182, January 2006.
[6] "Digital video broadcasting (DVB); multimedia home platform (MHP) specification 1.2," DVB Document A107, 2007, http://www.mhp.org/mhp_technology/mhp_1_2/.
[7] U. Ladebusch and C. Liss, "Terrestrial DVB (DVB-T): a broadcast technology for stationary portable and mobile use," *Proc. IEEE*, 94(1): 183–193, January 2006.
[8] G. Faria, J. A. Henriksson, E. Stare, and P. Talmola, "DVB-H: Digital Broadcast Services to Handheld Devices," *Proc. IEEE*, 94(1): 194–209, January 2006.
[9] "Digital video broadcasting (DVB); transmission system for handheld terminals (DVB-H)," ETSI EN 302 304 V1.1.1 (2004-11), European Telecommunications Standards Institute.

[10] "Digital video broadcasting (DVB); transmission to handheld terminals (DVB-H)," Validation Task Force Report VTF, ETSI TR 102 401 V1.1.1 (2005-04), European Telecommunications Standards Institute.

[11] "Digital video broadcasting (DVB), DVB-H implementation guidelines," ETSI TR 102 377 V1.1.1 (2005-02), European Telecommunications Standards Institute.

[12] "Radio broadcasting system: digital audio broadcasting (DAB) to mobile, portable and fixed receivers," ETSI EN 300 401 v1.3.3, May 2003.

[13] "Digital multimedia broadcasting," Telecommunications Technology Association in Korea, 2003SG05.02-046, 2003.

[14] V. H. S. Ha, S.-K. Choi, J.-G. Jeon, G.-H. Lee, and W.-S. Shim, "Portable receivers for digital multimedia broadcasting," *IEEE Trans. Consum. Electron.*, 50(2): 666–673, May 2004.

[15] S.-J. Lee, S. Lee, K.-W. Kim, and J.-S. Seo, "Personal and mobile satellite DMB services in Korea," *IEEE Trans. Broadcasting*, 53(1): 179–187, March 2007.

[16] M. Uehara, M. Takada, and T. Kuroda, "Transmission scheme for the terrestrial ISDB system," *IEEE Trans. Consumer Electron.*, 45(1): 101–106, January 1999.

[17] H. Asami and M. Sasaki, "Outline of ISDB systems," *Proc. IEEE*, 94(1): 248–250, January 2006.

[18] Ma. Takada and Ma. Saito, "Transmission systems for ISDB-T," *Proc. IEEE*, 94(1): 251–256, January 2006.

[19] W. Y. Zou and Yi. Wu, "COFDM: an overview," *IEEE Trans. Broadcasting*, 41(1): 1–8, March 1995.

[20] "Generic coding of moving pictures and associated audio information – Part 1: systems," Int. Telecommun. Union-Telecommun. (ITU-T) and Int. Standards Org./Int. Electrotech. Comm. (ISO/IEC) JTC 1, Recommendation H.262 and ISO/IEC 13818-1 (MPEG-2 Systems), November 1994.

[21] L. Longfei, Y. Songyu, and W. Xingdong, "Implementation of a new MPEG-2 transport stream processor for digital television broadcasting," *IEEE Trans. Broadcasting*, 48(4): 348–352, December 2002.

[22] S. Levi, "Designing encoders and decoders for mobile terrestrial broadcast digital television systems," in *Proc. TI Developer Conf.*, April 2006.

[23] M. Papaleo, R. Firrincieli, G. E. Corazza, and A. Vanelli-Coralli, "On the application of MPE-FEC to mobile DVB-S2: performance evaluation in deep fading conditions," in *Proc, Inter. Workshop on Satellite and Space Communications (IWSSC)*, pp. 223–227, September 2007.

[24] "Draft ITU-T recommendation and final draft international standard of joint video specification ITU-T Rec. H.264 | ISO/IEC 14496-10 AVC," Joint Video Team (JVT) of ISO/IEC MPEG and ITU-T VCEG, JVTG050, 2003.

[25] MPEG-4 AAC (information technology – coding of audiovisual objects – Part 3: Audio), ISO/ IEC IS 14496-3, 2001.

[26] "Coding of audio-visual objects, MPEG-4 Part 11: scene description and application engine (BIFS)," ISO/IEC 14496-11, http://www.chiariglione.org/mpeg/technologies/mp04-bifs/index.htm.

[27] K.-T. Lee, Y.-S. Park, S.-H. Park, J.-H. Paik and J.-S. Seo, "Development of portable T-DMB receiver for data services," *IEEE Trans. Consumer Electron.*, 53(1): 17–22, February 2007.

[28] S. Lee, M. Choi, J. Kim, D. Kim, N. Eum, and H. Jung, "The MPEG-4 BSAC audio decoder in terrestrial DMB receiver," in *Proc. IEEE Int. Conf. on Consumer Electronics (ICCE'06)*, pp. 257–258, 2006.

[29] M. S. Richer, G. Reitmeier, T. Gurley, G. A. Jones, J. Whitaker, and R. Rast, "The ATSC digital television system," *Proc. IEEE*, 94(1): 37–43, January 2006.

[30] S. Vernon, "Design and implementation of AC-3 coders," *IEEE Trans. Consumer Electron.*, 41(30): 754–759, August 1995.

[31] "Generic coding of moving pictures and associated audio information – Part 2: Video," Int. Telecommun. Union-Telecommun. (ITU-T) and Int. Standards Org./Int. Electrotech. Comm. (ISO/IEC) JTC 1, Recommendation H.262 and ISO/IEC 13 818-2 (MPEG-2 Video), November 1994.

[32] "Guide to the use of the ATSC digital television standard ATSC: A/54A", Advanced Television Systems Committee, Washington, DC, December 2003.

[33] C. Basile, A. P. Cavallerano, M. S. Deiss *et al.*, "The US HDTV standard," *IEEE Spectr.*, 32(4): 36–45, April 1995.

[34] J. Zdepski, R. S. Girons, P. Snopko *et al.*, "Overview of the Grand Alliance HDTV video compression system," in *Proc. 28th Conf. Signals, Systems and Computers 1994*, Vol. 1, pp. 193–197.

[35] K. Challapali, X. Lebegue, J. S. Lim, W. H. Paik, and P. A. Snopko, "The Grand Alliance system for US HDTV," *Proc. IEEE*, 83(2): 158–174, February 1995.

[36] H. Okuda, "Digital television as integrated information device," in *Proc. ITU-T Workshop on Home Networking and Home Services*, Tokyo, June 2004.

[37] W.-S. Cheong, J. Cha, S. Ahn, W.-H. Yoo, and K. A. Moon, "Interactive terrestrial digital multimedia broadcasting (T-DMB) player," *IEEE Trans. Consumer Electron.*, 53(1): 65–71, January 2007.

7 Multimedia quality of service of IP networks

Unlike stand-alone multimedia applications, in which the multimedia contents are originated and displayed on the same machine, multimedia networking has to enable multimedia data that originate on a source host to be transmitted through the IP networks (the Internet) and displayed at the destination host. The Internet, which uses IP protocols and packet switching, has become the largest network of networks in the world (it consists of a combination of many wide area and local area networks, WANs and LANs; see Figure 7.1. Multimedia over the Internet is fast growing among service providers and potential customers. Owing to the special requirements of audio and video perception, most existing and emerging real-time services need a high level of quality and impose great demands on the network. Real-time multimedia applications (e.g., live video streaming and video conferences), which are very sensitive to transmission delay and jitter and usually require a sufficiently high bandwidth, are a good example. To this end, various systems have been made available on network protocols and architectures to support the quality of service (QoS), such as the integrated services (IntServ) [22] [23] and differentiated services (DiffServ) [26] models. Multiprotocol label switching (MPLS) [21] is another technique often mentioned in the context of QoS assurance, but its real role in QoS assurance is not exactly the same as that of the IntServ and DiffServ models.

7.1 Layered Internet protocol (IP)

The IP protocol is the set (suite) of communications protocols that implement the protocol stack on which the Internet and most commercial networks run. It has also been referred to as the TCP/IP protocol, since two of the most important protocols in it are the transmission control protocol (TCP) and the Internet protocol (IP), which were also the first two networking protocols defined. The IP protocol can be viewed as a set of layers in which each layer solves a set of problems involving the transmission of data; generally, a protocol at a higher level uses a protocol at a lower level to help accomplish its aims. The IP suite uses encapsulation to provide abstraction of protocols and services. The upper layers are logically closer to the user and deal with more abstract data, relying on lower-layer protocols to translate the data into forms that can eventually be physically transmitted. The IP protocol is now commonly accepted as a top-down five-layer model, having application, transport, network, data-link, and physical layers (see Figure 7.2) [1].

7.1.1 Application layer

In the application layer, data is passed from the program in an application-specific format, then encapsulated into a transport-layer protocol. Popular application-layer formats are FTP,

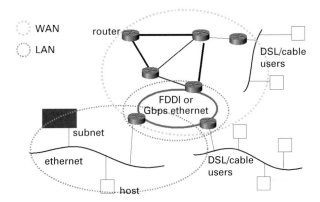

Figure 7.1 The Internet, which uses IP protocols and packet switching, has become the largest network of networks, consisting of a combination of many wide and local area networks.

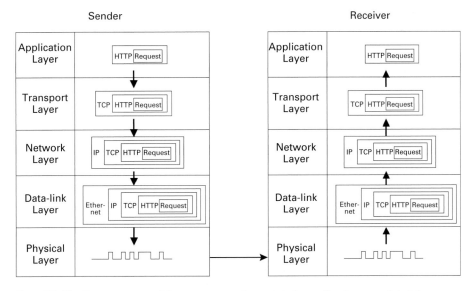

Figure 7.2 The Internet protocol is now commonly accepted as a five-layer model: it has application, transport, network, data-link, and physical layers.

Gopher, HTTP, NFS, RTSP, SIP, SMTP, SNMP, SSH, Telnet, etc. Servers have specific ports assigned to them (e.g., HTTP has port 80, FTP has port 21, etc.) while clients use short-lived ports whose numbers are automatically allocated from a predefined range by the TCP/IP stack software, typically to provide the port for the client end of a client–server communication. This allocation is temporary and is valid for the duration of the connection opened by the application using the protocols.

7.1.2 Transport layer

The transport layer is responsible for end-to-end data transfer that is independent of the underlying network, along with error control, fragmentation, and flow control. Since in an end-to-end data transmission the source and destination hosts might not be directly connected, the routing of packets is needed. Unfortunately packets might be lost (corrupted at the physical level or dropped because of full router buffers) or not be delivered successfully (destination unreachable, loops). Since IP provides only a best-effort delivery, the transport layer is the first layer of the TCP/IP stack to offer reliability through the following two protocols, the transmission control protocol (TCP) and the user datagram protocol (UDP).

The TCP is a connection-oriented protocol (with logical association) that has several mechanisms for providing a reliable byte stream, e.g., the data arrives in-order, it has minimal error (i.e., correctness), duplicate data is discarded, lost or discarded packets are resent, traffic congestion control is included. However, UDP is a connectionless datagram protocol and thus a best-effort or unreliable protocol. It is widely used in multimedia communication to ensure the real-time dissemination of audio and video data whose on-time arrival is more important than reliability. Since UDP is used without maintaining a connection, it does not guarantee delivery, preserve sequences, or protect against duplication. The reliability of UDP is simply addressed through error detection using a weak checksum algorithm. Both TCP and UDP are used to carry a number of higher-level applications. The applications at any given network address are distinguished by their TCP or UDP ports. In both TCP and UDP, each packet header will specify a source port and a destination port, each of which is a 16-bit unsigned integer (i.e., ranging from 0 to 65 535). A process listens for incoming packets whose destination port matches that port number, and/or sends outgoing packets whose source port is set to that port number. Processes may also bind to multiple ports.

Since UDP cannot provide good reliability for audio and video data, the real-time transport protocol (RTP), which is also a datagram protocol, was designed for the transmission of real-time data such as streaming audio and video. The RTP facilitates end-to-end real-time data delivery services by providing a timestamp and a sequence number on top of UDP; therefore both RTP and UDP contribute to the transport protocol functionality. It should be noted that the RTP is mainly used for monitoring the transmission and does not address resource reservation or QoS guarantees. It does not provide reliability in packet delivery, neither can it be used to recover from packet loss. Since RTP information is attached on top of UDP, it does not need to have a standard TCP or UDP port by which it communicates. The real-time transport control protocol (RTCP) is an accompanying protocol providing control information for an RTP flow. It partners RTP in the delivery and packaging of multimedia data but does not transport any data itself. It is used periodically to transmit control packets to participants in a streaming multimedia session. The primary function of the RTCP is to provide feedback on the quality of service information derived from the RTP. It gathers statistics on a media connection and information such as bytes sent, packets sent, lost packets, jitter, feedback, and round trip delay. An application may use this information to increase the quality of service, perhaps by limiting flow or using a low-compression codec instead of a high-compression codec. The RTCP is used for QoS reporting and normally uses the odd port next higher than the corresponding even port used in the UDP. The RTCP offers feedback on the quality of the data delivery service in forms of sender reports (SRs) and receiver report (RRs), and the data traffic associated with these reports is limited to 5% of the session bandwidth (SRs use 25% and RRs use 75%). Video and audio payloads are commonly transmitted as separate RTP sessions and RTCP packets are transmitted for

4	4	8	16	16	3	13	8	8	16	32	32	...	bits
VERS	HLEN	TOS	Total Length	ID	Flags	Frag. Offset	TTL	Protocol	Header Checksum	SA	DA	IP Options	Data

Figure 7.3 The IP header format of an ICMP message [2].

each medium using two different UDP port pairs and/or multicast addresses. This separation allows participants to decide to join either of these two sessions or both. Despite the separation, synchronized playback of a source's audio and video can be achieved using the timing information carried in RTP.

7.1.3 Network layer

In the Internet protocol suite, the network-layer protocol has the basic task of getting packets of data from end to end (source to destination), whereas the data-link layer is responsible for node-to-node (hop-to-hop) packet delivery. The network-layer protocol provides the functional and procedural means of transferring variable-length data sequences from a source to a destination via one or more networks. Some protocols carried by IP, such as the Internet control message protocol (ICMP) [2] and the Internet group management protocol (IGMP) [3] are typical protocols used in the network layer.

The ICMP is one of the core network-layer protocols; it is mainly used in response to errors in IP datagrams or for diagnostic or routing purposes in unicast connections, to indicate, for instance, that a requested service is not available or that a host or router could not be reached. The IGMP is another network-layer protocol used to manage the membership of IP multicast groups. There are two versions of ICMP: ICMPv4 for IPv4 and ICMPv6 for IPv6. ICMP messages are constructed at the network layer, usually from a normal IP datagram that has generated an ICMP response. The network layer encapsulates the appropriate ICMP message with a new IP header and transmits the resulting datagram in the usual manner. More specifically, according to IETF RFC 792 an IP data packet is arranged as in Figure 7.3, where the datagram fields are defined as follows.

- *VERS* is the IP version number (currently this is binary 0100, version 4, but in future it can also be version 6). All nodes must use the same version.
- *HLEN* is the header length in 32-bit (4-byte) words, so if header length is 6 then there are 6×32-bit words in the header, i.e., 24 bytes. The maximum size is 15×32-bit words, which is 60 bytes. The minimum size is 5×32-bit words or 20 bytes.
- *Type of service* (ToS) indicates how the datagram should be used, e.g., for delay, precedence, reliability, minimum cost, throughput, etc. This TOS field is now used by Differentiated Services and is called the Diff Serv code point (DSCP).
- *Total length* is the number of bytes that the IP datagram takes up including the header. The maximum size that an IP datagram can be is 65 535 bytes.
- *Identification* is a unique number assigned to a datagram fragment to help in the reassembly of fragmented datagrams.
- *Flags* Bit 0 is always 0 and is reserved. Bit 1 indicates whether a datagram can be fragmented (0) or not (1). Bit 2 indicates to the receiving unit whether the fragment is the last one in the datagram (1) or if there are still more fragments to come (0).

- *Frag offset* In units of eight bytes (64 bits), this specifies a value for each data fragment in the reassembly process. Different-sized maximum transmission units (MTUs) can be used throughout the Internet.
- *TTL* is the time that the datagram is allowed to live (exist) on the network. A router that processes the packet decrements this by one. Once the value reaches 0 the packet is discarded.
- *Protocol* This gives the layer-4 protocol sending the datagram: UDP uses the number 17, TCP uses 6, ICMP uses 1, IGRP uses 88 and OSPF uses 89.
- *Header checksum* gives the error control for the header only.
- *IP options* This field is for testing, debugging, and security.
- *Padding* is added sometimes just to make sure that the datagram is confined within a 32-bit boundary in multiples of 32 bits.

Many commonly used network utilities are based on ICMP messages. The "traceroute" command is implemented by transmitting UDP datagrams with specially set IP TTL header fields and looking for ICMP "time to live (TTL) exceeded in transit" and "destination unreachable" messages generated in response. Another related "ping" utility is implemented using the ICMP "echo request" and "echo reply" messages. For example, every machine (such as an intermediate router) that forwards an IP datagram has to decrement the TTL field of the IP header by one; if the TTL reaches 0 then an ICMP TTL-exceeded-in-transit message is sent to the source of the datagram. Each ICMP message is encapsulated directly within a single IP datagram and thus, like UDP, ICMP does not guarantee delivery.

Furthermore, according to the IP network header as shown in Figure 7.3, the network layer provides an unreliable service (i.e., only best-effort delivery), which means that the network makes no guarantees about packet loss, data corruption, out-of-order packet arrival, duplicate arrival, etc. In terms of reliability all the network layer does is to ensure that the IP packet's header is error free through the use of a checksum. An upper-layer protocol must be used to address any of these reliability issues. For example, to ensure in-order delivery the upper layer may have to buffer data until it can be passed up in order.

Since the Internet is a network of heterogeneous networks, given two networked nodes, A and B, which are not on the same LAN, a packet from A to B has to be routed. All routing protocols, such as the routing information protocol (RIP, IETF RFC 1058) [4] and the open-shortest-path-first protocol (OSPF, IETF RFC 2328, v2) [5], are part of the network layer because their payload is totally concerned with management of the network layer. More specifically, a router only knows about its surrounding network (in terms of the updated tree-image of the network topology), not all the hosts, and each entry in the routing table is a pair consisting of the destination network address and the next-hop router address, (Dest_Net, Next_Hop), as shown in Figure 7.4. Any destination network address can be further represented in terms of an IP address and mask (P, M). Some typical destination network addresses are shown below.

- Class-based network: $128.29 \rightarrow (P = 128.29.0.0, M = 255.255,0,0)$
- Subnetting: $128.29.3 \rightarrow (P = 128.29.3, M = 255.255.255.0)$
- Supernetting: $192.44.16\text{--}31 \rightarrow (P = 192.44.16.0, M = 255.255.240.0)$

When a packet with destination IP address X arrives at the router, the router goes over the entries in its routing table, and for each routing table entry $((P, M), Z)$, where Z denotes the

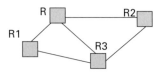

Destination Network	Next Hop
Net1	R1
Net2	R2
Net3	R3

Figure 7.4 A router only knows about its surrounding network (in terms of the updated tree-image of the network topology) and each entry of the routing table is denoted as a pair of the destination network address and the next-hop router address.

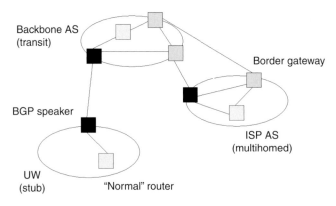

Figure 7.5 An autonomous system (AS) is a set of networks subject to a common authority.

next-hop router IP address, it checks whether the following condition is valid:

$$X \otimes M = P. \tag{7.1}$$

Here $\otimes$ denotes the logical AND operation: if Eq. (7.1) is satisfied then P is the (sub- or super-) network where X belongs, and the packet is forwarded to Z.

An autonomous system (AS) is a set of networks subject to a common authority. The Internet can be regarded as a set of ASs, connected by border gateways, which are routers connecting one AS to another. As shown in Figure 7.5, there are various kinds of ASs, as follows.

(1) *Stub AS*   This operates through a single connection to another AS, most traffic is local traffic.
(2) *Multihomed AS*   This operates through multiple connections to other ASs, most traffic is local traffic.
(3) *Transit AS*   This operates through multiple connections to other ASs, most traffic is transit traffic.

A typical example of an AS can be a university campus network, a company network, or an Internet service provider (ISP) network. Routing inside an AS is called intradomain routing. Popular intradomain routing protocols include RIP and OSPF. If, however, the routing has to be carried out across many different ASs then the border gateway protocol (BGP) is used for security and economics reasons, since there are more than 50 000 ASs in the Internet and building a routing table is too expensive. It should be noted that one border gateway among others from each AS is designated as the BGP speaker, and the BGP is executed among all BGP speakers. The BGP speakers then distribute routing information to other border gateways inside their AS.

The routing information protocol (RIP) [4] is one of the most commonly used intradomain routing protocols within an AS. The RIP helps routers adapt dynamically to changes in network connection by communicating information about which networks each router can reach and how far away those networks are. The RIP constructs its routing table using the Bellman–Ford algorithm, i.e., given a weighted graph and a destination node D, this algorithm finds the shortest path from each node in the graph to D. More specifically, each router exchanges a distance vector with neighboring routers, e.g., R1 : 3, R2 : 7, R3 : 5, as in Figure 7.4, and the Bellman–Ford algorithm can thus update the routing table using the received distance vectors. Although RIP is still actively used, it is generally considered to have been made obsolete by routing protocols such as OSPF owing to the slow convergence of the Bellman–Ford algorithm and, moreover, long routing loops can be formed as a result of routing table inconsistency and cause packets to be forwarded from router to router and never reach the destination.

The OSPF protocol [5] is another hierarchical intradomain protocol for routing in the Internet protocol; it uses a link-state [6] in the individual ASs that make up the hierarchy. A computation based on Dijkstra's algorithm is used to calculate the shortest path-tree inside each AS. In OSPF, a router periodically sends updated link-state advertisement (LSA) packets which describe the sender's cost to its neighbor, e.g., R1, R2 : 7, to all other routers in the same routing domain (see Figure 7.6). The LSA cost is determined on the basis of the following factors:

- path distance; with a weight 1 for all links (the cost of the path is defined to be the hop count);
- link latency (queuing and propagation delays);

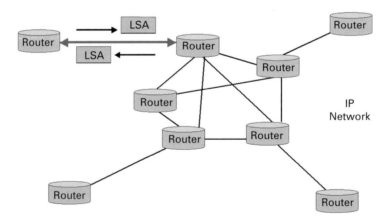

Figure 7.6 In OSPF, a router periodically sends updated LSA packets which describe the sender's cost to its neighbor to all other routers in the same routing domain.

- link capacity (the maximum throughput of the link in terms of bitrate);
- link reliability (packet-loss statistics).

Thus each router learns the current state of the whole network and runs Dijkstra's algorithm to build its own routing tables, i.e., given a weighted graph and a source node A, Dijkstra's algorithm finds the shortest path from A to each other node in the graph. Routers in the same AS domain form adjacencies when they have detected each other. This detection occurs when a router "sees" itself in a hello packet forwarded from other routers. This is called a two-way state and is the most basic relationship. The OSPF system uses both unicast and multicast to send "hello packets" and link-state updates. The multicast addresses 224.0.0.5 and 224.0.0.6 are reserved for OSPF. In contrast with the RIP or the BGP, OSPF does not use TCP or UDP; it uses IP directly. OSPF handles its own error detection and correction so does not need the TCP or UDP functions.

7.1.4 Data-link layer

The data-link layer is layer 2 of the five-layer TCP/IP reference model. It responds to service requests from the network layer and issues service requests to the physical layer. This is the layer which transfers data between adjacent network nodes in a wide area network (WAN) or between nodes on the same local area network (LAN) segment. The data-link layer provides the functional and procedural means to transfer data between network entities and can provide the means to detect and possibly correct errors that may occur in the physical layer. Since the data link provides data transfer across the physical link and the transfer might or might not be reliable, many data-link protocols do not provide acknowledgments of successful frame reception and acceptance, and some data-link protocols might not even have a checksum to check for transmission errors. That is why in the IP protocol suite, higher-level (such as network-layer or transport-layer) protocols are used to provide flow control, error checking, acknowledgments, and retransmission.

In IEEE 802 local area networks (LANs), the data-link layer is split into logical link control (LLC) and media access control (MAC) sublayers. The LLC is the upper sublayer of the data-link layer and is applicable to the various physical media (such as IEEE 802.3 Ethernet, IEEE 802.5 token ring, and IEEE 802.11 wireless LAN). The LLC sublayer is primarily responsible for multiplexing protocols running atop the data-link layer and *optionally* provides flow control, the detection and retransmission of dropped packets, acknowledgment, and error recovery. However, the MAC sublayer is used to determine who is allowed to access the media at any time (e.g., CSMA/CD in IEEE 802.3 and CSMA/CA in IEEE 802.11) in either a distributed or centralized control manner. That is why MAC is sometimes referred as a multiple access control protocol, since the MAC sublayer provides the protocol and control mechanisms that are required for a certain channel access method. This makes it possible for several stations connected to the same physical medium to share it. More specifically, the MAC provides addressing and channel access control mechanisms that make it possible for several terminals or network nodes to communicate within a multipoint network, typically a LAN or metropolitan area network (MAN, e.g., IEEE 802.16). A MAC protocol is not required in full-duplex point-to-point communication. The MAC sublayer also acts as an interface between the LLC sublayer and the network's physical layer. The MAC layer provides an addressing mechanism called physical address or MAC address, which is a unique serial number assigned to each network adapter, making it possible to deliver data packets to a destination within a subnetwork, i.e., a physical network without routers, for example an Ethernet network.

7.1.5 Physical layer

The physical layer is responsible for encoding, transmission, and reception of data over a physical data link connecting network nodes. It operates with data, in the form of raw bits rather than packets, that are sent from the physical layer of the sending (source) device and received at the physical layer of the destination device. Consequently, no packet headers or trailers are added to the data by the physical layer. The bitstream may be grouped into codewords or symbols and converted to a physical signal, which is transmitted over a physical transmission medium. Many major functions and services are performed by the physical layer; some typical examples are standardized interfaces to physical transmission media, such as the mechanical specification of electrical connectors and cables, the electrical specification of transmission-line signal level and impedance, etc. For wireless, the physical layer can include radio interfaces, such as the electromagnetic spectrum frequency allocation and the specification of signal strength, analog bandwidth, modulation, equalization filtering, training sequences, pulse shaping, and other signal processing information. Moreover, forward error correction (FEC), bit-interleaving, and other types of channel coding are also taken care of in the physical layer.

7.2 IP quality of service

An end-to-end IP-based service is defined by ITU as "a service provided by the service plan to an end user, e.g., a host (end system) or a network element, and utilizes the IP transfer capabilities and associated control and management functions for delivery of the user information specified by the service level agreements" [7]. In the networking community, quality of service (QoS) commonly refers to the service level offered by the network to applications or users to match their performance needs in terms of network QoS metric parameters, including delay or latency, delay variation (delay jitter), throughput or bandwidth, and packet loss or error rate, etc. [8]. These parameters describe the treatment experienced by packets while passing through the network. They can be translated into particular parameters of the network-architecture components used to ensure QoS. They are finally mapped into the configuration of network elements. They are also closely connected with protocols used in the network and equipment capabilities.

7.2.1 Delay or latency

One-way delay (or one-way trip time, OTT) has a direct impact on users' satisfaction in multimedia communication. Real-time applications require the delivery of information from the source to the destination within a certain period of time. According to the ITU-T one-way VoIP delay recommendation, an OTT below 150 ms is quite satisfactory and not noticeable; however, when the OTT exceeds 250 ms then the quality is not tolerable and can easily create user frustration during interactive tasks [9].

Several sources contribute to the end-to-end OTT for the data traffic, which is carried across a series of components in the communication system that interconnects the source and the destination. More specifically, in addition to the delay introduced by the network transmission, there are also source-processing (digitization, compression, and packetization) delays, introduced by the source that converts the captured raw data into compressed bitstreams and then packets and destination-processing delays introduced by the processing required at the destination to perform packet reconstruction and data decoding. These two processing delays depend strongly on the source and destination hosts' hardware configuration and load. With today's rapidly growing CPU power, these two delays are playing a less critical factor in the QoS.

Let us assume there are N links from the sender to the receiver, $i = 1, \ldots, N$. The capacity of link i is C_i. The sender sends out a packet of size S; the network transmission delay D, which is one critical component of OTT, is the sum of the delays at each link along the transmission path. More specifically, the delay at a single link consists of the propagation delay S/C_i, the queuing delay, and the processing delay at the router [10]:

$$D = \sum_{i=1}^{N} \left(\frac{S}{C_i} + d_i + \sigma_i \right), \tag{7.2}$$

where the processing delay σ_i is caused by the communication protocols executed at the different network components such as routers, gateways, and network interface cards. The processing delay depends on the particular protocols, the load of the network, and the configuration of the hardware that executes protocol processing. The queuing delay d_i^k is caused by the time a packet spends in the outgoing link queue at a network component. For example, such a delay can be incurred at an intermediate router output queue. The delay depends on the network congestion, the configuration of the hardware, and the link speed.

7.2.2 Delay variation (delay jitter)

Delay variation, commonly called delay jitter, is a QoS metric that refers to the fluctuation or variation of end-to-end delay from one packet to the next packet within the same packet stream or connection or flow. Since each packet in the network travels through different paths and since the network conditions for each packet can be different, the end-to-end delay varies. Given a pair of packets (Pkt$_i$, Pkt$_{i+1}$) with corresponding transmission and receiving timestamps $\{(T_i, T_{i+1}), (R_i, R_{i+1})\}$, the delay jitter is defined to be the difference between the transmission time gap $T_{i+1} - T_i$ and the receiving time gap $R_{i+1} - R_i$ (see Figure 7.7),

$$delay\_jitter = |(R_{i+1} - R_i) - (T_{i+1} - T_{i1})|. \tag{7.3}$$

According to the IETF VoIP recommendation, the quality is regarded as good if $delay\_jitter$ is less than 40 ms, and the quality is becoming unacceptable if $delay\_jitter$ is greater than 75 ms.

For data generated at a constant rate, the delay jitter distorts the time synchronization of the original traffic. For example, the receiver plays back the signal as soon as the packets arrive. The playback point might be changed from the original timing reference, and this introduces distortion into the playback signal. However, the receiver plays back the signal using the original timing reference. Late packets that miss the playback point will be ignored, which also introduces distortion. A better alternative is to use a de-jittered buffer. All packets are stored in the buffer and held for some time (the offset delay) before they are retrieved by the receiver with the original timing reference. The fidelity of the signal will be maintained as long as there are packets available in the buffer. A large amount of

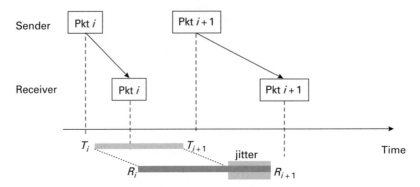

Figure 7.7 Delay jitter is defined to be the difference between the transmission time gap $T_{i+1} - T_i$ and the receiving time gap $R_{i+1} - R_i$.

delay jitter requires large buffer space to hold the packets and smooth out the jitter. However, a large buffer itself introduces large delays, which will be eventually constrained by the application's delay requirement. In summary, there is a tradeoff between the following three factors: de-jittered buffer space, delay requirement, and fidelity of the playback signal.

7.2.3   Throughput or bandwidth

In communication networks, the throughput or bandwidth is the amount of digital data per time unit that is delivered over a physical or logical link or that is passing through a certain network node. The throughput is usually measured in bits per second (bits/s or bps). The system throughput, or aggregate throughput, is the sum of the data rates that are delivered to all terminals in a network.

The maximum throughput of a point-to-point or point-to-multipoint physical transmission medium is equal to or near the channel capacity. This is affected by the modulation method and physical-layer protocol overheads such as error correction coding, bit synchronization, and equalizer training sequences. If the communication is mediated by several links in series with different bitrates, the maximum throughput is lower than or equal to the lowest (bottleneck) bitrate. In a computer network, the throughput that is achieved from one computer to another (end-to-end) is normally lower than the maximum throughput and lower than the network access channel capacity, because the traffic load may be lower than the maximum throughput, the channel capacity may be shared by other users, protocol overhead affects the perceived speed, or packets are lost owing to congestion or noisy channels, etc.

7.2.4   Packet loss or error rate

Packet loss directly affects the perceived quality of the application. It compromises the integrity of the data or disrupts the service. At the network level, packet loss can be caused by network congestion, which results in dropped packets. Another cause of loss is the presence of a noisy communication channel, especially in a wireless channel. There are several techniques for recoverery from packet loss or error, such as packet retransmission at the transport layer, error correction at the physical layer, or the use of a codec at the application layer that can compensate or conceal the loss.

7.3 QoS mechanisms

Based on either the quantitative or qualitative requests of QoS requirements, network and application providers need to respond to these requests by supplying QoS services using a number of QoS mechanisms. The bandwidth is the most important network resource that can be managed so as to satisfy the QoS requirements. There have been several QoS mechanisms based on bandwidth management, including traffic classification, packet scheduling, resource reservation, channel access, and traffic policing.

7.3.1 Traffic classification

For a network to provide differentiated services to various applications, the network requires a traffic classification mechanism that can differentiate between the various applications. The classification mechanism identifies and separates the traffic into flows or groups of flows (aggregated flows or classes). Therefore, each flow or each aggregated flow can be handled selectively. The classification mechanism can be implemented in the various network devices (i.e., the end hosts and intermediate devices such as switches, routers, and access points). To identify and classify the traffic, the traffic classification mechanism requires some form of tagging or marking of packets. A number of traffic classification approaches are implemented in the different IP layers.

7.3.1.1 Data-link layer classification

The data-link layer, layer 2, can be used to classify the traffic using the tag or field available in the layer-2 header. An example of layer-2 classification is the IEEE 802 user priority protocol, whose header includes a 3-bit priority field that enables eight priority classes. It aims to support service differentiation on a layer-2 network such as a LAN. The end host or intermediate host associates application traffic with a class (on the basis of the policy or the service that the application expects to receive) and tags the packets' priority field in the IEEE 802 header.

A classification mechanism identifies packets by examining the priority field of the IEEE 802 header and forwards the packets to the appropriated queues. The IEEE recommends mapping the priority value and the corresponding service as shown in Table 7.1. One such example of MAC layer traffic classification is the IEEE 802.11e standard [11], which is an extension for QoS improvements of the 802.11 WLAN formed by adding priorities to the MAC layer so as to provide improvements of the MAC access mechanisms. Eight user priorities, (see Table 7.1), defined in the 802 standard for various traffic classes, are mapped down to four access categories (ACs) in 802.11e. Each AC has its own output queue for transmission on the channel, but before a frame is put into one of these queues a mapping from a traffic class to an AC takes place in the 802.11e node. The four ACs are called AC voice (AC_VO), AC video (AC_VI), AC background (AC_BK), and the AC best-effort (AC_BE). The access category AC_VO has the highest priority and is meant for voice traffic with strict latency, jitter, and bandwidth requirements; AC_VI is

Table 7.1 The recommended mapping of the priority values and the corresponding services in IEEE 802 headers

Priority	Services
0	Default, assumed to be best-effort service
1	Less than best-effort service
2	Reserved
3	Reserved
4	Delay sensitive, no bound
5	Delay sensitive, 100 ms bound
6	Delay sensitive, 10 ms bound
7	Network control

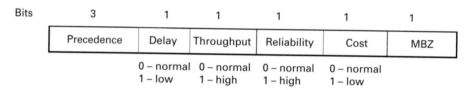

Figure 7.8 The type-of-service (TOS) fields defined in IPv4 are used in the network-layer service classification.

meant for video traffic that has strict bandwidth demands, but looser latency and jitter demands than the former category; AC_BK is meant for background traffic (such as FTP or web browsing traffic); and AC_BE is intended for best-effort traffic (such as email traffic).

7.3.1.2 Network-layer classification

Traffic classification can also be carried out in the network layer, i.e., layer 3, which classifies packets using the IP header to enable service differentiation. As shown in Figure 7.8, the type-of-service (TOS) fields defined in IPv4 are used in the network-layer service classification (IETF RFC 1349). The type-of-service field consists of a 3-bit precedence subfield, a 4-bit TOS subfield, and the final bit, which is unused and is set to 0. The 4-bit TOS subfield enables 16 classes of service. In an IPv6 header there is an 8-bit class of service field. The TOS field was never really used despite being part of TCP/IP for a long time. It has now been redefined as the differentiated services (DS) field and consists of the first six bits (called DS code points, DSCPs) and two that are unallocated, enabling 64 classes of service. The DS fields are now supported by most routers as a part of the DiffServ service [13], to be discussed later.

The 3-bit precedence fields in TOS are defined as follows:

- **000** (0)–Routine
- **001** (1)–Priority
- **010** (2)–Immediate
- **011** (3)–Flash
- **100** (4)–Flash Override
- **101** (5)–Critical

- **110** (6)–Internetwork Control
- **111** (7)–Network Control

The remaining TOS bits are also defined (note that bits set to 1 basically help speed up the packet flow):

- **Delay** – when set to "1" the packet requests low delay;
- **Throughput** – when set to "1" the packet requests high throughput;
- **Reliability** – when set to "1" the packet requests high reliability;
- **Cost** – when set to "1" the packet has a low cost;
- **MBZ** – checking bit;

7.3.1.3 Transport-layer classification

Using the network-layer traffic classification, the QoS differentiation at core routers depends solely upon the DS field in the IP header, yielding only coarse-grained service differentiation and resource isolation. This classification is based solely on the characteristics of each individual packet without consideration of the corresponding application or end-to-end connection flow, which can be uniquely identified by the 5-tuplet IP header (source IP, destination IP, source port, destination port, and protocol IP). This brings us a transport-layer traffic classification that can further differentiate TCP and UDP traffic to provide a finer-grain service differentiation and supports per-flow QoS service [14]. For service differentiation, transport-layer switching techniques have been proposed [15] in which routing decisions are made on the basis of the destination address as well as the header fields at the transport or higher layer.

7.3.1.4 Application- or user-layer classification

The application or user can also be uniquely identified by using user or application identification (ID). For the connection classification (signaling), there is a central station or entity in the network that is responsible for making the decision whether to allow a new session to join the network. First, the application or user sends the connection request to the central station. Then, if the new connection is admitted, it will be assigned a unique ID number. Packets from the application will be associated with this ID number.

7.3.2   Packet traffic management

Traditionally, data traffic was handled on a first-in first-out (FIFO) queuing basis, which forwards the packets in the same order as that in which they arrive at the interface (see Figure 7.9). Today, with the performance requirements for transmitting multimedia traffic across a data infrastructure, congestion management techniques based on queuing methods are available to prioritize traffic on the basis of its classifications or characteristics. Some well-known queuing methods for congestion management are priority queuing, custom queuing, and weighted fair queuing. Congestion management can also take the form of congestion avoidance, which prepares for and prevents congestion on typical bottlenecks throughout the data network. These techniques all involve dropping traffic during times of congestion, but the differences between these methods are the determining factors behind which packets to drop. Some of the common congestion avoidance techniques are tail drop, random early detection (RED), and weighted random

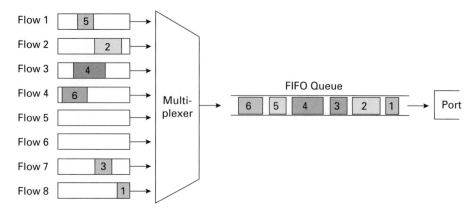

Figure 7.9 First-in first-out (FIFO) queuing forwards the packets in the same order as that in which they arrive at the interface.

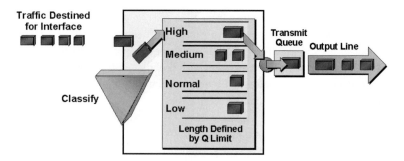

Figure 7.10 Priority queuing (PQ) places each packet in one of four queues using an assigned priority [16].

early detection (WRED). Within some of these methods, low-priority traffic has a high drop probability during times of congestion.

7.3.2.1 Queuing management

Several packet routing functions are related to queuing for congestion management. As mentioned above, these include priority queuing (PQ), custom queuing (CQ), and weighted fair queuing (WFQ). The priority features allow the router to control which packets are sent first at an interface. If there are too many packets during congestion, the queuing management in effect also selects which packets get dropped.

Priority queuing (PQ) places each packet in one of four queues, i.e., high, medium, normal, or low, using an assigned priority (see Figure 7.10). Packets that are not classified by this priority-list mechanism fall into the "normal" queue. When the router is ready to transmit a packet, it searches the high queue for a packet. If there is one, it gets sent. If not, the medium queue is checked. If there is a packet, it is sent. If not, the normal, and finally the low-priority queues are checked. For the next packet, the process repeats. If there is enough traffic in the high queue, the other queues may get starved since they never get

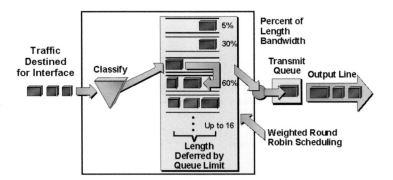

Figure 7.11 Custom queuing (CQ) is aimed at a fair division of bandwidth [17].

serviced. Priority queuing is regarded as a drastic scheme in which the high-priority traffic goes out of the interface at all costs, and any other traffic can be dropped. It is generally intended for use on low-bandwidth links.

Custom queuing (CQ) is aimed at a fair division of bandwidth. It uses 17 queues to divide up the bandwidth on an interface. Queue number 0 is a system queue, which is serviced first before any of the queues numbered 1 through 16 are processed. The system places high-priority packets, such as keep-alive and signaling packets, in queue 0. Other traffic cannot be configured to use this queue. The remaining traffic can be assigned to queues 1 through 16 (see Figure 7.11). These queues are serviced in a round-robin fashion. Packets are sent from each queue in turn. As each packet is sent, a byte counter is incremented. When the byte counter exceeds the default or configured threshold for the queue or when the queue is empty, transmission moves on to the next queue. Custom queuing ensures that no application achieves more than a predetermined pro-portion of the overall capacity when the channel is under stress. Like PQ, CQ is statically configured in advance and does not automatically adapt to changing network conditions.

Weighted fair queuing (WFQ) is a data packet scheduling technique useful for situ-ations in which it is desirable to provide a consistent response time to heavy and light network users alike without adding excessive bandwidth. It is a flow-based queuing algorithm that does two things simultaneously: it schedules interactive traffic to the front of the queue to reduce the response time, and it shares the remaining bandwidth fairly between high-bandwidth flows. Low-volume traffic streams, which comprise the majority of traffic, receive preferential service, transmitting their entire offered loads in a timely fashion. High-volume traffic streams share the remaining capacity proportionally between them.

Weighted fair queuing is a generalization of fair queuing (FQ). Both in WFQ and FQ, each data flow has a separate FIFO queue. In FQ, with a link data rate of R, at any given time the N active (non-empty) data flows are serviced simultaneously, each at an average data rate R/N. Since each data flow has its own queue, an ill-behaved flow, which has sent larger packets or more packets per second than the others since it became active, will be punished. In contrast with FQ, WFQ allows different sessions to have different service

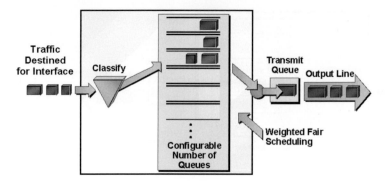

Figure 7.12 Weighted fair queuing (WFQ) is a data packet scheduling technique allowing different scheduling priorities to statistically multiplex data flows [18].

shares. If N data flows are currently active, with weights $w_1, w_2 \ldots, w_N$, data flow i will achieve an average data rate of (see Figure 7.12)

$$\frac{Rw_i}{w_1 + w_2 + \cdots + w_N}. \tag{7.4}$$

In WFQ, low-bandwidth traffic gets priority, with high-bandwidth traffic sharing what is left over. More specifically, WFQ is efficient in that it will use whatever bandwidth is available to forward traffic from lower priority flows if no traffic from higher-priority flows is present. This is different from time-division multiplexing (TDM), which on the one hand simply carves up the bandwidth and lets it go unused if no traffic is present for a particular traffic type. On the other hand, if the traffic is bursting ahead of the rate at which the interface can transmit, new high-bandwidth traffic gets discarded after the configured or default congestive-messages threshold has been reached. Weighted fair queuing uses some parts of the protocol header to determine flow identity. For IP, WFQ uses the type-of-service (TOS) bits, the IP protocol code, the source and destination IP addresses (if not a fragment), and the source and destination TCP or UDP ports.

7.3.2.2 Congestion avoidance

In the traditional *tail drop* algorithm, a router or other network component buffers as many packets as it can, and simply drops the ones it cannot buffer. If the buffers are constantly full, the network is congested. Tail drop distributes buffer space unfairly among traffic flows. It can also lead to TCP global synchronization as all TCP connections "hold back" simultaneously, and then step forward simultaneously. Networks become under-utilized and flooded by turn.

Random early detection (RED)   also known as random early discard or random early drop, is a queue management technique for congestion avoidance [19] , that has been proposed to address the difficulties of tail drop algorithms. It monitors the average queue size and drops packets on the basis of statistical probabilities. If the buffer is almost empty, all incoming packets are accepted. As the queue grows, the probability that an incoming packet will be dropped grows too. When the buffer is full, the probability reaches 1 and all incoming packets are dropped (see Figure 7.13). Random early detection is considered fairer than tail

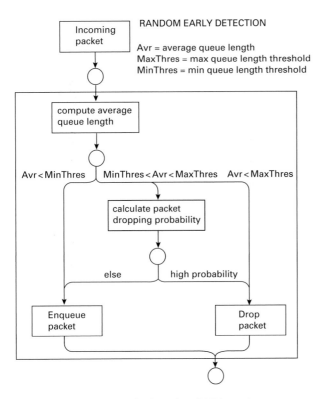

Figure 7.13 In random early detection (RED), as the queue grows the probability for dropping an incoming packet grows too. When the buffer is full, the probability reaches 1 and all incoming packets are dropped [19].

drop: the more a host transmits, the more likely it is that its packets are dropped. Early detection helps avoid global synchronization.

Weighted random early detection (WRED) is an extension to RED in which different queues may have different buffer occupation thresholds before random dropping starts, as well as different dropping probabilities; packets are classified into these queues according to priority information such as IP precedence or DSCP (for DiffServ). In this way QoS differentiation is made possible, since packets in queues with higher buffer occupation thresholds or lower dropping probabilities are effectively prioritized [20].

7.3.2.3 Traffic shaping

Traffic shaping, or rate limiting, algorithms are used to specifically shape the traffic so as to satisify some predefined data rate over the networks. Traffic shaping has the following key difference from traffic policing. As shown in Figure 7.14, when the traffic rate reaches the configured maximum rate, excess traffic is dropped (or remarked) in traffic policing. The result is an output rate that appears as a saw-tooth with crests and troughs. In contrast with policing, traffic shaping retains an instantaneous excess of packets in a queue and schedules the excess for later transmission over increments of time, resulting in a smoother packet output rate.

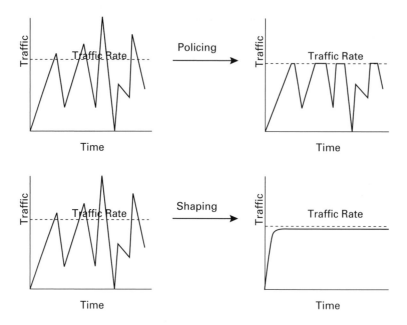

Figure 7.14 In contrast with policing, traffic shaping retains excess packets in a queue and then schedules the excess for later transmission over increments of time.

There are two predominant methods for shaping traffic, the token bucket and leaky bucket schemes [21]. Both schemes have distinct properties and are used for distinct purposes. They differ principally in that the token bucket scheme allows a certain amount of burstiness while imposing a limit on the average data transmission rate; the leaky bucket scheme imposes a hard limit on the data transmission rate.

The token bucket traffic shaping scheme is a control mechanism that dictates that traffic can be transmitted when there are "tokens" in the bucket. Each token can represent a number of bytes or a single packet. The tokens in the bucket are effectively "cashed in" (removed) in exchange for the ability to send a packet. The network administrator specifies how many tokens are needed per byte; when tokens are present, a flow is allowed to transmit traffic. If there are no tokens in the bucket, a flow cannot transmit its packets. Therefore, a flow can transmit traffic up to its peak burst rate if there are adequate tokens in the bucket and if the burst threshold is configured appropriately. The algorithm can be conceptually understood as follows using a simple example.

(1) Tokens are added to the bucket at a rate r, i.e., one is added every $1/r$ seconds (see Figure 7.15).
(2) The bucket can hold at the most b tokens. If a token arrives when the bucket is full, it is discarded (see Figure 7.16).
(3) When a packet of n bytes arrives, n tokens, say, are removed from the bucket, and the packet is sent to the network.
(4) If fewer than n tokens are available, no tokens are removed from the bucket, and the packet is considered to be non-conformant.

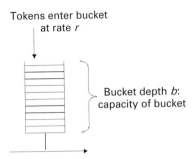

Figure 7.15 The token bucket scheme allows a certain amount of burstiness while imposing a limit on the average data transmission rate, while the leaky bucket scheme imposes a hard limit on the data transmission rate.

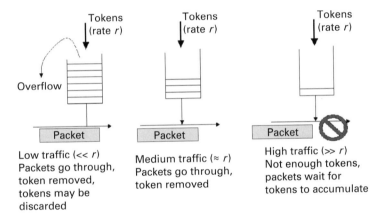

Low traffic (<< r)
Packets go through, token removed, tokens may be discarded

Medium traffic (≈ r)
Packets go through, token removed

High traffic (>> r)
Not enough tokens, packets wait for tokens to accumulate

Figure 7.16 The token bucket can hold at the most b tokens. If a token arrives when the bucket is full, it is discarded.

(5) The algorithm allows bursts of up to b bytes, but over the long run the output of conformant packets is limited to the constant rate r. Non-conformant packets can be treated in various ways:

- they may be dropped;
- they may be put in a queue for subsequent transmission when sufficient tokens have accumulated in the bucket;
- they may be transmitted but marked as being non-conformant, possibly to be dropped subsequently if the network is overloaded.

The leaky bucket scheme is another traffic shaping mechanism that is used to control the traffic rate sent to the network (see Figure 7.17). A leaky bucket provides a mechanism by which bursty traffic can be shaped to present a steady stream of traffic to the network, rather than traffic with erratic bursts of low-volume and high-volume flows. An appropriate analogy for the leaky bucket is a scenario in which four lanes of automobile traffic converge into a single lane. A regulated admission interval into the single lane of traffic flow helps the traffic move. The benefit of this approach is that traffic flow into the major arteries

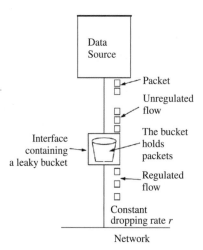

Data
Source

Packet

Unregulated
flow

The bucket
holds
packets

Interface
containing
a leaky bucket

Regulated
flow

Constant
dropping rate r

Network

Figure 7.17 In the leaky bucket mechanism, the bucket can hold at most b bytes. If a packet arrives when the bucket is full, the packet is discarded.

(the network) is predictable and controlled. The major liability is that when the volume of traffic is much greater than the bucket size, in conjunction with the drainage-time interval, inward traffic for the bucket that is beyond bucket capacity is discarded [22]. The algorithm can be understood conceptually as follows.

(1) Arriving packets are placed in a bucket with a hole in the bottom.
(2) The bucket can queue at most b bytes (limited by the memory of the system). If a packet arrives when the bucket is full, the packet is discarded.
(3) Packets drain through the hole in the bucket, into the network, at a constant rate of r bytes per second, thus smoothing traffic bursts.

7.3.3   Network resource management

Since today's Internet interconnects multiple administrative domains, or autonomous systems (ASs), it is the concatenation of domain-to-domain data forwarding that provides end-to-end delivery. Therefore, instead of dealing with packet-level traffic management, the IP QoS can also be handled through network resource management techniques for end-to-end QoS. There are two popular categories of resource management, one is *bandwidth over-provisioning* and the other is *explicit resource management* [8].

Bandwidth over-provisioning is a very simple QoS architecture which tries to allocate the network's bandwidth resources so that none of the resources become a bottleneck in the communication system. This can be made possible with today's new transmission technologies, which provide very high bandwidth everywhere. All traffic is served with the same QoS level on the basis of the provisioning mechanism, and there is no differentiation between users' flows since they belong to a single service class. The concern over this technique is that it can be less profitable for Internet Service Providers (ISPs), because of the use of higher than needed bandwidth and because there is no possibility of differentiating between the traffic of different users.

Explicit resource management techniques, however, are based on the concept of dividing all served flows into traffic classes that are served at various QoS levels. The specifications of traffic and its desired class of service can be given on a per-flow basis with a service-level agreement (SLA), which is a service contract between a customer and a service provider. In addition to the traffic specification, an SLA specifies all the aspects of packet forwarding treatment that a customer should receive from its service provider. There are two well-defined and standardized architectures for IP networks with class-based resource management: IntServ and DiffServ. The basic difference between these two is the level of granularity of independently treated flows in the network.

7.3.3.1 Integrated services (IntServ)

Integrated services (IntServ) is an explicit resource management QoS framework with dynamic resource reservation [22], which uses the resource reservation protocol (RSVP) as its working protocol to achieve end-to-end signaling [23]. In RSVP, resources are reserved at each router along the path between sender and receiver through explicit signaling for a flow that demands QoS. The benefit of end-to-end per-hop signaling is its ability to convert a connectionless Internet into a more reliable connection-oriented network. As shown in Figure 7.18, we have the following.

(1) The flow source of an RSVP session sends a PATH message to the intended flow receiver(s), specifying the characteristic of the traffic.
(2) As the PATH message propagates towards the receiver(s), each network router along the way records path characteristics such as available bandwidth.

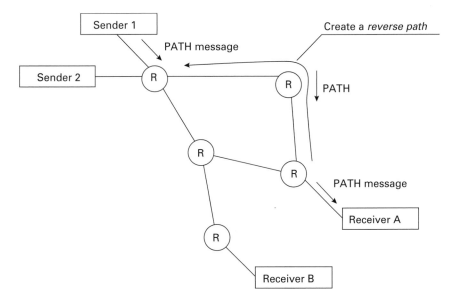

Figure 7.18 In RSVP, resources are reserved at each router along the path between sender and receiver through explicit signaling for a flow that demands QoS [24].

(3) Upon receiving a PATH message, the receiver responds with a RESV message to request resources along the path recorded in the PATH message in reverse order from the sender to the receiver.

(4) Intermediate routers can accept or reject the request of the RESV message. If the request is accepted, link bandwidth and buffer space are allocated for the flow, and the flow-specific state information is installed in the routers. Reservations can be shared along branches of the multicast delivery trees.

The RSVP takes the *soft state* approach, which regards the flow-specific reservation state at routers as cached information that is installed temporarily and should be periodically refreshed by the end hosts. A state that is not refreshed is removed after a timeout period. If the route changes, the refresh messages automatically install the necessary state along the new route. The soft state approach helps RSVP to minimize the complexity in the connection setup and improves robustness, but it can lead also to increased flow setup times and message overheads. The IntServ architecture adds two service classes to the existing best-effort model, guaranteed service and controlled load service.

(1) *guaranteed quality service* This is for applications with rigid end-to-end delay bounds.

(2) *controlled load service* This is for applications with looser performance criteria than guaranteed quality services but requiring higher attributes than normal best-effort services.

(3) *best-effort service* This is the service provided by the Internet at present.

By using per-flow resource reservation, IntServ can deliver fine-grained QoS guarantees. However, introducing flow-specific states into the routers represents a fundamental change to the current Internet architecture. Particularly in the Internet backbone, where a hundred thousand flows may be present, this may be difficult to manage as a router may need to maintain a separate queue for each flow. Although RSVP can be extended to reserve resources for aggregation of flows, many people in the Internet community believe that the IntServ framework is more suitable for intradomain QoS or for specialized applications such as high-bandwidth flows. IntServ also faces the problem that incremental deployment is only possible for controlled-load service, while ubiquitous deployment is required for guaranteed service, making it difficult for it to be realized across the network.

7.3.3.2 Differentiated services (DiffServ)

The main problem that prevents IntServ from being feasible is scalability. Maintaining thousands or millions of connections in an IntServ router is impractical. With large number of sessions existing in the backbone, the amount of computation complexity involved in using RSVP will simply overburden the routers. The IETF Differentiated Services (DiffServ) protocol [25] provides an alternative for better resource-reservation QoS. The basic assumption of DiffServ is to overcome the scalability issues in IntServ. More specifically, DiffServ completely eliminates the storage of session states in the router and eventually becomes a computationally inexpensive QoS solution for a router even in the backbone. In DiffServ, implementation of the concepts of aggregation of flows and per-hop behavior is applied to a network-wide set of traffic classes (see Figure 7.19). Flows are classified according to predetermined rules, so that many application flows can be aggregated to a limited set of class flows.

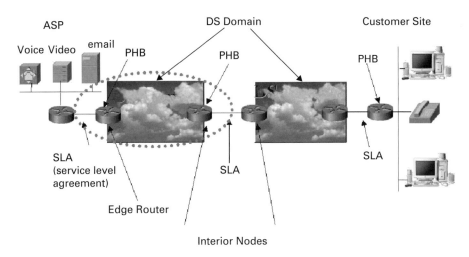

Figure 7.19 In DiffServ, flows (packets) served by the same class of service are aggregated and experience the same QoS level. Aggregated packet processing by a network node is called per-hop behavior (PHB).

Traffic entering the network domain at the border router is first classified for consistent treatment at each transit or core router inside the network. The classification is based on the IPv4 (or IPv6) type-of-service (TOS) 6-bit header field, also known as the differentiated services (DS) field, as discussed in Section 7.3.1.2 [26]. Currently two bits are used for standardization while the other six bits are reserved for later use. DiffServ, which can be either end-to-end or intradomain, provides a wide range of services through a combination of:

(1) setting DSCP bits in the DS field at network edges and administrative boundaries,
(2) using those bits to determine how packets are treated by the routers inside the network; and
(3) conditioning the marked packets at the network boundaries in accordance with the requirements of each service.

In the DiffServ model, independent flows select one of the predefined services and are served in the same way as other flows that choose the same service. Flows (packets) served by the same service are aggregated and experience the same QoS level. Aggregated packet processing by a network node is called per-hop behavior (PHB) and is shown in Figure 7.19. Each core router within the DiffServ domain will read the DS field of the packets and treatment will usually be applied by separating the traffic into queues according to the class of traffic. This type of framework is more scalable than IntServ, mainly due to its stateless design, flow aggregation, and minimization of signaling requirements.

Even though DiffServ resolves the scalability issue, it still lacks qualitative QoS guarantees. Unlike IntServ, once a packet enters into a DiffServ domain its behavior is unpredictable because of the absence of explicit connection-oriented end-to-end per-hop signaling. All the routers in DiffServ are expected to mark the packets trustfully and provide them with the required level of treatment. Therefore, in such cases it is left to the network operator to provide the best possible services. Defining service-level agreements (SLAs) between network providers implementing such services is considered a serious challenge. Such SLA requirements between two different domains implementing DiffServ necessarily

Table 7.2 Features of the DiffServ and IntServ architectures [8]

Features	IntServ	DiffServ
QoS assurance	Per flow	Per aggregate
QoS assurance range	End-to-end (application-to-application)	Domain (edge-to-edge) or DiffServ region
Resource reservation	Controlled by application	Configured at edge nodes using SLA
Resource management	Distributed	Centralized within DiffServ domain
Signaling	Dedicated protocol (RSVP)	Based on DSCP carried in IP packet header
Scalability	Not recommended for core networks	Scalable in all parts of network
Class of service (CoS)	GS, CL, BE	BE and a set of mechanisms for CoS design (EF and AF PHBs)

SLA, service level agreement; DSCP, DiffServ code point; GS, guaranteed service; CL, controlled load; BE, best-effort; EF, expedited forwarding; AF, assured forwarding

involve thousands of rules that govern their inter-working. These issues become more complicated if the network providers have different numbers of service classes. Effective QoS *mapping* between the service classes of one network provider and another is still an active research issue for DiffServ.

Per-hop behaviors may be specified in terms of their resources (e.g., buffer, bandwidth), their priority relative to other PHBs, or their relative observable traffic characteristics (e.g., delay, loss). Currently, two PHBs are defined for DiffServ.

(1) *Expedited forwarding PHB*   This is for real-time traffic and is related to the guaranteed service (GS) transfer capability.
(2) *Assured forwarding PHB*   This is for elastic traffic and is related to the controlled load (CL) service transfer capability.

Table 7.2 shows, for comparison, several features of the DiffServ and IntServ architectures [8]. In fact, the differences between the two can be complementary in functionality, therefore IETF has proposed an interoperability framework for the IntServ and DiffServ architectures [13]. The integrated IntServ–DiffServ model is used to provide QoS in the end-to-end relation. To avoid per-microflow servicing in the core, the proposed architecture uses DiffServ in the core to support aggregated IntServ microflows. The IntServ model is used in the access part of the network to provide the applications with the mechanisms necessary to express the QoS requirements. It provides a user-network QoS signaling interface and forwards requests for resources to the network. The QoS signaling is end to end, i.e., it takes place between the communicating terminals. Except for the per-flow resource reservation system, RSVP signaling and some extensions can be used to aid resource management in a DiffServ domain.

7.4   IP multicast and application-level multicast (ALM)

With the growing amount of multimedia, especially high-bandwidth video, disseminated over the Internet, the bandwidth requirement is getting higher and higher. End-to-end

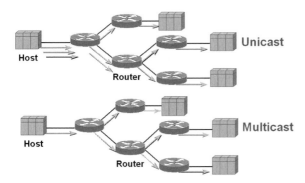

Figure 7.20 (a) IP unicast sends a separate copy of the multimedia data from the source to each destination over the Internet using the unicast path from the source to the destination. (b) IP multicast places receivers which desire the same stream into one multicast group and sends the stream only to that group.

delivery of live-streamed multimedia packets is mainly supported by the IP unicast, which sends a separate copy of the multimedia data from the source to each destination over the Internet using the unicast path from the source to the destination, as shown in the upper half of Figure 7.20. However, this is very inefficient since multiple copies of the same live-streamed multimedia stream traverse the same links in the network. For example, consider a video server using unicast to transmit a 300 kbps stream. To support 1000 receivers, the server needs a 300 Mbps access link. Receiver scalability is clearly an intrinsic issue since the bandwidth demand on the source grows linearly with the group size. To resolve effectively the matter of supporting a larger number of receivers for the same live-streamed media, the IP multicast delivery of UDP traffic was introduced; in this system the multicast receiver gets only the video content he or she desires, and this allows a significantly larger number of streams to be available. This is achieved by placing receivers who desire the same stream into one multicast group and sending the stream only to that group, as shown in the lower half of Figure 7.20. This is different from the traditional digital media broadcast delivery, where all streams are sent to all receivers, limiting the number of available streams.

7.4.1   IP multicast

IP multicast, mentioned above, is a technique for many-to-many communication over an IP infrastructure. IP multicast is receiver initiated, i.e., a receiver of a multicast stream expresses its desire to the network and the network will add it to the desired group. The sender, who may not even be in the group, does not know who is receiving the stream; this avoids implosion of join requests at the sources and improves the scaling in terms of the group size. In addition, this mechanism can potentially make it easier to support multiple sources. It scales to a larger receiver population by not requiring prior knowledge of who or how many receivers there are. Multicast utilizes network infrastructure efficiently by requiring the source to send a packet only once, even if it needs to be delivered to a large number of receivers. The nodes in the network take care of replicating the packet to reach multiple receivers only where necessary. Key concepts in IP multicast include an IP multicast group address, a multicast distribution tree, and receiver-driven tree creation.

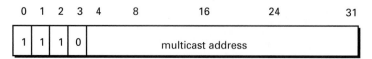

(IP addresses 224.0.0.0 through 239.255.255.255)

Figure 7.21 Each multicast group is identified by a class-D IP address, which corresponds to group addresses from 224.0.0.0 through 239.255.255.255.

According to IETF RFC 1112 [27], each multicast group is identified by a class-D IP address, which corresponds to group addresses from 224.0.0.0 through 239.255.255.255 (see Figure 7.21). More specifically, an IPv4 multicast address is 32 bits, of which the first four bits are always the same, leaving 28 bits. According to Internet Assigned Numbers Authority (IANA, http://www.iana.org/), the address 224.0.0.0 is guaranteed not to be assigned to any group; 224.0.0.1 is assigned to all IP hosts (including gateways) on a subnet and 224.0.0.2 to all routers on a subnet. These two are used to address only all multicast hosts on the directly connected subnet. There is no multicast address (or any other IP address) for all hosts on the total Internet as a whole.

Members (hosts) of a multicast group join or leave the group by indicating their desire to do so to the multicast routers, which listen to all multicast addresses and use multicast routing protocols to manage groups. In IP multicast, a host of a group can be a receiver without sending data to the group or the host can be a sender of data but not be a receiver or it can be both. When sourcing data, a host just sends the data and then maps the network-layer address to the link-layer address, so that routers can figure out where receivers are. When receiving data, a host within a local area network needs to perform two actions: one is to tell routers in which group this host is interested, and the other is to tell the associated LAN controller to receive the link-layer mapped MAC address, which can be derived from the network-layer IP multicast address. According to IANA, the 48-bit (6-byte) MAC multicast address always begins with the hex values "01, 00, 5E" in the first three bytes (24 bits), followed by another "0" in the next bit. The remaining 23 bits can then be derived as the least significant 23 bits from the IP multicast address. Figure 7.22 shows how an IP multicast address of "224, 65, 10, 154," (hex values "E0, 41, 0A, 9A") is converted to a MAC multicast address of "01, 00, 5E, 41, 0A, 9A" in hex values.

An IP multicast group address is used by sources and the receivers to send and receive content. Sources use the group address as the IP destination address in their data packets. Receivers use this group address to inform the network that they are interested in receiving packets sent to that group. For example, if some content is associated with group 228.1.2.3, the source will send data packets destined to 228.1.2.3. Receivers for that content will inform the network that they are interested in receiving data packets sent to the group 238.1.2.3. The receiver thus "joins" 228.1.2.3. The protocol used by receivers to join a group is called the Internet group management protocol (IGMP) [3]. Once the receivers join a particular IP multicast group, a multicast distribution tree is constructed for that group using one of various multicast routing protocols (MRPs) [28]. The most widely used MRP is protocol independent multicast (PIM), which sets up multicast distribution trees such that data packets from senders to a multicast group reach all receivers that have "joined" the group. There are variations of PIM, e.g., sparse mode (PIM-SM), dense mode (PIM-DM), source specific mode (PIM-SSM), and bidirectional mode (PIM-Bidir). Of these PIM-SM is the most widely deployed at present

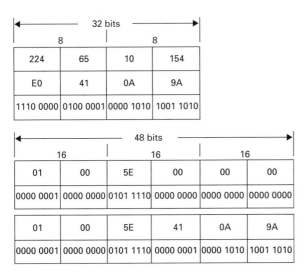

Figure 7.22 The MAC multicast address is 48 bits, of which the first 24 bits are always the same. One of the remaining bits is reserved, leaving 23 bits to specify the address within the LAN.

and PIM-SSM and PIM-Bidir are simpler and more scalable variations developed more recently and gaining in popularity. IP multicast does not require a source sending to a given group to know about the receivers of the group. The multicast tree construction is initiated by network nodes close to the receivers, i.e., it is receiver driven.

A multicast router does not need to know how to reach all other multicast trees in the Internet. It only needs to know about multicast trees for which it has receivers downstream of it. This allows it to scale to a large receiver population. In reality, most efforts at scaling multicast up to large networks have concentrated on the simpler case of single-source multicast, which seems to be more computationally tractable. Moreover, owing to the need for better-controlled usage metering and owing to discrepancies in the profit sharing between various Internet service providers (ISPs) and application service providers (ASPs), IP multicast is not widely used in the commercial Internet. Some limited applications of IP multicast are IPTV applications such as distance learning or the televising of company meetings. In stock exchanges it is used for distributing stock trading data. In content-delivery networks it is used to provide commercial television to a set of subscribers over the IP infrastructure.

The basic multicast functionalities of an IP multicast are as follows [29].

(1) *Group membership management* This learns and maintains information about members in the multicast group.
(2) *Data delivery path maintenance* This constructs a delivery path that reaches all the receivers requesting the stream.
(3) *Replication and forwarding* This enables interior nodes in the delivery path to replicate and forward the stream.

7.4.1.1 Internet group management protocol (IGMP)
Within the IP infrastructure, multicast membership management operations are supported by the Internet group management protocol (IGMP) [3], which operates between local hosts,

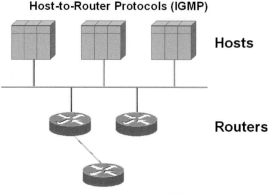

Host-to-Router Protocols (IGMP)

Hosts

Routers

Router-to-Router Protocols (MRP)

Figure 7.23 The router queries the local hosts for multicast group membership information and then "connects" active hosts to the multicast tree via the multicast routing protocol.

typically attached via a LAN, and the multicast router; its purpose is to support the creation of transient groups, the addition and deletion of members of a group, and the periodic confirmation of group membership. The router "queries" the local hosts for multicast group membership information and then "connects" active hosts to the multicast tree via the multicast routing protocol (see Figure 7.23). Local hosts respond with membership reports, that is to say, the first host which responds speaks for all. A host also issues "leave-group" messages to leave; this is optional since the router periodically polls anyway. The IGMP also specifies a "deadman timer" procedure whereby hosts periodically confirm their membership with the multicast agents. The IP module must maintain a data structure listing the IP addresses of all host groups to which the host currently belongs, along with each group's loopback policy, access key, and timer variables. This data structure is used by the IP multicast transmission service to discover which outgoing datagrams to loop back, and by the reception service to discover which incoming datagrams to accept. The purpose of IGMP and the management interface operations is to maintain this data structure.

Joining a multicast group   Multicast routers, commonly a chosen designated router (DR), periodically send IGMP host membership query messages (in brief, queries) to discover which host groups, say multicast group 224.3.4.5, have members on their attached local networks. Queries are addressed to the all-hosts group (via IP address 224.0.0.1) and carry an IP time to live (TTL) of 1. Hosts (Hs) respond to a query by generating host membership reports (in brief, reports), reporting each host group to which they belong on the network interface from which the query was received. Once the DR receives the reports, it will start forwarding packets for 224.3.4.5 to the network to which the reporting hosts belong (see Figure 7.24). In order to avoid an "implosion" of concurrent reports and to reduce the total number of reports transmitted, two techniques are used.

(1) When a host receives a query, rather than sending reports immediately it starts a report delay timer for each of its group memberships on the network interface of the incoming query. Each timer is set to a different, randomly chosen, value between zero and D seconds. When a timer expires, a report is generated for the corresponding host group. Thus, reports are spread out over a D-second interval instead of all occurring at once.

IGMP Membership Report

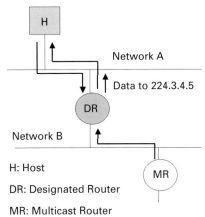

H: Host

DR: Designated Router

MR: Multicast Router

Figure 7.24 Hosts respond to a query by generating host membership reports, reporting each host group to which they belong on the network interface from which the query was received. Once the DR receives the reports, it will start forwarding packets to the network to which the reporting hosts belong.

(2) A report is sent with an IP destination address that is the same as the host group address being reported, and with an IP TTL of 1, so that other members of the same group on the same network can overhear the report. If a host hears a report for a group to which it belongs on that network, the host stops its own timer for that group and does not generate a report for that group. Thus, in the normal case, only one report will be generated for each group present on the network, by the member host whose delay timer expires first. Note that the multicast routers receive all IP multicast datagrams and therefore need not be addressed explicitly. Further, note that the routers need not know which hosts belong to a group on a particular network, only that at least one host belongs to that group.

Note that a multicast router keeps a list of multicast group memberships for each attached network, and a timer for each membership. The multicast group membership implies the presence of at least one member of a multicast group in a given attached network, not all the members. When a host joins a new group, it immediately transmits a report for that group rather than waiting for a query, in case it is the first member of that group on the network. To cover the possibility that the initial report is lost or damaged, it is recommended that it be repeated once or twice after short delays.

Leaving a multicast group   When a host (H) decides to leave a multicast group, say Group 224.3.4.5, the host has to send an IGMP leave-group message to IP address 224.0.0.2, which is the address for all multicast routers in the same subnet. Once the designated router (DR) receives this leave-group message, it will stop forwarding packets for 224.3.4.5 to network A (see Figure 7.25) if there are no more 224.3.4.5 group members on network A.

The IGMP version 2 (IGMPv2, IETF RFC 2236) allows group membership termination to be quickly reported to the routing protocol, which is important for high-bandwidth multicast groups and/or subnets with highly volatile group membership. Multicast routers also use IGMPv2 to learn which groups have members on each of their attached physical networks.

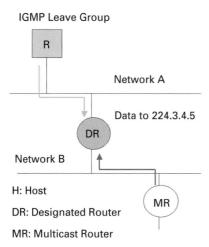

IGMP Leave Group

R

Network A

Data to 224.3.4.5

DR

Network B

H: Host

MR

DR: Designated Router

MR: Multicast Router

Figure 7.25 When a host decides to leave a multicast group, it has to send an IGMP Leave-Group message to IP address 224.0.0.2, to be received by the designated router.

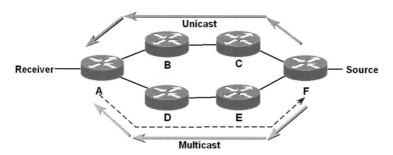

Unicast

Receiver — B C Source

A F

D E

Multicast

Figure 7.26 IP multicast uses the reverse path forwarding (RPF) algorithm to reduce drastically the overhead involved in forwarding multicast data.

7.4.1.2 Multicast routing protocols

IP multicast uses the reverse path forwarding (RPF) [30] algorithm, where a router forwards a multicast datagram on the basis of the path received at the interface used to send unicast datagrams to the source (see Figure 7.26), in order to reduce drastically the overhead involved in forwarding multicast data. More specifically, RPF forwards multicast traffic on a spanning tree, which spans all the hosts belonging to the specific multicast group, of all routers across the entire network; the root is the sender of the traffic. Upon receiving a multicast packet, leaf routers may prune themselves from the tree if no hosts on their network subscribe to the group. The pruning process proceeds up the tree from each leaf until a router is connected to a host that is subscribed to the group. This RPF algorithm in effect constructs a different tree for each sender, the so-called source tree (see Figure 7.27), where each source is the root of its own tree connecting to all the members. Even though the source trees can be constructed to get optimal paths from the source to all receivers with minimized delay, the associated routers need to keep track of all the states relating to all the different trees (which amounts to greater complexity and memory requirements), so that packets can flow from each source to all receivers.

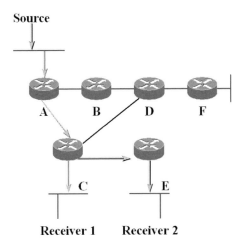

Figure 7.27 The RPF algorithm can be used to construct a different tree for each sender, a so-called source tree, where each source is the root of its own tree connecting to all the members.

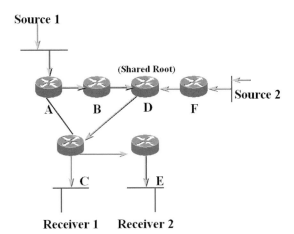

Figure 7.28 Core-based trees (CBTs) construct a single spanning tree for a group that accommodates all senders. The tree is constructed as routers contact a designated core (the so-called rendezvous point) to join the group.

Alternatively, core-based trees (CBTs) can construct a single spanning tree (with less complexity and memory in each router) for a group that accommodates all senders (see Figure 7.28). The tree is constructed as routers contact a designated core (the so-called rendezvous point) to join the group. Core-based trees commonly create sub-optimal paths from sources to receivers, with extra delays.

The RPF-based source tree construction algorithm works well for groups that are dense either within a local region or throughout the network, and CBT works well for multicast groups that are sparsely distributed throughout the network. Several multicast routing protocol standards have been developed using both these algorithms. The distance-vector

Figure 7.29 In the Mbone, upon receiving a packet the mrouted demon sends the packet using IP multicast on its local network if any host is listening for that address.

multicast routing protocol (DVMRP) [31], the multicast open-shortest-path-first (MOSPF) protocol [32], and the protocol-independent multicast-dense mode (PIM-DM) [33] use the RPF algorithm, while the protocol-independent multicast-sparse mode (PIM-SM) [34] uses an algorithm similar to CBT.

7.4.1.3 Multicast backbone (Mbone)

Several IP multicast technologies have been deployed. The multicast backbone or Mbone [35], started in 1992, is a virtual network on top of the Internet and connects routers and end hosts that are multicast capable. Each IP multicast island has one designated machine (end host or router) running the multicast routing (mrouted) demon. "Tunnels" are manually configured between mrouted machines to connect the multicast enabled islands, and the distance-vector multicast routing protocol (DVMRP) is used to route packets to different tunnel end points. Upon receiving a packet, the mrouted demon sends the packet using IP multicast on its local network if any host is listening for that address (see Figure 7.29). The Mbone has been used for many academic and technical conferences, working group meetings, university seminars, public radio and television programs, Space Shuttle missions, and live music performances. However, more than ten years after its initial deployment, the Mbone is still limited to a very small number of universities and research labs.

In the Mbone system, IP multicast packets are encapsulated for transmission through tunnels (see Figure 7.30), so that they look like normal unicast datagrams to intervening routers and subnets. A multicast router that wants to send a multicast packet through a tunnel will prefix another IP header, set the destination address in the new header to be the unicast address of the multicast router at the other end of the tunnel, and set the IP protocol field in the new header to be 4 (which means that the next protocol is IPv4). The multicast router at the other end of the tunnel receives the packet, strips off the encapsulating IP header, and forwards the packet within the multicast enabled island as appropriate.

7.4.2   Application-level multicast

As discussed previously, IP unicast (see Figure 7.31) does not scale well when multiple receivers are requesting the same multimedia contents from the media server, which therefore needs to support multiple high-data-rate UDP connections. This led us to the development of IP multicast (see Figure 7.32), which requires the source to send a packet only once, even if it needs to be delivered to a large number of receivers. The multicast

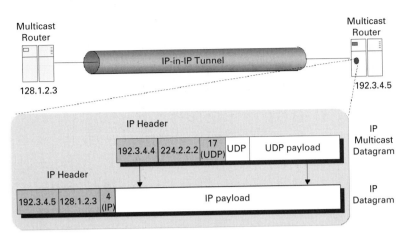

Figure 7.30 IP multicast packets are encapsulated for transmission through tunnels, so that they look like normal unicast datagrams to intervening routers and subnets.

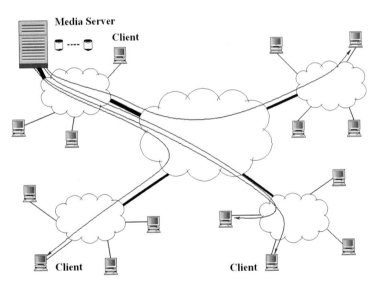

Figure 7.31 IP unicast does not scale well when multiple receivers are requesting the same multimedia contents from the media server.

routers in the network take care of replicating the packet to reach multiple receivers only where necessary. However, after a decade of research, there are still many hurdles in the deployment of IP multicast, such as the lack of higher-layer functionalities (e.g., reliable transport, flow, congestion, and admission control, and multicast security) and scalable interdomain multicast routing protocols. Therefore, application-level multicast (ALM) or overlay multicast, in which the multicast functionality is pushed up to the application layer, has recently been proposed for a number of Internet multicast applications. There are two different ALM architectures: one is the (proxy-based) content delivery network (CDN) and the other is the peer-to-peer (P2P) network.

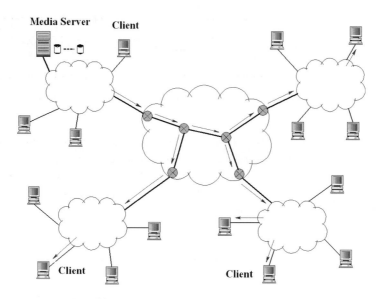

Figure 7.32 IP multicast requires the source to send a packet only once, even if it needs to be delivered to a large number of receivers.

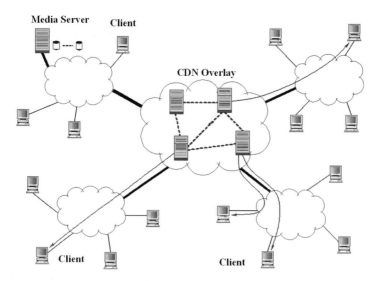

Figure 7.33 Content delivery networks are dedicated collections of servers located strategically across the wide-area Internet.

7.4.2.1 Content delivery networks (CDNs)

Content delivery networks (CDNs, see Figure 7.33) are dedicated collections of servers located strategically across the wide-area Internet. Content providers, such as websites or streaming video sources, contract with commercial CDNs to host and distribute content. Basically, a CDN is an *overlay network* constructed from a group of strategically placed and geographically distributed CDN servers. When a user requests some media content, the request is redirected to a nearby CDN server (the so-called user's local CDN server) through a certain *user request redirection* mechanism, so that content delivery takes place at the edge of the network where bandwidth is abundant. There are several ways to redirect the request from a user to its local CDN server. For example, a user-request redirection can be *non-transparent* (to the user), as in the explicit-user configuration, or, preferably, it can be *transparent*, so that it relies on network devices such as DNS servers, which have become popular because of their simplicity and generality. The local CDN server is the entry point for the user to access the CDN. It conducts *admission control* to either accept or block the request. If the request is accepted, the local CDN server serves the user if it has the content; otherwise, it performs *content routing* to locate and then forward the requested content to the user [36]. Multimedia contents delivery is commonly classified into two types: *on-demand* and *live streaming*. On-demand multimedia contents are normally requested by the users asynchronously. After locating a content, the local CDN server should preferably cache the content in order to serve future user requests. However, it is not feasible for a single CDN server to cache all the contents existing in the Internet because of its limited disk space; therefore, the *cooperative caching scheme* among CDN servers and content routing are closely related. Live multimedia contents cannot be cached in advance. But since live contents are synchronous, a CDN server can aggregate the user requests for the same live content that are redirected to it and become a multicast group member for that content. The multicast group members for the same content form an overlay multicast tree (or mesh) that delivers the content from the origin server to all multicast group members. If a CDN server that has not yet been a multicast group member for certain content receives a user's request for that content, it performs application-layer multicast content routing to add itself to the multicast tree.

In order to reach as many users as possible and to serve those users with a satisfactory QoS, the servers of a CDN need to be placed strategically on the Internet. These CDN servers are interconnected by the existing network infrastructure (e.g., the public Internet or transmission lines with dedicated bandwidth). They form a logical overlay network and function as overlay routers to deliver the contents from the origin servers to the users. For a large CDN consisting of hundreds or even thousands of geographically distributed CDN servers, such as the hierarchical topology of the IP routers on the Internet, a hierarchical overlay network topology is sometimes preferred for the CDN servers to perform content routing and content delivery efficiently [36]. There are several major differences between IP multicast and ALM as deployed in CDNs, as follows.

(1) *Multicast group identifier* In IP multicast, a class-D address is used to identify a multicast group. In ALM, a URL or other application-related key is commonly used to identify a multicast group.

(2) *Multicast group members* In IP multicast, in order to receive certain multicast content a user explicitly subscribes to its directly attached router via the IGMP. The IP routers of the same multicast group form a multicast tree and take responsibility for delivering the multicast content to their subscribed users. In ALM, the request from a user is redirected to the user's local CDN server. The CDN servers of the same multicast group form an overlay multicast tree (or mesh) and take responsibility for delivering the multicast content to their users.

(3) *Network topology*   In IP multicast, the topology of the IP routers exactly reflects the physical network topology. In ALM, however, the CDN servers form a logical overlay network on top of the underlying real physical network infrastructure.

(4) *Multicast routing*   Both IP multicast and ALM require a certain number of multicast routing protocols to construct a multicast tree for delivering the multicast content. IP multicast routing normally relies on the underlying unicast routing protocols, which are quasi-static and employ simple routing metrics such as *number-of-hops* or *delay*. On the contrary, as we will discuss later, the implementation of ALM routing is much more flexible, e.g., a *minimum-delay-path spanning tree* can be constructed to minimize the delay from the source to each of the nodes on the tree. However, for streaming applications to avoid annoying interruptions due to insufficient available bandwidth, a *widest-path spanning tree* that maximizes the available bandwidth of the tree links is preferred [37] [38].

7.4.2.2 Peer-to-peer (P2P) systems

Recently, as another alternative for application-level multicast (ALM), researchers have begun to focus on the application of the peer-to-peer (or P2P) concept to media streaming (see Figure 7.34), because P2P techniques can capitalize on receivers' bandwidth to provide services to other receivers and these techniques rely on IP unicast only. However, building an efficient P2P media streaming system faces the following challenges.

(1) *Peer management and distribution tree construction*   How can one get an efficient ALM distribution tree and manage this tree, while peers join and leave all the time, so as to deliver media content to large numbers of receivers

(2) *Robustness* A robust technique is needed to recover smoothly when failure occurs; otherwise there will be service interruption. Thus the unpredictable behavior of receivers needs to be accommodated.

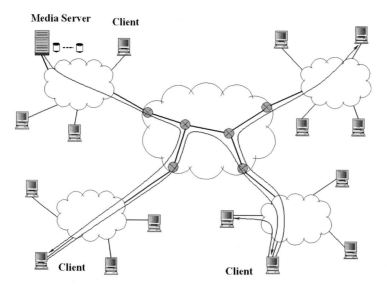

Figure 7.34 P2P techniques can capitalize on receivers' bandwidth to provide services to other receivers; they relays on IP unicast only.

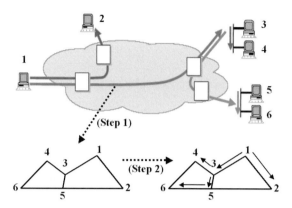

Figure 7.35 Narada constructs the distribution trees in a two-step process.

(3) *Dynamic adaptation*: An adaptive technique is needed to allow the streaming system to quickly adapt to changes in network conditions, e.g., the network may become congested and the loss rate may increase.

Research prototypes and commercially deployed P2P streaming systems have been made available to provide different solutions to the above challenges. These P2P live media streaming systems can be classified into three categories according to their data delivery mechanisms: tree–push, mesh–pull, and push–pull.

Tree–push P2P Most of the earliest P2P ALM protocols employed tree–push mechanisms, which mimic the distributed tree-building mechanisms used in most multicast routing protocols (MRPs) adopted by the IP multicast, for media data delivery [39] [40]. The media content is first delivered from the source peer (node) to several intermediate peers (nodes). The intermediate peers then forward the data to the rest of the peers. The content is actively pushed from the root of the tree to all other peers in the tree.

SpreadIt [40] and Narada [41] are representative examples of the tree–push P2P ALM mechanisms. However, they use different approaches to organize peers and build the distribution tree. More specifically, SpreadIt builds and maintains a single multicast distribution tree in a distributed manner over the set of peers, each peer running a peering layer that coordinates with other peers to establish and maintain the multicast tree. When a new receiver wants to join, SpreadIt traverses the tree nodes downward from the source until it finds one with unsaturated bandwidth. When a receiver wants to leave, if it has children then it will send messages asking its children to connect to its parent or source. The problem of SpreadIt is that it has to get the source involved whenever a failure occurs, thus it is vulnerable to disruption due to the severe bottleneck at the source.

Narada focuses on multi-sender multi-receiver streaming applications. It builds the distribution trees in a two-step process (see Figure 7.35). First, it constructs a richly connected application-level mesh by making connections among the peer pairs. The links in the mesh are monitored periodically to improve the quality of the mesh. In the second step, whenever a sender wants to transmit media content to a set of receivers, Narada runs a reverse path forwarding (RPF) algorithm on the mesh to construct an efficient multicast distribution tree. In Narada, every member maintains a list of all other members in the group and periodically

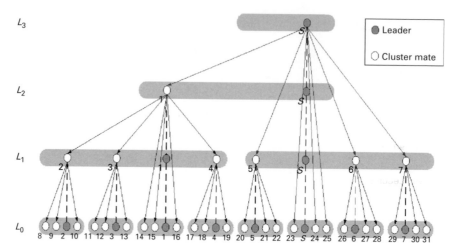

Figure 7.36 Once the leadership of cluster C in layer L_j of NICE is changed, the current leader has to remove itself from all layers higher than L_j. Each affected cluster in the higher layers has to choose a new leader.

exchanges its knowledge of group membership with its neighbors in the mesh. When a peer wants to join the group, it receives a list of group members by an out-of-band bootstrap mechanism and then sends them messages requesting to be added as a neighbor. When a peer wants to leave the group, it notifies its neighbor and this information is propagated to the rest of the group members along the mesh. Because of this feature, Narada cannot support large-scale networks. It is only employed for small P2P networks.

To support a larger quantity of P2P receivers on the Internet, Nice and Zigzag, which are hierarchical tree–push P2P mechanisms, were thus proposed for single-source media P2P streaming applications. To achieve better scalability, Nice and Zigzag organize peers into a multi-layer hierarchy of clusters recursively. Nice arranges end hosts into different clusters in an hierarchical structure. The number of nodes in each cluster is bounded by $[k, 3k - 1]$, where k is a predefined universal constant. If the cluster size is smaller than k or larger than $3k - 1$, subsequent cluster merging or splitting will be performed (see Figure 7.36). The Nice protocol defines basic operations to create and maintain its hierarchical structure. In Nice each cluster has a leader, which plays the important role of forwarding data to cluster members, as shown in the broken lines in Figure 7.36, where the data source is in the highest layer. To minimize the end-to-end delay between leaders and their cluster members, Nice selects the leader as the center node of the cluster, i.e., the node that has minimum distance to all other nodes in the cluster.

Leaders in layer L_i will join layer L_{i+1} after being selected. Therefore, an hierarchical structure is constructed. A cluster leader at layer L_j is also a cluster leader of a certain cluster at a lower layer L_i, where $L_i < L_j$. As members join or leave, the cluster leader, i.e., the center of a cluster, may change. Once the leadership of cluster C in layer L_j is changed, the current leader has to remove itself from all layers higher than L_j and each affected cluster in these higher layers has to choose a new leader. The multicast overlay data paths are implicitly defined on the basis of the hierarchical structure.

There are possible bottlenecks in this hierarchical architecture, because a node in a higher layer would need to support more nodes than a node at a lower layer. In particular, the node

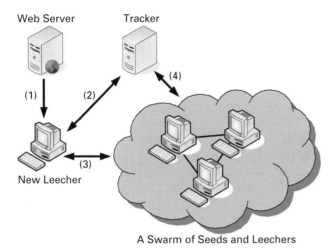

Figure 7.37 BitTorrent for file sharing.

at the highest layer needs to deliver data to $O(\log_k N)$ members. With increasing scale, this becomes more and more difficult to achieve. To mitigate the potential bottlenecks of high fan-out degrees of nodes in high layers in Nice, Zigzag was thus proposed [43]; it has a similar control topology to that of Nice in that it maintains a hierarchical structure for controlling and managing the membership. However, Zigzag defines a couple of connection rules (C-rules) and builds the multicast tree atop the administrative organization by applying these C-rules to achieve lower fan-out degrees than Nice.

One of the main focuses of the tree–push protocols mentioned above is to construct an efficient multicast tree for media data delivery. However, it is observed that even though the data delivery paths are efficient the intermediate nodes tend to be overloaded. In addition, most traditional architectures do not adapt well to occasional peer departures or failures. Once there are peer failures, the service could be seriously disturbed.

Mesh-pull P2P To overcome the adaptation issues of occasional peer failures or the frequent come-and-go of participating peers, many popular mesh–pull P2P live streaming systems make use of protocols modified from P2P file sharing protocols to deliver media content. The peers are randomly organized into a mesh structure and the selection of active peers for data exchange is commonly decided on the basis of the round-trip time (RTT) between peers. Peers exchange information about video chunks they possess before exchanging media data, which is also known as the gossiping process. After that, peers request video chunks which they do not have from other peers who have them, which is known as the pulling process.

To explain mesh–pull P2P media streaming applications, we first briefly describe the BitTorrent system, one of the most widely used P2P file sharing protocols. There are basically three key components within the BitTorrent protocol (see Figure 7.37). A *web server* keeps *torrent* files, i.e., the metadata of the sharing files. A *tracker* keeps track of the information of all participating peers. There is also a swarm of *seeds* and *leechers*, where a seed is a peer that has finished downloading the files but is still active for uploading the files to others and leechers are peers that are still downloading the files.

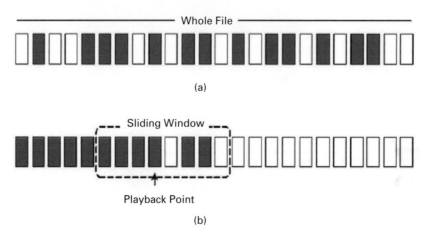

Figure 7.38 Buffer snapshot of (a) BitTorrent file sharing, and (b) mesh–pull P2P streaming.

When a new peer (most likely a leecher) tries to join the file sharing process, it first connects to the web server to download the *torrent* file, which includes the detailed file information and the tracker location. After that, the new peer connects to the tracker to register its information and get from the tracker an initial list of (~ 50) active peers which contain the requested file. After the leecher receives the initial peer list, it starts to connect to some (20–40) active peers selected from the peer list to ask for the exchange of pieces of files. As shown in Figure 7.38(a), the file to be P2P-shared is commonly broken into equal-size chunks (typically of 256 kB, and each chunk is further divided into 16 kB blocks as the smallest unit for peer-exchange purposes) and the peer connected can exchange blocks with one another. To update the tracker's global view of the system, active peers period-ically (e.g., every 30 minutes) report their state to the tracker or when joining or leaving the file sharing session. If a peer fails to maintain at least 20 connections, it re-contacts the tracker to obtain additional peers to which to connect. Although it has more than 20 maintained peer connections, a given peer normally limits the number of peers (to ~ 4) it is serving (unchoking) and being served by at each time instance on the basis of the corres-ponding uploading and downloading available bandwidth, which is evaluated every 10 seconds. A leecher preferentially sends data to peers that reciprocally send data to itself. This "tit-for-tat" strategy is used to encourage cooperation and avoid "free-riding." A leecher also performs optimistic unchoking to one of the four active peers it is serving regardless of the amount of data received from this peer, so that some peers can still receive some blocks of data even if they have nothing to give in return. However a seeder who is not looking for additional data will send data equally to all requesting peers. When a leecher requests blocks from other peers, it asks for blocks to be sent first on the basis of on the rarest-first policy, i.e., blocks that are rarely available from serving peers should be sent first.

A representative mesh–pull P2P live streaming architecture, e.g., Coolstreaming [54] or PPLive [55] (Figure 7.39), can be conceptually depicted as consisting of the following three major components [56].

(1) *The streaming peer node* This contains a streaming engine and a media player. The streaming engine pulls media chunks from other peer nodes or the channel streaming

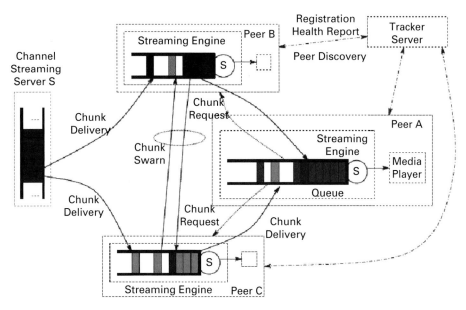

Figure 7.39 Conceptual diagram of PPLive [55] (© IEEE 2007).

server; these chunks are reassembled into the original media content and passed to the media player for playback.

(2) *The channel stream server* This converts the media content into small video chunks for efficient distribution among peers.

(3) *The tracker server* This provides information about streaming channels, peers, and video chunks for each peering join the network and pulling video chunks from multiple peers requesting the same media content in the system.

More specifically, when joining the protocol, peers first connect to the tracker to get the channel list. After they decide which channel to watch, they then register their information at the tracker server and download an initial peer list. While trying to get connected to peers in the initial peer list, they also exchange peer list information with other peers so as to collect more peers for best parent selection. In this manner, each peer maintains a list of other peers watching the same channel. The registration and peer discovery protocol commonly runs over UDP, while TCP may also be used if UDP fails, e.g., owing to the presence of a firewall. Utilizing this distributed gossip-like peer discovery protocol, the signaling overhead at the tracker server is considerably reduced; hence, a small number of tracker servers are able to manage possibly millions of streaming users.

Similar to BitTorrent in that it divides the shared file into chunks (pieces and blocks), a mesh–pull system also divides the streaming file into video chunks (around one second of video segment per chunk). The buffer map (BM) is used to record the existing video chunks available in a peer. Peers then exchange buffer map information before pulling data from other peers. The major difference between mesh–pull P2P streaming and BitTorrent file sharing lies in the way they manage their buffers. Figure 7.38(b) shows a buffer snapshot of a typical mesh–pull P2P streaming system: at any given time instant a peer buffers up to a few minutes' worth of chunks within a sliding window. When it has buffered a sufficient

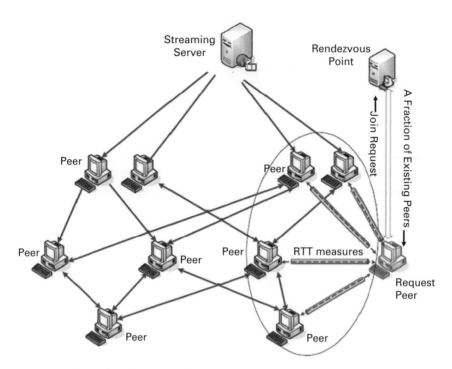

Figure 7.40 The push–pull system architecture of Gridmedia [56] (© IEEE 2006).

amount of video content, the media player begins to decode and render the video. Some of these chunks may have been recently played; the remaining chunks are scheduled to be played in the next few minutes. The continually exchanged buffer map message indicates which chunks a peer currently has buffered and can be shared. The buffer map message includes the offset (the ID of the first chunk), the length of the buffer map, and a string of zeroes and ones indicating which chunks are available (starting with the chunk designated by the offset).

As with the BitTorrent file sharing mechanism, a mesh–pull P2P peer may request chunks from tens of other peers simultaneously, and it also continually searches for new partners from which it can pull chunks. Different mesh–pull systems may differ significantly regarding their peer selection and chunk scheduling algorithms. More specifically, when a peer requests chunks it commonly gives higher priority to missing chunks that are to be played out first and also to rare chunks, that is, chunks that do not appear in many of its partners' buffer maps [56]. The chunks can be delivered over either TCP or UDP connections, depending on the tradeoff between reliable and on-time deliveries.

Push–pull P2P   As mentioned above, the advantage of the tree–push mechanism is the efficiency of the data delivery, since once the application-level multicast tree is built the media data is actively delivered from the parents to the children and additional information exchange (gossiping) is not necessary. However, the advantage of the more popular mesh–pull mechanism is its robustness to node failure and efficient bandwidth usage of all nodes. Push–pull P2P systems, such as Gridmedia [56], are proposed to leverage the advantages from both tree–push and mesh–pull mechanisms. Gridmedia was developed with the intention of reducing the incurred delay during information exchange. More specifically, once a Gridmedia peer finds that it persistently pulls media chunks from the same peer, it will then try to switch to the push mode to reduce the delay.

Figure 7.40 illustrates the system structure of Gridmedia, which includes the streaming server, the rendezvous point (RP) server, and peers. When the streaming server is connected to a request peer, it distributes the live content to the peer. The RP server is used to facilitate the login process of new arriving peers. Two main functions performed by peers are *unstructured overlay organization* and the *push–pull streaming schedule* [56]. Since the end nodes in Gridmedia are organized into an unstructured overlay network, an overlay manager in each node is used to find appropriate partnering peers by gossip protocol so that the application-layer overlay network can be effectively built up. During the startup, a new arriving node contacts the RP server first to get a candidate list (the so-called member table) of the nodes already in the overlay. This newly participating node selects some nodes with minimum measured round-trip time (RTT) as its initial partnering peers to reduce the chunk delivery delays. The rest of the initial partnering peers are then selected randomly from the given candidate list. During the streaming session, each actively participating node should maintain a list of partners from which it can fetch the streaming content. However, owing to the frequent joining and leaving of peers, the list should be updated from time to time. The update can be achieved by periodic exchanges of member tables among the partnering nodes. Each node in the member table has a modifiable field called *lifetime*, which denotes the elapsed time since the latest message was received from this member. When the lifetime of a node exceeds the predefined threshold, it will be removed from the member table. Once a node quits, it will flood a "quit message" via all its partners so that whoever receives this message will delete the corresponding node from its member table. Moreover, each node also periodically sends an "alive message" to all its partners to declare its existence. Thus, the failure of any partner can be detected as soon as possible.

Regarding the scheduling of push–pull operations, every peer in Gridmedia maintains a set of partners, whose end-to-end bandwidth and delay are dynamically varying. The use of the mesh–pull P2P mechanism can easily cause long average latency, owing to the need for the frequent exchange of buffer maps and chunk data requests. Moreover, the delay will be accumulate further owing to with the long hops of data delivery, causing a very big latency to be observed at the nodes. Gridmedia proposed a push–pull streaming mechanism in which the pull mode of a receiver and the push mode of a sender are used alternatively between partners. More specifically, each node uses the pull method as a startup and after that can relay a "pushing packet" to its partner as soon as the packet arrives, without explicit requests from the partner. According to the traffic condition of each partner, the node can also switch to pushing packets occasionally, e.g., in the case of lost packets.

7.5 Layered multicast of scalable media

With the fast deployment of Internet infrastructure, wired or wireless, the IP network is getting more and more heterogeneous. The heterogeneity of receivers (see Figure 7.41) under an IP multicast session significantly complicates the problem of effective data transmission. A major problem in IP multicast is the sending rate chosen by the sender. If the transmitted multimedia data rate is too high, this may cause packet loss or even a congestion collapse whereas a low transmission data rate will leave some receivers

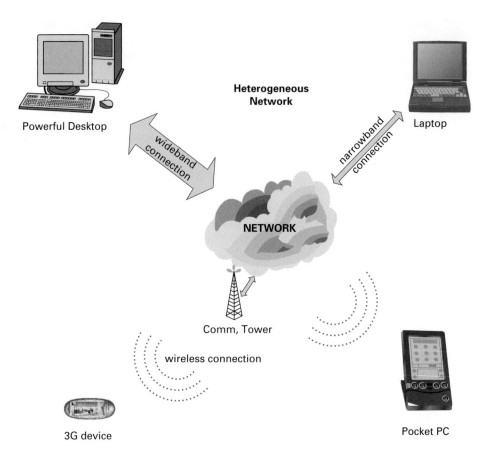

Figure 7.41 The heterogeneity of the receivers under an IP multicast session significantly complicates the problem of effective data transmission.

underutilized. This problem has been studied for many years and is still an active research area in IP multicast. To solve this issue, the transmission source should have a scalable rate, i.e., multirate, which allows transmission in a layered fashion. By using multirate, slow receivers can receive data at a slow rate while fast receivers can receive data at a fast rate. In general, multirate congestion control can perform well for a large multicast group with a large number of diverse receivers. This brings us to the scheme of layered multicast.

Basically, layered multicast is based on a layered transmission scheme. In a layered transmission scheme, data is distributed across a number of layers which can be incrementally combined, thus providing progressive refinement. The scalable video coding (SVC) discussed in Section 5.8 can easily provide such layered refinement. Thus, the idea of layered multicast is to encode the source data into a number of layers. Each layer is disseminated as a separate multicast group, and receivers decide to join or leave a group on the basis of the network condition (see Figure 7.42). It is assumed that the data that is going to be transmitted can be distributed into l multicast groups with bandwidths L_i, $i = 0$, ..., $l-1$. Receivers can adjust the transmission rates by using the cumulative layered transmission scheme. Now the adaptation to heterogeneous requirements becomes possible

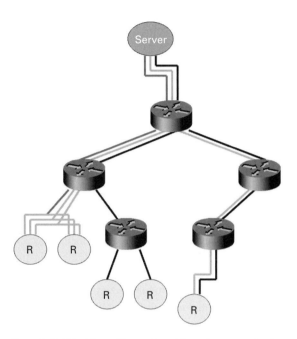

Figure 7.42 The idea of layered multicast is to encode the source data into a number of layers. Each layer is a separated multicast group; receivers decide to join or leave a group according to the network condition.

because it can be done independently in each receiver. On the basis of the network condition, a particular receiver can subscribe a bandwidth B_i by joining the $L_0, L_1, \ldots, L_i$ layers:

$$B_i = \bigcup_{j=0}^{i} L_j. \qquad (7.5)$$

The more layers the receiver joins, the better quality it gets. As a consequence of this approach, different receivers within a session can receive data at different rates. Also, the sender does not need to take part in congestion control.

To avoid congestion, end systems are expected to be cooperative by reacting to congestion and adapting their transmission rates properly and promptly. The majority traffic in the Internet is best-effort traffic. The transport control protocol (TCP) traffic uses an additive-increase multiplicative-decrease (AIMD) mechanism, in which the sending rate is controlled by a congestion window. The congestion window is halved for every window of data containing a packet drop and increased by roughly one packet per window of data otherwise. Similarly, IP multicast for UDP traffic needs a congestion control algorithm. However, IP multicast cannot simply adopt the TCP congestion control algorithm because acknowledgements can cause an "implosion problem" in IP multicast. Owing to the use of different congestion control algorithms in TCP and multicast, the network bandwidth may not be shared fairly between the competing TCP and multicast flows. Lack of an effective and "TCP friendly" congestion control is the main barrier for the wide-ranging deployment of multicast applications.

"Scalability" refers to the behavior of the protocol in relation to the number of receivers and network paths, their heterogeneity, and their ability to accommodate dynamically variable sets of receivers. The IP multicasting model provided by RFC 1112 is largely scalable, as a sender can send data to a nearly unlimited number of receivers. Therefore, layered multicast congestion control mechanisms should be designed carefully to avoid scalability degradation.

7.5.1   Receiver-driven layered multicast (RLM)

Receiver-driven layered multicast (RLM) [44] is the first well-known end-to-end congestion control algorithm based on layered multicast. In RLM, a receiver detects network congestion when it observes increasing packet losses. The receiver reduces the level of subscription if it experiences congestion. In the absence of loss, the receiver estimates the available bandwidth by doing so-called *join experiments* when the join-timer expires. A join experiment means that a receiver increases the level of subscription and measures the corresponding loss rate over a certain period. If a join experiment causes congestion then the receiver quickly drops the offending layer. Otherwise, another join-timer will be generated randomly and the receiver retains the current level of subscription and continues to do join experiments for the next layer once the newly generated join-timer has expired. To avoid the transient congestion that impacts the quality of the delivered signals, RLM performs join experiments less frequently when they are likely to fail, i.e., RLM implements this strategy by increasing the join-timer exponentially when congestion is detected and a layer is dropped. However, join experiments can interfere with each other. For example, a receiver can misinterpret the congestion caused by its join experiment and mistakenly reduce the level of subscription even though the congestion was induced by another receiver in a join experiment. To avoid this problem, RLM introduces a "shared learning" mechanism in which a receiver notifies the entire group by multicasting a message identifying the experimental layer before conducting a join experiment. A receiver can only do the join experiment for layers at or below the newly multicast experimental layer. In general, the subscription level can be increased or decreased in RLM according to the following rules.

(1) Before doing a join experiment, a receiver will perform "shared learning" by broadcasting a notification message to all receivers in the multicast group. By doing so, all the receivers are informed about which layer is currently participating in the join experiment.
(2) Join-timers are randomized to avoid a protocol synchronization effect. If a join-timer expires and no experiment at the same or a lower layer is in progress, the receiver will perform a join experiment to increase the level of subscription. Otherwise the current join-timer is ignored and a new one will be generated.
(3) If a packet loss is detected then, depending on the circumstances of the receiver, the following actions will be taken.
   - If the receiver is currently participating in a join experiment at the highest level, the receiver will drop the offending layer and back off the join-timer.
   - If the receiver is currently doing a join experiment but not at the highest level or if no experiment is being performed then RLM will measure the long-term congestion before dropping the offending layer.

Receiver-driven layered multicast is the first end-to-end layered multicast congestion control mechanism to avoid congestion collapse, and it can ensure that the link is fully utilized. However, there are a number of problems with RLM. More specifically, uncoordinated join experiments by downstream receivers create substantial problems. A receiver's join experiments can introduce packet losses at other receivers behind the same bottleneck link. This finally results in unfairness among the downstream receivers (known as "intra-session unfairness"). Several studies have reported that RLM has neither interprotocol nor intraprotocol fairness [45]. Join experiments *per se* are prone to packet losses when over-subscription causes bandwidth waste. As reported in [46], RLM is very slow in converging to an optimal rate. It can take several minutes to do a join experiment in order to discover the available bandwidth. Moreover, RLM has no support for error recovery. When receivers try to tackle congestion by unsubscribing from a layer, RLM can take several seconds to take effect, owing to the nature of the Internet group multicast protocol (IGMP) [3]. During this latency, multicast traffic still flows through the router. So, IGMP leave-latency can slow down bandwidth recovery and therefore causes congestion persistence.

7.5.2 Receiver-driven layered congestion control (RLC)

Receiver-driven layered multicast does not exhibit TCP-friendly or fairness behavior. To address this problem, another receiver-driven layered congestion (RLC) control was proposed [47]. This RLC control introduces *synchronized join experiments* and *burst test* techniques to adapt reception rates to network conditions. In RLC the layer rate uses a doubling scheme, i.e., the cumulative rates of the layers are 1, 2, 4, 8, ... Adding a layer will double the rate, while dropping a layer will multiplicatively decrease the rate. So, RLC is *doubling-increase multiplicative-decrease (DIMD)* rate adjustment.

Since the leave-delay of a multicast group is expensive, the subscription level can only be increased when it is certain that the attempt will be successful. To estimate the available bandwidth, the sender regularly generates short bursts of packets, the synchronization point being placed at the end of the burst. During the bursts, which have duration T_0 (this is the interpacket time in the base layer, as shown in the shaded rectangle in Figure 7.43), two back-to-back packets are sent at each transmission. Following the bursts, there is another interval T_0, during which transmission is suspended (the white rectangle in Figure 7.43). If no loss is experienced during bursts, the subscription level will be increased. It takes some time to complete the leaving phase of a multicast group. A receiver may continuously experience packet losses during the leaving phase and decrease the subscription many times

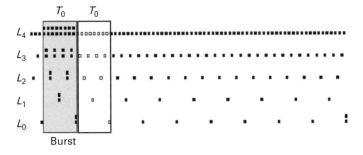

Figure 7.43 The doubling-increase multiplicative-decrease (DIMD) rate adjustment [49] in RLC (© IEEE 1998).

in response to a single failed join. To overcome this problem, RLC introduces a deaf period T_D. A receiver does not react to further losses for a time T_D after a loss is detected and the subscription level has been decreased. Basically, the congestion control mechanism is the following.

(1) Decrease the subscription if a loss is experienced.
(2) Increase the subscription at synchronization point if no loss is detected during the burst.
(3) Make no change if loss is detected only at burst or during the deaf period.

If the number of layers is large, RLC will take a long time to converge to steady state. Another weakness of RLC is that burst induces losses, which probably reduce the user's perceived quality. The rate doubling scheme can cause dramatic fluctuations in network bandwidth consumption and rapid queue build-up [48]. After a burst test, even if no loss is found it still cannot be assumed that adding a layer will be safe. So, its join experiments even with the burst test cannot ensure the avoidance of packet losses and oversubscription. It has been shown that RLC can be very sluggish before the optimal layer is reached, causing inefficiency of bandwidth usage. Although RLC improves some fairness issues, it was shown in [49] that RLC competes well with TCP when the round-trip time (RTT) of the TCP is much more than one second but not when the RTT of the TCP is much less than one second.

7.5.3   Packet-pair layered multicast (PLM)

Packet-pair receiver-driven layered multicast (PLM) [50] was proposed to improve RLC. It is also based on a receiver-driven cumulative layering scheme. However, to react appropriately to congestion PLM requires that all layers must follow the same multicast routing tree. Most existing rated-based protocols use the principle of "increasing rate when no congestion, decreasing rate under congestion" to adjust the sending rate (for layered multicast, joining or leaving a layer corresponds to increasing or decreasing the rate). Without knowledge of the available bandwidth, this rule results in rate oscillations and periodical packet loss even under smooth available bandwidth. To overcome these problems, the key mechanisms adopted in PLM are *receiver-side packet pair* probing and *fair queuing* at routers. More specifically, instead of relying on a join experiment technique such as in RLM or RLC, PLM uses a packet pair probing approach to infer the available bandwidth and avoid congestion.

The packet pair technique is a well-known procedure for measuring the capacity of a path. When a packet is transmitted in a link, it encounters a transmission or serialization delay due to the physical bandwidth limitations of the link and the hardware constraints of the transmitting equipment. In a link of capacity C_i and for a packet of size L, the transmission delay is $\tau_i = L/C_i$. A packet pair experiment consists of sending two packets back to back, i.e., with a spacing that is as short as possible, from the source to the sink. Without any cross traffic in the path, the packet pair will reach the receiver *dispersed* (spaced) by the transmission delay in the narrow link, $\tau_i = L/C_i$. So, the receiver can compute the capacity C from the measured dispersion Δ as $\tau_n = L/C_i$. Figure 7.44 illustrates the packet pair technique in the case of a three-link path, using a fluid analogy. Even though simple in principle, this technique can produce widely varying estimates and erroneous results. The main reason is that cross traffic in the path distorts the packet pair dispersion, increasing or decreasing the capacity estimates.

In the packet pair approach, a PLM source periodically sends a pair of its data packets as a burst to infer the bandwidth share of the flow. At the receiver side, the estimation of the

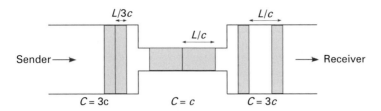

Figure 7.44 The packet-pair technique in the case of a three-link path, using the fluid analogy [50].

available bandwidth is calculated from the packet size divided by the inter-arrival gap. This estimated bandwidth is used to perform rate adaptation, which is done only once, at every regular *check period* interval, to avoid rate adaptation that is too oscillatory. To deal with the case when packet pairs are lost during severe congestion, PLM defines a *timeout period*. If no packet is received before the timeout, a layer will be dropped owing to the expectation of congestion. If some packets are received but no packet pair is received, and the loss rate is over a predefined *loss threshold*, a layer will be dropped. Then, to avoid overreaction to loss, PLM also waits for a predefined *blind period* before re-estimating the loss rate.

Packet-pair layered multicast assumes the deployment of a fair queuing mechanism in routers and relies on a fair scheduler to ensure fairness, including intraprotocol fairness, interprotocol fairness, and TCP-friendliness. The PLM system has a few advantages over RLC. First, it has a faster convergence for rate adaptation. It can quickly reach the optimal level of layer subscription. Furthermore, compared with the situation for join experiments, the packet-pair system can sense the bandwidth changing in the network before congestion becomes severe (i.e., before packet losses). Finally, the fair scheduler makes PLM very intraprotocol and interprotocol fair. However, it is still arguable whether it is feasible to implement the fair scheduler required in the router over the whole Internet. Packet-pair layered multicast has also been shown to be very scalable owing to the nature of receiver-driven layered multicast congestion control protocols.

7.5.4 Bandwidth inference congestion control (BIC) layered multicast

The one-way trip time (OTT) of a packet k, D_k (see Eq. (7.2)), is the sum of the delays at each link along the path, while the delay at a single link consists of the transmission, queuing, and processing delays at the router. Since an absolute value of the delay is difficult to obtain owing to a skew between the clocks at the sender and the receiver, it is common practice to use a variation called the *relative one-way trip time* (ROTT) to replace OTT in algorithms: ROTT is measured by the receiver as the time difference between the receiving time and the packet-sending timestamp recorded in the field of the multimedia UDP packet header, plus a fixed bias. This delay value is closely related to the queuing delay, which is mainly contributed by network congestion along the path. If the end-to-end available bandwidth is not enough to sustain the transmission rate of a sequence of probing packets, then the ROTTs $\{D_k\}$ will increase with time. Note that, for an increase trend, the increase may not necessarily be strictly due to stochastic cross traffic. At the receiver, a one-way delay can be detected by using the observed receiving time and the sending time encapsulated in the packet header. When there is no packet-size trend, i.e., the packet sizes of the probing sequence are approximately equal, the one-way delay trend can be effectively taken

Table 7.3 Layered bitrate representations

bw_i	Bitrate up to layer i
pbw_i	Bitrate up to layer i during probes

as a queuing delay trend. It is very important to detect correctly whether there are increase trends among the measured one-way delays. A good trend detection algorithm should have the following two properties. First, it should give correct trend results using as few probe packets as possible; because a lengthly probing may cause congestion and adversely affect the existing cross traffic, a short probe size is important to ensure that the probing is not intrusive. Second, the algorithm should not be sensitive to the algorithm parameters, in order to assist large-scale practical deployment. A trend detection algorithm called Fullsearch was proposed based on statistical tests to check the existence of a trend. Define $I(X)$ as 1 if X holds and 0 otherwise. Suppose that the measured one-way ROTT delays are D_k, $k = 1, \ldots, M$. The test used in Fullsearch is [51]

$$S_{Fullsearch} = \frac{\sum_{k=2}^{M} \sum_{l=1}^{k-1} I(D_k > D_l)}{\frac{1}{2} M(M-1)}. \tag{7.6}$$

As shown in Eq. (7.6), the Fullsearch algorithm reflects the statistical relationship between all pairs of measurements. The test gives a result between 0 and 1. If there is no trend among the measurements, the test result is around 0.5; if there is a strong increase trend, the test result approaches 1. A trend threshold t can be used to check whether there is an increasing trend: if the test result is larger than t, the measurements have an increasing trend, otherwise there is no increasing trend. The trend threshold is a major parameter in the available-bandwidth estimation process.

A rate-based bandwidth inference congestion (BIC) control for layered multicast was proposed in [52]; this uses a delay-trend detection technique to estimate the end-to-end available bandwidth and adjusts the sending rate accordingly. The BIC can achieve good stability, which is desirable for video transmission.

The BIC can work for arbitrary-layered data organization. Suppose that N is the total number of layers. In Table 7.3, the symbols representing the data bitrates in each layer i are shown. The sender is required periodically to send probes in each layer so that each receiver can use the statistics of the probe packets to estimate the spare capacity along the path. During the probe period the cumulative data bitrate up to layer i follows pbw_i instead of bw_i. The relationship between the two rates in Table 7.3 is seen from

$$pbw_i = bw_{i+1} \quad (i \neq N),$$
$$pbw_N = bw_N. \tag{7.7}$$

According to this relationship, a receiver subscribing to layer i receives data at a rate bw_i during normal periods and at a rate bw_{i+1} during probe periods. Two parameters control the behavior of the probes. One is the probe size P, i.e., the number of probe packets for each layer; the other one is the interval T between probes. The probe size should be as large as

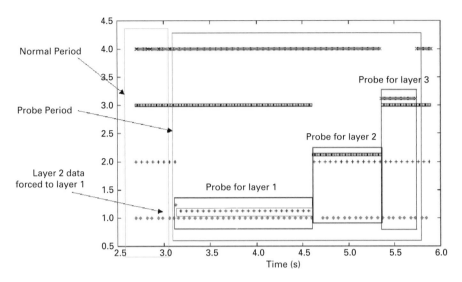

Figure 7.45 Output of the multi-layer token-bucket model [52] (© IEEE 2004).

required for delay trend detection. The probe interval T is configured so that the additional data bitrate (due to the probe bitrate) is less than 10% of the original data bitrate.

The multi-layer token-bucket model is used to reshape the source-generated raw data and to interleave packets from different layers so that the packets in each layer are as even as possible. Figure 7.45 shows an example of the output of this multi-layer token-bucket model for the first four layers. As shown in Figure 7.45, each point represents a packet. During the probe period for layer 1 (from 3.2 s to 4.6 s), the packets of layer 2 are sent through multicast group 1 so that the receivers subscribing to layer 1 receive packets up to layer 2, just as if they had subscribed to layer 2. The statistics at the receivers during the probe period are used to infer whether the receivers can join up to the next layer. From Figure 7.45, it can be seen that the probe duration is different for each layer, and the higher layer has a shorter probe duration (e.g., the probe period for layer 2 is from 4.6 s to 5.4 s). The reason is that for a given number of probe packets to be sent, higher layers require less time owing to the higher-data-rate assumption. These properties distinguish a source probe with the BIC infrastructure from that used in RLC.

In BIC, packet loss is used to detect congestion during normal periods in the same way as other protocols. However, delay trend detection is used during probe periods to infer the spare capacity. The basic idea is that when the data bitrate is larger or equal to the available bandwidth along the path, the observed packet delay at receivers will have an increase trend.

From the statistics of the probe packets, the delay-increase trend indicator $S_{fullsearch}$ defined in Eq. (7.6) (which has a value between 0 and 1; the larger this value, the more obvious that there is an increase trend) can be calculated. A threshold can be set for each layer so that the receiver can judge whether there is enough spare capacity to subscribe one more layer by comparing the trend detection result with the predefined threshold. If the delay trend result is lower than the threshold, it is probable that the path can support the probe rate, which is the rate at which the receiver subscribes to one more layer according to the source-layer organization described. If the path does not have enough spare capacity to support one more layer, a delay-increase trend will be detected. The router queue can absorb the limited insertion of probe packets (normally less than $P/2$ packets), thus it does not cause router queue overflow.

The join or leave decision in the BIC protocol is straightforward. Each receiver monitors the packet loss and leaves the current layer when the observed packet loss exceeds a predefined loss threshold. Each receiver also detects the delay trend during its probe period and joins one more layer when the delay trend is less than a predefined trend threshold (if loss is detected in the previous normal period or current probe period then the receiver keeps to its current subscription layer). Note that the probe also serves as a synchronization point, for all receivers in one session, for join action, which scales with the number of receivers in the same way as in RLC. A distinct feature of BIC is that no join experiment is required to be conducted.

Bandwidth inference congestion control adopts a different join or leave strategy during the startup phase. The receiver does not wait for the probe period to join the next layer in the startup phase. Instead, just when it has received P packets after the first packet of the current layer, it detects whether there is a delay-increase trend for these P packets. If there is no increase trend and no packet loss is observed then it joins the next layer; otherwise, it turns to a steady phase. A stricter trend threshold is applied for the startup phase to avoid oversubscription. This special startup-phase strategy enables BIC to converge quickly to the optimal subscription level. Intersession fairness can be achieved by choosing different loss thresholds and delay-trend thresholds for different layers. Stricter thresholds are used for higher layers, so that receivers with a lower subscription level are easier to reach.

References

[1] Douglas E. Comer, *Internetworking with TCP/IP, Principles, Protocols and Architectures*, fourth edition, Prentice Hall, 2000.

[2] "Internet control message protocol (ICMP)," IETF RFC 792, http://tools.ietf.org/html/rfc792.

[3] "Internet group management protocol (IGMP)," IETF RFC 988, RFC 1112, and RFC 2336.

[4] "Routing information protocol (RIP)," IETF RFC 1058, http://www.ietf.org/rfc/rfc1058.txt.

[5] "Open shortest path first (OSPF)," IETF RFC 2328, v2, http://www.ietf.org/rfc/rfc2328.txt.

[6] J. M. McQuillan, I. Richer, and E. C. Rosen, "The new routing algorithm for the ARPANet," *IEEE Trans. Commun.*, 28(5): 711–719, May 1980.

[7] "Support of IP-based services using IP transfer capabilities," ITU-T Rec. Y.1241, March 2001.

[8] J. Gozdecki, A. Jajszczyk, and R. Stankiewicz, "Quality of service terminology in IP networks," *IEEE Commun. Mag.*, 41(3): 153–159, March 2003.

[9] ITU-T recommendation G.114, February 1996.

[10] Q. Liu and J. N. Hwang, "End-to-end available bandwidth estimation and time measurement adjustment for multimedia QoS," in *Proc. IEEE Con. ICME 2003*, Baltimore MD, July 2003.

[11] "Draft supplement to standard for telecommunications and information exchange between systems – LAN/MAN specific requirements – Part 11: wireless medium access control (MAC) and physical-layer (PHY) specifications: medium access control (MAC) enhancements for quality of service (QoS)," IEEE 802.11 WG, IEEE 802.11e/D5.0, August 2003.

[12] "The type-of-service (TOS) field for network layer service classification," IETF RFC 1349, http://www.faqs.org/rfcs/rfc1349.html.

[13] Y. Bernet *et al.*, "A framework for differentiated services," Internet draft, http://www.ietf.org/internet-drafts/draft-ietf-diffserv-f~~mework-O2.txt, February 1999.

[14] H. Wang and K. G. Shin, "Transport-aware IP routers: a built-in protection mechanism to counter DDoS attacks," *IEEE Trans. Parallel and Distributed Process.*, 14(9): 873–884, September 2003.

[15] J. McQuillan, "Layer 4 switching," *Data Comm.*, October 1997.

[16] "Priority queuing, QoS," Cisco IOS Software, Cisco Systems, http://www.cisco.com/warp/public/732/Tech/pq/index.html.

[17] "Custom queuing, QoS," Cisco IOS Software, Cisco Systems, http://www.cisco.com/warp/public/732/Tech/cq/index.html.

[18] "Weighted fair queuing," Cisco IOS Software, Cisco Systems, http://www.cisco.com/warp/public/732/Tech/wfq/index.html.

[19] "Random early detection," Wikipedia, the free encyclopedia, http://en.wikipedia.org/wiki/Random_early_detection

[20] "Weighted random early detection," Wikipedia, the free encyclopedia, http://en.wikipedia.org/wiki/Weighted_random_early_detection

[21] J. Evans and C. Filsfils, *Deploying IP and MPLS QoS for Multiservice Networks: Theory and Practice*, Morgan Kaufmann, 2007.

[22] R. Braden, D. Clark, and S. Shenker, "Integrated services in the internet architecture: an overview," IETF RFC 1633, June 1994.

[23] R. Braden, ed., L. Zhang, S. Berson, S. Herzog, and S. Jamin, "Resource ReSerVation protocol (RSVP) – version 1 functional specification," IETF RFC 2205, September 1997.

[24] U. Payer, "DiffServ, IntServ, MPLS," Institute for Applied Information Processing and Communications, Technical University Graz, 2005.

[25] S. Blake *et al.*, "An architecture for differentiated services," IETF RFC 2475, December 1998.

[26] "Definition of the differentiated services field (DS field) in the IPv4 and IPv6 headers," http://www.ietf.org/internet-drafts/draft-ietf-diffserv-header-02.txt.

[27] S. Deering, "Host extensions for IP multicasting," IETF RFC 1112, August 1989, http://www.ietf.org/rfc/rfc1112.txt.

[28] R. Wittmann and M. Zitterbart, *Multicast Communication: Protocols, Programming, & Applications*, Morgan Kaufmann Series in Networking, June 2000.

[29] A. Ganjam and H. Zhang, "Internet multicast video delivery," *Proc. IEEE*, 93(1): 159–170, January 2005.

[30] Y. Dalal and R. Metcalfe, "Reverse path forwarding of broadcast packets," in *Proc. Conf. on Communications ACM*, December. 1978.

[31] S. Deering, "Multicast routing in internetworks and extended LANs," in *Proc. ACM SIG-COMM*, August 1988.

[32] J. Moy, "Multicast extensions to OSPF," IETF RFC 1584, March 1994.

[33] S. Deering, D. Estrin, D. Farinacci, V. Jacobson, C. Liu, and L. Wei, "An architecture for wide-area multicast routing," in *Proc. ACM SIGCOMM*, 1994.

[34] D. Estrin, D. Farinacci, A. Helmy *et al.*, "Protocol independent multicast-sparse mode (PIM-SM): protocol specification," IETF RFC-2117, June 1997.

[35] S. Casner and S. Deering, "First IETF internet audiocast," in *Proc. ACM Comput. Commun. Rev.*, pp. 92–97, 1992.

[36] J. Ni, D. H. K. Tsang, "Large scale cooperative caching and application-level multicast in multimedia content delivery networks," *IEEE Commun. Mag.*, 43(5): 98–105, May 2005.

[37] S. Y. Shi and J. S. Turner, "Routing in overlay multicast networks," in *Proc. IEEE INFOCOM 2002*, June 2002.

[38] S. Banerjee *et al.*, "Construction of an efficient overlay multicast infrastructure for real-time applications,"in *Proc. IEEE INFOCOM,* 2003

[39] V. Padmanabhan, H. Wang, and P. Chou, "Resilient peer-to-peer streaming" in *Proc. 11th IEEE Int. Conf. on Network Protocols (ICNP'03)*, Atlanta, November 2003.

[40] H. Deshpande, M. Bawa, and H. Garcia-Molina," Streaming live media over a peer-to-peer network," Technical Report, Stanford University, 2001.

[41] Y.-H. Chu, S. G. Rao, and H. Zhang, "A case for end systems multicast," in *Proc. ACM SIGMETRICS Conf.*, pp. 1–12, 2000.

[42] S. Banerjee, B. Bhattacharjee, and C. Kommareddy, "Scalable application layer multicast," Technical Report, UMIACS TR-2002-53 and CS-TR 4373, Department of Computer Science, University of Maryland, May 2002.

[43] D. Tran, K. Hua, and T. Do, "ZIGZAG: an efficient peer-to-peer scheme for media streaming," in *Proc. IEEE INFOCOM'03*, San Francisco, April 2003.

[44] S. McCanne, V. Jacobson, and M. Vetterli, "Receiver-driven layered multicast," in *Proc. ACM Sigcomm*, Palo Alto, pp. 117–130, August 1996.

[45] R. Gopalakrishnan, J. Griffioen, G. Hjalmtysson, and C. J. Sreenan, "Stability and fairness issues in layered multicast", in *Proc. NOSSDAV'99*, 1999.

[46] A. Legout and E. W. Biersack, "Pathological behaviors for RLM and RLC", in *Proc. of NOSSDAV'2000*, Chapel Hill, North Carolina, June 2000.

[47] L. Vicisano, J. Crowcroft, and L. Rizzo, "TCP-like congestion control for layered multicast data transfer", in *Proc. INFOCOM' 98*, San Francisco, March 1998.

[48] J. Byers, M. Frumin, G. Horn *et al.*, "FLID-DL: congestion control for layered multicast," in *Proc. of ACM NGC*, pp. 71–82, Palo Alto, November 2000.

[49] I. E. Khayat and G. Leduc, "A stable and flexible TCP-friendly congestion control protocol for layered multicast transmission," in *Proc. Int. Workshop of IDMS*, pp. 154–167, Lancaster, October 2001.

[50] A. Legout and E. W. Biersack, "PLM: fast convergence for cumulative layered multicast transmission", in *Proc. ACM SIGMETRICS Conf.*, pp. 13–22, Santa Clara, CA, June 2000.

[51] Q. Liu, J. Yoo, B.-T. Jang, K. Choi, and J.-N. Hwang, "A scalable VideoGIS system for GPS-guided vehicles," *Signal Process. Image Communi.*, 20(3): 205–208, March 2005.

[52] Q. Liu and J.-N. Hwang, "A scalable video transmission system using bandwidth inference in congestion control," in *Proc. IEEE Int. Symp. on Circuits and Systems (ISCAS)*, Vancouver, May 2004.

[53] Y. Liu, Y. Guo, and C. Liang, "A survey on peer-to-peer video streaming systems," *Peer-to-Peer Networking and Applications*, 1(1): 18–28, March 2008.

[54] S. Xie, B. Li, G. Y. Keung, and X. Zhang, "Coolstreaming: design, theory, and practice," *IEEE Trans. Multimedia*, 9(8): 1661–1667, December 2007.

[55] X. Hei, C. Liang, J. Liang, Y. Liu, and K. W. Ross, "A measurement study of a large-scale P2P IPTV system," *IEEE Trans. Multimedia*, 9(8): 1672–1686, December 2007.

[56] J. G. Luo, Y. Tang, M. Zhang, L. Zhao, and S. Q. Yang, "Design and deployment of a peer-to-peer based IPTV system over global Internet," in *Proc. First Int. Conf. on Communications and Networking in China (ChinaCom '06)*, October 2006.

8 Quality of service issues in streaming architectures

With the fast advances in computing and compression technologies, high-bandwidth storage devices, and high-speed networks, it is now feasible to provide real-time multimedia services over the Internet. This is evident from the popular use of the three most commonly used streaming media systems, i.e., Microsoft's Windows Media [1], RealNetwork's RealPlayer [2], and Apple's QuickTime [3]. The real-time transport of live or stored audio and video is the predominant part of real-time multimedia. In the download mode of multimedia transport over the Internet, a user downloads the entire audio/video file and then plays back the media file. However, full file transfer in the download mode usually suffers long and perhaps unacceptable transfer times. In contrast, in the streaming mode the audio and video content need not be downloaded in full but is being played out while parts of the content are being received and decoded. Multimedia streaming is an important component of many Internet applications such as distance learning, digital libraries, video conferencing, home shopping, and video-on-demand. The best-effort nature of the current Internet poses many challenges to the design of streaming video systems. Owing to its real-time nature, audio and video streaming typically has bandwidth, delay, and loss requirements. However, the current best-effort Internet does not offer any quality of service (QoS) guarantees to streaming media over the Internet.

The design of some early streaming media programs, such as VivoActive 1.0 [4], was based on the use of the H.263 video and G.723 audio protocols, with HTTP-based web servers, to deliver encoded media content. Since all HTTP server–client transactions are based on a very simple design using a guaranteed-delivery TCP transport protocol, they are not particularly suited for real-time streaming. For example, lack of control over the rate at which the web server pushes data through the network, as well as the use of the guaranteed-delivery transport protocol (TCP), caused substantial fluctuation in the delivery times for fragments of the encoded data. This is why the VivoActive player used a quite large *preroll buffer*, which was used to accumulate sufficient data before the received streaming data was played back; this was meant to compensate for the burstiness of such a delivery process. Nevertheless, if for some reason the delivery of the next fragment of data was delayed by more than the available *preroll time* (5–20 s), the player had to suspend rendering until the buffer was refilled. This so-called *rebuffering* process was a frequent cause of diminished user experience.

The first Internet-based streaming media system that featured both server and client components was RealAudio 1.0, introduced in March 1995 [5]. The communication between RealAudio server and RealAudio player was based on a suite of dedicated, TCP- and UDP-based network protocols, known as the progressive networks architecture (PNA). These protocols allowed transmission of the compressed audio packets via a low-overhead, unidirectional, UDP transport, retaining the use of TCP mainly for basic session control. While the use of UDP transport enabled better utilization of the available network bandwidth and made the transmission process much more continuous (in comparison with TCP traffic), it also introduced several problems such as lost, delayed, or delivered-out-of-order packets. Several mechanisms were used to combat the damage caused by these problems. For example, the automatic repeat-request (ARQ) mechanism was implemented to allow the client to request missing packets again from the server. If the retransmitted packets were successfully delivered within the available preroll time, the loss was recovered. In cases where ARQ failed, a frame-interleaving technique was used to minimize the perceptual damage caused by the loss of packets [6]. To satisfy the requirement of producing compressed data that can be streamed at some constant bitrate, a bandwidth-smoothing algorithm was implemented. Owing to the dynamic change of video content, such as scene changes, fast and slow motion, transitions, etc., it is likely that a very unequal distribution of bits between frames is created when they are encoded with the same level of distortion. With the availability of the preroll buffer, the constant bitrate requirement can be substantially relaxed and this eventually led to the design of the variable-bit-rate (VBR) rate-control (bandwidth-smoothing) algorithm in the RealVideo codec. Moreover, the RealVideo codec used a combination of forward error correction codes to protect the most sensitive parts of the compressed bitstream and various built-in *error concealment* mechanisms. Such a combination of techniques is commonly referred to as *unequal error protection* [7].

Since the advent of RealAudio and RealVideo, many real-time streaming systems, protocols, and architectures have been proposed and developed. In terms of multimedia streaming over IP networks, there are two common delivery mechanisms in which multimedia information can be distributed over the Internet, namely, *live streaming* and *on-demand streaming* [6].

Live streaming   A diagram illustrating various steps in the distribution of live content is presented in Figure 8.1 [6]. The source of live multimedia information (such as any standard

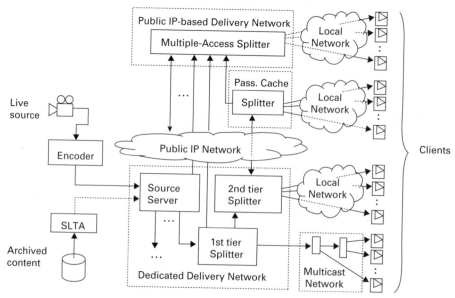

Figure 8.1 Delivery of live and/or simulated live media content [6] (© IEEE 2003).

video camcorder) is connected to the video and audio *encoders*. The encoding engine is responsible for capturing, digitizing, and compressing the incoming analog video and audio information and passing the resulting compressed bitstreams down to the *server*, which is the origin of the streaming content. Often, servers can be chained by *relay* configurations, as in tunneling. When the fan-out of a relay spans several destinations, the relay becomes a *splitter*. The splitting of media distribution allows a large set of users to be served and a CDN to be deployed. Alternatively, the server can receive such information from a *simulated live transfer agent* (*SLTA*), a software tool that reads pre-encoded information from an archive and sends it to a server as if it had just been encoded from a live source.

The server is responsible for disseminating the compressed bitstreams from the encoder to all connected *splitters* and/or *clients* who have joined the live streaming session. Splitters are additional servers that can be either part of a dedicated media delivery network or a public-IP-based multiple-access delivery network, or they can be embedded in network traffic caches, which simply pass the information through in the case of live streaming. The server (or splitter) commonly *unicasts* the encoded video information to each of the clients individually using a one-way data stream (in combination with some two-way streaming session control). In this case, the parameters of the connection between the server and each client can be estimated at the beginning of each session and can be systematically monitored during the broadcast. In the case where a network is equipped with multicast-enabled routers, the server needs to send only one *multicast* stream, which is automatically replicated to all subscribing clients on the network. Important limitations of multicasting are one-way transmission and the non-guaranteed delivery of information owing to the lack of two-way streaming session control. In addition, the server typically does not know how many clients are subscribed to the multicast session and/or their actual connection statistics.

On-demand streaming The second multimedia distribution mechanism, called on-demand streaming, is illustrated in Figure 8.2 [6], which has a major difference between the mechanism for live streaming (see Figure 8.1), i.e., it has no direct connection between the encoder and the server. Instead, a compressed video clip has to be recorded on disk first, and then the server will be able to use the resulting compressed file for distribution. The on-demand streaming structure also allows remote *proxy* servers to use their local storage to *cache* the most frequently used media clips. Server and client communication for delivering on-demand content is essentially the same as the unicast streaming of live content. The main difference is that with on-demand content a user is allowed to rewind and/or fast forward the presentation, while such controls are not available for live streaming. Server–proxy transfers can only be initiated by the client, i.e., at the time of the transfer the proxy may already have some information about the requested clip in its local storage. Using appropriate coding techniques, such information can be used to reduce the rate of the requested additional stream to the proxy.

8.1   QoS mechanisms for multimedia streaming

The streaming of video over the Internet faces many technical as well as business challenges, and new codecs, protocols, players, and systems need to be developed to address them. As shown in Figure 8.3 [8], there are six key components of a multimedia streaming system, namely multimedia compression, application-layer QoS control, continuous

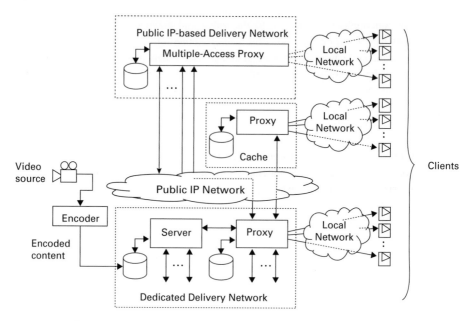

Figure 8.2 Delivery of on-demand media content [6] (© IEEE 2003).

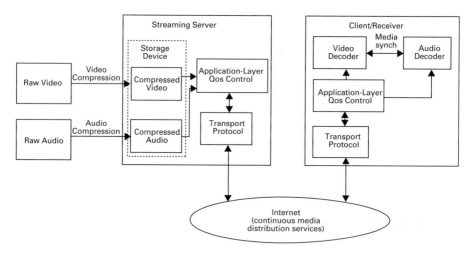

Figure 8.3 There are six areas in a generic architecture for multimedia streaming [8] (© IEEE 2003).

media distribution services, streaming servers, media synchronization mechanisms, and protocols for the streaming media. Each of the six components constitutes a basic block, and from these blocks an architecture for streaming multimedia can be built. The raw video and audio data are first precompressed by video or audio compression algorithms and then saved in storage devices. For live streaming cases, the media are real-time compressed and sent out while at the same time being saved in storage devices. Upon the client's request, a *streaming server* retrieves the compressed video and audio data from storage devices and then the *application-layer QoS control* module adapts the video or audio bitstreams according to the network status and QoS requirements. After this adaptation the transport *protocols* packetize the compressed bitstreams and send the video or audio packets to the Internet. Packets may be dropped or experience excessive delay inside the Internet owing to congestion or poor channel quality. To improve the quality of video and audio transmission, *continuous media distribution services* are deployed in the Internet. Packets that have been successfully delivered to the receiver have first passed through the transport layers and then been processed by the application layer before being decoded by the video or audio decoder. To achieve synchronization between the video and audio presentations, *media synchronization mechanisms* are required. From Figure 8.3, it can be seen that the six components are closely related and are coherent constituents of the video streaming architecture.

Video and audio compression Raw video and audio must be compressed before transmission to achieve efficiency. Video and audio compression schemes can be either scalable or non-scalable. Owing to the lower bandwidth demand of audio compression, non-scalable audio coding is the most widely used type. However, scalable video coding combined with receiver-driven layered multicast (see Section 7.5) is gaining more and more attention owing to its graceful degradation ability in coping with bandwidth fluctuations in the Internet.

Application-layer QoS control To cope with varying network conditions and the varying presentation quality requested by users, various application-layer QoS control techniques have been proposed. These application-layer QoS techniques include congestion control and error control. Congestion control is employed to prevent packet loss and reduce delay. Error control, however, is employed to improve video presentation quality in the presence of packet loss.

Continuous media distribution services In order to provide high-quality multimedia presentations, adequate network support to reduce transport delay and packet loss is crucial. Built on top of the Internet (IP protocol), continuous media distribution services provide several mechanisms for improving QoS and efficiency in streaming video and audio over the best-effort Internet.

Streaming servers Streaming servers play a key role in providing streaming services. To offer high-quality streaming services, streaming servers are required to process multimedia data under timing constraints and to support interactive control operations such as pause or resume, fast forward, and fast backward. Furthermore, streaming servers need to retrieve media components in a synchronous fashion.

Media synchronization mechanisms Media synchronization is a major feature that distinguishes multimedia applications from other traditional data applications. With media synchronization mechanisms, the application at the receiver side can present various media streams in the same way as they were originally captured. An example of media synchronization is that the movements of a speaker's lips match the played-out audio.

Protocols for streaming media Protocols are designed and standardized for communication between clients and streaming servers. Protocols for streaming media provide such services as network addressing, transport, and session control.

8.1.1   Audio and video compression

Owing to the lower bandwidth demands of audio compression, non-scalable audio coding is the most widely used coding in streaming systems. Nevertheless, scalable and non-scalable video coding are both available for streaming systems. For video streaming, congestion control tries to adapt the sending rate to the available bandwidth in the network; therefore scalable video is more adaptable to the varying available bandwidth in the network in comparison with non-scalable video coding. Most recent efforts on video compression for streaming video have been focused on scalable video coding. The primary objectives of on-going research on scalable video coding are to achieve high compression efficiency, high flexibility (bandwidth scalability), and/or low complexity. Moreover, streaming audio and/ or video (A/V) requires bounded end-to-end delay so that packets can arrive at the receiver in time to be decoded and displayed. If an A/V packet does not arrive in time for playout, it is useless and can be regarded as lost. Since the Internet introduces time-varying delays, to provide continuous playout a de-jitter buffer at the receiver is commonly used before the decoding process can start. The de-jitter buffer transforms the variable delay into a fixed delay. It holds the first sample received for a period of time before it plays it out. This holding period is known as the initial play-out delay [9].

It is essential to handle the de-jitter buffer properly. If A/V packets are held for too short a time, variations in delay can potentially cause the buffer to under-run and cause gaps in the A/V playback. If the sample is held for too long a time, the buffer can overrun and the dropped packets again cause gaps in the playback. Lastly, if packets are held for too long a time then the overall delay on the connection can rise to an unacceptable level. The optimal initial play-out delay for the de-jitter buffer should be approximately equal to the total variable delay along the end-to-end connection.

Packet loss is inevitable in the Internet and can damage perceptual quality; thus it is desirable that an A/V stream be robust to packet loss. Error-resilient techniques in the encoding process or error concealment techniques in the decoding process can be effectively applied to combat these packet losses. There is a growing use of handheld devices, such as smart phones and personal digital assistants (PDAs), which require only a low power consumption. Therefore, streaming A/V applications running on these devices must use special mechanisms in the design of A/V (mainly video) encoder and decoder.

8.1.2   Application-layer QoS control

The objective of application-layer QoS control is to avoid congestion and maximize video quality in the presence of packet loss. The application-layer QoS control techniques include congestion control and error control. These techniques are employed by the end systems and do not require any QoS support from the network. Most recent studies on source-based rate control of A/V communication have been focused on TCP-friendly adaptation [10]. A number of TCP-friendly adaptation schemes have been proposed, and have been demonstrated as achieving a certain degree of fairness among competing connections, including TCP connections. However, strictly TCP-like rate control may result in sharp reductions in the transmission rate, owing to its additive-increase multiplicative-decrease (AIMD) congestion control nature, and possibly unpleasant perceptual quality. Therefore, for TCP-like rate control, further investigation is needed on how to trade off responsiveness in detecting and reacting to congestion with smooth fluctuations in perceptual quality.

Error-control mechanisms include forward error correction (FEC), retransmission, error-resilient encoding, and error concealment. There are three kinds of FEC: channel coding, source-coding-based FEC, and joint source and channel coding. The advantage of all FEC schemes over retransmission-based schemes is the reduction in A/V transmission latency. Source-coding-based FEC can achieve lower delays than channel coding, while joint source and channel coding could achieve optimal performance in the rate–distortion sense. The disadvantages of all FEC schemes include an increase in the transmission rate and inflexibility to varying loss characteristics.

Unlike FEC, which adds redundancy regardless of correct receipt or loss, a retransmission-based scheme only resends the packets that are lost. Thus, a retransmission-based scheme is adaptive to varying loss characteristics, resulting in efficient use of network resources. The limitation of delay-constrained retransmission-based schemes is that their effectiveness diminishes when the round-trip time is too large. Currently, an important direction is to pursue combining FEC with retransmission [11]. In addition FEC can be used in layered video multicast, so that each client can individually trade off latency for quality on the basis of its requirements. Examples of such an FEC-protected multicast include hierarchical FEC and receiver-driven layered multicast [12]. Multiple description coding (MDC) is another mechanism for error-resilient encoding. The advantage of MDC is its robustness to loss. The cost of MDC is a reduction in compression efficiency. Current research efforts are geared toward finding a satisfactory tradeoff between compression efficiency and the reconstruction quality available from one description. Error concealment is performed by the receiver (when packet loss occurs) and can be used in conjunction with any other techniques (e.g., congestion control and other error control mechanisms).

8.1.2.1 TCP-friendly congestion control

Bursty loss and excessive delay have a devastating effect on multimedia presentation quality, and they are usually caused by network congestion. Thus, congestion-control mechanisms at end systems are necessary to helping reducing packet loss and delay. Because A/V data are delay-sensitive, most multimedia applications use rate-based congestion control algorithms that work by adjusting the sending rate. Reliable TCP transmission is considered inappropriate for these applications. Various rate-based protocols [13] [14] [15] have been proposed to enable these applications to adapt to the network dynamics. Many protocols use AIMD algorithms to achieve TCP-friendliness. Packet losses and round-trip delays are used to infer network congestion in these protocols. Audio and video data have two properties that affect the design of a congestion control in regard to its transmission. First, A/V data are loss tolerant (if the loss is low), since the receiver can use error-concealment techniques to hide the loss. Moreover, the source can add some redundancy to the data stream to make it more robust to loss. Second, the stability of the audio and video quality is important to the audience's overall subjective perception. Frequent quality changes within a short period can be very annoying. These two properties are not applicable for general TCP bulk data delivery. Designing a congestion control for multimedia applications should take them into consideration explicitly. The existing rate-based congestion control algorithms keep increasing the sending rate, when there is no congestion, until packet loss occurs; then the sending rate is decreased accordingly. This kind of oscillation is harmful to A/V perception and it is better avoided. Also these algorithms rely on the round-trip time (RTT) to collect the receiver's feedback information or use it directly in the sending rate adaptation as in equation-based protocols [15]. This constitutes unfairness among receivers with different RTTs: receivers with longer RTTs are allocated a lower bandwidth than receivers with shorter RTTs. This unfairness applies to all AIMD protocols including TCP. In rate-based applications, RTT should not play a dominant role in the bandwidth allocation, and all competing transmissions should grab an equal share of bandwidth regardless of their RTTs.

There are two issues regarding rate-based congestion control protocols. One is stability and the other is fairness. Most existing rate-based protocols (except PLM [16]) use the principle of "increasing rate when no congestion, decreasing rate under congestion" to adjust the sending rate (for layered multicast, joining or leaving a layer corresponds to increasing or decreasing the rate). Without knowledge of the available bandwidth this rule results in rate oscillations and periodical packet loss even under smooth available bandwidth. Many protocols have tried out mechanisms to avoid or relieve oscillation, for example, equation-based rate adjustment [15], thin layers [17], and dynamic join timer [18]. The rate oscillations should be greatly reduced if the path's available bandwidth can be accurately estimated. As shown in [16], PLM uses the packet-pair technique [19] to estimate the available bandwidth in a fair-queuing network, the stability performance does indeed improve significantly.

Many existing rate-based protocols try to be TCP-friendly, i.e., they do not grab more bandwidth than a TCP session under the same network environment. Some protocols, such as TCP-friendly rate control (TFRC) [15] and smooth multirate multicast congestion control (SMCC) [20], use the explicit TCP throughput equation for rate adjustment. To be TCP-friendly, a lot of effort (especially in the multicast case) is needed to acquire the RTT information, which is an important factor determining a TCP's throughput. However, TCP is known to be RTT-biased, i.e., a TCP session with a larger RTT achieves a lower throughput when competing in a bottleneck link with another TCP session having a smaller RTT. Note that the rate should be the dominant factor in determining the bandwidth share for rate-based protocols. In other words, rate-based protocols should not be RTT-biased. The bandwidth inference congestion (BIC) control protocol for layered multicast [21] [22], discussed in Section 7.5.4, uses a delay-trend detection technique to estimate the path's available bandwidth. When a BIC receiver judges that the path's available bandwidth cannot sustain another layer, it holds to its current layer without joining new layers. When the available bandwidth is smooth, a BIC session has no layer oscillations, as in PLM [16] (BIC differs from PLM in that BIC does not require fair queuing). The BIC protocol does not use RTT in its control loop for layer join or leave decisions. As a result, BIC is very simple, in that the sender and all receivers work independently without interaction. This simplicity helps BIC to scale to a large number of receivers and makes it easy for deployment. Moreover, BIC is not RTT-biased, which implies that BIC is not TCP-friendly for some RTTs, i.e., BIC may achieve a different throughput from TCP under some network conditions.

8.1.2.2 Congestion control by rate shaping (transcoding)

The objective of rate shaping is to match the rate of a precompressed video bitstream to the target rate constraint. A rate shaper, which performs rate shaping (or transcoding [23]), is required for source-based rate control. The reason is that the stored video may be pre-compressed at a certain rate which may not match the available bandwidth in the network. A straightforward realization of a transcoder is achieved by cascading a decoder and an encoder, i.e., the decoder decodes the compressed input video and the encoder re-encodes the decoded video into the target format. This is computationally very expensive. Therefore, reducing the complexity of a straightforward decoder–encoder implementation is a major driving force behind many research activities on transcoding. What makes transcoding different from video encoding is that the transcoding has access to many coding parameters and statistics are readily available from the input compressed video stream. The challenge of transcoding is then how intelligently to utilize the coding statistics and parameters extracted from the input to achieve the best possible video quality and the lowest possible computational complexity.

Figure 8.4 shows open-loop and closed-loop transcoding systems for rate conversion [23]. In the open-loop transcoding shown in Figure 8.4(a), the bitstream is variable-length decoded (VLD) to extract the variable-length codewords corresponding to the quantized DCT coefficients, as well as macroblock (MB) data corresponding to the motion vectors and other MB-level information. In this scheme, the quantized coefficients are inverse quantized and then simply requantized to satisfy the new output bitrate. Finally, the requantized coefficients and stored MB-level information are variable-length coded (VLC).

Open-loop systems are relatively simple since a frame memory is not required and there is no need for an inverse IDCT. However, open-loop architectures are subject to drift owing to the loss of high-frequency information in the requantization process (Q_2), i.e., high-frequency information is discarded by the transcoder to meet the new target bitrate. When a decoder receives the transcoded bitstream of a reference frame (an I- or a P-frame), it will decode with reduced quality and store it in its memory. When it is time to decode the next P-frame, the degraded reference frame is used as a predictive component and added to a degraded residual component for motion compensation (MC). Considering that the purpose of the residual is to represent accurately the difference between the original signal and the motion-compensated prediction and that now both the residual and predictive components

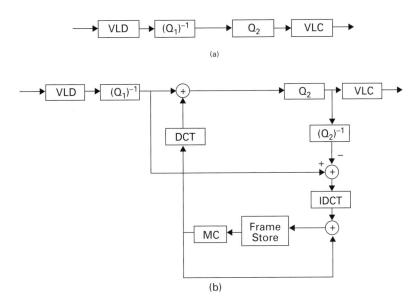

(a)

(b)

Figure 8.4 Simplified transcoding architectures for bitrate reduction: (a) open-loop, partial, decoding to DCT coefficients then requantizing (b) closed-loop drift compensation for the requantized data [23] (© IEEE 2003).

are different from what was originally derived by the encoder, errors will be introduced into the reconstructed frame. Such errors are a result of the mismatch between the predictive and residual components. As time goes on this mismatch progressively increases, with the result that the reconstructed frames become severely degraded, the so-called error drift problem.

A straightforward way of avoiding the drift problem is to use a cascaded decoder–encoder architecture, which decodes the video bitstream and fully re-encodes the reconstructed signal at the new rate. The best performance can thus be achieved by calculating new motion vectors and mode decisions for every MB at the new rate. However, a significant complexity saving can be achieved, while still maintaining acceptable quality, by reusing information contained in the original incoming bitstreams and also considering simplified architectures, e.g., performing MC in the DCT domain. Such a closed-loop transcoding architecture is shown in Figure 8.4(b); this is aimed to eliminate the mismatch between the predictive and residual components by approximating the cascaded decoder–encoder architecture [24]. The main difference in structure between the cascaded pixel-domain architecture and the architecture shown in Figure 8.4(b) is that reconstruction in the cascaded pixel-domain architecture is performed in the spatial domain, thereby requiring two reconstruction loops with one DCT and two IDCTs, whereas in the simplified structure shown in Figure 8.4(b) only one reconstruction loop is required, with one DCT and one IDCT. In this structure, some arithmetic inaccuracy is introduced owing to the non-linear way in which the reconstruction loops are combined. However, it has been found that this approximation has little effect on the quality [24]. With the exception of this slight inaccuracy, this architecture is mathematically equivalent to a cascaded decoder–encoder approach. Overall, though, in comparison with the open-loop architectures discussed earlier, drift is eliminated since the mismatch between the predictive and residual components is compensated for.

The closed-loop architecture described in Figure 8.4(b) provides an effective transcoding structure in which the MB reconstruction is performed in the DCT domain. However, since the MC requires spatial-domain pixels to be stored in the memory (frame store), the compressed-domain methods for MC proposed in [25] can be used. In this way, it is possible to reconstruct reference frames without decoding to the spatial domain; several architectures describing this reconstruction process in the compressed domain have been proposed [26]. It was found that decoding completely in the compressed domain could yield equivalent quality to spatial-domain decoding. However, this was achieved with floating-point matrix multiplication and proved to be quite costly. This computation can be simplified [23] by approximating the floating-point elements by power-of-2 fractions, so that shift operations can be used, or by matrix decomposition techniques.

Rate shaping can be accomplished by a frame-dropping filter, which distinguishes the frame types (e.g., I-, P-, and B-frame in MPEG) and drops frames according to their importance. For example, in MPEG the dropping order would be first B-frames, then P-frames, and finally I-frames. When reference frames such as I-frames and P-frames are dropped, the motion vectors pointing to the dropped reference frames need to be recalculated. The frame-dropping filter has been used with effect to reduce the data rate of a video stream by discarding a number of frames and transmitting the remaining frames at a lower rate. The frame-dropping filter can be used at the source [27] or used in the network.

8.1.2.3 Error control

Error control mechanisms include forward error correction (FEC), retransmission, error-resilient encoding, multiple descriptive coding (MDC), and error concealment.

Forward error correction The principle of FEC is to add redundant information so that the original message can be reconstructed in the presence of packet loss. For Internet applications, block-based FEC is the most widely used technique in terms of block codes since it does not need any modification of the compression algorithms, which are already standardized. Specifically, a media stream is first chopped into segments, each of which is packetized into packets; then, for each segment of K packets, a block code (e.g., the Tornado code [28] or a block erasure code [29] [30]) is applied to the K packets to generate an N-packet block. To recover a segment perfectly, a user needs only to receive any set of K packets in the N-packet block. For more details about block erasure codes, please refer to Section 10.1.2.

Delay-constrained retransmission Retransmission is not widely accepted as a method for recovering lost packets in real-time video since a retransmitted packet may miss its play-out time. However, if the one-way trip time is short with respect to the maximum allowable delay, a retransmission-based approach (called delay-constrained retransmission [31]) is a viable option for error control. The idea is based on effectively deriving the retransmission timeout (RTO) according to the statistical estimation of round-trip time (RTT), including error tolerance in estimating RTT, the sender's response time, and the receiver's decoding delay.

Error-resilient encoding The objective of error-resilient encoding is to enhance the robustness of compressed video to packet loss. The standard error-resilient encoding schemes include resynchronization marking, slice-based data partitioning, and data recovery [32].

When transporting video bitstreams over packet-based networks, it is necessary to fragment the bitstream before it is converted into a packet payload. When these compressed video packets are transmitted over noisy communication channels, errors are introduced into the bitstream due to packet loss. A video decoder that is decoding this corrupted bitstream will lose synchronization with the encoder (because it is unable to identify the precise location in the image where the current data belongs). The H.263 + protocol improves support for this fragmentation operation through the use of resynchronization markers [33] in the bitstream at various locations. When the decoder detects an error it can then hunt for this resynchronization marker and regain resynchronization.

Another error-resilient technique uses slice-based data partitioning. More specifically, as discussed in Chapter 5, in many video codecs (e.g., H.264/AVC), the macroblocks of a picture are organized into *slices*, which represent regions of the picture that can be decoded independently. Each slice is a sequence of macroblocks and is processed in the order of a raster scan, i.e., a scan from top left to bottom right; each slice is self-contained and basically can be decoded without the use of data from other slices of the picture. Slices can be used effectively for error resilience, as the partitioning of the picture allows spatial concealment within the picture and as the start of each slice provides a resynchronization point at which the decoding process can be re-initialized. Since some syntax elements in the H.264 bitstream are more important than others, data partitioning enables unequal error protection, according to the importance of the syntax elements. For example, the header information (e.g., picture size and quantizer) controls the whole slice, a picture, or even a whole sequence, whereas a loss of transform coefficients impairs only the block to which they belong or the rest of the slice, owing to error propagation. Data partitioning (DP) of H.264/AVC allows the partitioning of a normal slice into up to three parts (i.e., data partitions A, B, and C). Each part is then encapsulated into a separate NAL packet [34].

Data partition A contains the header information such as MB types, quantization parameters, and motion vectors, which are more important than the remaining slice data. With the loss of data in DP A, the data in the other two partitions becomes useless. Data partition B contains intra coded block patterns (CBPs) and transform coefficients of I-blocks. Because the intra frames and intra MBs are used as references, the loss of this part will severely impair the recovery of successive frames, owing to error propagation. Data partition C contains inter CBPs and coefficients of P-blocks. Compared with the data in partitions A and B, the data contained in DP C is less important. However, it is the biggest partition of a coded slice since a large number of frames are coded as P-frames. Both data partitions B and C depend on A, but B and C are independent of each other in terms of syntax. From this, we can safely draw the conclusion that A is the most important of the three data partitions. Thus, in practice, because of the small amount of data in A, it can be transmitted through out-of-band channel or protected with a higher level of FEC.

One problem with transmitting compressed video over error-prone channels is the use of variable-length codes (VLCs). During the decoding process, if the decoder detects an error while decoding VLC data then it loses synchronization and hence typically has to discard all the data up to the next resynchronization point. The reversible VLCs (RVLCs) adopted in MPEG-4 alleviate this problem and enable the decoder to isolate the error location better, by enabling data recovery in the presence of errors [35]. Reversible VLCs are special VLCs that have the prefix property when decoded in both the forward and reverse directions. Hence, they can be uniquely decoded in both the forward and reverse directions. The advantage of these codewords is that when the decoder detects an error while decoding the bitstream in the forward direction, it jumps to the next resynchronization marker and decodes the bitstream in the backward direction until it encounters another error. From the locations of these two errors the decoder can recover some data that would otherwise have been discarded. In general, the RVLC property reduces the coding efficiency of VLC tables. An RVLC, however, must satisfy the suffix condition for instantaneous backward decoding as well as the prefix condition for instantaneous forward decoding. The suffix condition is that no codeword should coincide with the suffix of any longer codeword, while the prefix condition expresses that there should be no coincidence with the prefixes of any longer codeword. The suffix condition is sufficient for instantaneous decoding in the backward direction just as the prefix condition is for the forward direction.

Multiple description coding   Multiple description coding (MDC) [36] is used for robust Internet video transmission, where a raw video sequence is compressed into multiple streams (referred to as descriptions), in which each description provides acceptable visual quality and combined descriptions provide a better visual quality in proportion. With the use of MDC, the streaming system can be made more robust to loss since even if a receiver receives only one description (all other descriptions being lost), it can still reconstruct video with acceptable quality. If a receiver receives multiple descriptions, it can combine them together to produce a better reconstruction than that produced from any one of them.

However, the advantages come with a cost. To make each description provide acceptable visual quality, each must carry sufficient information about the original video. This will reduce the compression efficiency in comparison with conventional single description coding (SDC). In addition, although more combined descriptions provide better visual quality, a certain degree of correlation between the multiple descriptions has to be embedded in each description, resulting in further reduction of the compression efficiency. Further investigation is needed to find a satisfactory tradeoff between compression efficiency and the reconstruction quality from a single description.

Error concealment Error-resilient encoding is executed by the encoding side to enhance the robustness of compressed video before packet loss actually happens, a so-called *preventive* approach. Error concealment, however, is performed by the decoding side when packet loss has already occurred; this is also called a *reactive approach*. Specifically, error concealment is employed by the receiver to conceal the loss of A/V data and to make the presentation less displeasing to human perception.

There are two basic approaches in error concealment for video, namely, spatial and temporal interpolation. In spatial interpolation, missing pixel values are reconstructed using neighboring spatial information. In temporal interpolation, the lost data is reconstructed from data in the previous frames. Typically, spatial interpolation is used to reconstruct the missing data in intra coded frames, while temporal interpolation is used to reconstruct the missing data in inter coded frames. In recent years, numerous error-concealment schemes have been proposed in the literature. Examples include maximally smooth recovery, projection onto convex sets, and various motion vector and coding mode recovery methods such as motion-compensated temporal prediction [37].

8.1.3   Continuous media distribution services

In order to provide high-quality multimedia presentations, adequate support from the network is critical to reduce transport delay and packet loss ratio. Streaming video and audio are classified as continuous media because they consist of a continuous sequence of media data (such as audio samples or video frames), which convey meaningful information only when presented in time. Continuous media distribution services are built on top of the best-effort Internet with the aim of achieving QoS and efficiency for media streaming. Some representative techniques include network filtering, application-level multicast, content replication, etc. A major topic of active research is how to build a scalable, efficient, cost-effective, and incremental deployable infrastructure for continuous media distribution.

8.1.3.1 Network filtering

As a congestion-control technique, network filtering aims to maximize media quality during network congestion. The network filter, which can be placed at the media server or in the network, can adapt the rate of media streams according to the network congestion status [38]. Commonly, the service provider places filters on the nodes that connect to network bottlenecks and multiple filters can be placed along the path from a server to a client. An example of filters in a network is shown in Figure 8.5, where the nodes labeled R denote routers that have no knowledge of the format of the media streams and may randomly discard packets. The "Filter" nodes receive the client's requests and adapt the stream sent by the server accordingly.

Typically, frame-dropping filters are used as network filters for video streaming. By the continuous measurement of the packet loss rate, the receiver can request a change in bandwidth of the video stream by asking the frame-dropping filters to increase or decrease the frame-dropping rate [38]. For example, if the packet loss rate is higher than a

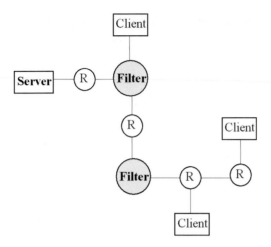

Figure 8.5 The nodes labeled R in a network filtering system denote routers that have no knowledge of the format of the media streams and may randomly discard packets. The "Filter" nodes receive the client's requests and adapt the stream sent by the server accordingly [8] (© IEEE 2003).

threshold, then the client will ask the filter to increase the frame-dropping rate, otherwise the receiver will ask the filter to reduce the frame-dropping rate. The frame-dropping filters inside the network are used to improve video quality and bandwidth efficiency. More specifically, when a video stream flows from an upstream link with larger available bandwidth to a downstream link with smaller available bandwidth, the use of a frame-dropping filter at the connection point (between the upstream link and the downstream link) can help improve the video quality by gracefully degrading the stream's quality instead of corrupting the flow through random packet loss. Since filtering can help to save network resources by discarding those frames that are less important, the bandwidth efficiency is also improved.

8.1.3.2 Application-level multicast

As discussed in Section 7.4.2, even though IP multicast is capable of providing efficient multipoint packet delivery, there are still many hurdles in the deployment of IP multicast, such as the lack of higher-layer functionalities (e.g., reliable transport, flow, congestion, and admission control, and multicast security) and scalable interdomain multicast routing protocols. To overcome these issues effectively, application-level multicast mechanisms were created, with the aim of building a multicast service on top of the Internet. This enables independent content service providers (CSPs), Internet service providers (ISPs), or enterprises to build their own Internet multicast networks and interconnect them into larger, worldwide, "media multicast networks" at the application level or at the streaming-media content layer. Media multicast networks can be built from an interconnection of multicast-capable nodes (called CDN servers [39]), which perform routing at the application layer. In addition, each CDN server is interconnected with one or more neighboring CDN servers through explicit configurations, which define the application-level overlay topology. Collectively, the CDN servers in a media multicast network employ a distributed application-level multicast routing algorithm to determine the optimal virtual paths for distributing content throughout the network.

When the underlying network fails or becomes overly congested, a media multicast network automatically and dynamically re-routes content via alternate paths according to application-level routing policies. In addition, CDN servers subscribe dynamically to multicast content when and only when a downstream client requests it. This capability replicates the IP multicast capability that ensures one and only one copy of the multicast content flows across any physical or virtual path independently of the number of down-stream clients; this saves network bandwidth. The advantage of application-level multicast is that it breaks through barriers, such as scalability, network management, and support for congestion control, which have prevented ISPs from establishing "IP multicast" peering arrangements.

8.1.3.3 Content replication

An important technique for improving the scalability of a media delivery system is content or media replication. Content replication takes two forms, namely, mirroring and caching, which are deployed by publishers, CSPs, and ISPs. Both mirroring and caching seek to place content closer to the clients and both share the advantages that they reduce bandwidth consumption on network links, reduce the load on streaming servers, reduce latency for clients, and increase availability.

In mirroring, the original multimedia files are placed on the main server while copies of the original multimedia files are placed on the duplicate servers scattered around the Internet. In this way, clients can retrieve multimedia data from the nearest duplicate server to achieve the best streaming performance (e.g., lowest latency). There are some disadvantages associated with mirroring, e.g., it is still rather expensive to establish dedicated mirrors, and the speed is slow. In addition, establishing a mirror on an existing server, while cheaper, is still an administratively complex process.

Caching is a more effective way of content replication; in this mechanism local copies are made of contents that the clients retrieve, assuming that different clients will request many of the same contents. When a client requests a video file, the local cache retrieves it from the origin server if this video file has not been requested before, storing a copy locally in the cache, and then passes it on to the requesting client. If other clients ask for video files which the cache has already stored, the cache will supply the local copy rather than going all the way to the origin server where the video file resides. In addition, cache sharing and cache hierarchies allow each cache to access files stored at other caches so that the load on the origin server can be reduced and network bottlenecks can be alleviated [40] [41]. To increase the cache hit-rate and reduce latency experienced by the clients, a smart network cache strategy based on "hints" for improved cache data scheduling was proposed in [42].

8.1.4 Streaming servers

To offer high-quality streaming services, streaming servers play a critical role. They are required to process multimedia data under strict timing constraints in order to provide synchronized A/V and data presentation as well as to prevent artifacts (e.g., jerkiness in video motion and pops in audio) during playback. In addition, streaming servers also need to support VCR-like control operations, such as stop, pause or resume, fast forward, and fast backward. A streaming server typically consists of the following three subsystems: (1) a *communicator* to handle the application-layer and transport protocols (see more details in

Chapter 7 and Section 8.1.6 below), implemented on the server; (2) a *real-time operating system* to satisfy real-time requirements for streaming applications; (3) a *storage system* to support continuous media storage and retrieval in the streaming services.

8.1.4.1 Real-time operating system

A real-time operating system (OS) is needed in a streaming service for efficient process management, resource management, and file management.

Process manager   To satisfy the timing requirements of continuous media, the *process manager* in the real-time OS is used to map each single process onto the CPU resource according to a specified scheduling policy. Most real-time scheduling techniques are variations of two basic algorithms for multimedia systems: earliest-deadline-first (EDF) [43] and rate-monotonic scheduling [44]. Each task in EDF scheduling is assigned a deadline and is processed in the order of increasing deadlines. However, each task in rate-monotonic scheduling is assigned a static priority according to its request rate and is processed in order of priority. Specifically, the task with the shortest period (or the highest rate) is given the highest priority, and the task with the longest period (or the lowest rate) gets the lowest priority. Both EDF and rate-monotonic scheduling are preemptive; that is, the schedulers can preempt the running task and schedule a new task for the processor on the basis of its deadline or priority. Execution of the interrupted task will resume at a later time. The difference between EDF and rate-monotonic scheduling is as follows. An EDF scheduler is based on a one-priority task queue and the processor runs the task with the earliest deadline. A rate-monotonic scheduler, however, is a static-priority scheduler with multiple-priority task queues. That is, the tasks in the lower-priority queue cannot be executed until all the tasks in the higher-priority queues are served.

Resource manager   The *resource manager* in a real-time OS is responsible for admission control and allocation of limited resources, which include CPUs, memories, and storage devices. Specifically, before admitting a new client, a multimedia server must perform an admission control test to decide whether a new connection can be admitted without violating performance guarantees already given to existing connections. If a connection is accepted, the resource manager allocates resources required to meet the QoS for the new connection. Admission control algorithms can be classified into two categories: deterministic admission control [45] and statistical admission control [46]. Deterministic admission control algorithms, which are simple, with strict assurance of quality, can provide hard guarantees to clients but with lower utilization of server resources. However, statistical admission control algorithms provide statistical guarantees to clients, i.e., the continuity requirements of at least a fixed percentage of media units are guaranteed to be met. Therefore statistical admission control improves the utilization of server resources by exploiting human perceptual tolerances as well as the differences between the average and the worst-case performance characteristics of a multimedia server. As for admission control algorithms, resource allocation schemes can be categorized as either deterministic or statistical. On the one hand, deterministic resource allocation schemes make reservations for the worst case, e.g., they reserve bandwidth for the longest processing time and the highest rate that a task might ever need. On the other hand, statistical resource allocation schemes achieve higher utilization by allowing temporary overloading and a small percentage of QoS violations.

File manager The *file manager* provides protocol access and control functions for file storage and retrieval [47]. To support continuous media in file systems, it is common to organize A/V files on distributed storage such as disk arrays. The disk throughput can be improved by scattering or striping each A/V file across several disks, and disk seek-times can be reduced by disk-scheduling algorithms. To design high throughput, large capacity, and fault-tolerant storage systems to accommodate the specifications imposed by the file manager, several storage system architectures have also been proposed. They are discussed in the following section.

8.1.4.2 Storage system for streaming systems

Data stripping In order to support a higher number of connection clients in a streaming system, the most widely used mechanism to increase throughput is based on the data-stripping technique, in which a multimedia file is scattered across multiple disks and the disk array can be accessed in parallel [48]. As shown in Figure 8.6, blocks 1, 2, and 3 of file A can be read in parallel, resulting in increased throughput. Load balancing is an important issue in the design of a data-stripping scheme so that the most heavily loaded disks can avoid overload situations while keeping latency small. It is a common practice to use tertiary storage (e.g., an automated tape library or a CD-ROM jukebox) to keep the storage cost down.

Hierarchical storage architecture To reduce the overall cost, an hierarchical storage architecture is typically used. Under such a storage architecture, a small portion only of frequently requested video files are kept on disks for quick access; the remainder resides in the automated tertiary tape library. To deploy large-scale streaming services, a storage area network (SAN) architecture [49] [50] can be used to provide high-speed data pipes between storage devices and hosts at far greater distances than those in a conventional host-attached small-computer-systems interface (SCSI). With these high-speed connections, an SAN is able to provide a many-to-many relationship between heterogeneous storage devices (e.g., disk arrays, tape libraries, and optical storage arrays), and multiple servers and storage

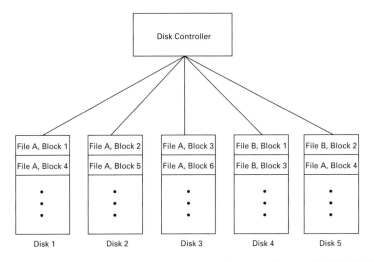

Figure 8.6 Data stripped in multiple disks and accessed in parallel [8] (© IEEE 2003).

clients. Another approach to deploying large-scale storage is network attached storage (NAS) [51], which can be attached to a local area network (LAN) or a wide area network (WAN) directly using popular protocols such as the hypertext transfer protocol (HTTP), the network file system (NFS), TCP, UDP, or IP. By separating the data from the application server, both NAS and SAN mechanisms can achieve the following advantages over the traditional storage: (1) the simplification of storage management by centralizing storage; (2) scalability; and (3) fault tolerance.

Fault-tolerant storage   In order to ensure uninterrupted service even in the presence of disk failures, a server must be equipped with fault tolerant abilities. There are two common fault-tolerant techniques: parity encoding and file mirroring [52]. Parity encoding data adds a small overhead on storage but requires the synchronization of data bits receiving additional processing time to decode lost information. In contrast, the mirroring of A/V files does not require the synchronization of reads or additional processing time to decode lost information, and this significantly simplifies the design and implementation of video servers. However, file mirroring incurs at least twice as much storage volume as a non-fault-tolerant technique. As a result, there is a tradeoff between reliability and complexity (i.e., cost). It was shown that for the same degree of reliability, mirroring-based schemes always outperform parity-based schemes in terms of per-stream cost as well as restart latency after disk failure [52].

8.1.5   Media synchronization

A major feature that distinguishes multimedia streaming from other traditional data transport applications is media synchronization, which refers to the maintaining of temporal relationships within one data stream and between various media streams. There are three levels of synchronization, corresponding to three semantic layers of multimedia data, i.e., intramedia (e.g., between audio and video streams), intermedia, and interobject (including texts and images) synchronization [53].

8.1.5.1   Timestamping

With the best-effort IP networks, media streams may lose synchronization after transportation from the server to the client since the end-to-end connection routes inevitably introduce delays and delay variations. The incurred delays and delay variations could disrupt intramedia, intermedia, and interobject synchronization. Therefore, media synchronization mechanisms are required to ensure the proper rendering of the multimedia presentation at the client. An essential part of any media synchronization mechanism is specification of the temporal relations within a medium and between media. A widely used specification method for continuous media is timestamping, i.e., at the source, a stream is timestamped with respect to other streams to keep the temporal information within the stream and; at the destination, the application presents the streams according to their temporal relation. Timestamping can be done either automatically or manually. In the case of A/V recording and playback, the timestamping is done automatically by the recording device. In the case of presentations that are composed of independently captured or otherwise created media, timestamping has to be done manually.

8.1.5.2 Synchronization mechanisms

In the absence of real-time synchronization support from the current IP network, most synchronization mechanisms are implemented on the application layer of the end systems. These synchronization mechanisms can be either preventive or corrective [54].

Preventive mechanisms These are designed to minimize synchronization errors, such as transmission latency and jitter, as data is transported from the server to the end users. Preventive mechanisms require a coordinated effort between several schemes, e.g., disk-reading scheduling algorithms, network transport protocols, operating systems, and synchronization schedulers. Disk-reading scheduling helps to organize and coordinate the retrieval of data from the storage devices. Network transport protocols provide a means for maintaining synchronization during data transmission over the Internet. Operating systems achieve the precise control of timing constraints by using EDF or rate monotonic scheduling. A synchronization scheduler can specify an end-to-end schedule for the delivery of the media streams to the client by the servers (the delivery schedule) and the presentation of these media streams to the user by the client application (the presentation schedule). This scheduler can be centralized (i.e., entirely located at the client) or distributed (the delivery scheduling functionalities are shared between the server and the client).

Corrective mechanisms These are designed to recover synchronization in the presence of synchronization errors, which are unavoidable due to the random delays and jitters introduced by the Internet. Corrective mechanisms are thus needed at the receiver to compensate the timing discrepancies when synchronization errors occur. An example of a corrective mechanism is the stream synchronization protocol (SSP) [55], where an "intentional delay" is used by the various streams in order to adjust their presentation times to recover from network delay variations. More specifically, at the client side a client-end monitor compares the real arrival times of data with those predicted by the presentation schedule and notifies the scheduler of any discrepancies. These discrepancies are then compensated by the scheduler, which delays the display of data that are ahead of other data, allowing the late data to catch up.

8.1.6 Protocols for multimedia streaming

Several standardized protocols are available for communication between a streaming server and clients. According to their functionalities, the protocols directly related to Internet streaming video can be classified into the following three categories.

(1) *Network-layer protocols* provide basic network service support, such as network addressing. The IP serves as the network-layer protocol for Internet video streaming.
(2) *Transport protocols* provide end-to-end network transport functions for streaming applications. Transport protocols include the user data protocol (UDP), the transport control protocol (TCP), the real-time transport protocol (RTP), and the real-time control protocol (RTCP); the UDP and the TCP are lower-layer transport protocols while the RTP and the RTCP [56] are upper-layer transport protocols, which are implemented on top of UDP or TCP.

(3) *Session control protocols* define the messages and procedures in the application layer needed to control the delivery of the multimedia data during an established session. The real-time streaming protocol (RTSP) [57] and the session initiation protocol (SIP) [58] are examples of such protocols.

These three types of protocol interact with one another as follows. At the sending side, the compressed A/V data is retrieved and packetized at the RTP transport layer. The RTP-packetized streams provide timing and synchronization information, as well as sequence numbers. The RTP-packetized streams are then passed to the UDP or TCP transport layer and the IP network layer. The resulting IP packets are delivered over the Internet. At the receiver side, the media streams are processed in the reverse manner before their playback presentations. For control purposes, RTCP packets and RTSP packets are multiplexed at the UDP or TCP layer and the application layer to facilitate many functionalities of packet transmission over the Internet.

8.1.6.1 Transport protocols

The transport protocol family for media streaming includes the UDP, TCP, RTP, and RTCP. The UDP and TCP provide basic transport functions while the RTP and RTCP run on top of UDP or TCP. The UDP and TCP support such functions as multiplexing, error control, congestion control, or flow control. These functions can be briefly described as follows. First, UDP and TCP can multiplex data streams from different applications running on the same machine with the same IP address. Second, for the purpose of error control, TCP and most UDP implementations employ a checksum to detect bit errors. If a single bit error or multiple bit errors are detected in the incoming packet, the TCP or UDP layer discards the packet so that the upper layer (e.g., RTP) will not receive the corrupted packet. However, unlike the UDP, the TCP uses retransmission to recover lost packets. Therefore, TCP provides reliable transmission while UDP does not. Third, TCP employs congestion control to avoid sending too much traffic, which may cause network congestion. This is another feature that distinguishes TCP from UDP. Lastly, TCP employs flow control to prevent the receiver buffer from overflowing whereas UDP does not have any flow control mechanism. Since TCP retransmission introduces delays that are not acceptable for streaming applications with stringent delay requirements, UDP is typically employed as the transport protocol for video streams. In addition, since UDP does not guarantee packet delivery, the receiver needs to rely on the upper layer (i.e., RTP) to detect packet loss.

The Internet standard protocol RTP, defined in IETF RFC 1889, was designed to provide end-to-end transport functions to support real-time applications. More specifically, RTP is used to indicate how encoded media data are to be packetized over UDP. Real data transport (RDT) is a proprietary protocol from RealNetworks and is used in place of RTP, allowing the client to issue retransmission requests. The companion protocol to RTP, RTCP is designed to provide QoS feedback to the participants of an RTP session. In other words, RTP is a data transfer protocol while RTCP is a control protocol. The RTP does not guarantee QoS or reliable delivery but, rather, provides the following information in support of media streaming.

(1) *Timestamping* The RTP provides timestamping to synchronize different media streams. Note that RTP itself is not responsible for synchronization, which is left to the applications.

(2) *Sequence numbering*   Since packets arriving at the receiver may be out of sequence, RTP employs sequence numbering to place the incoming RTP packets in the correct order. The sequence number is also used for packet loss detection.

(3) *Payload type identification*   The type of payload contained in an RTP packet is indicated by an RTP-header field called the payload type identifier. The receiver interprets the content of the packet from the payload type identifier. Certain common payload types, such as MPEG audio and video, have been assigned payload type numbers. For other payloads, this assignment can be made using session control protocols.

(4) *Source identification*   The source of each RTP packet is identified by an RTP-header field called the synchronization source identifier (SSRC), which provides a means for the receiver to distinguish different sources.

The real-time control protocol, as defined in IETF 3550, is the control protocol designed to work in conjunction with the RTP. In an RTP session, participants periodically send RTCP packets to convey feedback on the quality of data delivery and information about membership. Basically, RTCP provides the following services.

(1) *QoS feedback*   This is the primary function of RTCP. It provides feedback to an application regarding the quality of the data distribution. The feedback is in the form of sender reports (sent by the source) and receiver reports (sent by the receiver). The reports can contain information on the quality of reception such as: (a) the fraction of RTP packets lost since the last report; (b) the cumulative number of lost packets, since the beginning of reception; (c) the packet inter-arrival jitter; and (d) the delay since the last sender's report was received. The control information is useful to the senders, the receivers, and third-party monitors. Using the feedback, the sender can adjust its transmission rate; the receivers can determine whether congestion is local, regional, or global, and the network managers can evaluate the network performance for media distribution.

(2) *Participant identification*   A source can be identified by the SSRC field in the RTP header. Unfortunately, the SSRC identifier is not convenient for human users. To remedy this problem, the RTCP provides a human-friendly mechanism for source identification. Specifically, RTCP SDES (source description) packets contain textual information called canonical names as globally unique identifiers of the session participants. They may include a user's name, telephone number, email address, and other information.

(3) *Control packets scaling*   To scale the RTCP control packet transmission with the number of participants, a control mechanism is designed as follows. The control mechanism keeps the total control packets to 5% of the total session bandwidth. Among the control packets, 25% are allocated to the sender reports and 75% to the receiver reports. To prevent control packet starvation, at least one control packet is sent within five seconds to the sender or receiver.

(4) *Intermedia synchronization*   The RTCP sender reports contain an indication of real-time and the corresponding RTP timestamp. This can be used in intermedia synchronization, such as lip synchronization in video.

(5) *Minimal session control information*   This optional functionality can be used for transporting session information such as the names of the participants.

8.1.6.2 Real-time streaming protocol (RTSP)

The real-time streaming protocol (RTSP), as defined in IETF RFC 2326, is an application-layer session control protocol by which users can request streaming media over the Internet [57]. This protocol uses a request–response clear-text syntax, somewhat similar to SMTP and HTTP, but servers do maintain state information about each client connection. A main function of RTSP is to support VCR-like control operations such as stop, pause or resume, fast forward, and fast backward. In addition, RTSP also provides a means for choosing delivery channels (e.g., UDP, multicast UDP, or TCP), and delivery mechanisms based upon RTP. It works for multicast as well as unicast. Another main function of RTSP is to establish and control streams of continuous audio and video media between the media servers and the clients. Specifically, RTSP provides the following operations.

(1) *Media retrieval* The client can request a presentation description and ask the server to set up a session to send the requested media data
(2) *Adding media to an existing session* The server or the client can notify each other when any additional media become available to the established session.

In RTSP, each presentation and media stream is identified by an RTSP uniform resource identifier (URI, IETF RFC 3986). The URIs used in RTSP are generally uniform resource locators (URLs) as they give a location for the resource. As URLs are a subset of URIs, they are commonly referred to as URIs to cover cases when an RTSP URI would not be a URL. The overall presentation and the properties of the media are defined in a presentation description file, which may include the encoding, language, RTSP URLs, destination address, port, and other parameters. The presentation description file can be obtained by the client using HTTP, email, or other means.

Different kinds of RTSP request are named *methods*; some popular methods are as follows.

- DESCRIBE is sent by clients asking for the composition of an RTSP URI. The server responds by sending back the session description protocol (SDP), which describes the kind of RTP media to be used.
- ANNOUNCE carries the SDP metadata which describe an RTSP URI media composition, now available or changed.
- SETUP is sent by clients and creates session state information about a server, whose identifier is returned in the SETUP response. If the SDP returned in response to DESCRIBE contains media formats that the client cannot receive, the latter should refrain from proceeding to the SETUP phase.
- PLAY is sent by clients, and starts transmission of the media associated with the URI referenced in the SETUP phase;
- TEARDOWN is sent by clients, and destroys session state information at the server.

Microsoft Media Server (MMS) is a proprietary protocol that provides an alternative to RTSP for the transfer of unicast data. It can be transported via UDP or TCP, with default port UDP or TCP 1755. Microsoft depreciated MMS in favor of RTSP in 2003 with the release of the Windows Media Services 9 Series, but continued to support the MMS for some time in the interests of backwards compatibility. Support for the protocol was finally dropped in Windows Media Player 11 and Windows Media Services 2008.

8.1.6.3 Session initiation protocol (SIP)

The session initiation protocol (SIP) [58], as defined in IETF RFC 2543, is another application-layer session control protocol mainly used for multipoint interactive streaming (such as for multimedia conferencing over IP). This protocol can establish, modify, or terminate multimedia sessions such as Internet-telephony calls (i.e., voice over IP, VoIP) using a text-based protocol. As with RTSP, SIP can also create and terminate sessions with one or more participants. Unlike RTSP, however, SIP supports user mobility by proxying and redirecting requests to the user's current location. Although SIP is designed to be independent of the transport-layer protocol, typically it runs over UDP rather than over TCP; TCP's connection setup and acknowledgment routines introduce delays, which are annoying and must be avoided in voice transmission. By adopting UDP, however, the timing of messages and their retransmission can be controlled by the application layer.

Figure 8.7 illustrates the architecture of an SIP network. An SIP client is any network element, such as a PC with a headset attachment or an SIP phone, that sends SIP requests and receives SIP responses. An SIP server is a network element that receives requests and sends back responses which accept, reject, or redirect the request. So SIP is a client–server protocol. Note that "clients" and "servers" are logical entities which last only for the duration of a certain transaction, i.e., a client might also be found on the same platform as a server. For example, SIP enables the use of proxies, which act as both client and server for the purpose of making requests on behalf of other clients. There are four different types of server, i.e., proxies, user agent servers (UASs), redirect servers, and registrars (see Figure 8.7).

Proxy servers are application-layer routers that are responsible for receiving a request, determining the location of the user, and then sending it there. To other entities it appears

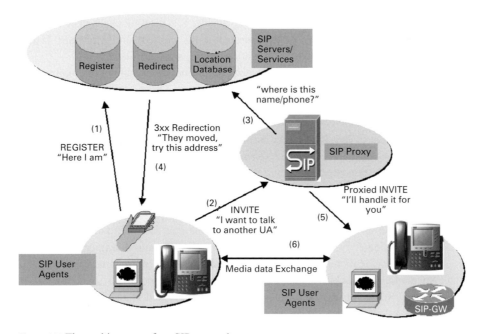

Figure 8.7 The architecture of an SIP network.

as if the message is coming from the proxy rather than from some entity hidden behind the proxy. A proxy must implement both the client and server requirements of a specification. It is also useful for enforcing policy and for traversing firewalls (which block selected network packets). A UAS is a logical entity that generates a response to an SIP request and contacts the user. In reality, an SIP device (such as an SIP-enabled telephone) will function as both a user agent client (UAC) and as a UAS, which enables SIP to be used for end-to-end communication. A redirect server is a server that accepts SIP requests, maps the destination address to a set of one or more addresses, and returns the new routing information to the originator of the request. Thereafter, the originator of the request can send a new request to the address(es) returned by the redirect server. A redirect server does not issue any SIP requests of its own. A registrar acts as a front end to the location service for a domain, reading and writing mappings based on the contents of the REGISTER messages. Session initiation protocol proxy servers, which are responsible for sending a request to the current host at which the person called is to be reached, need to consult this location service, provided by a registrar. Note that the distinction between SIP server types is logical only, not physical. Typically, a registrar is combined with a proxy or redirect server in a real network. In summary, SIP enables the following:

(1) determination of the location of the target end point. SIP supports address resolution, name mapping, and call redirection;
(2) determination of the media capabilities of the target end point. Using the session description protocol (SDP), SIP determines the lowest level of common services between the end points. Conferences are established using only the media capabilities that can be supported by all end points;
(3) determination of the availability of the target end point. If a call cannot be completed because the target end point is unavailable, SIP determines whether the party called is already on the phone or did not answer in the allotted number of rings. It then returns a message indicating why the target end point was unavailable;
(4) establishment of a session between the originating and target end point. If the call can be completed, SIP establishes a session between the end points. It also supports mid-call changes, such as the addition of another end point to the conference or the changing of a media characteristic or codec;
(5) handling of the transfer and termination of calls. It supports the transfer of calls from one end point to another. During a call transfer, SIP simply establishes a session between the transferee and a new end point (specified by the transferring party) and terminates the session between the transferee and the transferring party. At the end of a call, SIP terminates the sessions between all parties.

In addition to SIP, there are other protocols that facilitate voice over IP (VoIP). The H.323 standard was the first for call control for VoIP and was originated as an International Telecommunications Union (ITU) multimedia standard for both packet telephony and video streaming. The H.323 standard incorporates multiple protocols, including Q.931 for signaling, H.245 for control signaling and negotiation, and H.225 for registration admission and status (RAS) used in session control. Owing to its lack of scalability and extensibility, this complicated multiprotocol is more or less obsolete for use in video conferencing systems. Most of its functionalities have been replaced by SIP in more recent real-time interactive conferencing or messaging systems.

8.2 Windows Media streaming technology by Microsoft

Microsoft Windows Media Services 9 Series is the server component of the Windows Media 9 Series platform, and works in conjunction with Windows Media Encoders and Windows Media Players to deliver audio and video streaming content to clients over the Internet or an intranet. These clients might be other computers or devices that play back the content using a player, such as Windows Media Player, or they might be other computers running Windows Media Services (called Windows Media servers) that are proxying, caching, or redistributing content. Clients can also be custom applications that have been developed using the Windows Media Software Development Kit (SDK).

Windows Media Services can deliver a live stream or pre-existing content, such as a digital media file. The container file format is known as the *advanced systems format*, and the files have extensions .WMV, .WMA, or .ASF, depending on whether they contain video and audio (WMV), only audio (WMA), or content that is not coded by Windows Media codecs (ASF). To stream live content, a broadcast publishing point needs to be configured and then connected to an encoding software, such as Windows Media Encoder, that is capable of compressing a live stream into a format supported by the server. A Windows Media Server can handle HTTP, RTSP, and MMS control protocols and is limited to the streaming of ASF, WMV, and WMA container formats. To stream pre-existing content encoded by Windows Media Encoder, Microsoft Producer for PowerPoint 2002, Windows Movie Maker, Windows Media Player, or many other third-party encoding programs can be used. The pre-existing content is streamed from an on-demand publishing point.

Some technical features have been uniquely developed for Windows Media streaming systems, such as the fast streaming and dynamic content delivery technologies [60].

8.2.1 Fast streaming

Fast streaming refers to a set of features in Windows Media Services that significantly improves the quality of the A/V streaming experience over a variety of networks – even when network connections are unreliable. Fast streaming is possible because of these four components: fast start, fast cache, fast recovery, and fast reconnect. We now discuss these in turn.

8.2.1.1 Fast start

Fast start provides instant playback with no buffering delay, whether playing a single piece of content, or switching between on-demand clips or broadcast channels. Before it can start playing content, Windows Media Player must buffer a certain amount of data. When streaming to clients who use Windows Media Player, fast start is used to provide data directly to the buffer at speeds higher than the bitrate of the content requested. This enables users to start receiving content more quickly. After the initial buffer requirement is fulfilled, on-demand and broadcast content then streams at the bitrate defined by the content. Fast start is useful when the available network bandwidth of the client exceeds the required bandwidth of the content, when the network connectivity is intermittent or has high latency (as for wireless networks), or when the quality of the content received is of paramount importance (e.g., pay-per-view movie services).

Fast start enables users to have a better experience when playing back the content, e.g., fast-forwarding or rewinding takes place without additional delay and rebuffering: a player that connects through broadband networks starts playing the content more quickly, making the experience much more like viewing a television program or listening to a radio broadcast. Content delivered from the server by using server-side playlists switches smoothly and seamlessly between content items. Additionally, the prebuffering of data makes the player resistant to playback errors due to lost packets or other network problems.

8.2.1.2 Fast cache

For the effective use of the fast start mechanism, fast cache is needed to allow the streaming of content to the Windows Media Player cache as fast as the network will allow, so reducing the likelihood of an interruption in play due to network issues. For example, using fast cache the server can transmit a 128 kbps stream at 700 kbps. The stream is still rendered in Windows Media Player at the specified 128 kbps data rate, but the client is able to buffer a much larger portion of the content before rendering it. This allows the client to handle variable network conditions with no perceptible impact on the playback quality of either the on-demand or the broadcast content.

8.2.1.3 Fast recovery

Fast recovery works in conjunction with forward error correction (FEC) to provide redundant packets of information to clients that are using wireless connections. By providing the necessary amount of redundant packets, it can be ensured that no data is lost as a result of connectivity disruptions. Because of FEC, Windows Media Player can usually recover lost or damaged data packets without having to request that the data be resent by the Windows Media server. In environments that are subject to latency problems, such as satellite networks and other wireless networks, this process of receiving data is much more efficient.

8.2.1.4 Fast reconnect

Fast reconnect automatically restores live or on-demand player-to-server and server-to-server connections, if they become disconnected during a streaming session, to ensure an uninterrupted viewing experience. If the client was connected to an on-demand streaming server, the client restarts playback at the point at which the connection was lost by synchronizing itself with the content timeline. If the client was connected to a live broadcasting streaming server, the client reconnects to the broadcast in progress with a gap in the broadcast.

8.2.2   Dynamic content delivery

With Windows Media Services 9 Series, the distribution of users' content can be customized using server-side playlists and advertisements. The Windows Media server-side playlist is based on the synchronized multimedia integration language (SMIL) 2.0 standard, which provides a robust mechanism for assembling content for playback on personal computers and portable devices. Both broadcast and on-demand streaming servers can stream content from a playlist that executes on the server. A server-side playlist can contain live or pre-encoded content and be delivered using unicast or multicast transmission.

Several additional features are provided in Windows Media Services to ensure better QoS of A/V streaming. For example, cache or proxy solutions are used to conserve network bandwidth, decrease network-imposed latency, and decrease the load on Windows Media origin servers. Some improved protocols, based on RTSP and HTTP, are used to provide support between servers, for both unicast and multicast streaming in IPv4 and IPv6. Moreover, flexible distribution between servers uses combined UDP and TCP transports for streaming in mixed environments.

8.3   SureStream streaming technology by RealNetworks

RealMedia is a streaming technology, offered by RealNetworks, which runs on Windows and Linux. Both the container format and the codecs are proprietary, and the files have extensions .RA, .RAM, .RM, and .RPM. A RealMedia server called Helix Universal Server can handle RTSP and MMS control protocols and is able to deliver a large variety of container formats, including RealMedia, Windows Media (WM), Apple's QuickTime (QT), MPEG, and others. It can also accept connections from WM, QT, and MPEG-4 encoders, and from any of these players. It runs on the Linux and Windows platforms. Stream splitting is obtained either by *push* or *pull* techniques. In pull mode, no data will flow until a client makes a request.

To match the actual bandwidth and loss statistics of the channel, it is necessary to have an *adaptive* channel (or joint source and channel) encoding of the streaming content. Since today's streaming media servers are designed to be capable of serving thousands of clients per CPU, only very simple types of processing can be done at the bitstream level. Instead of using the more computationally demanding types of encoding such as scalable video coding (SVC) or multiple description coding (MDC), RealSystem G2 offers an extensive set of tools and public APIs, known in 1998 as the SureStream technology [64], to provide a comprehensive media streaming solution which is more scalable and more adaptive. The key idea of SureStream is to use the encoder to produce multiple representations (or *streams*) of the original content, optimized for various channel conditions. These encoded streams are then stored in a single SureStream file, in a form that facilitates their efficient retrieval by a server. During the streaming session, a client (RealPlayer G2) monitors the actual bandwidth and loss characteristics of its connection and instructs the server to switch to the stream whose transmission over the current channel would yield the minimum distortion in the reconstructed signal.

The use of client-side (receiver-driven) processing in SureStream offers at least two major benefits. First, it greatly reduces the complexity of server-side processing needed to support stream selection and thus increases the number of simultaneous connections that the server can maintain. Second, it allows a very simple extension of SureStream mechanism for *multicast* delivery, i.e., if encoded streams are assigned to different multicast groups then each client can subscribe and unsubscribe dynamically to a different multicast group using a rate-distortion minimization process. Since the implementation of SureStream services in RealSystem G2 is not tied to any particular file format or video coding algorithm, it can be used to take advantage of various *scalable video coding* techniques for channel adaptation. RealSystem G2 also marked an important phase in the development of Internet streaming infrastructure, being the first system built on the IETF and W3C standards for Internet multimedia (see Figure 8.8). More specifically, RealSystem G2 used the standard RTSP protocol for session control and supported the RTP standard for the framing and transporting of data packets. RealSystem G2 was also one of the first systems that embraced the W3C SMIL standard [63] for multimedia presentations.

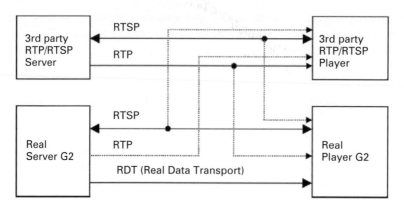

Figure 8.8 RealSystem G2 uses the standard RTSP protocol for session control, and supports the RTP standard for framing and transporting of data packets [6] (© IEEE 2003).

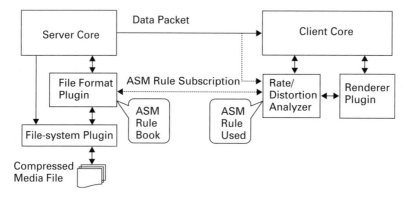

Figure 8.9 Adaptive stream management in RealSystem 8 [6] (© IEEE 2003).

The key technological components of RealSystems G2 are:

(1) *an adaptive stream management (ASM) protocol;*
(2) *SureStream* file format access and rendering mechanisms;
(3) *source and channel coding algorithms.*

These components will now be discussed in turn.

8.3.1   Adaptive stream management (ASM) protocol

The adaptive stream management protocol is a mechanism that allows the client (RealPlayer) to communicate and request efficiently the appropriate type of encoded SureStreams with the server in order to minimize the distortion of the received information. The structure of the server's and client's components involved in the ASM process is shown in Figure 8.9, where it can be seen that compressed media files are accessed by the server with the help of the *file system* and *file format* application programming interface (API) in RealSystems 8. The file format API has knowledge about the way data are compressed and stored in the media file, and is capable of producing various combinations of the encoded streams as they are requested by the client. To produce such combinations, the file format API uses the so-called *ASM rules*, which are based on a sophisticated fully programmable syntax. These ASM rules can be used to describe various means of channel adaptation ranging from assigning simple priorities to different packets to producing expressions describing the effects of various combinations of bandwidth, packet loss, and other types of loss on the reconstructed signal that can be assessed by the client. At the initial phase of the communication, the ASM rules are transferred to the client. In turn, the client collects information about the channel, parses the ASM rules, and sends the server a request to *subscribe* to a rule or combination of rules that match current statistics in the channel. When the server receives the request to subscribe to a rule, it passes it to the file format API, which in turn begins to generate the SureStream data according to its knowledge of their structure.

8.3.2 The structure of the RealVideo 8 algorithm

The overall structure of the RealVideo 8 encoding process to create various SureStreams is presented in Figure 8.10, where digitized and captured video frames (or fields) along with their timestamps are passed to a set of *input filters*. These filters are used mainly to remove the noise and specific artifacts introduced by edits and conversions of the video signal.

The output of the filtering engine is connected to a *spatial resampler*, which is used to *downscale* input frames to a set of spatial resolutions that are suitable for encoding at various output bitrates. The optimal selection of such resolutions depends not only on the set of target bitrates (which is typically known), but also on the type of content and type of distortion (e.g., smoothness versus clarity of individual frames) that the expected audience will be more willing to tolerate. For this reason, RealVideo 8 allows content creators to select one of four modes, "smooth motion," "normal motion," "sharpest image," and "slide show" when encoding a video clip. Later, this selection is used to deduce the desired tradeoff between spatial and temporal resolutions for each target bitrate.

After passing the *resampler*, video frames of various resolutions are forwarded to a set of actual video *codecs*, which are configured to produce encoded *SureStreams* for a given set of target bitrates. Potentially, each codec can work independently from the others after receiving the data. However, since some encoding operations (such as *scene-cut detection*, *motion compensation*, etc.) that are performed for each stream are generic to different streams, these codecs are designed to share their intermediate data and use them to reduce the complexity of the overall encoding process.

The *CPU scalability control* module (see Figure 8.10), which tracks the actual processing times at various points in the RealVideo 8 engine, sends signals of various

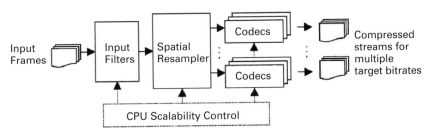

Figure 8.10 The structure of the RealVideo 8 encoding module [6] (© IEEE 2003).

resolutions to the appropriate algorithms informing them whether they should switch to lower-complexity modes. This type of control is essential to maintain the best possible quality level for encoding live presentations, especially when the encoding machine is being used for other needs.

8.3.3   Technical components of RealVideo 8 encoding

Some technical components of the RealVideo 8 encoding algorithm are described as follows [6].

8.3.3.1 Pre-filtering

The input video filters in RealVideo 8 are used to remove noise and other artifacts contained in the original input video. In addition, these filters can also remove special types of *irrelevant* (and, in some cases, also *redundant*) information, thus helping the main video compression algorithm to achieve its goals. For example, if a video signal is captured from a digital source, such as a USB camera, it is already presented in *progressive* form and may only need to be passed through a filter that removes low-energy spatial noise. However, if the signal is being captured from an analog NTSC (or PAL/SECAM) source, such as a TV tuner, camcorder, or VCR, which involves an *interlaced* and potentially edited video signal, additional filters can be applied. Several of these filters are designed to remove artifacts in the capturing process that become obvious when the captured frames are displayed in a progressive mode (such as on a computer monitor). The *de-interlace* filter is designed to intelligently combine information from odd and even *fields* of an NTSC (or PAL/SECAM) video signal to produce progressive video frames without introducing jagged artifacts. The *inverse telecine* filter is designed to remove the effects of the telecine process. Film is a progressive medium that is composed of 24 frames per second (fps), while NTSC video is an interlaced media having a rate of 29.97 fps (or 59.94 fields per second). The telecine process injects a redundant 5.97 fps (or 11.94 fields per second), and this should be removed before the encoding process. Also, owing to possible edits of the NTSC-converted film, the regular pattern of the inserted frames (fields) can be changed. For this reason, the inverse telecine filter detects all changes in order of fields and passes this information to the de-interlace filter to make sure that it combines them in the proper order.

8.3.3.2 Core compression algorithm

The core single-rate video compression algorithm in RealVideo 8 is essentially a motion-compensated hybrid scheme. To achieve better coding gain than other well-known standard codecs, some sophisticated proprietary features are incorporated in the core compression algorithms, such as motion prediction [65], adaptive transform sizes, and advanced statistical models.

8.3.3.3 Rate control

Rate control for RealVideo has two major modalities, single-pass and two-pass rate control. The rate control used in the single-pass mode, which is always used for live encoding, aims to pick the number of frames to skip until the next frame is encoded and also the "correct" quality and type (for example, intra or inter coded) for that frame. Knowledge of both the current and previous encoded and unencoded frames is used. For two-pass encoding,

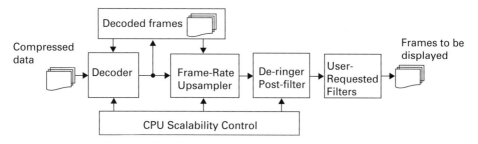

Figure 8.11 Client-side video processing in RealSystem 8 [6] (© IEEE 2003).

knowledge of future frames and the efficiency with which they can be coded are also used in the rate-control process.

There are commonly several, sometimes conflicting, requirements that drive the rate-control choices for streaming video encoding. For example, users should have the ability to join live streams easily and the coding should be resilient to loss in a streaming session. However, these requirements must be balanced against generic video coding requirements such as maximizing the quality of the content, i.e., minimizing the short-term variability of the frame rate and quality level. RealVideo 8 allows content creators to adjust some of the fundamental parameters of rate control such as the temporal depth of the preroll buffer, which plays a significant role in streaming performance. As discussed earlier, the preroll buffer limits the amount of bit averaging that is allowed in the encoding process. A larger buffer allows a coder to encode transient scenes, such as a pan or fade, with no degradation in video quality. The downside of a larger buffer is increased startup latency of a streaming presentation. A core rate-control algorithm should be able to make all these tradeoff decisions (in terms of explicitly specifying a cost function and then attempting to minimize it) while taking into account the content creators' desires (such as the ability to support a larger preroll buffer) and yielding a significant improvement in video coding quality.

8.3.3.4 Client-side video post-processing

In addition to its basic video compression and decompression services, RealVideo 8 also offers several post-processing filters aimed at improving the subjective quality of the video playback. The structure of these filters is presented in Figure 8.11. After decompression, video frames and their motion vectors are sent to the *frame rate upsampler*, which is a special temporal filter that attempts to interpolate intermediate frames; this is quite critical for low-bitrate encoded content where original frames are regularly skipped. A de-ringing filter is also used to reduce the ringing artifacts that are common in most transform-based codecs. Moreover, RealPlayer also offers a variety of additional post-processing tools to improve the decoding quality, such as a sharpening filter, color controls, etc.

8.4 Internet protocol TV (IPTV)

The technological advances in the past decades for VoIP, last-mile IP connections, video streaming, and video conferencing over IP can now be applied to an emerging technology, Internet protocol TV (IPTV), which will provide a more function-rich, user-interactive form of TV to consumers. The IPTV provides several additional functionalities based on the packetized two-way TV system. The primary roadblocks to the deployment of IPTV over the existing IP infrastructure are the need for sufficient bandwidth and quality of service (QoS). As a rule of thumb, it requires approximately 2 Mbps to transmit a broadcast-quality video stream using MPEG-2 compression; a DVD-quality TV signal takes 4 to 5 Mbps; and high-definition television (HDTV) requires approximately 9 Mbps. These TV programs can now be delivered digitally to the ordinary household with the use of advanced DSL (ADSL2+ and VDSL) or fiber-to-the-home (FTTH) technologies. The next generation network (NGN), which promises better manageable IP capabilities with broadband connectivity, can bring us another opportunity for supporting various real-time broadband multimedia services. ITU-T finished the development of NGN Release 1 in July 2006 regarding its requirements and functional architectures including QoS, resource control and security. Influencing these NGN standards developments was a pressing request from the industries involved, including operators, to initiate the standards of IPTV.

Internet protocol TV is formally defined as multimedia services, such as audio and video, delivered over IP-based networks that are managed in such a way as to provide the required level of QoS and quality of experience (QoE), security, interactivity, and reliability. The services are not limited to the delivery of traditional broadcasting TV (to TV, PC, PDA, or mobile phones) and include advanced services linked with TV programs and/or additions of new features during an IPTV session.

8.4.1   Swisscom IPTV

Figure 8.12 shows an IPTV system, which is built on top of Bluewin's DSL system, that operates through the same physical network as Swisscom's telephone system [66]. In a DSL system, packets of data flow in and out of a home computer from or to a phone system device (digital subscriber line access multiplexer, DSLAM) no more than 5000 meters away. The DSLAM aggregates data packets from a number of users before shuttling them into Swisscom's backbone network, and from there they are sent through two separate 10 Gbps interconnection points in Zurich to the Internet as a whole. The DSLAM also divides up the various signals coming from a household, i.e., the regular telephone service goes out to the public telephone system while video and DSL data go out to the Internet.

Bluewin's IPTV set-top boxes run Windows CE, a version of Microsoft's ubiquitous operating system designed for consumer electronics devices. The IPTV streaming is based on Windows Media 9 technology. In addition, these set-top boxes can also play TV shows and other media files and run Microsoft's own program guide software. They interact with media servers set up by Bluewin across a sophisticated data network.

A big challenge in IPTV is the single-channel problem, i.e., when viewers are watching one channel of digitally encoded and compressed video programming, only the media data associated with that specific channel is disseminated from an IP connection; all the other available channels are not. This is unlike cable TV, where all the channels are sent to the household at once but only one channel is tuned in for viewing. For a limited last-mile bandwidth IPTV, a channel change is a somewhat large challenge. More specifically, when a channel-surfer at home starts clicking away, the set-top box builds a new UDP connection with the data center, kilometers away. The IPTV server there stops sending the old channel selection and sends the new one using UDP data flow. This channel switching is done in less than 200 ms, so fast that the user has no idea that the video programming was not there in the set-top box all along, waiting for its chance to leap onto the screen. Transmitting data

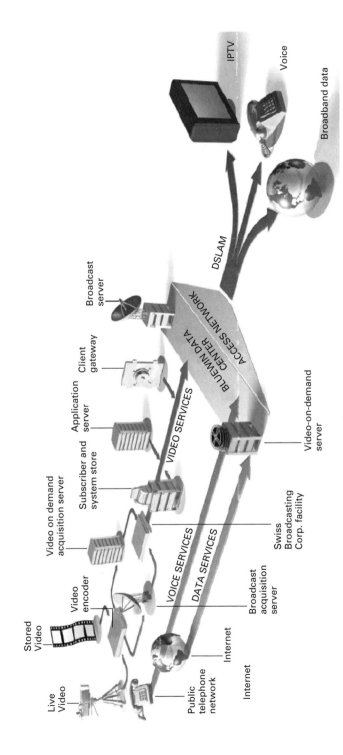

Figure 8.12 An IPTV system built on Bluewin's DSL system operates through the same physical network as Swisscom's telephone system [66] (© IEEE 2005).

across an Internet in 200 ms might not seem to be a problem, since even email can be sent and received as quickly as that, sometimes. Doing it with bulky video data, though, and doing it reliably enough to create a TV experience that can be watched for hours on end, is something that Microsoft's Windows Media fast streaming technology is targeted to resolve. More specifically, the Windows Media platform is used to provide the framework for IPTV systems. In addition to the fast streaming features described earlier, some useful features in Windows Media Services have been developed to deliver a high-quality experience to IPTV viewers, as follows.

(1) *Unicast, multicast, video on demand (VoD), and broadcast* Windows Media Services can deliver broadcast and on-demand streams, using unicast and multicast protocols.
(2) *Intelligent streaming* When digital media is streamed, the player detects network conditions and sends feedback to the Windows Media server, which then adjusts the properties of the streams to maximize quality so that the highest-quality stream can be delivered to the viewer regardless of network conditions. Typically, this intelligent streaming makes use of multiple-bitrate streams.
(3) *Advanced fast start* This feature adds to fast start capabilities by allowing the player to begin playing content as soon as its buffer receives a minimum amount of data, further reducing the amount of time a user has to wait to begin receiving the stream.
(4) *Advanced FF/RW* This feature improves fast-forward and rewind ("trick mode") functionality for the video portion of encoded files and stabilizes network-bandwidth availability by smoothing the rate at which data is sent. Potential server performance bottlenecks are reduced because the server reads less presentation data from the source content disk while delivering a seamless experience to clients.
(5) *Play while archiving* Archived files can be made available for on-demand requests or for rebroadcast, even before a broadcast that is being archived has finished.

8.4.2 An IPTV reference architecture and functionality modules

Architectural study in ITU-T IPTV starts with the analysis of the relevant domains involved in the provision of an IPTV service. As shown in Figure 8.13, there are four main domains involved in an IPTV reference architecture [70].

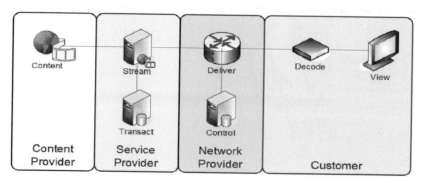

Figure 8.13 A reference architecture for running IPTV services [70] (© IEEE 2007).

(1) The *content provider* is the entity that owns or is licensed to sell content or content assets.
(2) The *service provider* is the entity that provides the IPTV service to the customer. Typically, the service provider acquires or licenses content from content providers and packages this into a service that is sold to the customer.
(3) The *network provider* is the entity providing the platform that connects the customers and the service providers.
(4) The *customer* is the entity that consumes and pays for the IPTV service.

From the functionality aspects, the IPTV reference architecture consists of the following three functionality modules: *customer premises*, *video headend*, and *transport networks* [67].

8.4.2.1 Customer premises; the set-top box (STB)

A set-top box (STB) is a device on the customer side that interfaces with the user terminal (e.g., a television, PC, or laptop) with DSL or cable wiring. An STB is usually installed with middleware client software to obtain the program-guide data, decode MPEG-2 or MPEG-4 video data, and enable display on the screen. Alternatively, a web browser can obtain the program-guide data from a central server. An STB can be integrated with a DSL or cable modem or even with an IEEE 802.11 switch for home Internet access networking.

8.4.2.2 Video headend

Video encoders and video on demand servers are the major sources of video content for IPTV services. The video headend is composed of the following components.

Video encoder   This can encode real-time analog video signals from a content provider or a live event location to a digital format based on a given video compression technology such as MPEG-2/4 or H.264/AVC. The encoder also deals with on-demand content stored or redistributed at different video on demand servers after encoding and other processing, such as digital rights and encryption (see Chapter 11).

Live video broadcast server   This is in charge of reformatting and encapsulating video streams in the case where video streams with different formats from a video encoder or pre-encoded video file are received. The server also interfaces the core network and transmits the video signal over the core network toward the access network.

Video on demand (VoD) server   This houses on-demand content and has streaming engines and a large storage capacity.

Content and subscriber management   The system essentially facilitates the operation of IPTV services, in terms of subscription account setup or tear-down, for client control, billing, and authentication for back-office control as well as digital rights and security issues for content and delivery control.

8.4.2.3 Transport networks

There are two major parts of the transport network, the *core* and *access* networks.
Core networks   These networks connect the access networks to customer premises and can be simply a single national distribution network running Gigabit Ethernet or IP plus various regional distribution networks running carrier-grade Ethernet. The managed content is

usually centralized and processed within the national distribution network before being delivered to different access networks. However, a wider range of choices for unmanaged content from other content providers can be made, and this unmanaged content is fed into the national distribution network to customers through the Internet.

Access networks   These networks serve as a critical part of the transport network and are used to reach each individual customer at his or her home through the STB. The technologies available today are mainly xDSL and coaxial hybrid fiber cable (HFC) or fiber techniques such as fiber-to-the-home (FTTH), which extended the reach to customer communities before the advent of xDSL or cable wiring. Because the bandwidth of access networks usually is very limited, to cater for all customers needing simultaneous access to TV channels, multicasting has been widely adopted to enable a scalable delivery of video data for IPTV.

8.4.3   ATIS IPTV Interoperability Forum (IIF)

The TV industry as a whole is moving aggressively into the next generation of digital programming: high-definition television. The Alliance for Telecommunications Industry Solutions (ATIS, www.atis.org), participated in by more than 350 global communications companies, is a technical planning and standards development organization that is committed to rapidly developing and promoting technical and operations standards for the communications and related information technologies industry worldwide using a pragmatic, flexible, and open approach.

The IPTV Interoperability Forum (IIF) was established by ATIS in June 2005 at the recommendation of an exploratory group established by the ATIS Board of Directors to assess the technical and operational opportunities and challenges surrounding the deployment of IPTV and the status of the related standardization activity. It is widely believed that the IPTV market is expanding rapidly already and that carriers need to use a standard process that is more akin to Internet development than that in the traditional telecommunications industry. More and more service providers deploying IPTV, including AT&T and Verizon, are increasingly concerned about standards. The exploratory group, meeting over a 45-day period, created an IPTV definition both for current models and for future implementations enabled by NGN, identified issues impeding the widespread adoption of IPTV, and surveyed the work of other organizations related to IPTV. The conclusion of the exploratory group was that the adoption of IPTV could be advanced by the creation of a committee specifically focused on IPTV applications.

The key issues (i.e., work items) discussed by IIF are related to IPTV architectures, digital rights management, and quality of service metrics. The mission of IIF is to enable the interoperability, interconnection, and implementation of IPTV systems and services by developing ATIS standards and facilitating related technical activities. The forum places an emphasis on the needs of North American and ATIS Member Companies in coordination with other regional and international standards development organizations. The forum's initial focus is: the creation of an overall industry reference architecture for IPTV; content delivery (quality of experience); digital rights management (DRM); interoperability standards and testing requirements for components; the reliability and robustness of service components;

and establishing user expectations. The definition of IPTV developed by the exploratory group and being used by the IIF is as follows [71].

IPTV is defined as the secure and reliable delivery to subscribers of entertainment video and related services. These services may include, for example, Live TV, Video On Demand (VOD) and Interactive TV (iTV). These services are delivered across an access agnostic, packet switched network that employs the IP protocol to transport the audio, video and control signals. In contrast to video over the public Internet, with IPTV deployments, network security and performance are tightly managed to ensure a superior entertainment experience, resulting in a compelling business environment for content providers, advertisers and customers alike.

8.4.4 High-level requirements of IPTV

The long-term goal of the IIF is to specify fully all interfaces in the IPTV "ecosystem" and identify and specify appropriate metrics for the quality of IPTV service across these interfaces. The IPTV ecosystem is very large and complex, however, and the IIF has recognized that it will not be able to specify the whole ecosystem at a detailed level all at once. Therefore, as an initial effort, several focus groups (FG) under ITU-T ATIS IIF are working on the development of high-level requirements for various important aspects: IPTV architecture, QoS and performance, security and content protection, network and control, end systems and middleware, and public interest. On the basis of these high-level requirements, the working groups in FG IPTV can further develop more detailed requirements. The high-level requirements for various aspects are summarized below [70] [71]:

8.4.4.1 Requirements for IPTV architectural aspects
The key features of the IPTV architectural requirements are as follows:

(1) to support the display of multiple supplementary video streams and layouts to provide service management capability, enabling service providers to integrate different contents into a content bundle or arrange a schedule for a program;
(2) to provide capability to support service and content protection, as well as easy integration with existing services through a unified service platform;
(3) to allow the seamless provision and operation of IPTV services across different administrative domains;
(4) to adapt dynamically to possible changes in wireless network characteristics (e.g. bandwidth, packet loss rate, etc.) when operating over a mobile network;
(5) to provide a mechanism that allows service signaling messages to be routed on the basis of the capabilities of the end-user device and/or the application servers available to provide services;
(6) to provide the ability to remotely configure and administer network and service elements;
(7) to allow private entities (e.g., residential users) to act as content providers for the purpose of sharing their own content. This mechanism may need to deal with issues related to dynamic IP-address assignment, non DNS registered hosts, servers located on private networks behind NAT and firewalls;
(8) to support Emergency Alert Services as per relevant national regulations and provide appropriate interfaces to support the ingestion of suitably authorized Emergency Alert information.

8.4.4.2 Requirements for QoS and performance aspects

Requirements on QoS and performance are being developed in IPTV FG in conjunction with QoE and traffic management aspects. Key features of these requirements are as follows:

(1) to allow the delivery of IPTV services with a defined QoE for the IPTV end user and provide mechanisms for supporting appropriate resilience in the service-provider infrastructure to maintain a high QoE for video services;

(2) to support, if possible, the delivery of multiple services, either from the same or multiple providers, over the common IP transport with a manageable IP QoS;

(3) to provide the capability to monitor video quality and perhaps to support the ability to adjust QoS or QoE parameters dynamically;

(4) to provide a mechanism by which network operators can integrate IPTV QoS management functions into a common framework with other NGN services and applications;

(5) to provide a means to support flexible channel switching times as a tradeoff with improved efficiency;

(6) to support a mechanism for assigning packet priority, and to support the relevant mechanisms for IPTV traffic identification, classification and marking, policing and conditioning, scheduling and discarding.

(7) to provide mechanisms for dynamic IPTV traffic load balancing so as to accommodate dynamically the network load and congestion conditions at any given time, so as to deliver a set of IPTV services to end users with the relevant level of quality.

8.4.4.3 Requirements for security and content protection aspects

Requirements on security and content protection aspects are being developed in IPTV FG including authentication. The key features of these requirements are as follows:

(1) to comply with the service and content protection requirements found in ATIS-0800001 (www.itu.int/md/T05-FG.IPTV-IL-0046/, "IPTV DRM interoperability requirements";

(2) to provide mechanisms to enable the application of appropriate DRM on content and provide mechanisms to support secure storage for prepositioned content;

(3) to provide mechanisms to establish authentication and authorization before service delivery;

(4) to provide mechanisms to control unauthorized access by unsubscribed end users to the service, and perhaps to redirect unsubscribed end users to a mechanism by which they may subscribe;

(5) to provide mechanisms to allow the authentication of a device in the provisioning and operation of the IPTV service;

(6) to provide a mechanism for the service provider to authenticate the source of content.

8.4.4.4 Requirements for network and control aspects

Requirements on network and control aspects are being developed in IPTV FG, including the home network aspects. The key features of these requirements are as follows:

(1) to provide the ability to support both multicast and unicast transmission schemes with supporting bidirectional communications to many IPTV terminal devices to provide interactive control;

(2) to support the personal video recording (PVR) and digital video recording (DVR) functions residing in either the end user or service provider networks, which support content streaming control;

(3) to define traffic management mechanisms for the differential treatment of IPTV traffic and to provide a means of signaling end users of the occurrence of an Emergency Alert Notification (EAN) regardless of the services currently being consumed by the end users;

(4) to support static and dynamic address allocation schemes for both IPv4 and IPv6 and automated configuration capabilities for IPTV devices;

(5) to support at least one *intelligent peripheral interface* (IPI-4) in the home network (HN) with sufficient bandwidth and QoS for IPTV service, including the simultaneous transport of multiple video streams in addition to voice and data traffic;

(6) to support the ability for the *delivery network gateway function* (DNGF, FG IPTV-C-0276) to present the entire home network through one or more WAN (public) IP addresses, which are used for all external communication.

8.4.4.5 Requirements for end systems and middleware aspects

Requirements for end systems and middleware including interoperability aspects are being developed in IPTV FG. The key features of these requirements are as follows:

(1) to provide API to and be able to communicate with IPTV service providers to implement media transmission and media control functions;

(2) to provide API for DRM functions to communicate with DRM systems;

(3) to provide API to communicate with IPTV service providers in order to implement media transmission and media control functions;

(4) to provide a negotiation mechanism between clients and servers at any service session time;

(5) to perform self-diagnostics of information such as power status, boot status, memory allocation, software version, middleware version, network address, and network status;

(6) to support the capability for the end user to switch between multiple supplementary video streams within a program;

(7) to provide the end user with the ability to access different IPTV content on different IPTV terminal devices if the delivery network allows the simultaneous delivery of several TV channels to the end user premises;

(8) to provide the end user with the ability to identify easily the TV channels delivering the content selected and the relative progress position in the delivery of this content;

(9) to provide the end user with the ability to capture high-value premium content.

8.4.4.6 Requirements for public interest aspects

Requirements for public interest aspects are also being developed in FG IPTV, including the provision of accessibility for people with disabilities and people with special needs, to meet the minimum regulatory requirements. The key features of these requirements are as follows:

(1) to provide mechanisms to support closed captioning, to select and receive two (related) video sources (e.g., one with sign language translation) and to provide the ability to select and receive two audio sources (e.g., one with audio description);

(2) to design applications and equipments using principles of universal design, so that a wider population with users of varying capabilities can access such applications and equipments;

(3) to include appropriate accessibility features in the recordings made by equipment, e.g., to record all accessibility features streams and link them to the service as a whole;

(4) to provide the end user with the ability to be notified and receive regulatory information services (unassociated with TV channels) whenever the IPTV terminal device is active (e.g. displaying an EPG or a TV channel);

(5) to support the Emergency Alert Service as per the relevant national regulations;

(6) to support a mechanism for customers to select IPTV network providers, IPTV service providers, and IPTV content providers according to their preferences.

8.4.5   High-level functional components in the IPTV architecture

The *IPTV high-level architecture standard* (ATIS-0800007) [69] specifies a reference architecture (see Figure 8.13) for further IIF work and is designed to enable an end-to-end IPTV system to be deployed using hardware or software from multiple vendors. This will serve as a reference architecture for the IPTV functional specifications to be defined in separate IIF specification documents. On the basis of the IPTV reference architecture shown in Figure 8.13, five functional components are identified in terms of IPTV provisioning: content provisioning, IPTV control, content delivery, end system and IPTV system management and security. Figure 8.14 shows this high-level architecture model, where the information between functional components is as follows [70]:

- A, content stream
- B, content request, Descriptive Metadata/Content Info etc.
- C, rights management interaction (e.g., DRM and the conditional access system (CAS))

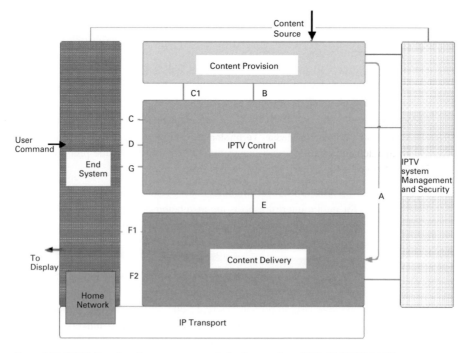

Figure 8.14 IPTV functional components and the interactions [70] (© IEEE 2007).

- C1 rights management
- G authentication and authorization, conditional access system
- D service interaction message
- B content location information, billing information, content control command, etc.
- F1 play control signal,
- F2 content stream

Content provision This component converts the ingested content from the content source into the designated format and encrypts it for rights management. The converted contents are then transferred to the content delivery component.

IPTV control This component is responsible for IPTV service preparation control and for offering controls such as packing the contents into services, generating content distribution policy, and then publishing these deliverable services through navigating and electronic program guide (EPG) modules. IPTV control is also responsible for issuing the content licence to the users according to their subscription, if the empowerment of content provision is agreed.

Content delivery This component delivers contents which are packed in IPTV services to the end system. In order to support services such as video on demand (VoD) and personal video recoding (PVR), and for the sake of transport efficiency, contents are stored or cached within the content delivery component. When the end user requests a piece of content, the IPTV control functions will direct the content delivery functions to obtain this content data.

End system The end system component collects control commands from users and interacts with the IPTV control component to obtain service information (e.g., EPG), content licenses and keys for decryption. The end system component also provides a capability for content acquisition, content decrypting, and decoding.

IPTV system management and security The system management and security functions are in charge of system overall status monitoring, configuration, and security aspects.

References

[1] "Microsoft Windows Media Technologies," http://www.microsoft.com/windows/window-smedia/default.mspx.
[2] RealNetworks RealPlayer, http://www.realplayer.com/.
[3] Apple QuickTime, http://www.apple.com/quicktime/.
[4] "VivoActive software and documentation," Vivo Software website, 1997, http://www.vivo.com/help/index.html.
[5] "Progressive Networks launches the first commercial audio-on-demand system over the Internet (press release)," Progressive Networks 1995, http://www.realnetworks.com/company/press-room/pr/pr1995.
[6] G. J. Conklin, G. S. Greenbaum, K. O. Lillevold, A. F. Lippman, and Y. A. Reznik, "Video coding for streaming media delivery on the Internet," *IEEE Trans. Circuits Syst. Video Technol.*, 11(3): 269–281, March 2003.
[7] A. E. Mohr, E. A. Riskin, and R. E. Ladner, "Unequal loss protection: graceful degradation of image quality over packet erasure channels through forward error correction," *IEEE. J. Selected Areas in Commun.*, 18(6): 819–828, June 2000.

[8] D. Wu, Y. T. Hou, W. Zhu, Y.-Q Zhang, and J. M. Peha, "Streaming video over the Internet: approaches and directions," *IEEE Trans. Circuits Syst. Video Technol.*, 11(3): 282–300, March 2003.

[9] "Understanding delay in packet voice networks," Document ID: 5125, Cisco Systems, http://cisco.com/warp/public/788/voip/delay-details.html#dejitterdelay.

[10] W. Tan and A. Zakhor, "Real-time Internet video using error resilient scalable compression and TCP-friendly transport protocol," *IEEE Trans. Multimedia*, 1: 172–186, June 1999.

[11] Q. Guo, Q. Zhang, W. Zhu, and Y.-Q. Zhang, "Sender-adaptive and receiver-driven video multicasting," in *Proc. IEEE Int. Symp. on Circuits and Systems (ISCAS 2001)*, Sydney May 2001.

[12] P. A. Chou, A. E. Mohr, A. Wang, and S. Mehrotra, "Error control for receiver-driven layered multicast of audio and video," *IEEE Trans. Multimedia*, 3 (1), March 2001.

[13] D. Sisalem and H. Schulzrinne, "The loss-delay based adjustment algorithm: a TCP-friendly adaptation," in *Proc. Workshop on Network and Operating System Support for Digital Audio and Video*, July 1998.

[14] R. Rejaie, M. Handley, and D. Estrin, "RAP: an end-to-end rate-based congestion control mechanism for realtime streams in the Internet," in *Proc. IEEE INFOCOM 99*, Vol. 3, pp. 1337–1345, 1999.

[15] S. Floyd, M. Handley, J. Padhye, and J. Widrner, "Equation-based congestion control for unicast applications," in *Proc. SIGCOMM'00*, August 2000.

[16] A. Legout and E. W. Biersack, "PLM: fast convergence for cumulative layered multicast transmission," in *Proc. ACM SIGMETRICS Conf.*, Santa Clara CA, pp. 13–22, June 2000.

[17] L. Wu, R. Sharma, and B. Smith, "Thin streams: an architecture for multicasting layered video," in *Proc. NOSSDAV'97*, St Louis MO, pp. 173–182, May 1997.

[18] S. McCanne, V. Jacobson, and M. Vetterli, "Receiver-driven layered multicast," in *Proc. ACM Sigcomm. Conf.*, Palo Alto CA, pp. 117–130, August 1996.

[19] S. Keshav, "The packet pair flow control protocol," Technical Report 91–028, Berkeley, California, May 1991.

[20] G. I. Kwon and J. W. Byers, "Smooth multirate multicast congestion control," in *Proc. IEEE INFOCOM '03*, April 2003.

[21] Q. Liu and J.-N. Hwang, "A scalable video transmission system using bandwidth inference in congestion control," in *Proc. IEEE Int. Conf. on Circuits and Systems*, Vancouver, May 2004.

[22] Q. Liu, J. Yoo, B.-T. Jang, K. Choi, and J.-N. Hwang, "A scalable VideoGIS system for GPS-guided vehicles," *Signal Process. Image Commun.*, 20 (3): 205–208, March 2005.

[23] A. Vetro, C. Christopoulos, and H. Sun, "Video transcoding architectures and techniques: an overview," *IEEE Signal Process. Mag.*, March 2003.

[24] P. AssunÓno and M. Ghanbari, "Post-processing of MPEG-2 coded video for transmission at lower bit-rates," in *Proc. IEEE Int. Conf. on Acoustics, Speech and Signal Processing*, Atlanta GA, pp. 1998–2001, 1996.

[25] S. F. Chang and D. G. Messerschmidt, "Manipulation and compositing of MC-DCT compressed video," *IEEE J. Selected Areas in Commun.*, 13: 1–11, January 1995.

[26] H. Sun, A. Vetro, J. Bao, and T. Poon, "A new approach for memory-efficient ATV decoding," *IEEE Trans. Consum. Electron.*, 43: 517–525, August 1997.

[27] Z.-L. Zhang, S. Nelakuditi, R. Aggarwa, and R. P. Tsang, "Efficient server selective frame discard algorithms for stored video delivery over resource constrained networks," in *Proc. IEEE INFOCOM'99*, pp. 472–479, March 1999.

[28] A. Albanese, J. Blömer, J. Edmonds, M. Luby, and M. Sudan, "Priority encoding transmission," *IEEE Trans. Inform. Theory*, 42: 1737–1744, November 1996.

[29] L. Rizzo and L. Vicisano, "A reliable multicast data distribution protocol based on software FEC techniques," in *Proc. 4th IEEE Workshop on the Architecture and Implementation of High Performance Communication Systems (HPCS'97)*.

[30] R. E. Blahut, *Theory and Practice of Error Control Codes*, Addison Wesley, MA, 1984.

[31] J. Cai, W. Zhan, Z. He, "Optimal retransmission timeout selection for delay-constrained multimedia communication," in *Proc. IEEE Int. Conf. on Image Processing*, pp. 2035–2038, 2004.

[32] "Information technology – coding of audio-visual objects, part 1: Systems, part 2: Visual, part 3: Audio," ISO/IEC JTC 1/SC 29/WG 11, FCD 14 496, December 1998.

[33] "Video coding for low bitrate communication," Draft ITU-T Recommendation H.263 Version 2, ITU Telecom. Standardization Sector of ITU, September 1997.

[34] T. Wiegand, G. J. Sullivan, G. Bjntegaard, and A. Luthra, "Overview of the H.264/AVC video coding standard," *IEEE Trans. Circuits and Syst. Video Technol.*, 13(7): 560–576, July 2003.

[35] Y. Takishima, M. Wada, and H. Murakami, "Reversible variable length codes," *IEEE Trans. Commun.* 43(2,3,4): 158–162, February/March/April 1995.

[36] Y. Wang, M. T. Orchard, and A. R. Reibman, "Multiple description image coding for noisy channels by pairing transform coefficients," in *Proc. IEEE Workshop on Multimedia Signal Processing*, pp. 419–424, June 1997.

[37] Y. Wang and Q.-F. Zhu, "Error control and concealment for video communication: a review," *Proc. IEEE*, 86: 974–997, May 1998.

[38] M. Hemy, P. Steenkiste, and T. Gross, "Evaluation of adaptive filtering of MPEG system streams in IP networks," in *Proc. IEEE ICME Conf.*, pp. 1313–1317, July 2000.

[39] J. Ni and D. H. K. Tsang, "Large scale cooperative caching and application-level multicast in multimedia content delivery networks," *IEEE Commun. Mag.*, 43(5): 98–105, May 2005.

[40] L. Fan, P. Cao, J. Almeida, and A. Z. Broder, "Summary cache: a scalable wide-area web cache sharing protocol," *IEEE Trans. Networking*, 8: 281–293, June 2000.

[41] Z. Miao and A. Ortega, "Proxy caching for efficient video services over the Internet," in *Proc. Packet Video'99*, New York, April 1999.

[42] R. Kermode, "Smart network caches: localized content and application negotiated recovery mechanisms for multicast media distribution," Ph.D. thesis, MIT Media Laboratory, June 1998.

[43] C. L. Liu and J. W. Layland, "Scheduling algorithms for multiprogramming in a hard real-time environment," *J. Assoc. Comput. Mach.*, 20 (1): 46–61, January 1973.

[44] J. Y. Chung, J. W. S. Liu, and K. J. Lin, "Scheduling periodic jobs that allows imprecise results," *IEEE Trans. Comput.*, 19: 1156–1173, September 1990.

[45] J. Gemmell and S. Christodoulakis, "Principles of delay sensitive multimedia data storage and retrieval," *ACM Trans. Information Syst.*, 10 (1): 51–90, January 1992.

[46] H. M. Vin, P. Goyal, A. Goyal, and A. Goyal, "A statistical admission control algorithm for multimedia servers," in *Proc. ACM Multimedia'94*, pp. 33–40, October 1994.

[47] J. Gemmell, H. M. Vin, D. D. Kandlur, P. V. Rangan, and L. A. Rowe, "Multimedia storage servers: a tutorial," *IEEE Comput. Mag.*, 28; 40–49, May 1995.

[48] P. Shenoy and H. M. Vin, "Efficient striping techniques for multimedia file servers," *Perform. Eval.*, 38 (3,4): 175–199, December 1999.

[49] D. H. C. Du and Y.-J. Lee, "Scalable server and storage architectures for video streaming," in *Proc. IEEE Int. Conf. Multimedia Computing and Systems*, pp. 62–67, June 1999.

[50] A. Guha, "The evolution to network storage architectures for multimedia applications," in *Proc. IEEE Int. Conf. Multimedia Computing and Systems*, pp. 68–73, June 1999.

[51] G. A. Gibson and R. V. Meter, "Network attached storage architecture," *Commun. ACM*, 43 (11): 37–45, November 2000.

[52] J. Gafsi and E. W. Biersack, "Performance and reliability study for distributed video servers: mirroring or parity?," in *Proc. IEEE Int. Conf. Multimedia Computing and Systems*, pp. 628–634. June 1999.

[53] G. Blakowski and R. Steinmetz, "A media synchronization survey: reference model, specification, and case studies," *IEEE J. Selected. Areas in Commun.*, 14: 5–35, January 1996.

[54] J. P. Jarmasz and N. D. Georganas, "Designing a distributed multimedia synchronization scheduler," in *Proc. IEEE Int. Conf. Multimedia Computing and Systems*, pp. 451–457, June 1997.

[55] N. D. Georganas, "Synchronization issues in multimedia presentational and conversational applications," in *Proc. Pacific Workshop on Distributed Multimedia Systems (DMS'96)*, June 1996.

[56] H. Schulzrinne, S. Casner, R. Frederick, and V. Jacobson, "RTP: a transport protocol for real-time applications," Internet Engineering Task Force RFC 1889, January 1996.

[57] H. Schulzrinne, A. Rao, and R. Lanphier, "Real time streaming protocol (RTSP)," Internet Engineering Task Force RFC 2326, April 1998.

[58] M. Handley, H. Schulzrinne, E. Schooler, and J. Rosenberg, "SIP: session initiation protocol," Internet Engineering Task Force RFC 2543, March 1999.

[59] "H.323: packet-based multimedia communication Systems," ITU, itu.int/rec/T-REC-H.323/e.

[60] "Technical overview of Windows Media Services 9 Series," Microsoft Windows Server 2003, Microsoft Corporation, March 2005.

[61] "RealNetworks announces RealSystem G2, the next generation streaming media delivery system (press release)," RealNetworks. (1998). www.realnetworks.com/company/pressroom/pr/pr1998.

[62] A. Lippman, "Video coding for multiple target audiences," in *Proc. IS&T/SPIE Conf. on Visual Communications and Image Processing*, San Jose CA, pp. 780–784, January 1999.

[63] P. Hoschka, "The application/SMIL media type," 1999, draft-hoschka-smilmedia- type-04.txt.

[64] "RealSystem G2 SDK and documentation," RealNetworks, Seattle, WA, www.realnetworks.com/devzone/downlds/index.html.

[65] K. O. Lillevold, "Improved direct mode for B pictures in TML," in *Proc. ITU-T Conf.*, Portland OR, ITU-T SG16 (Q15), Doc. Q15-K-44, August 2000.

[66] S. Cherry, "The battle for broadband," *IEEE Spectrum*, pp. 24–29, January 2005.

[67] J. She, F. Hou, P.-H. Ho, and L.-L. Xie, "IPTV over WiMAX: key success factors, challenges, and solutions," *IEEE Commun. Mag.*, 45(8): 87–93, August 2007.

[68] "Universal plug and play," www.upnp.org/specs/qos/UPnP-qos-Architecture-v2-20061016.pdf .

[69] "IPTV high level architecture," ATIS-0800007, Secretariat Alliance for Telecommunications Industry Solutions, Approved March 2007.

[70] C.-S. Lee, "IPTV over next generation networks in ITU-T," in *Proc. 2nd IEEE/IFIP Int. Workshop on Broadband Convergence Networks (BcN'07)*, pp. 1–18, May 2007.

[71] "Status report on the work of the ATIS IPTV Interoperability Forum (IIF)," ITU Document 34–E, Geneva, April 2006.

9 Wireless broadband and quality of service

With the gradual paradigm shift from analog to digital media, from push-based media broadcasting to pull-based media streaming, and from wired interconnectivity to wireless interconnectivity, wireless broadband access with provisioned quality of service (QoS) for digital multimedia applications to mobile end users over wide area networks is the new frontier of the telecommunications industry. The shift from wired to wireless Internet is also coming as a strong wave. The great success of broadband wireline services (either based on cable or DSL interconnectivity) and short-range portable wireless data services (based on Wi-Fi) has created a strong consumer demand for wireless broadband Internet anytime and anywhere. Wireless technology describes telecommunications in which electromagnetic waves carry the signal over part of or the entire communication path without cables. Wireless broadband is an extension of point-to-point wireless communication for the delivery of high-speed and high-capacity pipe that can be used for voice, multimedia, and Internet access services. Though there are many technologies available for providing broadband wireless access to the Internet, the main focus is on 3G, Wi-Fi, and WiMAX owing to their potential benefits (see Figure 9.1). The wireless LAN (WLAN or the so-called Wi-Fi standards [1]) technologies IEEE 802.11a/b/g and the next generation very-high-data-rate (> 200 Mbps) IEEE 802.11n are being deployed everywhere for Internet access (the so-called hotspot) with very affordable installation costs. Almost all newly shipped computer products and more and more consumer electronics also come with WLAN receivers for Internet access. To provide mobility support for Internet access, cellular-based technologies such as 3G networking [2] [3] are being deployed aggressively with more and more multimedia application services from the traditional telecommunication carriers. However, mobile wireless microwave access (WiMAX) serves as another powerful alternative for mobile Internet access from data communication carriers. Fixed or mobile WiMAX (IEEE 802.16d and 802.16e [4] [5]) can also serve as an effective backhaul for WLAN whenever it is not easily available, such as in remote areas or in moving vehicles with compatible IP protocols.

All 3G, Wi-Fi and WiMAX are access or network-edge technologies, which offer alternatives to last-mile wireline networks. Beyond the last mile, all rely on similar network connections and transmission-support backbone infrastructure. They all support broadband data services with different distance ranges (Wi-Fi has a range of about 100 m, 3G and WiMAX have ranges of several km). The data rates offered by Wi-Fi (11/54 Mbps) and WiMAX (several Mbps depending on the mobility) are substantially higher than the several hundred kbps expected from 3G services. The emergence of mobile WiMAX as a competitor to 3G has tended to enlarge the mobile broadband market with a reasonably level playing field. Mobile WiMAX (802.16e) implementations have been shown to be superior competitors to the currently deployed 3.5G mobile systems (such as HSDPA and EV-DO implementations resulting from 3GPP and 3GPP2 within the IMT 2000 framework). In fact,

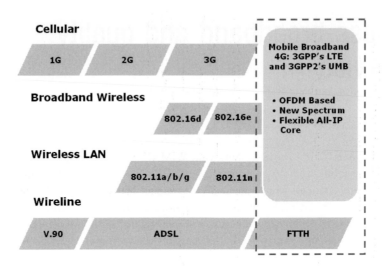

Figure 9.1 The technological evolution of 3G, Wi-Fi, and WiMAX for broadband wireless access [7] (© IEEE 2007).

WiMAX is regarded as the next generation (4G) wireless system (see Section 9.1.3) owing to its higher and more reliable data rates along with all the IP support. While 3G licenses have been awarded in a number of markets at a high cost, there is only limited growth with respect to service deployment. However, Wi-Fi networking equipments and systems have been installed rapidly, for example, as in home and office networks. It is expected that WiMAX will have a similar momentum of deployment growth owing to the strong endorsement of the computer and IC industry in integrating WiMAX customer premise equipments (CPEs) into laptops and many portable devices in 2008. This can be best evidenced by Wi-Fi's rapid large-scale market penetration well beyond the applications initially targeted.

A key distinction between 3G, Wi-Fi, and WiMAX is that the 3G and WiMAX technologies use a licensed spectrum, while Wi-Fi uses an unlicensed shared spectrum. This has important implications for cost of service (CoS), quality of service (QoS), congestion management, and industry structure. For example, both 3G and WiMAX are designed to provide a higher level of reliability and quality of service (especially in voice service) than Wi-Fi. In terms of a business model, 3G represents an extension of the mobile service provider model whereas Wi-Fi and WiMAX come out of the data communications industry, which is a by-product of the computer industry. WiMAX has great advantages over ADSL and cable in terms of reach and backhauling, not to mention mobility support; it can therefore become an extension to ADSL and cable in rural areas when the deployment cost of WiMAX becomes comparable with that of ADSL, estimated as sometime in 2009.

Even though these wireless technologies are competing with one another, they also complement one another in several services and applications. The presence of two or more wireless network systems in the same area is becoming more and more popular, as fourth generation systems will be composed of heterogeneous networks. Various interworking and handover strategies have been presented in the literature for the interworking between 3G, Wi-Fi, and WiMAX. The combination of 3G and Wi-Fi [6] has been viewed as quite complementary because the former provides wide area coverage with high mobility and medium rates and the latter provides local coverage and low mobility with high data rates.

However, with the emergence of the WiMAX networks, operators are increasingly interested in the interworking between Wi-Fi and WiMAX. More specifically WiMAX, which was designed for metropolitan area networks (MANs), complements Wi-Fi by extending its local area networking coverage reach and providing a similar experience in a larger area. Another important synergy between Wi-Fi and WiMAX is in the provisioning of nomadic broadband, which is a hybrid of fixed and mobile broadband. It can now enable mobility within the coverage area of a WiMAX base station, hot zone, i.e., which corrects for the inability of hotspots to hand over at vehicular speeds. The most successful nomadic technology today is Wi-Fi. By integrating with WiMAX, nomadic broadband can make a difference by enabling users to take laptops and subscriptions to other locations and yet continue to receive true broadband service. The difference between a Wi-Fi hotspot and the nomadic broadband enabled by WiMAX is that the consumer does not need to pay extra for the service. It would be available in places not usually served by public hotspots, the connection would be more secure, and the bandwidth would be higher.

9.1   Evolution of 3G technologies

Third generation (3G) mobile telephone systems combine high-speed mobile access with Internet-protocol(IP)-based services. The first generation mobile service was called the advanced mobile phone service (AMPS) and was based on analog frequency-division multiple access (FDMA) technology (see Figure 9.2), which allocates one available frequency channel in response to one requested cellular call and offers its service in the 800–900 MHz bands. In the 1990s, mobile services were upgraded to digital mobile technologies; these are known as the second generation (2G) of mobile services. Within these mobile phone networks, all portable mobile phones receive or make calls through a base station (a transmitting tower). To achieve increased service capacity, reduced power usage, and better coverage, the mobile networks in a large geographic area, which represents the coverage range of a service provider, are split up into smaller and slightly overlapping radio cells to deal with the large number of active users in an area (see Figure 9.3). In cities, each cell site has a range of up to approximately 0.5 mile while in rural areas the range is approximately 5 miles. That is why mobile phone networks are also called cellular phone networks.

The first US 2G system used time-division multiple access (TDMA) technology (see Figure 9.2) based on circuit switching, and was also once known as North American digital cellular (NADC). Time-division multiple access technology was also used to introduce the

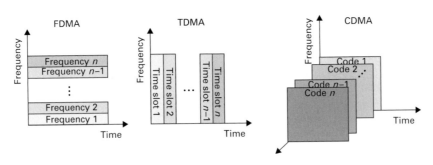

Figure 9.2 Various multiple access technologies: FDMA, TDMA, and CDMA.

global-system-for-mobile(GSM)-based 2G cellular system to Europe in the early 1990s and is now available in more than 100 countries. Code-division multiple access (CDMA) technology (see Figure 9.2) was later adopted to became another popular digital 2G system, with the US introduction of Interim Standard IS-95, also called CdmaOne. The TDMA system alleviated the channel capacity problem by dividing a single radio channel into time slots and then allocating a time slot to a user's digitized and compressed voice data. For example, the US TDMA system had three time slots per 30 kHz channel while the GSM system, also called IS-136, had eight time slots per 200 kHz channel. Instead of allocating time slots, CDMA systems assign each digital voice call with a unique code before it is added to the radio channel. The process is often called noise (spread-spectrum) modulation because the resulting signal looks like background noise. In comparison with TDMA, CDMA offers better capacity at essentially the same or better quality.

To further support higher-bandwidth wireless digital data communications, a vision of the next generation (i.e., third generation, 3G) cellular networks for public land mobile telecommunications systems, called international mobile telecommunications 2000 (IMT-2000) [8], was proposed by the Telecommunication Standardization Section of ITU. The 3G systems are intended to work in a universally acceptable spectrum range and provide voice,

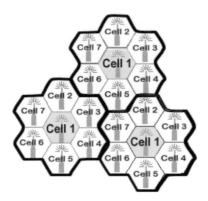

Figure 9.3 Mobile networks are split up into smaller and slightly overlapping radio cells to deal with the large number of active users in an area.

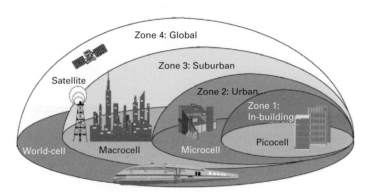

Figure 9.4 The 3G mobile services are divided into different sizes of cell, picocell, microcell, macrocell, and world-cell.

data, and multimedia services. As shown in Figure 9.4, for a technically stationary user operating in a picocell (100 m coverage area, within a building), the data rate would be up to 2.048 Mbps. For a pedestrian user operating in a microcell (0.5 km coverage area, in an urban neighborhood), the data rates would be up to 384 kbps. For a user with vehicular mobility operating in the macrocell (5–10 km metropolitan coverage area), the data rates would be up to 144 kbps. A critical part of the 3G system is the provision of packet-switched data services. The evolution from 2G to 3G begins with the creation of robust, packet-based, data services.

9.1.1 Wideband CDMA-based 3G

The 3G wireless system evolved from GSM is called wideband CDMA (WCDMA) and is also referred to as the universal mobile telecommunications system (UMTS) in the European market. To evolve to the 3G system from a basic GSM system, the general packet radio service (GPRS) was invented as the first step towards introducing a packet-switched data service that is more sophisticated than the short message service (SMS). If all time slots of each GSM's 200 kHz channel were allocated to data services, then theoretically GPRS can deliver a 115.2 kbps peak data rate, which falls short of the minimum 144 kbps required in 3G services. With overheads, the effective rates of GPRS are somewhere in the 20–40 kbps range, which is why GPRS is often referred to as a 2.5G service. Another effort moving GSM toward 3G in the US market was enhanced data rates for GSM evolution (EDGE), which can provide data rates up to 384 kbps. This technology uses the same 200 kHz channel with eight time slots and gets its improved speed by the use of a more efficient modulation (8-PSK) scheme to achieve a 384 kbps peak data rate, even though the actual rates are in the 64–128 kbps range. The strength of EDGE is that it uses a traditional channel size, thus requiring no additional spectrum. At the same time most other markets decided to move directly to WCDMA using the new 3G spectrum, which uses an at least 5 MHz channel to deliver data at rates of up to 2 Mbps. The WCDMA variations include frequency-division duplex (FDD), time-division duplex (TDD), and time-division CDMA (TDCDMA). China has its own standard, time-division synchronous CDMA (TD-SCDMA). The production of UMTS handsets designed to roam between CDMA-based WCDMA and TDMA-based GSM networks provides an additional challenge. Network operators with their own networks (Vodaphone, for example) have resolved this problem; subscribers can now easily roam between WCDMA and GSM networks without a significant number of dropped calls.

To make a globally applicable 3G mobile phone standardization on radio, core networks, and the service architecture of evolved GSM specifications within the scope of the ITU IMT-2000 standard, the Third Generation Partnership Project (3GPP) was formed in December 1998 to establish a collaboration between groups of telecommunications associations, including the European Telecommunications Standards Institute (ETSI), the Association of Radio Industries and Businesses and the Telecommunication Technology Committee (ARIB/TTC) of Japan, the China Communications Standards Association, the Alliance for Telecommunications Industry Solutions of North America, and the Telecommunications Technology Association of South Korea.

Each release of 3GPP incorporates hundreds of individual standards documents, each of which may have been through many revisions. Systems within 3GPP are deployed across much of the established GSM market, specified primarily as Release 99 systems. But, as of 2006, the growing interest in high-speed downlink packet access (HSDPA) is driving the adoption of Release 5 and its successors, as shown in the evolution roadmap of 3GPP in Figure 9.5 [9]. High-speed downlink packet access (HSDPA) provides a smooth evolutionary path for UMTS networks to higher data rates and higher capacities, in the same way as EDGE does in the GSM world. It allows time and code multiplexing among different users and consequently is well matched to the bursty nature of packet-based traffic in a multi-user environment. Downlink dedicated channels and downlink shared channels can share the same frequency spectrum by using code multiplexing. The introduction of shared channels for different users in HSDPA guarantees that channel resources are used efficiently in the packet domain and are less expensive for users than dedicated channels for mobile services, such as Internet access and file download, that require high-data-rate packet transport on the downlink. It is well adapted to the urban environment and to indoor deployment. Some key technological advances adopted in HSDPA include: (1) adaptive modulation and coding schemes (e.g., QPSK and 16-QAM); (2) a hybrid automatic repeat request (HARQ) retransmission protocol; and (3) fast packet scheduling controlled by the medium [10]. With these key technologies, assuming comparable cell sizes, HSDPA has the possibility of achieving peak data rates of about 10 Mbps, with a maximum theoretical rate of 14.4 Mbps.

The 3GPP release 6 high-speed uplink packet access (HSUPA), which is similar to HSDPA, is another 3G mobile telephony protocol in the HSPA family with uplink speeds up to 5.76 Mbps, which considerably improves on today's maximum uplink data rate of 384 kbps. More specifically, coupled with Quality of Service (QoS) techniques and reduced latency, HSUPA uses a packet scheduler to achieve higher uplink rates. It operates on a *request-grant* principle, where the users request permission to send data and the scheduler decides when and how many users will be allowed to do so. A request for transmission contains data about the state of the transmission buffer and the queue at the user and its available power margin. The HSUPA protocol almost doubles cell capacity, resulting in lower cost per bit and, together with HSDPA, provides a complete wireless system supporting an entire range of broadband applications, from faster email synchronization and the uploading of bulky attachments to real-time online gaming, mobile video conferencing, and VoIP.

We now clarify the information relating to the six footnote indicators (on Rev A etc.) in Figure 9.5 [9].

[1]The labels EV-DO Rev A and Rev B show that OFDM has been incorporated for multicasting.

[2]The data rates are based on 64-QAM and a 2×20 MHz FDD band allocation and are scalable with the number of carriers assigned: up to 15 carriers, up to 4.9 Mbps per carrier.

[3]The multiple modes supported are CDMA, TDM, OFDM, OFDMA, and LS-OFDM. The new antenna techniques used are 4×4 MIMO and SDMA. The leverages are EV-DO protocol stack.

[4]The data rates are based on 2×20 MHz FDD band allocation and 4×4 MIMO; they depend on the level of mobility.

[5]The upper range of DL peak data rates for Release 7 and Release 8 enhancements is based on 64-QAM, 2×2 MIMO.

[6]The initial requirements are based on OFDMA in the DL and on SC-FDMA in the UL: they are FDD, 64-QAM, and 2 TX MIMO in DL and a 16-QAM single TX stream in UL.

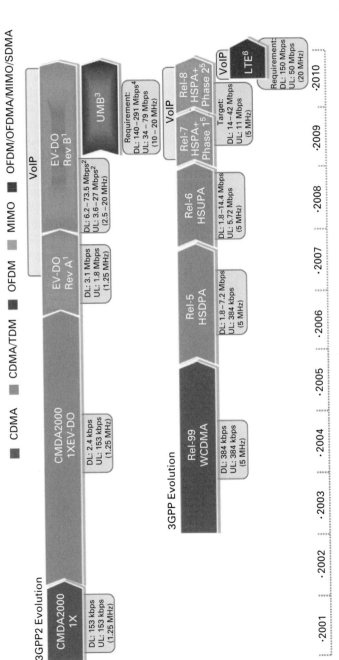

Figure 9.5 The evolution roadmap of 3GPP [9]. Note that the timeline depicts the initial commercial availability of each technology. Those introduced beyond 2008 are under standardization and are subject to variation (© IEEE 2007).

9.1.2   CDMA2000-based 3G

Unlike the rest of the world, which was taking the GSM evolution towards 3G WCDMA, the US decided to take a different route owing to a lack of wireless spectrum; this route was from IS-95 toward 3G CDMA2000, which is structured in a way that allows some 3G service levels to meet IMT-2000 requirements in the traditional 1.25 MHz IS-95 carrier channel. The CDMA2000 $1\times$ technology (see Figure 9.5), originally developed by Qualcomm in 1999, is the air interface technology that immediately follows IS-95 and is backward compatible with IS-95. The term $1\times$ is an abbreviation of $1\times$ radio transmission technology ($1\times$RTT), which denotes one times the IS-95-channel radio carrier size (1.25 MHz) and provides 153 kbps peak data rate for both downlink and uplink transmission.

As a direct evolution of $1\times$RTT technology, the first CDMA2000-based 3G services, referred to as CDMA2000 $1\times$ evolution-data only or $1\times$ evolution-data optimized ($1\times$EV-DO, also called ITU TIA-856) Release (Rel) 0, was approved in August 2001. As the name suggests, EV-DO is a data-centric technology that allows its channels to carry only data traffic [11]. South Korea made an early decision to move to CDMA2000 and deployed the first worldwide mobile broadband technology in 2002, based on $1\times$EV-DO Rel 0 technology. The $1\times$EV-DO Rel 0 provides a peak data rate of 2.4 Mbps in the downlink and 153 kbps in the uplink in a single 1.25 MHz FDD carrier. In commercial networks, Rel 0 delivers an average throughput of 300–700 kbps in the forward link and 70–90 kbps in the reverse link. It offers an always-on user experience and supports IP-based network connectivity with many broadband data applications such as broadband Internet or VPN access, MP3 music downloads, 3D gaming, TV broadcasts, and video and audio downloads. Beyond $1\times$EV-DO, one gets into the realm of multichannel CDMA2000. At its full 3G capability, CDMA2000 can use a 3.75 MHz channel, three times the traditional channel width, and is called $3\times$RTT. The $3\times$RTT is a 3.75 MHz carrier channel implemented in 5 MHz of spectrum; the remaining 1.25 MHz is used for upper and lower guard bands.

Since $1\times$EV-DO is a data only service and no voice is allowed in the channel, the $1\times$EV-data/voice ($1\times$EV-DV) was later developed to offer a true multimedia channel. Continuous work within the standards body on $1\times$EV-DV ceased and instead work became focused on two further enhancements to the first implementation, Rel 0, of EV-DO. The first such enhancement was EV-DO Revision A (TIA-856-A), which was the first upgrade of Release 0; it was approved in March 2004 and commercial services began around 2006. The second enhancement EV-DO Revision B, followed Revision A logically, and was standardized in 2006. Through Revision B, all further EV-DO revisions will be fully backward and forward compatible. Ultimately, there could be several "phases" of Revision B, each introducing greater functionality and richer features.

The EV-DO Rev A implementation improves the performance of Rel 0 by offering faster packet establishment on both the forward and reverse links along with air interface enhancements that reduce latency and improve data rates [11]. Therefore EV-DO Rev A is ready for real-time VoIP applications. Revision A also incorporates OFDM technology for multicasting. In addition to an increase in the maximum burst downlink rate from 2.45 Mbps to 3.1 Mbps, Rev A has a significant improvement in the maximum uplink data rate, from 153 kbps to a maximum uplink burst rate of 1.8 Mbps; typical data rates in fact average below 1 Mbps.

The EV-DO Rev B implementation builds on the efficiencies of Rev A by introducing the concept of dynamically scalable bandwidth [11]. Through the aggregation of multiple

1.25 MHz Rev A channels, Rev B enables data traffic to flow over more than one carrier and hence improves user data rates and reduces latencies on both forward and reverse links. Peak data rates are proportional to the number of carriers aggregated. When 15 channels are combined within a 20 MHz bandwidth, Rev B delivers peak rates of 46.5 Mbps (or 73.5 Mbps with 64-QAM modulation incorporated) in the downlink and 27 Mbps in the uplink. Typical deployments are expected to include three carriers for a peak rate of 14.7 Mbps. Through bundling multiple channels together to achieve a higher data rate, user experience is enhanced and new services such as high-definition video streaming are made possible. To achieve this performance the 1.25 MHz carriers do not have to be adjacent to one another, thus giving operators the flexibility to combine blocks of spectrum from different bands. This is a unique benefit of Rev B that is not available in WCDMA or HSDPA. In addition to supporting mobile broadband data and OFDM-based multicasting, Rev B uses statistical multiplexing across channels to further reduce latency, enhancing the experience for delay-sensitive applications such as VoIP, push-to-talk, video telephony, concurrent A/V, and multiplayer online gaming. Revision B also provides efficient support for services that have asymmetric download and upload requirements (i.e., different data rates are required in each direction) such as file transfers, web browsing, and broadband multimedia content delivery.

As in the case of 3GPP, the Third Generation Partnership Project 2 (3GPP2), established in December 1998, is a collaboration between several CDMA2000 associations (rather than WCDMA in the case of 3GPP) specifically to create globally applicable 3G mobile phone system specifications within the scope of the ITU's IMT-2000 project. The participating associations include ARIB and TTC of Japan, the China Communications Standards Association, the Telecommunications Industry Association of North America, and the Telecommunications Technology Association of South Korea.

9.1.3 Moving towards 4G wireless

Even with today's fast growing use of 3G mobile services, there are still important disadvantages in comparison with those of wired networks. One main concern is that the bandwidth is still not as large and as stable as for fixed wired networks. Therefore a much higher wireless bandwidth is still highly desirable, since if the subscriber data rate can be improved then the experience of a single subscriber would be similar to that delivered by a fixed wired network. This would allow many fixed wired network services to be shifted to mobile networks, according to so-called fixed mobile convergence (FMC). Currently, it is still a challenging task for 3G services to remain online permanently, implying that radio and core networks maintain continuous sessions including push-to-talk over cellular (PoC) presence and other IP-based services. The next generation wireless system should not only improve communication efficiency but also enable subscribers to enjoy more diversified and personalized services, such as browsing, searches, location management, information exchange, purchasing, and banking, as well as a range of entertainment multimedia services, for example, IPTV, content/multimedia, VOD, and online gaming. These considerations call for the following three 4G wireless systems: the long term evolution (LTE) of 3GPP, the ultra mobile broadband (UMB) of 3GPP2, and the IEEE 802.16e WiMAX.

There are two main goals of 4G wireless systems. First of all, higher and more stable bandwidth is required as just discussed. Second, and more importantly, 4G networks will no longer have a circuit-switched subsystem as do the current 2G and 3G networks. Instead, the network is based purely on the Internet protocol (IP). The main challenge of this design is how to support the stringent requirements of voice calls for constant bandwidth and delay. Having sufficient bandwidth is a good first step. Efficient mobility and QoS support for a voice and video connection is clearly another. Currently, 3G networks are transforming into 3.5G networks, as carriers add technologies such as HSDPA and high HSUPA to UMTS. Similar activities can be observed in the EV-DO revisions. Staying with the UMTS example, such 3.5G systems are realistically capable of delivering about 6–7 Mbps in a 5 MHz band. Higher-bitrate designs have been reported too. However, these speeds can only be reached under ideal conditions (very close to the antenna, no interference, etc.) which are rarely found in the real world. To actually achieve the speeds specified for 4G wireless systems, some technological breakthroughs are needed, more specifically: orthogonal frequency-division multiplexing access (OFDMA), to replace the CDMA used in 3G: scalable channel bandwidths up to 20 MHz; and multiple-input multiple-out (MIMO) smart antenna technologies.

9.1.3.1 OFDM, OFDMA, and MIMO

Orthogonal frequency-division multiplexing (OFDM), also referred to as multicarrier or discrete multitone modulation, is a technique which divides a high-speed serial information signal into multiple lower-speed subsignals, which the system transmits simultaneously at different frequencies in parallel. It has been successfully adopted in digital A/V broadcasting (DAB/DVB) standards. The benefits of OFDM are high spectral efficiency, resilience to RF interference, and lower multipath distortion. More specifically, in multicarrier modulation, the signal bandwidth is divided (decimated) into parallel subcarriers, i.e., narrow strips of bandwidth, which overlap. As shown in Figure 9.6(a), the signal is passed through a series-to-parallel converter, which splits the data stream into K low-rate parallel subchannels. The data symbols mapped from the bit sequence of each subchannel are applied to a modulator; there are K modulators whose carrier frequencies are $\{f_0, f_1, \ldots, f_K\}$. The K modulated carriers are then combined to generate the OFDM signal. The OFDM system uses parallel overlapping subcarriers, from which information can be extracted individually [12]. This is possible because the frequencies (subcarriers) are orthogonal, meaning that the peak of one subcarrier coincides with the null of an adjacent subcarrier, as shown in Figure 9.6(b). The orthogonal nature of OFDM allows subchannels to overlap, and this has a positive effect on spectral efficiency. Each subcarrier transporting information is just far enough apart from the others to reduce intersymbol interference caused by adjacent carriers; this arises from the interleaving of the signals forwarded to each modulator. Since the duration of each time-interleaved symbol is long, it is feasible to insert a guard interval between the OFDM symbols to reduce further intersymbol interference. Moreover, the computational complexity required in the K subchannels' modulation can be greatly reduced by employing an FFT-based implementation.

Because transmissions are at a lower data rate, multipath-based delays are not nearly as significant as they would be with a single-channel high-rate system. The use of multicarrier OFDM significantly reduces this problem. Multipath distortion can also cause intersymbol interference, which occurs when one signal overlaps with an adjacent signal. Orthogonal frequency-division multiplexing signals typically have a time guard of 800 ns, which is good enough for overcoming intersymbol interference except in the harshest environments.

The OFDM system, as mentioned above, uses a large number of evenly spaced subcarriers for modulation to increase the efficiency of data communications by increasing data throughput. This system allows only one user on the channel at any given time. To accommodate multiple users, the multi-user OFDM access (called OFDMA) was

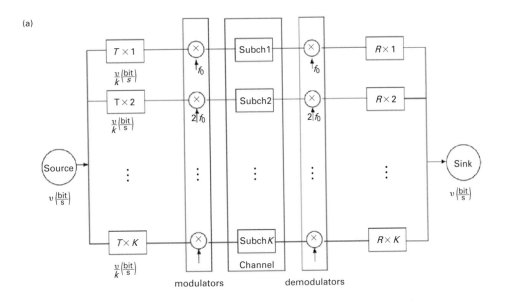

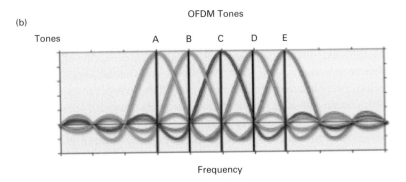

Figure 9.6 (a) A simplified block diagram of the OFDM scheme [12] (© IEEE 2007). (b) The subcarrier frequencies are orthogonal, meaning that the peak of one subcarrier coincides with the null of an adjacent subcarrier.

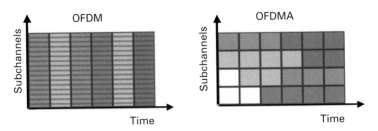

Figure 9.7 Optimizing subchannels with OFDMA, which divides a channel into subchannels. The coding and modulation schemes of each subchannel are assigned separately, in contrast with OFDM. Each subchannel is optimized according to its link condition.

introduced. Like OFDM, OFDMA employs multiple closely spaced subcarriers, but the subcarriers are further divided into groups of subcarriers. Each group is named a subchannel (see Figure 9.7). The subcarriers that form a subchannel need not be adjacent. The OFDMA system can accommodate many users in the same channel at the same time. Moreover, subcarrier-group subchannels can be matched to each user to provide the best performance, i.e. meaning the fewest problems with fading and interference, on the basis of the location and propagation characteristics of each user. Another alternative is to assign a subchannel to a different user, who may have better channel conditions for that particular subchannel; this allows users to concentrate transmitted power on specific subchannels, resulting in improvements to the uplink budget and providing a greater range. This technique is known as space-division multiple access (SDMA). With OFDMA, the client device is assigned subchannels on the basis of geographical location, with the potential of eliminating the impact of deep fades.

Multiple-input multiple-output (MIMO), or smart antenna, systems take advantage of *multipath interference*, which occurs when transmitted signals bounce off objects and create reflected signals that take multiple paths to their destination (see Figure 9.8 [14]). With standard antennas, signals arriving out of phase can interfere with and cancel out one another; MIMO systems, however, use multiple receiving and sending antennas that can fit onto a wireless device. By resolving data from the multiple flows, MIMO can use multipath channels as additional data paths rather than just redundant carriers of the original signal, thereby increasing bandwidth and transmission range. More specifically, using multiple antennas, MIMO uses the spectrum more efficiently without sacrificing reliability. It uses multiple diverse antennas tuned to the same channel, each transmitting with different spatial characteristics. Every receiver listens for signals from every transmitter, enabling path diversity where multi-path reflections (which are normally disruptive to signal recovery) may be recombined to enhance the desired signals. Another valuable benefit that MIMO technology may provide is spatial-division multiplexing (SDM). Spatial-division multiplexing spatially multiplexes multiple independent data streams (essentially virtual channels) simultaneously within one spectral channel of bandwidth; MIMO SDM can significantly increase data throughput as the number of resolved spatial data streams is increased. Each spatial

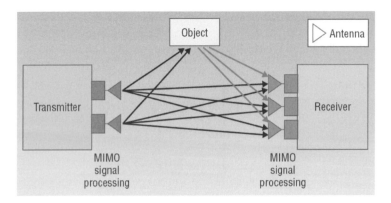

Figure 9.8 Multiple-input multiple-output (MIMO) systems use multiple receiving and sending antennas to resolve multiple data flows so that they can use multipath channels as additional data paths rather than just redundant carriers of the original signal [14] (© IEEE 2006).

stream requires its own transmit and receive (TX/RX) antenna pair at each end of the transmission.

9.1.3.2 Long-term evolution (LTE) of 3GPP

The long-term evolution (LTE) of 3GPP is the next project designed to improve the UMTS mobile phone standard in order to cope with future requirements. As of 2008 LTE is still an ongoing project, but it will be ratified as the new evolved Release 8 of the UMTS standard at the end of 2008 or in early 2009; this will include additional extensions and modifications of the UMTS system. The main objective of the LTE project is to upgrade UMTS to a so-called fourth generation (4G) mobile communications technology, essentially a wireless broadband Internet system with voice and other services built on top. There has been a large amount of work aiming at simplifying the architecture of the system, as it transitions from the existing UMTS circuit plus packet-switching combined network to an all-IP 4G system. Some specific performance targets of LTE include: (1) download rates of 100 Mbps, and upload rates of 50 Mbps for every 20 MHz of spectrum; (2) at least 200 active users in every 5 MHz cell; (3) sub-5 ms latency for small IP packets; (4) increased spectrum flexibility, with spectrum slices as small as 1.25 MHz (and as large as 20 MHz) supported; (5) an optimal cell size of 5 km, a 30 km size with reasonable performance, and up to a 100 km size supported with acceptable performance; (6) co-existence with legacy standards so that users can interoperably start a call or transfer of data in an area using an LTE standard and, should coverage be unavailable, continue the operation without any action on their part using GSM and GPRS or WCDMA-based UMTS.

It is important to note that Release 8 is intended for use over any IP network, including WiMAX and WiFi, and even over wired networks. The proposed system uses OFDMA for the downlink (tower to handset) and single-carrier FDMA (SC-FDMA) for the uplink and employs multiple-input multiple-output (MIMO) with up to four antennas per station. The channel coding scheme for the transport blocks is turbo coding and a contention-free quadratic permutation polynomial (QPP) turbo code internal interleaver. The use of OFDMA in LTE has a link spectral efficiency greater than the older CDMA-based systems that dominated 3G and, when combined with modulation formats such as 64-QAM and techniques such as MIMO, the LTE system should be considerably more efficient than WCDMA with HSDPA and HSUPA. Companies such as Siemens, Nokia, Nortel, and Verizon started some demonstrations of LTE prototyping systems in 2006 and 2007.

9.1.3.3 Ultra-mobile broadband (UMB) of 3GPP2

The ultra-mobile broadband (UMB) is the next generation of ultra-fast mobile broadband networks within 3GPP2 designed to improve the CDMA2000 mobile phone standard for next generation applications. Ultra-mobile broadband is expected to deliver economically IP-based voice, multimedia, broadband, information technology, entertainment, and consumer electronic services within most kinds of devices. In September 2007, 3GPP2 announced the official release of UMB 2.0 in San Diego, USA, the industry's first 4G technical standard. The UMB system also employs OFDMA technology along with MIMO and space-division multiple access (SDMA) techniques to provide peak rates of up to 291 Mbps for downlink and 79 Mbps for uplink using an aggregated 20 MHz channel. To provide ubiquitous and universal access, UMB supports intertechnology handoffs with other technologies, including the existing CDMA2000 $1\times$ and $1\times$EV-DO systems. The UMB standardization was completed in 2007, and commercialization is expected to take place around mid-2009.

Ultra-mobile broadband has a 3–4 times higher downlink spectrum efficiency and a 2–3 times higher uplink spectrum efficiency than those of existing 3G technologies (such as EV-DO). The flattened architecture can further reduce air interface delays so that the subscriber-side one-way delay can be less than 17 ms, which remarkably improves user experience and raises operators' competitive strengths. The UMB network can also remain online permanently, so that radio and core networks can maintain continuous sessions in most IP-based multimedia services.

9.1.3.4 Competition between 4G technologies

As all three types of 4G network (WiMAX, LTE, and UMB) have similar technologies and properties, the time-to-market will be an essential component in the overall competitiveness of a standard. The network that is set to enter the market first is 802.16e Mobile WiMAX, since the air interface part of the standard was approved by the members of the IEEE standards body in February 2006. The 3GPP has also started its activities around 4G and the UMTS LTE standardization is well on track. It is widely believed that work on the WiMAX standards is 12 to 24 months ahead of the LTE work. According to Alamouti [7], Intel believes that Mobile WiMAX can meet the needs of mobile operators for their data-centric mobile broadband systems, and 3GPP LTE does not have any fundamental technological advantage over Mobile WiMAX. Mobile WiMAX already has a ratified standard with products in the market and large-scale deployment. In fact, the timelines of LTE are similar to the evolved version of Mobile WiMAX for 2009+ deployment, i.e., IEEE 802.16m WiMAX Release 2, which has even higher performance (see Table 9.1). The work on 3GPP2 UMB seems to be even further behind WiMAX; the reason might be that the 3GPP2 group is still working on EV-DO Rev B, the multicarrier extension to the current EV-DO Rev A networks.

The LTE and UMB projects might gain some advantages due to their backward compatibility with a predecessor technology which is already in place. Thus, handsets and other mobile devices will work not only in LTE or UMB networks but also in existing 3GPP or 3GPP2 networks. This is especially important in the first few years of network deployment when coverage is still limited to big cities. WiMAX, however, is not backward compatible with any previous wireless network standard. Thus, it remains to be seen whether devices will also include a 3G UMTS or EV-DO chip with internetworking capabilities. It is expected that to be able to deploy in a large-scale economical wireless infrastructure, WiMAX will be closely internetworking with Wi-Fi for indoor internet access as well as in backhauling for Wi-Fi environments. Moreover, WiMAX also needs to be internetworking closely with the existing 2G networks in order to handle most of the voice traffic with minimal cost. This is not only a question of technology but also a question of strategy. On the one hand, if a company with a previously installed 2G or 3G network deploys WiMAX then they will surely be keen to offer such handsets. New alternative operators without an already existing network, on the other hand, might be reluctant to offer such handsets as they would have to partner with an already existing network operator. They might not have much choice, though, if they want to reach a wider target audience.

As 4G network technology is based on IP only and includes no backward compatibility for circuit-switched services, current operators do not necessarily have to select the evolution path of the standard that they are currently using. The only consideration would be compatibility with their existing core network infrastructure, billing systems, and services. Since current 3.5G networks offer enough capacity for a number of years to come, 3GPP or

Table 9.1 Performances of 3GPP LTE, 3GPP UMB, and 802.16m Mobile WiMAX [7] (© IEEE 2007)

Feature	3GPP LTE (Source: 3GPP RAN1)	3GPP2 UMB (Source: Qualcomm)	802.16m Mobile WiMAX R2
Duplexing modes	TDD, FDD	TDD, FDD	TDD, FDD
Channel bandwidths	1.25, 1.6, 2.5, 5, 10, 15, 20 MHz	1.25 to 20 MHz	5, 10, 20, 40 MHz
Peak data rates (per sector at 20 MHz)	DL: 288 (4×4) UL: 98 (2×4)	DL: 250 Mbps (4×4) UL: 100 Mbps (4×4)	DL: > 350 Mbps (4×4) UL: > 200 Mbps (2×4)
Mobility	Up to 350 km/hr	Up to 250 km/hr (350 km/hr per SRD)	Up to 350 km/hr
Latency	Link-layer access: < 5 ms Handoff: < 50 ms	Link-layer access: < 10 ms Handoff: < 20 ms	Link-layer access: < 10 ms Handoff: < 20 ms
MIMO configuration	DL: 2×2, 2×4, 4×2, 4×4 MIMO UL: 1×2, 1×4, 2×2, 2×4 MIMO	DL: 2×2, 2×4, 4×2, 4×4 MIMO UL: 1×2, 1×4, 2×2, 2×4 MIMO	DL: 2×2, 2×4, 4×2, 4×4 MIMO UL: 1×2, 1×4, 2×2, 2×4 MIMO
Average sector throughput @ 20 MHz TDD (DL:UL = 2:1)	DL: ~ 32 Mbps UL: ~ 6 Mbps	DL: 32 Mbps UL: 7.1 Mbps	DL: > 40 Mbps UL: > 12 Mbps
Spectral efficiency (per sector)	Peak DL: 14.4 bps/Hz (4×4) UL: 4.8 bps/Hz (2×4) Sustained DL: ~ 2.4 bps/Hz UL: ~ 0.9 bps/Hz	Peak DL: 13 bps/Hz (4×4) UL: 5 bps/Hz (4×4) Sustained DL: ~ 2.3 bps/Hz UL: ~ 1.0 bps/Hz	Peak DL: > 20 bps/Hz (4×4) UL: > 10 bps/Hz (2×4) Sustained DL: > 3 bps/Hz UL: > 1.5 bps/Hz
Coverage (km)	5/30/100 km (Optimal performance at 5 km)	5/10/30/100 km	1/5/30 km
Number of VoIP active users	> 80 users/sector/FDD MHz	> 64 users/sector/FDD MHz (100 per 3GPP2 SRD)	> 100 users/sector/FDD MHz > 50 users/sector/TDD MHz

3GPP2 operators are currently in no hurry with 4G technologies. However, WiMAX might have a chance, since some operators may try a different game to see whether they can gain a competitive advantage. For example Sprint, by collaborating with ClearWire (www.clear-wire.com), in the USA made a decision and announced that they had chosen mobile WiMAX as their 4G technology; then started a rollout of network trials in early 2008. The disadvantage of WiMAX in not having a network legacy could in the end be a major advantage. It will allow new companies to enter the market more easily and thus increase competition, network coverage, and services and, it is to be hoped, decrease prices.

9.2   Wi-Fi wireless LAN (802.11)

Thanks to the authorization of the industrial, scientific, and medical (ISM) frequency bands by the Federal Communications Commission (FCC) of the US in 1985, the development of wireless local area networks (WLANs) was accelerated. In 1989, the IEEE 802.11 Working Group began elaborating on WLAN medium access control (MAC) and physical (PHY) layer specifications. The final draft was ratified on 26 June 1997. Since then 802.11-based WLANs have been rapidly accepted and recently deployed widely in many different environments, including campuses, offices, homes, and hotspots. In the communication architecture of an IEEE 802.11, all components that can connect into a wireless medium in a network are referred to as stations (STAs). All stations are equipped with wireless network interface cards (WNICs). Wireless stations fall into one of two categories, access points and clients. An access point (AP) is the base station for the wireless LAN. It transmits and receives the radio frequencies at which wireless enabled devices can communicate. The transmitter of a WLAN AP sends out a wireless signal that allows wireless devices to access it within a circle of roughly 100 meters. The zone around the transmitter is known as a hotspot. A wireless client can be a mobile device such as a laptop, a personal digital assistant (PDA), an IP phone, or a fixed device such as a desktop or workstation equipped with a WNIC. In the network configuration of a WLAN, a basic service set (BSS) is defined as a set of wireless stations that can communicate with each other. Two types of configuration (mode) are specified for a BSS in the standard, i.e., ad hoc BSS and infrastructure BSS [15]. An ad hoc BSS (also referred to as an independent BSS, IBSS), see Figure 9.9(a), contains no APs, which means it cannot connect to any other basic service set. In an infrastructure BSS, see Figure 9.9(b), a station can communicate with other stations, whether or not they are in the same basic service set, only through APs.

The ad hoc mode of a BSS enables mobile stations to interconnect with each other directly (peer-to-peer) without the use of an AP; all stations are usually independent and equivalent in an ad hoc network. Stations may broadcast and flood packets in the wireless coverage area without accessing the Internet. Like the addressing system in wired LAN (802.3), each device in the WLAN (802.11) system, including stations, access points, and routers, has its own unique 48-bit local area network MAC address (also known as a link address). Each data packet to be exchanged in a WLAN is called a frame, which has the same payload as the packet but with some WLAN MAC headers and QoS control components. The ad hoc configuration can be deployed easily and promptly when the users involved cannot access or do not need a network infrastructure. For instance, participants at a conference can configure their laptops as a wireless ad hoc network and exchange data without much effort. However, in many instances, the infrastructure network configuration

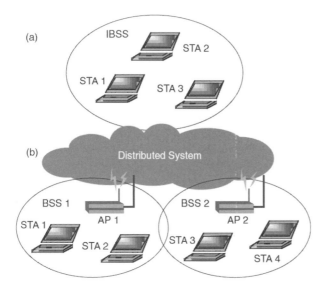

Figure 9.9 (a) Ad hoc and (b) infrastructure WLAN architectures [15] (© IEEE 2003).

is adopted. In the infrastructure mode there are access points which bridge wireless stations and the wired network. Basic service sets can be connected by a distributed system, which is normally a LAN, to form an extended service set (ESS) that may contain multiple APs, as shown in Figure 9.9(b). The coverage areas of these BSSs in the ESS usually overlap. Handoff will happen when a station moves from the coverage area of one AP to another AP. Although the radio range of a BSS limits the movement of wireless stations, seamless roaming among BSSs can help to construct a campus-wide wireless network service. It is also possible to move with slow velocity (nomadic users) without breaking the network connection while connected to a Wi-Fi network.

9.2.1 Various IEEE 802.11 standards

Wi-Fi's popularity really took off with the growth of high-speed broadband Internet access in homes and business offices. It was, and remains, the easiest way to share a broadband link between several computers spread over a home or office. The growth of hotspots, free and fee-based public access points, has added to Wi-Fi's popularity. There were several standards under the umbrella of IEEE 802.11 Wi-Fi as listed in Table 9.2 [16]. The original version of the standard was called Legacy 802.11, released in 1997 and ratified in 1999, which specified two raw data rates of 1 and 2 Mbps to be transmitted in the Industrial Scientific Medical (ISM) frequency band at 2.4 GHz on the basis of the frequency-hopping spread spectrum (FHSS) and the direct sequence spread spectrum (DSSS). Legacy 802.11 was rapidly supplemented and popularized by 802.11b. The 802.11b standard is the longest, most well-supported, stable, and cost effective standard; it runs in the same ISM band of 2.4 GHz range, which makes it prone to interference from other devices (e.g., microwave ovens, cordless phones, etc.) and also has security disadvantages. The number of 802.11b APs in an ESS is limited to three. It has 11 channels, three of which are non-overlapping, to avoid frequency interference with other devices and neighboring WLANs. The 802.11b uses direct-sequence spread-spectrum technology and supports rates from 1 to 11 Mbps. A more flexible design that is completely different from 802.11b is 802.11a, which runs in the Unlicensed National Information Infrastructure band of 5 GHz range and has less interference from other devices. It is more flexible because multiple channels can be combined for faster throughput and more access points can be collocated. It uses frequency-division multiplexing technology and has a shorter range of radio coverage than 802.11b (or 802.11g). The 802.11a standard has 12 channels, eight of which are non-overlapping, and supports rates from 6 to 54 Mbps. A new variant, 802.11g, which uses the more advanced OFDM technology, is backward compatible with 802.11b and allows a smooth transition from 802.11b to 802.11g. The 802.11g standard uses the 2.4 GHz band and can achieve speeds of up to 54 Mbps with a shorter range of radio coverage than 802.11b.

Presently, 802.11a/b/g WLANs provide adequate performance for today's networking applications where the convenience of a wireless connection is the chief consideration. Next generation wireless applications will require higher WLAN data throughput and people will begin to demand more range. In response to these needs, IEEE 802.11n was introduced and has been on the market since 2008. The objective of the task group working on 802.11n was to define modifications to the PHY layer and MAC layer that can deliver a minimum of 100 Mbps throughput at the MAC SAP (service access point). As is the case with its predecessor, the new 802.11n Wi-Fi operates in the 2.4 GHz frequency range and is backward compatible with 802.11b and 802.11g. Its minimum throughput requirement approximately quadruples the WLAN throughput performance in comparison with today's 802.11a/g networks. Its over-the-air throughput is targeted to exceed 200 Mbps in order to meet the 100 Mbps MAC SAP throughput requirement. Other necessary improvements include the range at given throughputs, robustness to interference, and an improved and more uniform service within the coverage of a BSS. To meet these demands and so increase the physical transfer rate, 802.11n adds MIMO smart antenna systems for both the transmitter and the receiver on top of the 802.11g technology.

More specifically, 802.11n bonds two or more of MIMO's 20 MHz channels together to create even more bandwidth. Thus, 802.11n has a theoretical maximum throughput of 600 Mbps and a transmission range of up to 50 meters. For security, 802.11n relies on earlier technologies such as the 802.11i-based Wi-Fi protected access.

Some concerns over the use of WLANs are their power consumption and security protection. Even though Wi-Fi networks have limited radio coverage range, e.g., a typical Wi-Fi home router using 802.11b or 802.11g has a range of 150 ft (46 m) indoors and 300 ft (92 m) outdoors, the power consumption is fairly high compared with other networking standards, making battery life and heat a matter of concern. Wi-Fi commonly uses the *wired equivalent privacy* (WEP) protocol for protection, which can be easily breakable even when properly configured. Newer wireless solutions are using the superior Wi-Fi protected access (WPA) protocol in their implementation of the 802.11i protocol, though many systems still employ WEP. Adoption of the 802.11i protocol makes available a better security scheme for future use when it is properly configured.

9.2.2   MAC layer of a WLAN

The IEEE 802.11 standard [1] specifies the PHY layer and the MAC layer [17] of a WLAN. The IEEE 802.11 MAC specifies two different MAC mechanisms in WLANs: the

Table 9.2 Various IEEE 802.11 standards and their characteristics [16] (© IEEE 2003)

Standard	Spectrum	Maximum physical rate	Layer-3 data rate	Transmission	Compatible with	Major advantages	Major disadvantages
802.11[a]	2.4 GHz	2 Mbps	1.2 Mbps	FHSS/DSSS	None	Higher range	Limited bitrate
802.11a	5.0 GHz	54 Mbps	32 Mbps	OFDM	None	Higher bitrate in less crowded spectrum	Smallest range of all 802.11 standards
802.11b	2.4 GHz	11 Mbps	6–7 Mbps	DSSS	802.11	Widely deployed, higher range	Bitrate too low for many emerging applications
802.11g	2.4 GHz	54 Mbps	32 Mbps	OFDM	802.11/802.11b, due to narrow spectrum	Higher bitrate in 2.4 GHz spectrum	Limited number of collocated WLANs

[a] Legacy standard.

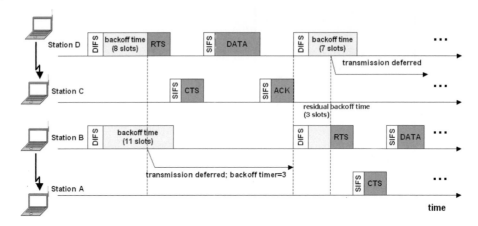

Figure 9.10 The CSMA/CA protocol for 802.11 MAC DCF.

contention-based distributed coordination function (DCF) and the polling-based point coordination function (PCF). At present, only the mandatory DCF is implemented in 802.11-compliant products. The 802.11 PHY layers provide multiple data transmission rates by employing different modulation and channel coding schemes. In the MAC layer, the DCF is known as the carrier sense multiple access with collision avoidance (CSMA/CA) protocol. This protocol is specified in all the stations of both the ad hoc and the infrastructure configurations. All 802.11 stations can deliver MAC service data unit (MSDU) frames of arbitrary lengths up to 2304 bytes, with the help of the MAC protocol data unit (MPDU), after detecting that there is no other transmission in progress on the channel. If two or more stations detect the medium as being idle at the same time, they may initiate their transmissions at the same time, and inevitably a collision occurs. To reduce the probability of collisions, the DCF applies a collision avoidance (CA) mechanism, where stations perform a so-called *backoff procedure* before initiating a transmission together with optional RTS/CTS control functions.

According to CSMA/CA (see Figure 9.10), when the wireless medium is sensed as being idle, the STA waits for a time interval called the DCF interframe space (DIFS) and then enters a backoff procedure. A slotted *backoff time* is generated randomly from a *contention window (CW)* size:

$$backoff\_time = rand[0; CW] \times slot\_time$$
$$= rand[0; CW_{min} \times 2^n] \times slot\_time, \tag{9.1}$$

where CW is set equal to a minimum value, CW_{min}, at the first transmission attempt ($n = 0$), and n denotes the number of consecutive unsuccessful transmissions. As indicated in Eq. (9.1), CW is doubled after each unsuccessful transmission until reaching a maximum value, CW_{max}. It is reset to CW_{min} after successful transmission. The IEEE 802.11 MAC implements a network allocation vector (NAV), which stores the backoff time that indicates to a station the amount of time that remains before the medium will become available. The backoff time is decremented by each slot when the medium is sensed as being idle for that slot. It is frozen if the medium becomes busy and resumes after the medium has been sensed idle again for another DIFS period. Only when the backoff time reaches zero is the STA authorized to access the medium. If two or more STAs finish their backoff procedures and detect the medium as idle at the same time, they may transmit frames simultaneously; thus, a collision may occur. As an optional feature, the 802.11 standard also includes the request to send (RTS) and clear to send (CTS) functions to control station access to the medium. If the RTS or CTS feature is enabled at a particular station, it will refrain from sending a data frame until the station completes an RTS or CTS handshake with another station, such as an AP. More specifically, a station initiates the process by sending an RTS frame. The AP receives the RTS and responds with a CTS frame. The station must receive a CTS frame before sending the data frame. The CTS contains a time value that alerts other stations to avoid accessing the medium while the station initiating the RTS transmits its data. The RTS or CTS handshaking provides positive control over the use of the shared medium. A positive acknowledgement (ACK) is used to notify the sender that the frame has been successfully received. The time duration between a data frame and its ACK is the short interframe space (SIFS). If an ACK is not received within a time period *ACKTimeout*, the sender assumes that there is a collision and schedules a retransmission by entering the backoff process again until the maximum retransmission limit is reached.

The CSMA/CA mechanism can also solve the *hidden node* and *exposed node problems*, as illustrated in Figure 9.11. As shown in part (a) of the figure, node B can reach both node A and node C, but node A and node C cannot hear each other due to the separating distance. If node A is transmitting data to node B, node C will not detect the transmission. Thus node C might also send data to node B. Thus, a data collision will occur in node B. This is called the *hidden node problem* [15] since node A and node C are *hidden* from each other. The exchange of RTS and CTS frames between nodes A and B will prevent node C from sending data to node B, since node C will detect the CTS frame sent by node B.

In Figure 9.11(b), we assume that nodes B and C intend to transmit data only without receiving data. When node C is transmitting data to node D, node B is aware of the transmission, because node B is within the radio coverage of node C. Without exchanging

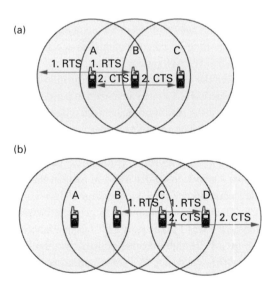

Figure 9.11 (a) Hidden node and (b) exposed node problems [15]. The large circles show the radio coverage of the nodes (© IEEE 2003).

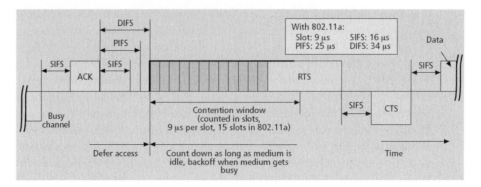

Figure 9.12 A summary of the CSMA/CA backoff procedure (© IEEE 2003).

RTS and CTS frames, node B will not initiate data transmission to node A because it will detect a busy medium. The transmission between nodes A and B, therefore, is blocked even if both are idle. This is referred to as the *exposed node problem*. Also, in Figure 9.11(b) the RTS frame sent by node C will reach node B, but the CTS frame sent by node D will not propagate to node B because node B is not within the radio coverage of node D. Thus, node B knows that the transmission from node C to node D might not interfere with its transmission to node A. Hence, node B will try to initiate transmission by sending an RTS to node A if it has data to transmit. This RTS/CTS strategy overcomes the exposed node problem and enhances the radio efficiency. However, when the data frame size is small, the overheads of RTS/CTS are considerable and it is recommended that this option be disabled. A summary of the CSMA/CA backoff procedure and the corresponding CW_{min} and various IFS parameter values for the 802.11a DCF are shown in Figure 9.12 [18].

Another access control mechanism, called the point coordination function (PCF), was included in the original 802.11 MAC standard to support multimedia transmissions but can only be used if a WLAN operates in an infrastructure mode. It is an optional MAC function because the hardware implementation of PCF was thought to be too complicated at the time the standard was finalized. The PCF is a polling-based contention-free access scheme, which uses an AP as a point coordinator. When a WLAN system is set up with PCF enabled, the channel access time is divided into periodic intervals called *beacon intervals*. A beacon interval is composed of a contention-free period (CFP) and a contention period (CP). During a CFP the AP maintains a list of registered STAs, and it polls them according to the list. Only after a STA is polled can it start transmission. The size of each data frame is bounded by the maximum MAC frame size (2304 bytes). If the PHY data rate of every STA is fixed, the maximum CFP duration for all STAs, *CFP_max_duration*, can then be decided by the AP. However, the link adaptation (multirate support, to be discussed in Section 9.2.3) capability makes the transmission time a frame variable and may induce large delays and jitters that reduce the QoS performance of PCF. The time used by an AP to generate beacon frames is called the target beacon transmission time (TBTT). The next TBTT is announced by the AP within the current beacon frame. To give PCF higher access priority than the DCF, the AP waits for an interval, shorter than the DIFS, called the PCF interframe space (PIFS), before starting PCF. But PCF is not allowed to interrupt any ongoing frame transmissions in DCF. Although the PCF mechanism was intended to support QoS, several problems have been identified in its original specification and it has not been widely adopted.

9.2.3   Link or rate adaptation of 802.11

As discussed previously IEEE 802.11b, based on direct sequence spread-spectrum (DSSS) technology, can have data rates up to 11 Mbps in the 2.4 GHz band and IEEE 802.11a, based on orthogonal frequency-division multiplexing (OFDM) technology, can have data rates up to 54 Mbps in the 5 GHz band. Moreover, the IEEE 802.11g standard, which extends the 802.11b PHY layer, can support data rates up to 54 Mbps in the 2.4 GHz band. There are many reasons for the highly volatile nature of the wireless medium used by the IEEE 802.11 standard: fading, attenuation, interference from other radiation sources, interference from other 802.11 devices in an ad hoc network, etc. To accommodate this volatile nature of the wireless medium, IEEE 802.11 stations support multiple transmission rates, which are used for frame transmissions in an adaptive manner depending on the underlying channel condition. For example, the IEEE 802.11b supports four transmission rates, namely, 1, 2, 5.5, and 11 Mbps, and the high-speed 802.11a supports eight transmission rates, from 6 to 54 Mbps. In nomadic systems such as WLANs without high mobility, the channel condition is often dominated by the distance between the transmitting and receiving stations. In such a case, the further the distance between two stations the lower the transmission rate typically employed. For example, in IEEE 802.11b WLAN, as a station moves away from its AP it decreases the transmission rate it uses for transmissions to the AP from 11 Mbps to 1 Mbps. A mechanism for selecting one out of multiple available transmission rates is referred to as *link adaptation* (also called *rate adaptation*) and is the process of dynamically switching data rates to match the wireless channel conditions, with the goal of selecting the rate that will give the optimum throughput for the given channel conditions. For example, owing to the heuristic and conservative nature of the link adaptation schemes implemented in most 802.11 devices, the current 802.11 systems are likely to show low bandwidth utilization when the wireless channel presents a high degree of variation.

The dynamic nature of transmission-quality variations can be categorized into either transient short-term modifications to the wireless medium or durable long-term modifications to the transmission environment. Typically, if someone walks around, closes a door, or moves big objects, this will have an effect on the transmission medium for a very short time. Its throughput capacity might drop sharply but not for long. If a user decides to move to another office, thus approaching the AP, the attenuation will decrease and this will have a longer lasting effect on the energy of the radio signal that will probably decrease the bit error rate (BER). This, in turn, will allow higher application-level throughput since the packet error rate (PER) is lower. Algorithms that adapt the transmission parameters to the channel conditions can be designed to optimize mainly either power consumption and/or throughput. More specifically, mobile devices which implement 802.11 radios usually have a fixed energy budget (owing to finite battery life), therefore it is of quite high importance to minimize power consumption. Furthermore, even though higher 802.11 transmission rates can provide a potentially higher throughput they usually have higher BERs, and this results in more retransmissions to obtain error-free transmission and thus decreases the effective throughput.

There are two aspects to rate adaptation: channel-quality estimation and rate selection [19]. Channel-quality estimation involves measuring the time-varying state of the wireless channel for the purpose of generating predictions of future quality. Issues include: which metrics should be used as indicators of channel quality (e.g., signal-to-noise ratio, signal strength, symbol error rate, bit error rate), which predictors should be used, whether predictions should be short term or long term, etc. It is common practice to perform channel-quality estimation in the receiver, which determines whether a packet can be received. Furthermore, it is also advantageous to minimize the delay between the time of the estimate and the time the packet is transmitted. Rate selection involves using the channel-quality estimation and predictions to select an appropriate rate. A common technique is threshold selection, where the estimated values of channel quality are compared with a list of threshold values to determine the appropriate data rate.

9.2.3.1 Autorate fallback

Autorate fallback (ARF) [20] was the first published rate adaptation algorithm. It was designed to optimize the application throughput in Lucent Technologies' WaveLan-II devices, which implemented the 802.11 DSSS standard. In ARF, each sender attempts to use a higher transmission rate after a fixed number of successful transmissions at a given rate and switches back to a lower rate after one or two consecutive failures. Specifically, the original ARF algorithm decreases the current rate and starts a timer when two consecutive transmissions fail in a row. When either the timer expires or the number of successfully received per-packet acknowledgments (ACKs) reaches 10, the transmission rate is increased to a higher data rate and the timer is reset. When the rate is increased, the first transmission after the rate increase (commonly referred to as the probing transmission or probing packet) must succeed or the rate is immediately decreased and the timer is restarted rather than trying the higher rate a second time. If the channel conditions change very quickly, ARF cannot adapt effectively owing to the requirement that one or two packet failures are needed for the rate to be decreased and up to 10 successful packet transmissions are needed to increase it; thus it cannot be synchronized with fast-changing channel conditions.

9.2.3.2 Receiver-based autorate (RBAR)

The receiver-based autorate (RBAR) protocol [19] is a rate adaptation algorithm whose goal is to optimize the application throughput. The RBAR was motivated to develop a more timely and reliable receiver-based channel-quality estimation to facilitate the rate selection strategy. This algorithm requires incompatible modification to the IEEE 802.11 standard: the interpretation of some MAC control frames is changed and each data frame must include a new header field. While this algorithm is of little practical interest because it cannot be deployed in existing 802.11 networks, it is of important theoretical interest because it can be used as a performance reference.

The core idea of RBAR is to allow the receiver to select the appropriate rate for a data packet during the RTS and CTS packet exchange. More specifically, a pair of RTS and CTS control frames are exchanged between the source and the destination nodes prior to the start of each data transmission. The receiver of the RTS frame calculates the transmission rate to be used by the upcoming data frame transmission on the basis of the signal-to-noise ratio (SNR) of the received RTS frame and on a set of SNR thresholds calculated with reliable knowledge of an a priori wireless channel model [21]. The rate to be used is then sent back to the source in the CTS packet. The RTS, CTS, and data frames are modified to contain information on the size and rate of the data transmission in order to allow all the nodes within the transmission range to correctly update their NAVs. More specifically, instead of carrying the duration of the reservation, the RTS and CTS packets in RBAR carry the modulation rate and size of the data packet. This modification serves the dual purpose of providing a mechanism by which the receiver can communicate the chosen rate to the

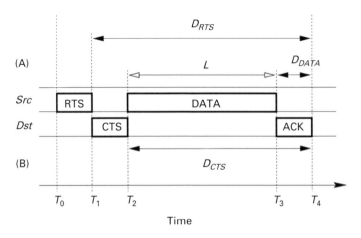

Figure 9.13 Timeline showing changes to the DCF protocol as needed for the receiver-based autorate protocol [19] (© IEEE 2001).

sender while still providing neighboring nodes with enough information to calculate the duration of the requested reservation. The protocol can be summarized as follows.

Referring to Figure 9.13 [19], the sender *Src* chooses a data rate based on some heuristic factor (such as the most recent rate that was successful for transmission to the destination *Dst*), and then stores the rate and the size of the data packet in the RTS. Node A, overhearing the RTS, calculates the duration of the requested reservation, DRTS, using the rate and packet size carried in the RTS. This is possible because all the information required to calculate DRTS is known to A, which then updates its NAV to reflect the reservation. While receiving the RTS, the receiver *Dst* uses information available to it about the channel conditions to generate an estimate of the conditions for the impending data packet transmission. *Dst* then selects the appropriate rate based on that estimate and transmits it and the packet size in the CTS back to the sender. Node B, overhearing the CTS, calculates the duration of the reservation DCTS using a procedure similar to that of A, and then updates its NAV to reflect the reservation. Finally, *Src* responds to the receipt of the CTS by transmitting the data packet at the rate chosen by *Dst*.

If the rates chosen by the sender and receiver are different then the reservation DRTS calculated by A is referred to as a tentative reservation. Such a reservation serves only to inform neighboring nodes that a reservation has been requested but that the duration of the final reservation may differ. Final reservations are confirmed by the presence or absence of a special subheader, called the reservation subheader (RSH), in the MAC header of the data packet. The reservation subheader consists of a subset of the header fields that are already present in the 802.11 data packet frame, plus a check sequence that serves to protect the subheader. The reservation subheader is used as follows. Referring again to Figure 9.13, if the tentative reservation DRTS is incorrect then *Src* will send a data packet with a special MAC header containing the RSH subheader. Node A, overhearing the RSH, will immediately calculate the final reservation DRSH and then update its NAV to account for the difference between DRTS and DRSH. Note that, for A to update its NAV correctly, it must know what contribution DRTS has made to its NAV. One way in which this can be done is to maintain a list of the end times of each tentative reservation, indexed according to the

Table 9.3 The eight modes of IEEE 802.11a PHY. Here BpS stands for bytes per OFDM symbol [22] [23] (© IEEE 2002)

Mode	Modulation	Code rate	Data rate	BpS
1	BPSK	1/2	6 Mbps	3
2	BPSK	3/4	9 Mbps	4.5
3	QPSK	1/2	12 Mbps	6
4	QPSK	3/4	18 Mbps	9
5	16-QAM	1/2	24 Mbps	12
6	16-QAM	3/4	36 Mbps	18
7	64-QAM	2/3	48 Mbps	24
8	64-QAM	3/4	54 Mbps	27

sender, receiver pair. Thus, when an update is required, a node can use the list to determine whether the difference in the reservations will require a change in the NAV.

9.2.3.3 MiSer: a minimum-energy transmission strategy

The MiSer algorithm [22] is based on the 802.11a and 802.11h standards; the latter offers dynamic frequency selection (DFS) and transmit power control (TPC) to the 802.11a MAC. MiSer was proposed for optimization of the local power consumption rather than the application throughput. The 802.11a PHY is based on OFDM and provides eight PHY rates with different modulation schemes and convolutional codes at the 5 GHz U-NII band. As shown in Table 9.3, the OFDM system provides a WLAN with capabilities of communicating at 6 to 54 Mbps. A key feature of the IEEE 802.11a PHY is that it provides eight PHY modes with different modulation schemes and coding rates, making the idea of link adaptation feasible and important. The supported modulation schemes are BPSK, QPSK, 16-QAM, and 64-QAM. Support for transmitting and receiving at data rates 6, 12, and 24 Mbps is mandatory in the IEEE 802.11a PHY. Owing to the use of forward error correction (FEC) to protect the actual data payload transmitted over a noisy wireless channel, the data payload rate can only be a fraction (i.e., the code rate) of the real transmission data rate. The FEC is performed by bit interleaving and rate-1/2 convolutional coding. The higher code rates 2/3 and 3/4 are obtained by puncturing the original rate-1/2 code.

MiSer is a simple table-driven approach. The basic idea is that the wireless station first computes offline a rate–power combination table indexed by the data transmission status. Each entry in the table is the optimal rate–power combination $< R^*, P^* >$ in the sense of maximizing the energy efficiency, which is defined as the ratio of the expected delivered data payload to the expected total energy consumption, under the corresponding *data transmission status*. The data transmission status is characterized by a quadruplet (l, s, SRC, LRC), where l is the data payload length, s is the path loss from the transmitter to the receiver, and the retry counts of a data frame are denoted as (SRC, LRC), which means that there have been SRC unsuccessful RTS transmission attempts and LRC unsuccessful data transmission attempts for this frame. This table is then used at runtime to determine the proper PHY rate and transmit power for each data transmission attempt.

Once the rate–power combination table has been established and is ready for runtime use, the retry counts *SRCcurr* and *LRCcurr* for the frame at the header of the data queue are both set to 0. At runtime, the wireless station estimates the path loss between itself and the receiver and then selects the rate–power combination $< Rcurr, Pcurr >$ for the current data transmission attempt by a simple table lookup. Note that the rate–power selection is made before the RTS frame is transmitted, so that the duration and ID information carried in the RTS frame can be appropriately set according to the PHY rate selection. If an RTS and CTS frame exchange successfully reserves the wireless medium and an ACK frame is received correctly within the ACK timeout to indicate the successful transmission attempt of previous data, both retry counts are reset to 0. Otherwise, either $SRCcurr$ or $LRCcurr$ is increased and the wireless station will reselect the rate–power combination for the next transmission attempt of the data frame. If the data frame cannot be successfully delivered after $SRCmax$ medium reservation attempts or $LRCmax$ data transmission attempts, the frame will be dropped and both $SRCcurr$ and $LRCcurr$ are reset to 0 for the next frame waiting in the data queue. One important aspect of MiSer is that it shifts the computation burden offline and, hence, simplifies the runtime execution significantly. Therefore, embedding MiSer at the MAC layer has little effect on the performance of higher-layer applications, which is a desirable feature for any MAC-layer enhancement.

Two problems were identified in the MiSer algorithm (other than mandating the use of the RTS and CTS protocol) [21]: one is the requirement for a choice of a priori wireless channel model for the offline table calculation and the other is the requirement for a priori knowledge of the number of contending stations on the wireless network.

9.2.3.4 Adaptive autorate fallback (AARF)

Autorate fallback (ARF) was designed for a low-latency system with reasonably good performance in handling the short-term variations of the wireless medium characteristics in an infrastructure WLAN; unfortunately, however, it fails to handle efficiently stable channel conditions. Typically, higher rates can achieve a higher application-level throughput but their higher BERs and PERs generate more retransmissions, which then decreases the application-level throughput. Autorate feedback can recognize this best rate and use it extensively but it also tries constantly (every 10 successfully transmitted consecutive packets) to use a higher rate in order to react to channel condition changes. This process can be costly since the regular trail transmission failures generated by ARF decrease the application throughput.

To avoid the scenario described above, an obvious solution is to increase the threshold for an increase in the current rate. While this approach can indeed improve performance in certain scenarios, it does not work in practice since it completely disables the ability of ARF to react to short-term channel condition changes. This problem led to the idea that formed the basis of adaptive autorate fallback (AARF) [21], i.e., the threshold is continuously changed at runtime to better reflect the channel conditions. This adaptation mechanism increases the amount of history available to the algorithm, which helps it to make better decisions. In AARF, a binary exponential backoff (BEB) is used to determine the adaptive threshold. If the transmission of the probing packet fails, the wireless system switches back immediately to the previous lower rate (as in ARF) but the number of consecutive successful transmissions (with a maximum bound set to 50) required to switch to a higher rate is multiplied by 2. This threshold is reset to its initial value of 10 if the rate is decreased because of two consecutive failed transmissions. The effect of this adaptation mechanism is to increase the period between successive failed attempts to use a higher rate. Fewer failed transmissions and retransmissions improves the overall throughput.

9.2.3.5 Link adaptation via effective goodput analysis

When a wireless station is ready to transmit a data frame, its expected effective *goodput* is defined as the ratio of the expected delivered data payload to the expected transmission time. Clearly, depending on the data payload length and the wireless channel conditions, the expected effective goodput varies with different transmission strategies. The more robust the transmission strategy, the more likely it is that the frame will be delivered successfully within the frame retry limit, with, however, less efficiency. So, there is a tradeoff and the key idea of link adaptation is to select the most appropriate transmission strategy for the frame to be successfully delivered in the shortest possible transmission time.

The effective goodput performance of an 802.11a DCF system can be analytically expressed as a closed-form function [23] of the data payload length l, the frame retry limit n_{max}, the wireless channel conditions $\hat{s}$ during all the potential transmission attempts, and the transmission strategy $\hat{m}$. Here, $\hat{s}$ is a vector of receiver-side SNR values that quantify the wireless channel conditions, $\hat{m}$ is a vector of 802.11a PHY mode selections (see Table 9.3), and both $\hat{s}$ and $\hat{m}$ are of length n_{max}.

On the basis of the above effective goodput analysis it was observed that, to deliver a data frame over a wireless channel, the higher the PHY modes that are used, the shorter will be the expected transmission time but the less likely it is that the delivery will succeed within the frame retry limit. So, for any given set of wireless channel conditions, there exists a corresponding set of PHY modes that maximizes the expected effective goodput. Such a set of PHY modes, denoted by $\hat{m}^*$, is called the best transmission strategy for the data frame delivery under the given wireless channel conditions. A link adaptation scheme based on the MAC protocol data unit (MPDU) for the 802.11a systems was proposed in [23]. This MPDU-based scheme is a simple table-driven approach and the basic idea is to establish a best-PHY-mode table in advance, offline, by applying the dynamic programming technique. The best-PHY-mode table is indexed by a system status triplet that consists of the data payload length, the wireless channel condition, and the frame retry count. At runtime, a wireless station determines the most appropriate PHY mode for the next transmission attempt by a simple table lookup, using the most up-to-date system status as the index, and each entry of the table is the best PHY mode in the sense of maximizing the expected effective goodput under the corresponding system status.

Before running the program, the wireless station computes the best PHY mode for each set of values of the data payload length l, the SNR s, and the frame retry count n. A best-PHY-mode table is thus pre-established and ready for runtime use. The counts *succ_count* for delivered frames and *fail_count* for dropped frames are both reset to 0 and the retry count n_{curr} for the frame at the header of the data queue is set to 1. At runtime, the wireless station monitors the wireless channel condition and determines the current system status. Then, the wireless station selects the best PHY mode $m_{n_{curr}}$ for the next transmission attempt by a simple table lookup. Whenever an ACK frame is received correctly within the ACK timeout, n_{curr} is reset to 1 and *succ_count* is increased; otherwise n_{curr} is incremented by 1. If a frame cannot be successfully delivered after n_{max} transmission attempts, it will be dropped; *fail_count* is increased and n_{curr} is reset to 1 for the next frame waiting in the data queue. Notice that since the best-PHY-mode table is computed offline, there is no extra runtime computational cost for the proposed MPDU-based link adaptation scheme.

9.2.3.6 Link adaptation via fragmentation

The 802.11 MAC also supports a concept called fragmentation, which provides for flexi-bility in transmitter and receiver design and can be useful in environments with RF inter-ference. An 802.11 transmitter optionally breaks an IP packet into smaller fragments for sequential transmission. A receiver can more reliably receive the shorter data bursts because the shorter duration of each transmission fragment reduces the chance of packet loss due to signal fading or noise. Moreover, smaller fragments have a better chance of escaping burst interference such as that from a microwave source. The 802.11 standard mandates that all receivers support fragmentation but leave such support optional on transmitters. By appropriately and adaptively incorporating this fragmentation in the MAC layer, link adaptation based on adaptive frame size control in response to different noisy channel condition can further improve the QoS in using 802.11.

A source station uses fragmentation to divide 802.11 frames into smaller pieces (frag-ments, with typical size 256 to 2048 bytes), which are sent separately to the destination. Each fragment consists of a MAC layer header, a frame check sequence (FCS), and a fragment number indicating its ordered position within the frame. Because the source station transmits each fragment independently, the receiving station replies with a separate acknowledgement for each fragment.

During the interval when a station sends a fragmented packet, there are no RTSs between packet fragments, so a given wireless terminal keeps sending its packet fragments as long as it is receiving the corresponding ACKs. Meanwhile, all other stations are "quiet." This leads to almost the same data rate shares as if there were no fragmentation, unless there is fragment loss (and thus a new RTS) due to a noisy channel for example. This fragmentation technique limits the risk that a packet will have to be retransmitted and thus improves overall performance. The MAC layer is responsible for the reconstitution of the received fragments, the processing being thus transparent for protocols at higher levels.

9.2.4 Performance anomaly due to link adaptation in multirate 802.11

The link adaptation, which dynamically switches data rates to match the wireless channel conditions of a specific STA, also causes degradation of the overall performance of mul-tirate IEEE 802.11 WLANs with geographically scattered stations. More specifically, when many stations are associated with the AP in a BSS, some stations are far away from the AP. These stations may have bad channel quality and the received signal strength can be weak. In such a case, it is more efficient to modulate the signal with a lower transmission data rate based on the link adaptation. However, if the rate becomes low then it takes more time for the same amount of data to be transmitted; this results in an unequal sharing of time between stations. Therefore, low-rate stations can degrade the overall performance of the BSS sig-nificantly. As pointed out analytically in [24], the (aggregate) throughput performance of high-rate stations is heavily affected by low-rate stations, and the former suffer from throughput degradation although they are near the AP. This is basically caused by the fact that the 802.11 MAC provides a fair channel-access (throughput fairness) among con-tending stations by giving the same channel access opportunities. Then, assuming that the same amount of data is to be transmitted over the wireless channel, stations with lower rates, i.e., stations far from the AP, tend to grab the wireless channel for longer than those with higher rates, i.e., stations near to the AP. Accordingly, stations do not obtain a throughput linearly proportional to their transmission rate. This phenomenon is often referred to as the "performance anomaly" of IEEE 802.11; this unexpected performance degradation of the BSS arises because the overall system performance is dominated by stations with the lowest rates. This, of course, causes a serious resource leakage problem that needs to be avoided or mitigated. Considering that hotspot areas with many stations are being deployed at an ever increasing pace, this problem becomes even more important.

It has been shown that the degradation of the aggregate throughput due to multirate WLAN performance anomaly can be mitigated via an appropriate control of the initial contention window size, CW_{min} (see Eq. 9.1); this has been shown to be the most effective MAC parameter for the control of stations' channel access [24]. The CSMA/CA system tends to give all stations fairness from the viewpoint of channel-access probability. However, each station has a different transmission time with a different transmission rate for the same payload size. In order to improve this unfairness, the opportunities of the low-rate stations must be reduced, i.e., they should have a larger CW_{min} value than those for high-rate stations. As was presented in [25], the CW_{min} value affects the performance of a station significantly. Controlling CW_{min} can have another advantage. As the number of stations increases, the collision probability also increases. In such a case, increasing the CW_{min} of low-rate stations can reduce the collision probability as well.

To minimize the anomaly effect, the frame size of low-rate stations can also be reduced so that it will take more frame transmissions to finish the same amount of payload and thus reduce the service opportunities for low-rate stations. In order to do this, the fragmentation threshold parameter can be adjusted. Note that how to make a reduction in the frame size of low-rate stations needs to be treated with care since there is a close relation between system utilization and frame size, i.e., reducing the frame size might decrease the transmission efficiency of IEEE 802.11 MAC.

9.3   QoS enhancement support of 802.11

The most important functions of the MAC layer for a wireless network include controlling channel access, maintaining quality of service (QoS), and providing security. Wireless links have characteristics that differ from those of fixed links, such as high packet loss rate, bursts of packet loss, packet re-ordering, and large packet delay and delay variation. Furthermore, the wireless link characteristics are not constant and may vary in time and place. As discussed in Section 9.2, two medium-access coordination functions are defined in the original 802.11 MAC: a mandatory distributed coordination function (DCF) and an optional point coordination function (PCF). Both medium-access mechanisms suffer from some QoS limitations, as follows.

In DCF, only best-effort service is provided. Delay-sensitive multimedia applications (e.g., voice over IP, video conferencing) require certain bandwidth, delay, and jitter guarantees. With DCF, all the stations compete for the channel with the same priority. There is no differentiation mechanism to provide a better service for real-time multimedia traffic than for data applications. As explained in an example in [26], suppose that there are a variable number of wireless stations located within an area where there are no hidden terminals and operating in ad hoc mode. The PHY-layer data rate of each station is set to 36 Mbps. Each station transmits three types of traffic flow (voice, video, and

Table 9.4 The average delays of the three types of traffic vs. the number of stations (STAs), including queuing delays [26] (© IEEE 2005)

STAs	Voice (ms)	Video (ms)	Background (ms)
4	0.20	0.20	0.23
6	0.47	0.48	0.47
8	0.77	0.78	0.77
10	3.77	3.71	3.77
12	179.75	179.75	178.89
14	296.47	298.17	296.47
16	373.66	371.29	373.66
18	419.44	419.56	419.87

background data) using UDP as a transport-layer protocol. The MAC-layer queue size is set to 50. The voice flow is chosen as a 64 kbps pulse code modulated (PCM) stream. The transmission rate of a video flow is 640 kbps with packet size 1280 bytes. The transmission rate of the background traffic is 1024 kbps. The load rates were varied from 9.6 to 90 percent by increasing the number of stations from 2 to 18. Shown in Table 9.4 are the average delays of the three types of traffic vs. the number of stations, including the queuing delays. The mean delays of the voice, video, and background flows are lower than 4 ms when the channel load is less than 70 percent (i.e., the number of stations is below 10). However, when the number of stations exceeds 10, the mean delays for the three types of flows increase up to about 420 ms and are almost the same for the different types of flow. Note that 802.11 does not offer any admission control to guarantee QoS. These sharply increased delays are caused by the well-known "saturation throughput" problem [25] of 802.11, which is a fundamental performance figure defined as the limit reached by the system throughput as the offered load increases and represents the maximum load that the system can carry in stable conditions. The throughput saturation causes frequent retransmission and increasing contention-window size in response to collisions resulting from congestion. This experiment demonstrates that a serious QoS problem for multimedia applications can occur when no service differentiation between the different types of flow is available under a high-traffic-load situation for 802.11 WLANs.

Even with the use of PCF to support delay-sensitive multimedia applications, this mode has some major problems that lead to poor QoS performance [26]. For example, PCF defines only a single-class round-robin scheduling algorithm, which cannot handle the various QoS requirements of different types of traffic. Moreover, stations in PCF mode are allowed to transmit even if the frame transmission cannot finish before the next TBTT. The duration of the beacon to be sent after the TBTT defers the transmission of data frames during the following CFP, which introduces (on average 250 μs [27]) delays to those data frames. Finally, a polled station is allowed to send a frame of any length between 0 and 2304 bytes, which may introduce unpredictable variable transmission times. Furthermore, the PHY data rate of a polled station can change according to varying channel conditions. Thus, the AP is not able to predict transmission times in a precise manner. This prevents it from providing guaranteed delay and jitter performance for other stations present in the polling list during the rest of the CFP interval.

9.3.1   Service differentiation techniques of 802.11 DCF

To improve the QoS and the overall system performance of 802.11, there have been several efforts [28] to introduce service differentiation for the IEEE 802.11 based on the DCF MAC mechanism:

(1) varying the DIFS and backoff time;
(2) limiting the maximum frame length;
(3) varying the initial contention-window size;
(4) blackburst;
(5) distributed fair schedling.

These techniques will now be discussed in turn.

9.3.1.1 Varying the DIFS and backoff time

According to the 802.11 CSMA/CA standard, after a channel is sensed as idle, the total amount of time during which a station should wait before it seizes the channel is the sum of the IFS and the random backoff time. Therefore, the DIFS length and the backoff time can be used to determine the priority of the corresponding station, i.e., which station should send first [29]. More specifically, instead of waiting for the DIFS time after the ACK is received following the successful receipt of a correct packet, the higher-priority stations wait only for the PCF interframe spacing (PIFS) time, which is shorter than the DIFS, to decide whether the medium is busy or idle. As a result, such a higher-priority station has a higher chance of seizing the channel. Another important factor in determining the priority is the backoff time. For higher-priority and lower-priority stations the backoff time is modified as follows:

$$
\begin{aligned}
high\_priority, \ backoff\_time &= \tfrac{1}{2} \, rand[0; CW_{min} \times 2^n] \times slot\_time, \\
low\_priority, \ backoff\_time &= \tfrac{1}{2} CW \times 2^n \times slot\_time \\
&\quad + \tfrac{1}{2} \, rand[0; CW_{min} \times 2^n] \times slot\_time,
\end{aligned}
\tag{9.2}
$$

where n denotes the number of consecutive unsuccessful transmission attempts, as in Eq. (9.1). Fairness can be achieved by including the packet size in the backoff interval calculation, with the result that flows with smaller packets are sent more often.

9.3.1.2 Limiting maximum frame length

Another mechanism that can be used to introduce service differentiation into IEEE 802.11 is to limit the maximum frame length used by each station. This can be achieved by either

(1) dropping packets that exceed the maximum frame length assigned to a given station (or simply configuring it to limit its packet lengths), or
(2) fragmenting packets that exceed the maximum frame length; as discussed in Section 9.2.3.6, this mechanism can also be used to increase transmission reliability and is used for service differentiation.

9.3.1.3 Varying the initial contention-window size

As discussed in Section 9.2.4, the initial contention-window size, CW_{min}, is an effective MAC parameter for controlling the backoff time and thus the airtime share. An extended DCF scheme involves varying the value of CW_{min} as a way of giving higher priority to some stations than to others. Assigning a small CW_{min} to those stations with higher priority ensures that in most (though not all) cases, higher-priority stations will be able to transmit ahead of lower-priority stations.

9.3.1.4 Blackburst

Blackburst (a channel-jamming technique) [30] [31] is proposed to minimize the delays for real-time traffic by imposing certain requirements on the traffic to be prioritized. Blackburst requires that all high-priority stations try to access the medium at constant intervals, $Tsch$ (this interval has to be the same for all high-priority stations). Further, Blackburst also requires the ability to jam the wireless medium for a period of time. When a high-priority station wants to send a frame, it senses the medium to see whether it has been idle for a PIFS time and then sends its frame. However, if the medium is found to be busy then the station waits until the channel has been idle for a PIFS time and then enters a blackburst contention period. The station now sends a so-called blackburst by jamming the channel for a period of time. The length of the blackburst is determined by the time the station has been waiting to access the medium and is calculated as a number of *black slots*. After transmitting the blackburst, the station listens to the medium for a short period of time (less than a black slot) to see whether some other station is sending a longer blackburst. This would imply that the other station has waited longer and, thus, should access the medium first. If the medium is idle then the station will send its frame, otherwise it will wait until the medium becomes idle again and enter another blackburst contention period. By using slotted time, and imposing a minimum frame size on real-time frames, it can be guaranteed that each blackburst contention period will yield a unique winner [30]. After the successful transmission of a frame, the station schedules the next access instant (when it will try to transmit the next frame) $Tsch$ seconds in the future. This has the useful effect that real-time flows will synchronize and share the medium in a time-division multiplex (TDM) fashion. This means that unless there is a transmission by a low-priority station when an access instant for a high-priority station occurs, very little blackbursting will have to be done once the stations have synchronized. Low-priority stations use the ordinary DCF access mechanism of IEEE 802.11.

There are two different modes of operation of Black burst: with and without feedback from the MAC layer to the application [31]. There are two different modes of operation of Blackburst: with and without feedback from the MAC layer to the application [31]. If the application is not Blackburst-aware and a mode without feedback is used then a slack time δ is used to ensure the stability of the system. The access intervals are scheduled at a time δ before the packet is expected to arrive at the MAC layer. This is to ensure that delayed access caused by interfering traffic does not make the system unstable.

9.3.1.5 Distributed fair scheduling

It is not always desirable to sacrifice completely the performance of low-priority traffic in order to give very good service to high-priority traffic. Often it is sufficient to provide relative differentiation, for example, specifying that one type of traffic should get twice as much bandwidth as some other type of traffic [32]. An access scheme called distributed fair scheduling (DFS), which applies the ideas behind fair queuing in the wireless domain, was proposed in [33]. There exist several fair queuing schemes that provide the fair allocation of bandwidth between different flows on a node. In this context, "fair" means that each flow gets a bandwidth proportional to some *weight* that has been assigned to it. These schemes are centralized in the sense that they run on a single node, which has access to all the information about the flows. Since different weights can be assigned to the flows, this can be used for differentiation between flows.

The DFS scheme is based on fair queuing plus the backoff mechanism of IEEE 802.11 to determine which station should send first. Before transmitting a frame the backoff process is always initiated, and the backoff time is calculated as proportional to the size of the packet to be sent and inversely proportional to the assigned weight of the station. Since the lower the weight of the sending station, the longer the backoff time, differentiation is achieved. Further, fairness is achieved by using the size of the packet to be sent in the calculation of the backoff time. This will cause larger packets to be given longer backoff intervals than smaller packets, allowing a station with small packets to send more often so that the same amount of data is sent.

If a collision occurs, a new backoff time is calculated using the standard backoff algorithm (which involves doubling the CW) of the IEEE 802.11 standard with a quite small CW_{min} value. The reason for choosing such a small CW_{min} even though a collision has occurred is that DFS tries to maintain fairness among nodes, and thus a node that was scheduled to send a packet should be able to send it as soon as possible. Otherwise fairness would suffer.

9.3.2   802.11e: enhanced QoS support for WLANS

The IEEE 802.11e standard [33], which was finalized in 2003, is an approved amendment to the IEEE 802.11 standard that defines a set of QoS enhancements for WLAN applications through modifications to the MAC layer. The enhanced MAC-layer standard can be incorporated into any of the 802.11 a/b/g MAC designs and is considered of critical importance for delay-sensitive applications such as voice over wireless IP and streaming multimedia. The IEEE 802.11e standard supports QoS on the basic of a hybrid coordination function (HCF), which defines two medium-access mechanisms: one is contention-based channel access and the other is controlled channel access (it includes polling). Contention-based channel access is referred to as enhanced distributed channel access (EDCA) and controlled channel access as HCF controlled channel access (HCCA) [18]. Note that channel access is used in 802.11e as a synonym for medium access.

To achieve better QoS, 802.11e uses multiple queues for the prioritized and separately handled access categories (ACs), which are labeled according to their target applications: AC_VO (voice), AC_VI (video), AC_BE (best-effort), and AC_BK (background).

9.3.2.1   Hybrid coordination function (HCF) of 802.11e

With 802.11e, there may still be two phases of operation within a frame, i.e., the contention period (CP) and the contention-free period (CFP). Enhanced distributed channel access is used in the CP only, while HCF-controlled channel access is used in both phases. The hybrid coordination function (HCF), as its name befits, combines the methods of the PCF and DCF. The *hybrid coordinator* (HC), which resides within an 802.11e AP, operates as the central coordinator for all the other 802.11e stations within the same QoS-supporting basic service set (QBSS), i.e., a BSS that includes an 802.11e-compliant HC is referred to as a QBSS. Since there are multiple backoff processes operating in parallel within one 802.11e station (owing to the existence of multiple ACs), each is referred to as a *backoff entity* that, instead of a station, attempts to deliver MSDUs.

The AP and stations that implement the 802.11e HCF are called a QoS-enhanced AP (QAP) and QoS-enhanced or 802.11e stations (QSTAs) respectively. An important new feature of HCF is the concept of transmission opportunity (TXOP). The TXOP denotes a time duration during which a QSTA is allowed to transmit a burst of data frames. Thus, the problem of the unpredictable transmission time of a polled station encountered in PCF (as mentioned in Section 9.2.2) is solved. A TXOP is defined by its starting time and duration. In 802.11e, a backoff entity that obtains medium access must be within a TXOP, during which a backoff entity has the right to deliver MSDUs. A TXOP is called an EDCA-TXOP when it is obtained by a backoff entity after a successful EDCA contention and an HCCA-TXOP when it is obtained by a backoff entity on receiving a QoS poll frame from the QAP. In order to control the delay, the maximum value of a TXOP is bounded by a value called $TXOP_{limit}$, which is determined by the QAP. A QSTA can transmit multiple frames within its TXOP allocation. This new feature also tends to provide time-based fairness between QSTAs, and this can help to remedy the performance anomaly of multirate 802.11. The QAP allocates an *uplink* HCCA-TXOP to a QSTA by sending a QoS poll frame to it, while no specific control frame is required for a *downlink* HCCA-TXOP. More details of these two 802.11e HCF subfunctions, EDCA and HCCA, will be discussed below [18].

9.3.2.2 Enhanced distributed channel access (EDCA)

The QoS support in 802.11e EDCA is provided by the introduction of access categories (ACs) and multiple independent backoff entities. MAC service data units are delivered by parallel backoff entities within one 802.11e station, where backoff entities are prioritized using AC-specific contention parameters, called an EDCA parameter set. As just mentioned there are four ACs; thus, four backoff entities exist in every 802.11e station. As shown in Figure 9.14(a), there is only one backoff entity with one single priority for each 802.11 station, while there are four backoff entites with four different priorities that can be identified for each 802.11e station: one entity represents one AC as shown in Figure 9.14(b).

The EDCA parameter set defines the priorities in medium access by setting individual interframe spaces, contention windows, and many other parameters per AC. Each AC queue works as an independent legacy 802.11 distributed coordination function (DCF) station with its own set of EDCA contention parameters (see an example of an EDCA contention parameter set in Table 9.5) as defined in the hybrid coordinator inside the AP. Such a set of contention parameters could consist of the arbitration interframe space number *AIFSN*, CW_{min}, and CW_{max}; the AIFSN per AC times the slot time added to a short interframe space (SIFS) gives the arbitration interframe space (AIFS) per AC:

$$AIFS = SIFS + AIFSN \times slot\_time. \tag{9.3}$$

As shown in Figure 9.15, multiple backoff entities contend in parallel for medium access with different priorities and the earliest possible medium access time after the medium has been busy is *DIFS*.

Note that the EDCA parameter set can also be modified over time by the HC and is announced via information fields in beacon frames. The same EDCA parameter set is used by the backoff entities of the same AC in different stations: it is essential that the same values for the parameters should be used by all these backoff entities. Each backoff entity within a station independently contends for a TXOP.

Before starting the transmission, the terminal picks the backoff time by selecting a random number between 0 and *CW*, which is initially set to CW_{min}. Transmission starts

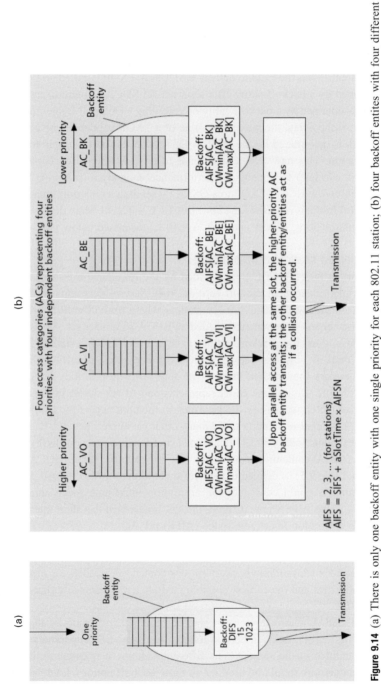

Figure 9.14 (a) There is only one backoff entity with one single priority for each 802.11 station; (b) four backoff entites with four different priorties can be identified for each 802.11e STA, one entity representing one AC [18] (© IEEE 2003).

Table 9.5 An example of an EDCA contention parameter set for 802.11e

Access categories	AC_VO	AC_VI	AC_BE	AC_BK
AIFSN[AC]	2	2	3	7
CWmin[AC]	7	15	31	31
CWmax[AC]	15	31	1023	1023

when the medium has been idle for at least a time $AIFS$ and the backoff counter reaches 0. The counter stops when the medium is busy and resumes again when the medium has been idle for a time $AIFS$. Because of the difficulty of detecting a collision or fading loss from the channel, each transmitted frame needs an acknowledgement (ACK) that follows immediately, to be considered as a successful transmission. After an unsuccessful transmission with the number of attempts no more than the retransmission limit for a frame, an exponential backoff process is performed with a double-sized CW_{new}, up to the value CW_{max} (see the discussion after Eq. (9.1)):

$$CW_{new} = (CW_{old} + 1) \times 2 - 1 \tag{9.4}$$

where CW is reset to CW_{min} in the case of a successful transmission.

It is obvious that the smaller the value of CW_{max} per AC, the higher the medium-access priority. However, a small value of CW_{max} may increase the collision probability. Furthermore, it should be highlighted that there are retry counters (as in legacy 802.11) that limit the number of retransmissions. The 802.11e protocol also defines a maximum MSDU lifetime per AC, which specifies the maximum time a frame may remain in the MAC. Once the maximum lifetime has passed since a frame arrived at the MAC, the frame is dropped without being transmitted. This feature can be useful since transmitting a frame too late is not meaningful in many real-time multimedia applications.

In addition to the backoff parameters, the value of $TXOP_{limit}$ is defined for a particular AC as part of the EDCA parameter set. The larger this value is, the larger the share of capacity for this AC. Once a TXOP is obtained using a backoff, a backoff entity may continue to deliver more than one MSDU consecutively during the same TXOP, which may take a time up to $TXOP_{limit}$. This important concept in 802.11e is referred to as the continuation of an EDCA-TXOP.

9.3.2.3 HCF controlled channel access (HCCA)

The HCF controlled channel access (HCCA) protocol further extends the EDCA access rules by allowing the highest-priority medium access to the HC during the CFP and CP [18]. More specifically, a TXOP can be obtained by the HC via controlled medium access. The HC may allocate TXOPs to itself to initiate MSDU deliveries whenever it requires, after detecting the medium as being idle for a time $PIFS$, and without backoff. To give the HC higher priority over legacy DCF and EDCA access, the parameter $AIFSN$ must be selected such that the earliest time of medium access for EDCA stations is $DIFS$ for any AC.

During CP, each TXOP of an 802.11e station begins either after a time $AIFS$ plus the random backoff time when the medium is determined as being available, or when a backoff entity receives a polling frame, the QoS CF-poll, from the HC. The latter can be transmitted after an idle period $PIFS$, without any backoff, by the HC. During CFP, the starting time and

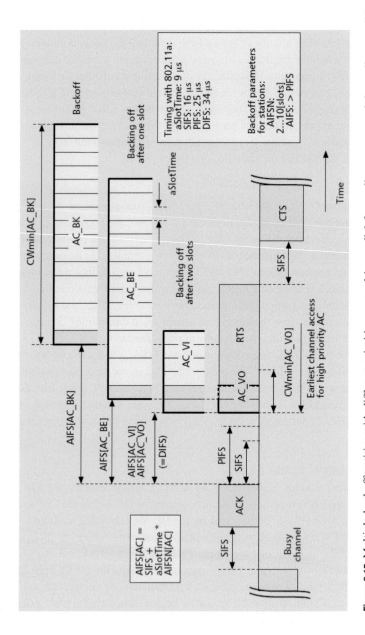

Figure 9.15 Multiple backoff entities with different priorities contend in parallel for medium access, and the earliest possible medium access time after the medium has been busy is DIFS [18] (© IEEE 2003).

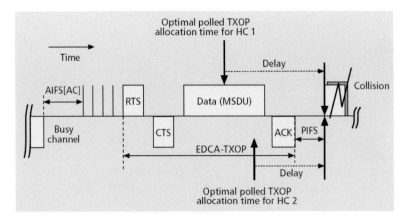

Figure 9.16 An example of polled TXOP allocation, where any 802.11e frame exchange will not take longer than a time $TXOP_{limit}$, which is the limit for all EDCA-TXOPs and under control of the HC [18] (© IEEE 2003).

maximum duration of each TXOP is also specified by the HC, again using the QoS CF-poll frames. During CFP, 802.11e backoff entities will not attempt to access the medium without being explicitly polled, hence only the HC can allocate TXOPs, by transmitting QoS CF-poll frames or by immediately transmitting downlink data. During a polled TXOP, a polled station can transmit multiple frames that the station selects to transmit according to its scheduling algorithm, with a time gap *SIFS* between two consecutive frames, as long as the entire frame exchange duration is not over the allocated maximum $TXOP_{limit}$.

Polled TXOP allocations may be delayed by the duration of an EDCA-TXOP, as illustrated in Figure 9.16. The HC controls the maximum duration of EDCA-TXOPs within its QBSS by announcing the $TXOP_{limit}$ value for every AC via the beacon. Therefore, it is able to allocate polled TXOPs at any time during the CP and the optional CFP. When very small MSDU delivery delays are required, CF-polls may be transmitted a time duration $TXOP_{limit}$ earlier than the optimal polled TXOP allocation time to avoid completely any MSDU delivery delay being imposed by EDCA-TXOPs. However, the largest $TXOP_{limit}$ of the four ACs must be considered.

9.3.3 WLAN mesh

To build a small-to-large-scale wireless distribution system based on the WLAN infrastructure, the IEEE 802.11s extended service set (ESS) was proposed for the installation, configuration, and operation of a WLAN multi-hop mesh [35]. A WLAN mesh consists of a set of devices interconnected with each other via wireless links, resulting in a mesh of connectivity, which is distinct from the BSS defined in Section 9.2. The implementation of 802.11s is to be atop the existing PHY layer of IEEE 802.11a/b/g/n operating in the unlicensed spectrum of the 2.4 GHz and 5 GHz frequency bands. The specification includes extensions in topology formation to make the WLAN mesh self-configure as soon as the devices are powered up. A path selection protocol is specified in the MAC layer instead of in the network layer, for routing data in the multi-hop mesh topology. This standard is expected to support MAC-layer broadcasts and multicasts in addition to

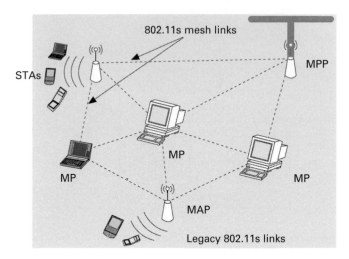

Figure 9.17 In a WLAN mesh architecture, any IEEE 802.11-based entity (either AP or STA) that partially or fully supports a mesh-relay function is defined as a mesh point (MP) [35]. (© IEEE 2006)

unicast transmissions. This standard will also accommodate devices that are able to support multichannel operations or are equipped with multiple radios, with the aim of boosting the capacity of the overall network. The IEEE 802.11s Mesh Networking Task Group approved the Draft D1.0 for the standard in early 2008. Current efforts aim to resolve some remaining issues and so reach agreement on Draft D2.0 (see http://www. ieee802.org/11/Reports/tgs_update.htm).

9.3.3.1 Proposed IEEE 802.11s mesh architecture

Figure 9.17 shows a proposed WLAN mesh architecture, in which any IEEE 802.11-based entity (either AP or STA) that partially or fully supports a mesh-relay function is defined as a mesh point (MP). Minimal MP operations include neighbor discovery, channel selection, and forming an association with neighbors. Besides this, MPs can directly communicate with their neighbors and forward traffic on behalf of other MPs via bidirectional wireless mesh links. The proposed WLAN mesh also defines a mesh access point (MAP), which is a specific MP but acts as an AP as well. The MAP may operate as part of the WLAN mesh or in one of the legacy 802.11 modes. A mesh portal (MPP) is yet another type of MP, through which multiple WLAN meshes can be interconnected to construct networks of mesh networks. An MPP can also co-locate with an IEEE 802.11 portal and function as a bridge or gateway between the WLAN mesh and other networks in the distribution system. To uniquely identify a WLAN mesh, a common mesh ID is assigned to each MP; this is similar to the use of a service set identifier (SSID) to represent an entity in legacy 802.11 BSS networks.

9.3.3.2 Media access coordination function (MCF)

The major components of media-access coordination function (MCF) in the proposed 802.11s are shown in Figure 9.18. Built on top of the legacy 802.11 physical-layer specification, the 802.11s MCF explicitly provides the following WLAN mesh services: topology learning, routing and forwarding, topology discovery and association, path selection

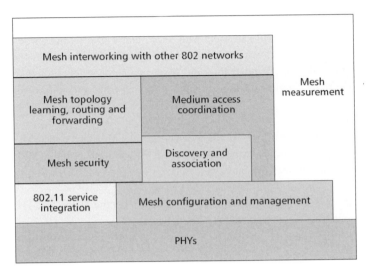

Figure 9.18 The major media-access coordination function (MCF) components of the proposed 802.11s [35] (© IEEE 2006).

protocols, medium access coordination, mesh configuration and management, mesh measurement, interworking, and security functions [35].

Topology learning, routing, and forwarding   This service function component focuses on the peer-to-peer discovery of MPs. It enables automatic topology learning, establishes links, and eventually forms a dynamic data delivery path across the WLAN mesh.

Topology discovery and association   A new node (a candidate MP) can discover and associate itself with neighboring MPs that belong to the active WLAN mesh, either by active scanning (i.e., sending probe messages) or passive listening (i.e., receiving periodic beacons).

Path selection protocols   A layer-2 path selection protocol is proposed for handling unicast and broadcast or multicast data delivery in a WLAN mesh. Moreover, a hybrid scheme using the ad hoc on-demand distance vector (AODV) together with the optimized link state routing (OLSR) protocol is proposed to support a wide range of application scenarios [36] [37]. In addition, radio-aware metrics that reflect actual link conditions are used to make the routing protocols more robust against link failures. More specifically, an airtime metric [36] is used to reflect the costs of channel, path, and packet error rate. Moreover another metric, called the weighted radio and load aware (WRALA) metric [37], is used to indicate the protocol overhead at the MAC and PHY layers, the frame size, bitrate, link load, and error rate.

Medium access coordination   The medium access coordination of the 802.11s proposal [36] [37] is based on the enhanced distributed channel access (EDCA) mechanism used in 802.11e [34]. The proposed media access coordination mechanisms facilitate congestion control, power saving, synchronization, and beacon collision avoidance. On the basis of the proposed mechanisms, it is possible to enable multiple channel operations in multiradio or single radio environments as well as in mixed environments.

Mesh configuration and management   Since unmanaged deployment of WLAN mesh networks is possible, autonomic management modules are required to minimize the burden of manual configuration for the service provider and to ensure the smooth operation of the network. Since any available MP can route packets, a failure of a particular device is not likely to affect the network as a whole. However, the system should still be able to report the malfunctioning of devices. Support for radio frequency autoconfiguration is expected to be provided for efficient multi-hop transmission, power saving, and improving the total capacity.

Internetworking   For interworking of the WLAN mesh with other networks, the IEEE 802.1D MAC bridge standard is to be incorporated in the mesh portals (MPPs), which define the interworking framework and service access interface across all 802 standards.

Security   Similarly, security architecture is to be based on the IEEE 802.11i standard, which specifies security features for all WLAN networks.

9.4   Worldwide interoperability for microwave access (WiMAX)

The IEEE 802.16 standard, commonly referred to as worldwide interoperability for microwave access (WiMAX) [4] [5], specifies the air interface, including the MAC and PHY layers, for the next generation wireless broadband access. First published in 2001, the IEEE 802.16 standard specified a frequency range 10–66 GHz with a theoretical maximum bandwidth of 120 Mbps and maximum transmission range of 50 km. However, the initial standard only supports line-of-sight (LOS) transmission and thus does not seem to favor deployment in urban areas. A variant of the standard, IEEE 802.16a-2003, approved in April 2003, can support non-LOS (NLOS) transmission and adopts OFDM at the PHY layer. It also adds support for the 2–11 GHz range. With a focus on several main profiles and on defining interoperability testing for WiMAX equipment, the IEEE 802.16 standard evolved to the 802.16–2004 standard (also known as 802.16d) [4]. The latter provides technical specifications for the PHY and MAC layers for fixed wireless access and addresses the first-mile or last-mile connection in wireless metropolitan area networks (WMANs). To further support mobility, which is widely considered to be a key feature in wireless networks, the new IEEE 802.16e (also known as 802.16–2005) [5] added mobility support and is generally referred to as mobile WiMAX. Several significant technological enhancements were included in 802.16e, including the use of advanced antenna diversity schemes and hybrid automatic repeat request (hARQ) for improved NLOS coverage, the use of dense sub-channelization of OFDMA to increase system gain and improve indoor penetration, the use of MIMO and adaptive antenna systems to improve coverage, and the introduction of a downlink subchannelization scheme to enable better coverage and capacity tradeoff.

With its improved data rate and mobility support, WiMAX is available to a wide variety of applications (see Figure 9.19 [13]). Furthermore, with its large range and high transmission rate, WiMAX can serve as a backhaul for 802.11 hotspots when connecting to the Internet, i.e., an 802.11 hotspot can now be a moving spot (such as a bus or a subrailway) owing to the mobility support, and users inside the moving spot can still enjoy Internet access using 802.11 customer premise equipment (CPE) devices. Alternatively, users can also connect mobile devices such as laptops and handsets with 802.16 CPE directly to WiMAX base stations.

Throughout WiMAX's development, the industry-led and non-profit WiMAX forum, which comprises a group of industry leaders (Intel, AT&T, Samsung, Motorola, Cisco, and

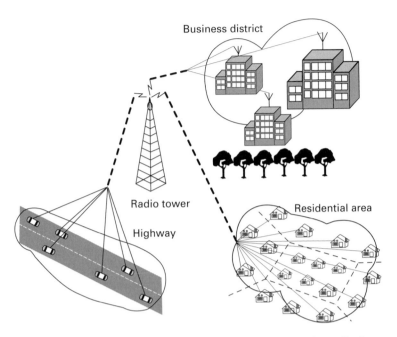

Figure 9.19 A WiMAX base station can serve terminals in a business district or residential area and on moving vehicles [13]. (© IEEE 2006)

others), has closely supported, promoted, and certified the compatibility and interoperability of WiMAX technologies. The group's workforce is divided along multiple working groups that focus on technical, regulatory, and marketing aspects. The certification working group has developed a WiMAX product certification program, which aims to ensure interoperability between WiMAX equipment from vendors worldwide. The certification process also considers interoperability with the high-performance radio metropolitan area network (HiperMAN), which is the MAN standard of the European Telecommunications Standards Institute (ETSI). Such interoperability is needed because 802.16 and HiperMAN were modified to include features from one another. In fact, they now share the same PHY-layer and MAC-layer specifications. The WiMAX forum, through its regulatory working group, is also in discussion with governments worldwide about spectrum regulations.

9.4.1   Protocol architecture of WiMAX

Figure 9.20 shows a common protocol architecture of the WiMAX standard [38] [39], where the MAC layer consists of three sublayers: the service-specific convergence sublayer (CS), the MAC common part sublayer (MAC CPS), and the security sublayer. For the PHY layer, the standard supports multiple PHY specifications, each handling a particular frequency range. The interface between different PHY specifications and the MAC is called the transmission convergence sublayer, which hides the detailed PHY technologies from the MAC. The CS sublayer is specified to facilitate various types of applications such as IP, asynchronous transfer mode (ATM), Ethernet, and point-to-point protocol (PPP) services. More specifically, the main functionality of the CS is to transform or map external data from the upper layers into appropriate MAC service data units (MSDUs), which are the basic data

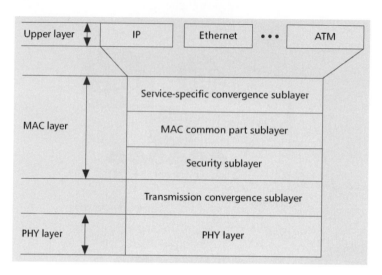

Figure 9.20 A common protocol architecture of the WiMAX standard [38] (© IEEE 2007).

units exchanged between two adjacent protocol layers, for the MAC CPS. This includes the classification of external data using the appropriate MAC service flow identifier (SFID) and connection identifier (CID). The MAC CPS provides the core functionality for system access, allocation of bandwidth, connection establishment and maintenance, and the QoS aspect of data transmission. The security sublayer provides functionalities such as authentication, secure key exchange, and encryption.

9.4.1.1 PHY layer of WiMAX

The IEEE 802.16d standard (fixed WiMAX) uses OFDM in the PHY layer on the basis of the OFDM 256 FFT mode, which means that there are 256 subcarriers available for use in a single channel. Multiple access on one channel is accomplished using TDMA or FDMA. The key technologies in 802.16e on the PHY level are OFDMA and an enhanced version called scalable OFDMA (SOFDMA). The OFDMA 128/512/1024/2048 FFT modes are all supported in IEEE 802.16e (mobile service), the default mode being the 2048 FFT. As discussed in Section 9.1.3.1, OFDMA uses a multicarrier modulation in which the carriers are divided among the users to form subchannels. For each subchannel the coding and modulation are adapted separately, allowing channel optimization on a smaller scale (rather than using the same parameters for the whole channel). Different levels of modulation scheme, including binary phase shift keying (BPSK), quaternary PSK (QPSK), 16-quadrature amplitude modulation (16-QAM), and 64-QAM, can be chosen depending on channel conditions. In the earlier versions of line-of-sight deployed WiMAX, the 10–66 GHz band with channel bandwidths of 20, 25 (the typical US allocation), or 28 MHz (the typical European allocation) can be used. For newer commercial versions of WiMAX (16d and 16e), used to take advantage of better multipath propagation and thus reduce the deployment cost, the 2–11 GHz frequency bands are used (e.g., 3.5 GHz licensed and 5.8 GHz license-exempt bands), with a flexible channel bandwidth allocation between 1.25 MHz and 20 MHz. Furthermore, optional features of intelligent adaptive antenna systems are also allowed to optimize the use of spectrum resources and enhance indoor coverage by assigning a robust scheme to vulnerable links. The OFDMA technology is only an option in 802.16d for fixed access; however, OFDMA is necessary in 802.16e devices and is required for certification. The SOFDMA technology is an enhancement of OFDMA that scales the number of sub-carriers in a channel with possible values of 128, 512, 1024, and 2048. It has been widely approved by standards makers and manufacturers. Observing the successful use of SOFDMA in the Korean standard WiBro (Wireless Broadband), IEEE is convinced that it should enable interoperability between WiBro and 802.16e. In addition, Intel announced that SOFDMA will be the PHY layer of choice for its indoor and mobile equipment.

9.4.1.2 MAC Layer of WiMAX

Owing to the growing popularity of multimedia applications with wireless broadband access, IEEE 802.16 is expected to provide the capability to offer new wireless services such as multimedia streaming, real-time surveillance, voice over IP (VoIP), and multimedia conferencing. Quality of service (QoS) provisioning with differentiated service require-ments becomes one of the most important issues in the WiMAX MAC (CPS) layer defin-ition. The IEEE 802.16 MAC provides two modes of operation: point-to-multipoint (PMP) and multipoint-to-multipoint (mesh).

Point-to-multipoint mode The PMP operational mode, supported both by 802.16d and 802.16e, fits a typical fixed access scenario in which multiple service subscribers are served by a centralized service provider. In the PMP mode, uplink transmissions from a subscriber; this can be a static or mobile station station (SS) to a base station (BS) occur in separate timeframes. In the downlink subframe, the BS can transmit a burst of MAC protocol data units (MPDUs). Since the downlink transmission is broadcast, an SS listening to the data transmitted by the BS is only required to process MPDUs addressed to itself or explicitly intended for all SSs. Subscriber stations share the uplink to the BS on a demand basis. Depending on the class of service utilized the SS may be issued a continuing right to transmit, or the right to transmit may be granted by the BS after receipt of a request from the user. Downlink and uplink subframes are duplexed either using frequency-division duplex (FDD) or time-division duplex (TDD). Subscriber stations can be either full-duplex or half-duplex.

An example of the MPDU frame structure in the TDD mode of 802.16e is illustrated in Figure 9.21, where an OFDMA frame consists of a downlink (DL) subframe for trans-mission from the BS to SSs and an uplink subframe for transmissions in the reverse direction [49]. The transmitting–receiving (Tx–Rx) transition gap (TTG) and the Rx–Tx transition gap (RTG) are specified to allow SS terminals to switch between reception and transmission, i.e., between the downlink and uplink subframes. A DL subframe starts with a preamble, which helps SSs to perform synchronization and channel estimation. The first two subchannels in the first *data* OFDMA symbol in the downlink is called the frame control header (FCH) and specifies the length of the immediately succeeding downlink MAP (DL-MAP) message and the modulation coding used for DL-MAP. In the downlink subframe, both the downlink MAP (DL-MAP) and uplink MAP (UL-MAP) messages are transmitted and comprise the bandwidth allocations for data transmission in the downlink and uplink directions, respectively. On the basis of the schedule received from the BS (as well as the location, format, and length of each OFDMA zone in the DL-MAP and UL-MAP), each SS can determine when (from the OFDMA symbols) and where (from the subchannels) it should receive from and transmit to the BS.

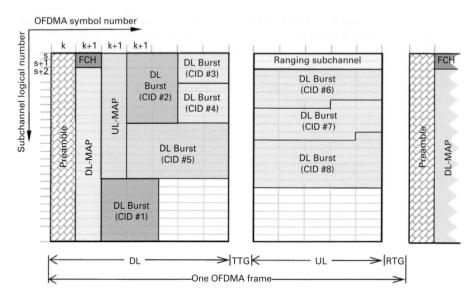

Figure 9.21 An example of an MPDU OFDMA frame structure in the TDD mode for a WiMAX PMP mode [49] (© IEEE 2007).

To handle the bandwidth asymmetry issue, that the downstream traffic is higher than the upstream in Internet applications, the lengths of the uplink and downlink subframes are determined dynamically by the BS and are broadcast to the SSs through UL-MAP and DL-MAP messages at the beginning of each frame. Therefore, each SS knows when and how long to receive data from and transmit data to the BS. Another important management message, which is associated with UL-MAP, is called an uplink channel descriptor (UCD) and can be periodically transmitted in the downlink subframe. The values of the minimum backoff window, W_{min}, and the maximum backoff window, W_{max}, are defined in this message; they are used for the contention-resolution algorithm and are explained in detail in Section 9.4.4. The uplink subframe contains transmission opportunities scheduled for the purpose of sending bandwidth-request (BW-REQ) messages, which allows for SSs to indicate to the BS that they need UL bandwidth allocation. The BS controls the number of transmission opportunities for both BW-REQ messages and data packet transmission through the UL-MAP message. An SS may establish multiple connections with a BS. While one BW-REQ request should be made for each connection, each station is only awarded a certain *grant*, the grant per subscriber station (GPSS), in the 802.16 standard. By choosing the GPSS aggregation method, a WiMAX system is supposed to reduce the granting workload of the BS when there are many connections per SS, and the winning SS should redistribute bandwidth among multiple connections to maintain the agreed QoS levels.

Multipoint-to-multipoint mode   In the mesh mode of WiMAX, nodes are organized in an ad hoc fashion. Unlike the PMP mode, there are no explicitly separate downlink and uplink subframes in the mesh mode. Each station is able to establish direct communication with a number of other stations in the system. However, in a typical scenario there can be certain nodes that provide the BS function of connecting the mesh network to the backhaul links [39].

Note that in the current IEEE 802.16e standard, the mesh mode is not supported, since mesh mode is mainly intended for stationary scenarios.

The IEEE 802.16 defines two mechanisms for scheduling data transmissions in the mesh mode; centralized and distributed scheduling. In centralized scheduling a designated BS-like node works like a cluster head, similarly to the BS in the PMP mode, and determines how SSs are to share the channel in different time slots. Because all the control and data packets need to go through the BS-like node, the scheduling procedure is simple but with a potentially long connection setup. The key difference of this centralized mesh mode from the PMP mode is that in the mesh mode all SSs may have direct links with other SSs. In distributed scheduling, however, every node competes for channel access using a pseudo-randoms election algorithm based on scheduling information from its two-hop neighbors. Data subframes are allocated on the basis of a request–grant–confirm three-way handshaking protocol [39]. As is the case with the PMP mode, specific schedulers for the mesh mode are not defined in the IEEE 802.16 standard.

9.4.2   QoS differentiation in 802.16d

The quality of service provisioning in IEEE 802.16d and IEEE 802.16e is slightly different. In 802.16d, a service flow is defined as a one-way flow of MAC SDUs (MSDUs) on a connection associated with specific QoS parameters such as latency, jitter, and throughput. These QoS parameters are used for transmission and scheduling. Service flows are typically identified by SSs and BSs from their *SFID* value. There are three basic types of service flows, as follows.

(1) A *provisioned service flow* is defined in the system with an *SFID* value, but it might not have any traffic presence. It may be waiting to be activated for usage.
(2) An *admitted service flow* undergoes the process of activation. In response to an external request for a specific service flow, the BS or SS will check for available resources, using the QoS parameters to see whether it can support the request. If there are sufficient resources, the service flow will be deemed to be admitted. The resources assigned to this service flow may still be used by other services.
(3) An *active service flow* is an admitted service flow with all the resources allocated. Packets will flow through the connection allocated to the active service flow.

The use of service flows is the main mechanism used in QoS provisioning. Packets traversing the MAC sublayer are associated with service flows as identified by the CID when QoS is required. Bandwidth grant services define bandwidth allocations using the QoS parameters associated with a connection. In downlink transmissions a BS has itself sufficient information to perform scheduling, but in uplink transmissions a BS performs the scheduling of various service transmissions using information gathered from SSs. In such cases an SS will request uplink bandwidth from the BS, and the BS will allocate bandwidth on an as-needed basis. For the proper allocation of bandwidth, four differentiated services are defined to support different types of data flow (see Table 9.6) [39], as follows.

(1) *unsolicited grant service (UGS)* is designed to support real-time constant bitrate (CBR) traffic, such as VoIP, i.e., a fixed-size data transmission at regular time intervals without the need for requests or polls.

Table 9.6 The five differentiated services defined, for proper allocation of bandwidth, to support different types of data flows in WiMAX [39]

QoS category	Applications	QoS specifications
UGS: unsolicited grant service	VoIP	Maximum sustained rate, maximum latency tolerance, jitter tolerance
rtPS: real-time polling service	Streaming audio or video	Minimum reserved rate, maximum sustained rate, maximum latency tolerance, traffic priority
ertPS: extended real-time polling service	Voice with activity detection (VoIP)	Minimum reserved rate, maximum sustained rate, maximum latency tolerance, jitter tolerance, traffic priority
nrtPS: non-real-time polling service	File transfer protocol (FTP)	Minimum reserved rate, maximum sustained rate, traffic priority
BE: best-effort service	Data transfer, Web browsing, etc	Maximum sustained rate, traffic priority

(2) A *real-time polling service (rtPS)* is designed to support variable bitrate (VBR) traffic, such as MPEG video. In this service the BS offers the SS periodic requests to indicate the required bandwidth.

(3) *Non-real-time polling service (nrtPS)* is designed for a delay-tolerant data service with a minimum data rate, such as FTP. The nrtPS allows an SS to use contention requests and unicast requests to determine the required bandwidth. Unicast polls are offered regularly to ensure that every SS has a chance to request bandwidth even in a congested network environment.

(4) A *best-effort (BE)* service does not specify any service-related requirements. Like nrtPS, it provides contention requests and unicast requests but does not provide bandwidth reservation or regular unicast polls.

While the concept of service flow is similar, to a certain extent, in both standards, IEEE 802.16e differs from IEEE 802.16d in its bandwidth grant services. In addition to the four data services listed above, IEEE 802.16e includes a new service known as *extended rtPS*, which provides a scheduling algorithm that builds on the efficiency of both UGS and rtPS. Like UGS, it is able to offer unsolicited unicast grants, such as VoIP. However, the size of the bandwidth allocation is dynamic, unlike in UGS, where the bandwidth allocation is a fixed size. The purpose of this service is to support real-time service flows that generate variable-size data packets on a periodic basis; one such example is VoIP with silence detection incorporated to produce a variable data rate.

9.4.3   Frame structure of 802.16 OFDMA

As shown in Figure 9.21, the DL-MAP in the TDD frame structure of an IEEE 802.16 system can present resource allocation information for each burst or each user. To support various types of physical channel condition, IEEE 802.16 OFDMA systems define several types of subchannel building methods; the two most popular are diversity subcarrier permutation, also called the partial usage subchannel (PUSC), and contiguous subcarrier

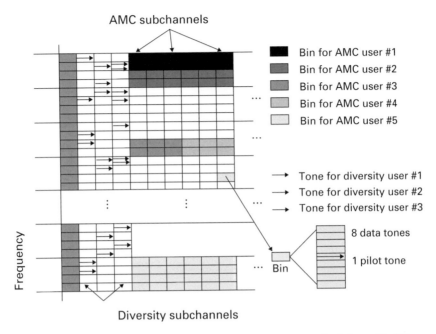

Figure 9.22 The IEEE 802.16 OFDMA system defines two types of subchannel building method, diversity subcarrier permutation (PUSC) and contiguous subcarrier permutation (AMC subchannels).

permutation, i.e., the adaptive modulation coding (AMC) subchannel. The ratio of these modes can be flexible in the IEEE 802.16 standard (see Figure 9.22).

Diversity subchannel (PUSC) mode The distributed subcarrier permutation mode draws subcarriers pseudo-randomly to form a subchannel. It provides frequency diversity and intercell interference averaging. It is a very useful scheme for the averaging of intercell interference and the avoidance of deep fading; this is achieved by the pseudo-random selection of subcarriers. Therefore, it is expected to be suitable for users with a high velocity and/or low signal-to-noise ratio (SNR). The basic resource units in the frequency domain of this mode are called *diversity subchannels*.

AMC subchannel mode In the adjacent subcarrier permutation mode, adjacent or contiguous subcarriers are grouped into clusters and are allocated to users. In this channel structure, the channel response can be seen as a flat fading channel; thus, the frequency selectivity of the channel cannot be exploited. This system can make better use of multi-user diversity as long as the channel state does not change significantly during the scheduling process. Therefore, it is expected to be suitable for users with low velocity and/or high SINR. The basic resource units in the frequency domain of this mode are called *band AMC subchannels*.

Figure 9.23 shows the time–frequency structure in the IEEE 802.16e OFDMA AMC subchannel [40]. A bin consists of nine contiguous subcarriers in a symbol, eight being assigned for data and one for a pilot. The position of the pilot is predefined. A slot, the minimum possible data allocation unit, is defined as a collection of bins of the type

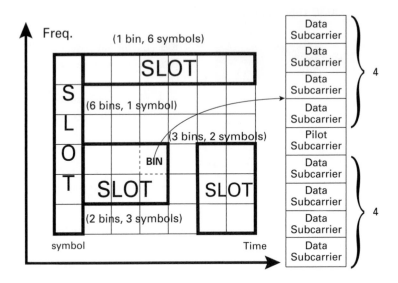

Figure 9.23 The time–frequency structure in the IEEE 802.16e OFDMA AMC subchannel [40].

$N \times M = 6$, where N is the number of contiguous bins and M is the number of contiguous symbols. The four allowed combinations for 802.16e are, as shown in Figure 9.23: 6 bins, 1 symbol; 3 bins, 2 symbols; 2 bins, 3 symbols; and 1 bin, 6 symbols. Note that the bandwidth of all subchannel equals the bandwidth of N contiguous bins. The standard defines four system bandwidths, 1.25 MHz, 5 MHz, 10 MHz, and 20 MHz with corresponding subcarrier numbers 128, 512, 1024, and 2048. Take for example, the 10 MHz system channel bandwidth and the 2 bins, 3 symbols combination: 160 subcarriers are reserved as guard subcarriers and a DC subcarrier and the remaining 864 are used for the data and the pilot. The modulation and coding scheme (MSC) can thus be determined from the AMC table using the instantaneous SNR based on the channel quality indicator (CQT) feedback [5]. More specifically, with respect to the different SNRs from the CQI feedback, the MCS provides seven modulation and coding alternatives (QPSK 1/2 and 3/4, 16-QAM 1/2 and 3/4, 64-QAM 1/2, 2/3, and 3/4).

9.4.4   Bandwidth-request (BW-REQ) mechanisms

A bandwidth request (BW-REQ) in WiMAX can be issued as either a stand-alone request or in an uplink data packet as a piggyback request, which is optional in the standard. When network access begins, an SS engages with a BS in a series of activities that include power leveling and signal ranging by using a series of ranging-request messages. The results of these activities are returned to the SS using ranging-response messages. The BS also may occasionally transmit ranging-response messages to the SS, requesting it to make adjustments to its power or timing settings. During the initial ranging process, the SS requests downlink service by transmitting a specific burst profile to the BS, which has the option of confirming or rejecting the request. In order to determine which SS is allowed to transmit its BW-REQ from multiple candidates, two schemes are suggested in the standard [38], as follows.

(1) *Contention scheme*   This is a contention-based random access, where SSs send BW-REQ during the contention period. Contention is resolved using backoff resolution.

(2) *Contention-free scheme*   This is a contention-free polling, where the BS polls each SS, which in turn replies by sending BW-REQ. This mode is more suitable for real-time applications owing to the predictable signaling delay of the polling scheme.

Instead of waiting for an explicit acknowledgment from BS regarding its BW-REQ, an SS expects to receive a grant within a special timeout interval after which it starts to use the allocated bandwidth for transmission. If the SS does not receive the grant within the timeout interval, it should conclude that its BW-REQ is corrupted either due to noisy channel or collision and then start a contention-resolution process. All bandwidth requests are made in terms of the number of bytes needed to carry the MAC header and payload, but not the PHY overhead. The actual time required to transmit the requested bandwidth depends on the modulation format used. The BS grants bandwidth on an aggregate basis for the entire CIDs associated with this SS, not to individual CIDs; this is the basis of the GPSS method discussed in Section 9.4.1.2.

In contention-based random access, an SS uses a random backoff mechanism to resolve contention among BW-REQ PDUs from multiple SSs. The mandatory method of random access contention resolution used in WiMAX is based on a truncated binary exponential backoff (BEB) scheme without carrier sensing, in contrast with the widely used CSMA/CA mechanism in IEEE 802.11 Wi-Fi networks. More specifically, before each transmission attempt of a BW-REQ, an SS chooses uniformly an integer number from the interval of $[0, W_i)$, where W_i denotes the current value of its backoff window after the ith collision. The chosen value, also referred to as a backoff counter, indicates the number of time slots during which the station has to wait before the transmission of a request. For the first transmission attempt, the backoff window size is at the minimum value W_{min}. Upon each transmission failure, a station doubles its backoff window value. Hence, the backoff window after the ith collision, W_i, will be $2 \times i \times W_{min}$; eventually the maximum value W_{max} $(= 2 \times m \times W_{min})$, where m is the allowed maximum transmission attempts, is reached. Both W_{min} and W_{max} are defined by the BS. On each successful transmission, the backoff window size should be reset to the minimum value W_{min}, assuming that the channel is free again.

Using a contention-free polling-based BW-REQ allocation system, the BS polls register SSs from a maintained list. Each SS is allowed to transmit a BW-REQ message only after it is polled. Actually, the poll schedule information for polling-based BW-REQ is conveyed by the UL-MAP and UCD in the downlink subframes, as shown in Figure 9.21. Note that the scheduling algorithms for polling are vendor-dependent and are not specified in the standard. One vendor may choose a simple round-robin scheduler to poll each SS sequentially in the polling list, but other, priority-based, polling mechanisms may also be used for BW-REQ scheduling if different QoS levels are required by different SSs. Furthermore, the polling allocation can be issued to a group of SSs. Allocation schedules to groups are also indicated in UL-MAP and UCD. This grouping mechanism is particularly important for specifically polling active groups of SSs for multicast and broadcast services (MBS) to save resource usage. Certain CIDs are reserved in MBS zones of an MPDU frame for multicast groups and broadcast messages, as specified in the standard.

9.4.5   Scheduling and resource allocation

Since WiMAX is a connection-oriented communication system, an SS must register at the BS before it can start to send or receive data. During the registration process, an SS can negotiate the initial QoS requirements with the BS, under a subscriber-station-based GPSS bandwidth granting method. These requirements can be changed later, and a new connection may be established on demand. The BS is responsible for providing the QoS guarantees in the WiMAX network by appropriately scheduling and resource-allocating for both the uplink and downlink directions. In other words, an algorithm at the BS has to map the QoS requirements of the chosen (scheduled) SSs into the appropriate number and location of slots in the time–frequency zone with appropriately chosen structure (see Figure 9.23). The algorithm can also account for the bandwidth request size that specifies the size of the SS input buffer. When the BS makes a scheduling decision, it informs all SSs about it by using the UL-MAP and DL-MAP messages at the beginning of each frame. These special messages define explicitly the slots that are allocated to each SS in both the uplink and downlink directions. The scheduling and resource allocation policy is not defined in the WiMAX specification and is open for alternative implementations.

Therefore, a critical scheduling and allocation issue is how a WiMAX BS centrally and efficiently allocates the channels in different slots to different SSs for uplink and downlink transmissions; in turn these resources are allocated to the various connections supported by the SSs at that particular time. Figure 9.24 shows a proposed MAC–PHY cross-layer scheduling and resource allocation mechanism for WiMAX [41]. In this system, it is supposed that the BS is aware of the PHY channel state (e.g., in terms of SNR values) of all subchannels for all the SSs. Thus the BS MAC can exploit channel-user diversity in allocating resources. The overall system throughput can be maximized by allocating a subchannel to the SS with the best channel state. However, this may not satisfy the QoS and fairness requirements of different SSs when there are insufficient resources to satisfy the QoS of all the users.

9.4.5.1   Fair packet scheduling and subchannel allocation in 802.16d

As discussed previously, 802.16d supports mainly fixed wireless services where the mobility of SSs is not supported. There are four QoS service classes in 802.16d, and each has different bandwidth request and grant mechanisms. A connection can be associated with any one of the service classes depending upon its QoS requirements. An SS requests the BS for bandwidth on a per-flow basis for uplink. The BS grants a total bandwidth for all the connections which belong to that SS, using the GPSS method. Then the SS redistributes this sum-total grant among its users according to the service class of a user's connection and its QoS requirements.

Given the requirements of different SSs and their channel conditions, the BS schedules the channels and slots to different SSs one frame at a time (or it can be several frames at a time). Owing to the use of TDD, the resource allocation of upstream and downstream traffic can be considered separately. Since real-time multimedia applications are delay sensitive, the requirements of real-time applications (UGS and rtPS) need to be met in the same frame, while the packets of non-real-time data can be deferred. Therefore it is good practice [42] for the BS first to try to satisfy the needs of the UGS applications in the upstream and downstream (normally the needed bandwidth is low, such as in the case of voice traffic). Next the requirements of the rtPS in both directions can be satisfied. After

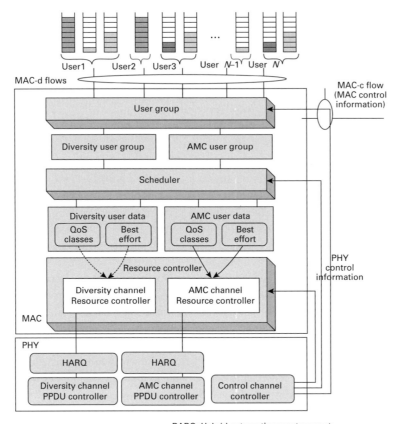

RARQ: Hybrid automatic repeat request
PPDU: Physical layer protocol data unit

Figure 9.24 A typical MAC–PHY cross-layer scheduling and resource allocation mechanism for WiMAX [41].

that the minimum requirements of the nrtPS can be satisfied, and finally the requirements of nrtPSs without a minimum rate requirement and of best-effort traffic are taken care of. Once the BS has done this calculation, it then divides the frame (since TDD is used) into upstream and downstream. The overall channel and slot allocation for every SS, under the GPSS principle, for upstream is broadcast to it at the beginning of the frame, so that an SS can be notified about its upstream channel and slot allocation in a single frame. The allocation of resources to different SSs can be formulated as an optimization problem.

Given a BS with m subchannels and n serviced SSs, assume that the ith subscriber station SS_i has a total demand of λ_i bytes for its UGS connection(s) for upstream in a given frame. Also, assume that the rate allocated on subchannel j is R_{ij} bytes per slot in this frame for SS_i and that the total number of slots allocated to SS_i in this frame on subchannel j is N_{ij}. Then, after this frame the resulting unsatisfied demand C_i of SS_i is [42]

$$C_i = \max\left(0, \lambda_i - \sum_{j=1}^{m} R_{ij}N_{ij}\right), \tag{9.5}$$

where C_i must be non-negative. The objective of optimal resource allocation is to minimize C, the total unsatisfied demand of the serviced SSs,

$$C = \sum_{i=1}^{n} C_i = \sum_{i=1}^{n} \left(\lambda_i - \sum_{j=1}^{m} R_{ij}N_{ij} \right), \tag{9.6}$$

subject to

$$\sum_{i=1}^{n} N_{ij} \leq \overline{N}_j, \quad \forall j; \qquad \lambda_i \geq \sum_{j=1}^{m} R_{ij}N_{ij}, \tag{9.7}$$

where $\overline{N}_j$ denotes the total number of slots in subchannel j available for data transmission.

To solve this optimization problem, which enables the BS to allocate channels and slots to different SSs in each of M frames, an integer programming optimization should be used since the cost function and the constraints of Eqs. (9.6) and (9.7) increase linearly with the integer-valued solution N_{ij}. To reduce the computational complexity, an heuristic algorithm was developed for 802.16d WiMAX [42], in which different subchannels are allocated to SSs one slot at a time. After allocation has been made for t slots, a subchannel is given to the SS that can transmit the maximum amount of data (out of its remaining data) on that subchannel in slot $t+1$. While this is a very intuitive algorithm, its performance has been shown also to be close to optimal.

It has been shown also that algorithms which optimize the overall system performance may not be fair to different SSs and users; therefore the scheduling and allocation algorithms developed should be modified to obtain fairer solutions with little sacrifice in overall system performance. This can be achieved by still using the minimization formulation in Eq. (9.6) but with a slight modification to the constraint given in Eq. (9.7), as follows:

$$\sum_{i=1}^{n} N_{ij} \leq \overline{N}_j, \quad \forall j; \qquad \lambda_i \geq \sum_{j=1}^{m} R_{ij}N_{ij} \geq \beta\lambda_i, \tag{9.8}$$

where a fixed $\beta \geq 0$ is used [42].

It is possible that for a given β this problem may not have a feasible solution. Therefore β can be started with a value close to 0 and then gradually increased until no feasible solution is reached. This will provide a good solution which is also proportionally fair, since it will ensure that each SS gets at least a fraction β of its demand. If β is small then the solution obtained will be more globally optimal but less fair. Thus β provides a tradeoff between global efficiency and fairness.

9.4.5.2 Joint packet scheduling and subchannel allocation for 802.16e

Of the two different types of OFDMA subchannel building scheme, the AMC scheme groups a block of contiguous subcarriers to form a subchannel which enables multi-user diversity by choosing the subchannels with the best frequency response. Multi-user diversity gain promotes system throughput substantially, while it requires the adaptive modulation coding scheduler to consider the PHY channel quality individual (CQI) as well as MAC QoS requirements. This kind of packet scheduling and subchannel allocation involves the MAC and PHY layers, therefore it is also called a cross-layer scheduling scheme.

Consider a downlink OFDM system with N subchannels and M time slots. Assume further that there are K users, J connections, and L packets in the system. The objective of resource allocation is to maximize the overall system throughput while guaranteeing the provision of QoS. The allocation task can thus be formulated as the following constrained optimization problem [40]: find the optimial arguments $A_i(m, n)$ and $g(j)$, i.e., find

$$\arg \max_{A_i(m,n), g(j)} \sum_{i=1}^{L} \sum_{m=1}^{M} \sum_{n=1}^{N} A_i(m, n) R_i(m, n) \tag{9.9}$$

subject to

$$\sum_i A_i(m, n) - 1 \le 0, \qquad A_i(m, n) \in \{0, 1\}, \quad \forall m, n, \tag{9.10}$$

$$W_i \le T_j \qquad \forall i, i \to j, \quad \text{(packet } i \text{ belongs to connection } j) \tag{9.11}$$

$$g(j) \ge \min\left(\frac{C_{\min}(j)}{d(j)}, P_j(t)\right), \qquad \forall j, \tag{9.12}$$

where $A_i(m, n)$ identifies packet i as allocated to slot (m, n) and $R_i(m, n)$ is the equivalent data rate obtained by packet i in this slot; T_j denotes the maximum allowed latency of the rtPS jth connection, W_i denotes the waiting time of packet i and $g(j)$ is the number of scheduled packets in connection j. The quantity $C_{\min}(j)$ represents the minimum reserved data rate of connection j, $d(j)$ being connection j's packet length and $P_j(t)$ the number of packets present in connection j.

Since the CQI of PHY is usually assumed to be constant for each frame, $R_i(m, n)$ can be simplified to $R_i(n)$. Equation (9.10) ensures that only one slot can be allocated to a given packet, while Eqs. (9.11) and (9.12) correspond to the QoS requirements in terms of delay and throughput. Owing to practical considerations such as the implementation complexity of the modulator and demodulator, the limited capacity of DL-MAP and UL-MAP messages defined in the standard to inform users on which slots their packets are loaded, and so on, solving Eq. (9.9) for the optimal solution is not viable in practice, and a practical scheme that allows a tradeoff between implementation complexity and spectrum efficiency was thus proposed in [40]. This practical scheme is divided into two phases.

(1) *Packet scheduling* defines the scheduling priority of each packet on the basis of its channel quality, QoS satisfaction, and service priority. Since packets have diverse channel qualities on different subchannels, the priorities of a given packet vary over subchannels. Thus for each subchannel one particular scheduling-priority queue should be obtained.

(2) *Packet allocation* searches for a valid subchannel which has free space and is not in deep fading. From the scheduling-priority queue of this valid subchannel, obtained in the above packet scheduling phase, the packet with the highest scheduling priority should be allocated to this subchannel. However, the packet may have highest priority on more than one subchannel. To maximize the system throughput, the best subchannel (in terms of the CQI) for this packet in the subchannel set in which the packet owns the highest priority should be chosen. Finally, the packet is allocated to this same subchannel.

This procedure is repeated until no packet remains to be transmitted or the valid subchannels are used up. More details of these two phases are discussed below.

Packet scheduling Owing to the constant-rate requirements of the unsolicited grant service (UGS) in the MAC, it can be presumed that fixed numbers of symbols are allocated for UGS ahead of scheduling and thus it is reasonable not to consider UGS service in the scheduling and allocation analysis. For an rtPS packet i, the scheduling priority $\varphi_{i,n}$ on subchannel n is defined as [40]

$$
\varphi_{i,n} = \begin{cases} \beta_{rtPS} R_i(n)/R_{\max} 1/F_i & \text{if} \quad F_i \geq 1, \quad R_i(n) \neq 0, \\ \beta_{rtPS} & \text{if} \quad F_i < 1, \quad R_i(n) \neq 0, \\ 0 & \text{if} \quad R_i(n) = 0, \end{cases} \tag{9.13}
$$

where

$$
F_i = \frac{T_j - W_i}{T_g} \tag{9.14}
$$

and β_{rtPS} denotes the priority of rtPS over other types of service, which is similar to priority queue (PQ) scheduling in a wireline system. The ratio $R_i(n)/R_{\max}$ is the normalized data rate, where R_{max} is the highest modulation and coding mode (i.e. the rate under 64-QAM 3/4); $R_i(n) = 0$ denotes that packet i is under deep fading on subchannel n and should not be scheduled; F_i represents the service satisfaction level of packet i and is defined as the ratio of the remaining allowable waiting time for packet i and the guard time. When $F_i \geq 1$, it is expected that the smaller the F_i is, the higher the scheduling priority should be. When $F_i < 1$, which indicates that the allowable waiting time of packet i is smaller than the guard time T_g, the priority of packet i is promoted to β_{rtPS}, enabling scheduling precedence over other packets. According to [40], T_g is set as the frame length, i.e., the interval between two scheduling times. In the case of very limited resources, so that it cannot get scheduled, the waiting time W_i of packet i will exceed the maximal allowed latency T_j; therefore this packet is deemed invalid and discarded at the time of the next scheduling.

Similarly, for an nrtPS packet i, $\varphi_{i,n}$ is defined as before,

$$
\varphi_{i,n} = \begin{cases} \beta_{nrtPS} R_i(n)/R_{\max} 1/F_i & \text{if} \quad F_i \geq 1, \quad R_i(n) \neq 0, \\ \beta_{nrtPS} & \text{if} \quad F_i < 1, \quad R_i(n) \neq 0, \\ 0 & \text{if} \quad R_i(n) = 0, \end{cases} \tag{9.15}
$$

but now

$$
F_i = \frac{\bar{C}_j(t)}{C_{\min}(j)}, \qquad i \to j, \tag{9.16}
$$

and

$$
\bar{C}_j(t+1) = \bar{C}_j(t)\left(1 - \frac{1}{t_c}\right) + \frac{C_j(t)}{t_c}, \tag{9.17}
$$

where β_{nrtPS} is the priority of the nrtPS service, F_i for nrtPS packet i is defined as the ratio of the average throughput of connection j and the prescribed minimum reserved data rate. The average throughput is estimated during a time window t_c. Obviously the packets in a single nrtPS connection have the same priority in a given subchannel and only need to be calculated once for this connection's packets. If the $F_i < 1$, i.e., the nrtPS connection's average throughput is less than its prescribed minimum reserved rate then the scheduling priorities of this connection's packets are upgraded to β_{nrtPS}.

In a design similar to that in the rtPS service, once the QoS is met the scheduling priorities are determined by packets' CQI. The larger the CQI value, the higher the priority. In fact, the design is a modification of proportionally fair scheduling, which has proved to be an optimal utility-based scheduling scheme [48]. Finally, for a BE packet i, $\varphi_{i,n}$ is defined as

$$\varphi_{i,\,n} = \beta_{BE} \frac{R_i(n)}{R_{max}} \tag{9.18}$$

where β_{BE} is the priority of the BE service. Since no QoS requirements are prescribed for BE service in the standard, its priority is related only to CQI and service type.

Packet allocation There are five steps in packet allocation to the subchannels [40], as follows.

(1) Calculate the scheduling priority $\varphi_{i,n}$, $n \in \{1,2,\dots,N\}$, for packet i on subchannel n and then create N priority queues $\{P_n\}$, corresponding to N subchannels for all the L packets:

$$P_n = \{i_1, i_2, \dots, i_L : \varphi_{i_1,n} \geq \varphi_{i_2,n} \geq \dots \geq \varphi_{i_L,n}\},$$

where first-in first-out (FIFO) is applied for packets with the same priority.
With respect to any user k, according to the CQI value we can also define a quality mapping Q_k, $k \in \{1, 2, \dots, K\}$, as an ordering from the best subchannel to the worst one:

$$Q_n = \{n_1, n_2, \dots, n_N : \gamma_{k,n_1} \geq \gamma_{k,n_2} \geq \dots \geq \gamma_{k,n_N}\},$$
$$\Delta j = M, \quad 1 \leq j \leq N; \qquad \Delta l = d(l), \quad 1 \leq l \leq L,$$

where Δj denotes the available space (in terms of slots) of subchannel j, Δl indicates the length (in terms of bytes) of the unscheduled part of packet l, and $d(l)$ is the initial length of packet l.

(2) Search for the first valid subchannel set V, where

$$V = \{n : P_n \neq \{\}, \Delta n \neq 0, R_{P_n(1)}(n) \neq 0\}, \qquad n = \min(V);$$

where $P_n \neq \{\}$ indicates that subchannel n has packets to transmit and $\Delta n \neq 0$ implies that free space is available on subchannel n. Then $P_n(1)$ denotes the packet with the highest priority on subchannel n and $R_{P_n(1)}(n) \neq 0$ excludes the possibility that subchannel n falls into deep fading.

(3) If valid subchannel n exists, search the best subchannel m^* for packet $P_n(1)$, where $m^* \in \{m: P_m(1) = P_n(1)\}$ and m^* queues the first in U_k if packet $P_n(1)$ belongs to user k.

(4) Denote the length of the untransmitted part of packet $P_n(1)$ as $\Delta P_n(1)$, and let Δm^* represent the free space of subchannel m^*. Finally, P_n denotes the priority queue. Denote as S the number of slots that packet $P_n(1)$ takes up. Then

if $S \leq \Delta m^*$ then $\Delta P_n(1) = 0$, $\Delta m^* = \Delta m^* - S$
else $\Delta P_n(1) = \Delta P_n(1) - \Delta m^* \times Th_{P_n(1)}(n)$, $\Delta m^* = 0$

Here, $Th_{P_n(1)}(n)$ is the slot unit capacity of packet $P_n(1)$ on subchannel n. The former condition corresponds to the situation where subchannel m^* has enough space for packet $P_n(1)$ while the latter condition corresponds to the opposite. Finally, delete all transmitted packets from the queue P_n, $n \in \{1, 2, \dots, N\}$.

(5) Search for the next valid subchannel. Update valid subchannel set V: if V is empty then the allocation is finished; if some elements in V are larger than n then n is updated to the first such element or else to the first element in V. Then go to step (3).

9.4.6   The 802.16j standard: multi-hop relay

Even though WiMAX offers a unique opportunity to provide users with a seamless broadband connection, there are many technical issues that must be carefully examined, including the deployment of infrastructure, data transmission, service guarantees, and scheduling. For instance, what is the tradeoff between using a single point with large coverage and using many points with smaller coverage, since the number of users that can be supported with sufficient throughput in the latter case can be significantly lower under single-point coverage? The IEEE 802.16j mobile multi-hop relay (MMR) standard, which is to be completed sometime in 2008, is proposed as providing an attractive solution for the coverage extension and throughput enhancement of IEEE 802.16e networks, as illustrated in Figure 9.25 [49].

As throughput enhancement is a major objective in 802.16j MMR networks, the ongoing research efforts on effective resource allocation are deemed to be a necessity for such networks. As shown in Figure 9.25, there are three kinds of network element in an 802.16j MMR network: the base station (BS), relay stations (RSs), and 802.16e compliant mobile stations (MSs), e.g., a notebook or a smart phone. The IEEE 802.16j standard is intended to support the multi-hop relaying function, whereby a BS and several RSs can form a multi-level tree topology; the footprint of such an 802.16 network is thus greatly expanded, in a highly economical manner. The purpose of enabling such a relay function is to enhance the coverage and throughput of a base station, and to enable low-power devices to participate in the network. The RS can be a fixed station or a station capable of mobility. It is assumed that the RS will not generate user traffic of its own, but it is expected that an RS will provide user access and will support the generation of control and management messages necessary for proper relay operation. The MMR infrastructure can support a hop-count number greater than or equal to 2, and it is also possible to establish

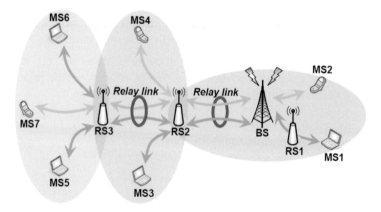

Figure 9.25 The IEEE 802.16j use multi-hop relay stations in an attempt to extend the coverage area and improve throughput at a feasible economical level [49].

multiple communication paths between a BS and an MS so as to improve transmission robustness.

As for WLAN mesh networks, an increase in the number of relay hops in an MMR adversely affects overall throughput; the latter is further challenged by the possibility of QoS degradation and scheduling complexity. With the potential for multimedia applications in such networks, there is also a need to look into ways of enhancing resource utilization and maximizing capacity in each coverage area. Maintaining end-to-end QoS is another challenge; in particular, mapping the QoS parameters from the BS across multiple hops to the last relay station needs to be addressed properly. This is especially so when there are many users with different QoS requirements under the coverage of different relay stations along the path. It will also be important to quantify the number of hops in QoS provisioning. Finally, there can be many interesting problems due to handover in MMR coverage; this has to preserve the QoS and timing requirements of potential multimedia services.

9.4.7   WiMAX modules for network simulator 2 (ns-2)

Two IEEE 802.16 simulation modules are available for public use: one is from the Networks and Distributed Systems Laboratory (NDSL) of the Chang Gung University (CGU) in Taiwan [43] and the other from the US National Institute of Standards and Technology (NIST) [44]. These modules are still in their early stages, and not all standardized features have been implemented.

The 802.16-based WiMAX simulation module by CGU, named as the Mac802_16 class, is in accordance with the specifications of the IEEE 802.16–2004 (802.16d) standard [4] and the ns-2 version 2.29 [5]. The proposed module is composed of a convergence sublayer (CS) and MAC sublayer. The CS sublayer has two major functions: (1) *IP-SFID mapping*, which transforms IP addresses to several SFIDs or vice versa; (2) *SFID-CID mapping*, which actually maps essential QoS parameters of the upper layers into QoS classes. The MAC sublayer implements functions such as MAC management, ranging, priority queue, and QoS scheduling with five messages, i.e., downlink channel description (DCD), uplink channel description (UCD), DL-MAP, UL-MAP, and bandwidth request (BR). An SS has to perform ranging to join a network, and packets will be queued in one of five QoS queues (UGS, rtPS, ertPS, nrtPS, or BE). The scheduler evaluates the available bandwidth according to a selection of modulation methods and distributes it to stations using a weighted round-robin policy. In its most updated version, 2.03, the mobility extension in 802.16e has not been included. A physical-layer component featuring OFDMA as well as an FDD mode is expected to replace the generic wireless physical component in ns-2. The module is fully tested with ns-2.29 and can be downloaded at http://ndsl.csie.cgu.edu.tw/wimax_ns2.php.

The NIST module, however, is currently based on the IEEE 802.16-2004 and 802.16e-2005 [5] standards. Three layers, CS, MAC, and PHY, are implemented to be relatively extensible. The supported functions are a configurable wirelessMAN-OFDM physical layer, a TDD mode, management messages, a round-robin (non-QoS) scheduler, scanning and handover extensions, and the fragmentation and reassembly of frames. Major work has been done on the wirelessMAN-OFDMA PHY, the FDD mode, QoS scheduling, periodic ranging, power adjustments, packing, and error correction. The module was fully tested with ns-2.29 and can be downloaded at http://www.antd.nist.gov/seamlessandsecure/download.html.

9.5   Internetworking between 802.16 and 802.11

The internetworking and interoperability between IEEE 802.16 and other wireless technologies, especially 802.11, constitute a critical task that warrants the provision of a more comprehensive wireless network. As discussed in Section 9.2, the MAC protocol of 802.11 uses contention access, in which each SS has to compete to access a wireless access point (AP). A subscriber distant from an AP can suffer performance degradation, which makes it difficult to maintain the service requirements for applications like VoIP and streaming video. However, IEEE802.16 MAC uses scheduling mechanisms for channel arbitration. It also has the flexibility to allocate different numbers of time and frequency slots to users with different needs.

From the MAC perspective, one of the main issues is how to allow the MAC information associated with a data frame to be mapped correctly across different wireless platforms. For example, one possible IEEE 802.16 deployment is to combine IEEE 802.16 and IEEE 802.11 to form a wireless network for both outdoors and indoors. This is because it may not be practical or economically feasible to use WiMAX for providing full coverage of an indoor environment, as obstructions and building materials can attenuate outdoor signals to a large extent. Therefore, to provide more complete coverage of indoor and outdoor environments, it seems natural to use a mix of IEEE 802.16 and IEEE 802.11. For multimedia traffic transmission with mobility, the best combination to use is IEEE 802.16e; and IEEE 802.11e; the latter is specified to support QoS over WLAN. Figure 9.26 [46] shows an integrated 802.16 and 802.11 system, where a WiMAX BS operating in a licensed band serves both WiMAX SSs and Wi-Fi APs or routers in its coverage area. The connection between the BS and an SS is dedicated to a single user, which is presumably shared among multiple sessions, while the connection between the AP or router and the BS is shared among the WLAN nodes. A potential application of WiMAX is to provide backhaul support for mobile Wi-Fi hotspots. Traditionally, a Wi-Fi hotspot is connected to the Internet via a wired connection (e.g., a digital subscriber line, DSL). However, by using an IEEE 802.16e–WiMAX-based backbone network to connect Wi-Fi hotspots to the Internet, costly wired infrastructure can be avoided and mobile hotspot services can also be provided (e.g., for Internet access on buses and for certain intelligent transportation system applications).

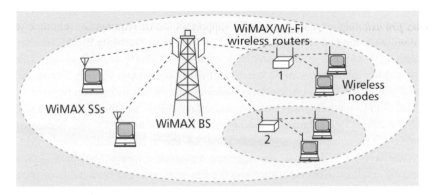

Figure 9.26 An integrated 802.16 and 802.11 system, where a WiMAX base station (BS) operating in a licensed band serves both WiMAX subscriber stations (SSs) and Wi-Fi access points or routers in its coverage area [46] (© IEEE 2007).

In one specific scenario, shown in Figure 9.26, an end user performs a handover from an IEEE 802.16e network to an IEEE 802.11e network. The core network represents the backbone of the overall network, which provides connectivity for BSs and APs. A key requirement is to ensure seamless handover with minimum service disruption to applications. In such a scenario there is a need to map the application-specific QoS parameters across different wireless platforms, which in this case are IEEE 802.16e and IEEE 802.11e. One of the challenges is to sufficiently map different QoS profiles using a limited set of QoS parameters and ensure that the QoS requirements are fulfilled as they map from one platform to another. Furthermore, the fact that the QoS mechanisms in two platforms are inherently different can bring considerable difficulty in ensuring proper end-to-end QoS provisioning.

Another problem is rerouting packets from the WiMAX BS to the 802.11e AP. The network must be able to recognize the handover and to reroute packets appropriately, so that there is minimum disruption to the traffic routing. In such cases the challenge is to ensure that the packet reroute time is sufficiently short to handle any QoS transmission latency caused by the handover. Thus, there is a need to design intelligent routing protocols that can work across different wireless platforms. Moreover, since the WiMAX and the Wi-Fi networks have different protocol architectures and QoS support mechanisms, there are many challenging issues in an integrated WiMAX–Wi-Fi network, such as protocol adaptation, quality of service (QoS) provisioning, and pricing for resource allocation, etc. [46].

Protocol adaptation   Protocol adaptation is critically required for two different wireless infrastructures, WiMAX and Wi-Fi, to internetwork with each other. Some protocol adaptation mechanisms originally proposed for internetworking between 3G and Wi-Fi can be extended for WiMAX–Wi-Fi integrated networks [47]. For example, with a layer-2 approach, adaptation would be required in the MAC layer for the WiMAX BS and Wi-Fi nodes. With a layer-3 approach, the adaptation would be performed at the IP layer, and a Wi-Fi user would interact only with the corresponding Wi-Fi AP or router. This layer-3 approach is preferred for the WiMAX–Wi-Fi integrated network, since Wi-Fi APs or routers can fully control bandwidth allocation among the nodes. Since a Wi-Fi AP or router is responsible for protocol adaptation up to the IP layer, modifications of the Wi-Fi user equipment and the WiMAX BS (in hardware and/or software) are not required.

QoS provisioning   Quality of service support would be required for real-time (e.g., video and voice) traffic in an integrated WiMAX–Wi-Fi network. While WiMAX networks have a predefined QoS framework, Wi-Fi networks also support QoS at the MAC layer on the basis of the IEEE 802.11e standard. The WiMAX QoS framework supports five major service types, UGS, rtPS, ertPS, nrtPS, and BE, which facilitates the implementation of traffic scheduling and resource management to achieve the target QoS performance. However, IEEE 802.11e also supports four traffic access categories (ACs): AC_VO (voice), AC_VI (video), AC_BE (best-effort), and AC_BK (background). A QoS framework with service mapping for WiMAX–Wi-Fi internetworking has been proposed [47], for the support of different types of traffic (e.g., constant bitrate, variable bitrate, best-effort). Moreover, the impacts of mobility on QoS also need to be taken into account.

Resource allocation pricing   The radio resource allocation mechanisms need to be developed accordingly. Specifically, optimal and adaptive bandwidth sharing mechanisms that satisfy both the WiMAX and Wi-Fi service providers need to be developed. While radio resource management schemes, such as traffic scheduling, rate and power allocation, and admission control, determine how the limited radio resource is used, the pricing scheme controls the amount of radio resource usage by network subscribers. The resource allocation and sharing mechanism in an integrated WiMAX–Wi-Fi network would be closely related to the pricing model used. A pricing scheme aims at maximizing the system utility, which is a function of the network QoS requirements. To this end, a game theory-based-optimal pricing scheme for bandwidth sharing in an integrated WiMAX–Wi-Fi network was proposed in [46].

References

[1] "Wireless local area networks," IEEE 802.11, http://www.ieee802.org/11/.
[2] "The Third Generation Partnership Project (3GPP)," http://www.3gpp.org/About/about.htm.
[3] "The Third Generation Partnership Project 2 (3GPP2)," http://www.3gpp2.org/Public_html/Misc/AboutHome.cfm.
[4] "IEEE standard for local and metropolitan area networks, Part 16: air interface for fixed broadband wireless access systems," IEEE Standard 802.16-2004 (revision of IEEE Standard 802.16-2001), 2004.
[5] "IEEE standard for local and metropolitan area networks, Part 16: air interface for fixed and mobile broadband wireless access systems, Amendment, 2: physical and medium access control layers for combined fixed and mobile operation in licensed bands, and Corrigendum 1," IEEE Standard 802.16e-2005 and IEEE Standard 802.16-2004/Cor 1-2005 (Amendment and Corrigendum to IEEE Standard 802.16-2004), 2006.
[6] K. Ahmavaara, H. Haverinen, and R. Pichna, "Interworking architecture between 3GPP and WLAN systems," *IEEE Communi. Mag.*, 41(11): 74–81, November 2003.
[7] S. M. Alamouti, "Mobile WiMAX: vision and evolution," plenary presentation in *IEEE Mobile WiMAX'07*, Orlando, Florida, March 2007.
[8] "International Mobile Telecommunications-2000 (IMT-2000)," ITU, http://www.itu.int/home/imt.html.
[9] R. Yallapragada, "Mobile broadband technologies for the present and the future," plenary presentation in *IEEE Mobile WiMAX'07*, Orlando, Florida, March 2007.
[10] B. Haberland, S. Bloch, and V. Braun, "3G evolution towards high speed downlink packet access," Technology White Paper, pp. 1–11, Alcatel Telecommunications Review – 4th Quarter 2003/1st Quarter 2004.
[11] M. W. Thelander, "The 3G evolution – taking CDMA2000 into the next decade," Signals Research Group, LLC, White Paper developed for the CDMA Development Group, October 2005.
[12] M. Jiang, L. Hanzo, "Multiuser MIMO-OFDM for next generation wireless systems," *Proc. IEEE*, 95(7): 1430–1469, July 2007.
[13] Z. Abichar, Y. Peng, and J. M. Chang, "WiMAX: the emergence of wireless broadband," *IT Proc., IEEE Computer Society*, 44–48, July/August 2006.
[14] S. J. Vaughan-Nichols, "Will the new Wi-Fi fly?," *IEEE Computer*, 16–18, October 2006.
[15] J.-H. Yeh, J.-C. Chen, C.-C. Lee, "WLAN standards,", *IEEE Potentials*, 16–22 October/November 2003.
[16] U. Varshney, "The status and future of 802.11-based WLANs," *IEEE Computer*, 102–105, June 2003.
[17] M. Ergen, "IEEE 802.11 tutorial," University of California, Berkeley, June 2002, http://wow.eecs.berkeley.edu/ergen/FILES/publications.htm.

[18] S. Mangold, S. Choi, G. R. Hiertz, O. Klein, B. Walke, "Analysis of 802.11e for QoS support in wireless LANs," *IEEE Wireless Commun.*, 40–50, December 2003.

[19] G. Holland, N. Vaidya, and P. Bahl, "A rate-adaptive MAC protocol for multi-hop wireless networks," in *Proc. ACM/IEEE Int. Conf. on Mobile Computing and Networking (MOBICOM'01)*, Rome, July 2001.

[20] A. Kamerman and L. Monteban. "WaveLAN-II: a high-performance wireless LAN for the unlicensed band," *Bell Laboratories Technical J.*, 118–133, Summer 1997.

[21] M. Lacage, M. H. Manshaei, and T. Turletti, "IEEE 802.11 rate adaptation: a practical approach," in *Proc. ACM Int. Symp. on Modeling, Analysis, and Simulation of Wireless and Mobile Systems (MSWiM)*, Venice, October 2004.

[22] D. Qiao, S. Choi, A. Jain, and K. G. Shin. "MiSer: an optimal low-energy transmission strategy for IEEE 802.11a/h." in *Proc. ACM MOBICOM*, pp. 161–175, September 2003.

[23] D. Qiao, S. Choi, and K. G. Shin. "Goodput analysis and link adaptation for IEEE 802.11a wireless LANs." *IEEE Trans. Mobile Computing*, 1(4): 278–292, October–December 2002.

[24] D.-Y. Yang, T.-J. Lee, K. Jang, J.-B. Chang, and S. Choi, "Performance enhancement of multirate IEEE 802.11 WLANs with geographically scattered stations," *IEEE Trans. Mobile Computing*, 5(7): 906–919, July 2006.

[25] G. Bianchi, "Performance analysis of the IEEE 802.11 distributed coordinated function," *IEEE J. Selected Areas Comm. (JSAC)*, 18 (3), 535–547, March 2000.

[26] Q. Ni, "Performance analysis and enhancements for IEEE 802.11e wireless networks," *IEEE Networks*, 21–27, July/August 2005.

[27] S. Mangold *et al.*, "IEEE 802.11e wireless LAN for quality of service," in *Proc. Euro. Wireless conf.*, Florence, February 2002.

[28] Y. Drabu, "A Survey of QoS techniques in 802.11," Dept. Computer Science of CS, Kent State University.

[29] D.-J. Deng, and R.-S. Chang, "A priority scheme for IEEE 802.11 DCF access method," *IEICE Trans. Commun.*, E82-B(1): 96–102, January 1999.

[30] J. L. Sobrinho and A. S. Krishnakumar, "Real-time traffic over the IEEE 802.11 medium access control layer," *Bell Laboratories Technical J.*, 172–187, 1996.

[31] J. L. Sobrinho and A. S. Krishnakumar, "Quality-of-service in ad hoc carrier sense multiple access networks," *IEEE J. selected Areas in Commun.*, 17(8): 1353–1368, 1999.

[32] A. Lindgren, A. Almquist, and O. Schelen, "Quality of service schemes for IEEE 802.11 wireless LANs – an evaluation," *J. of Mobile Networks and Applications*, 8: 223–235, 2003.

[33] N. H. Vaidya, P. Bahl and S. Gupta, "Distributed fair scheduling in a wireless LAN", in *Proc. 6th Annual Inte. Conf. on Mobile Computing and Networking*, Boston, 2000.

[34] "Amendment: medium access control (MAC) quality of service (QoS) enhancements," IEEE 802.11 WG, IEEE 802.11e/D13.0, January 2005.

[35] M. J. Lee, J. Zheng, Y.-B. Ko., and D. M. Shrestha, "Emerging standard for wireless mesh technology," *IEEE Wireless Communications*, 56–63, April 2006.

[36] M. Aoki *et al.*, "802.11 TGs simple efficient extensible mesh (SEE-mesh) proposal," IEEE 802 11–05/0562r01, 2005.

[37] M. Sheu *et al.*, "802.11 TGs MAC enhancement proposal," IEEE 802 11–05/0575r4, 2005.

[38] Q. Ni, A. Vinel, Y. Xiao, A. Turlikov, and T. Jiang, "Investigation of bandwidth request mechanisms under point-to-multipoint mode of WiMAX networks," *IEEE Commun. Mag.*, 45(5): 132–138, May 2007.

[39] B. Li, Y. Qin, C. Ping Low, and C. Lim Gwee, "A Survey on Mobile WiMAX," *IEEE Commun. Mag.*, 45(12): 70–75, December 2007.

[40] L. Wan, W. Ma, Zihua. Guo, "A cross-layer packet scheduling and subchannel allocation scheme in 802.16e OFDMA system," in *Proc. IEEE Wireless Communications and Networking Conf. (WCNC'07)*, pp. 1865–1870, March 2007.

[41] T. Kwon, H. Lee, S. Choi, *et al.*, "Design and implementation of a simulator based on a cross-layer protocol between MAC and PHY layers in a WiBro compatible IEEE 802.16e OFDMA system," *IEEE Commun. Mag.*, 43(12): 136–146, December 2005.

[42] V. Singh, and V. Sharma, "Efficient and fair scheduling of uplink and downlink in IEEE 802.16 OFDMA networks," in *Proc. IEEE Wireless Communications and Networking Conf. (WCNC'06)*, pp. 984–990, March 2006.

[43] J. Chen, C. Wang, F. C. Tsai *et al.*, "The design and implementation of WiMAX module for ns-2 simulator," in *Proc. Workshop on ns-2: the IP Network Simulator*, Pisa, 2006.

[44] "Seamless and Secure Mobility," US National Institute of Standards and Technology, http://www.antd.nist.gov/seamlessandsecure./download.html.

[45] "Network Simulator 2 (NS-2)," http://www.isi.edu/nsman/ns/.

[46] D. Niyato and E. Hossain, "Integration of WiMAX and WiFi: optimal pricing for bandwidth sharing," *IEEE Commun. Mag.*, 45(5):140–146, May 2007.

[47] A. Lera *et al.*, "End-to-end QoS provisioning in 4G with mobile hotspots," *IEEE Network*, 19(5): 26–34, September/October 2005.

[48] G. Song, and Y. Li, "Utility-based resource allocation and scheduling in OFDM-based wireless broadband networks," *IEEE Commun. Mag.*, 43(12): 127–134, December 2005.

[49] Z. Tao, A. Li, K. Teo, J. Zhang, "Frame structure design for IEEE 802.16j mobile multihop relay (MMR) networks," IEEE Globecom, 4301–4306, November 2007.

10 Multimedia over wireless broadband

The rapid growth of wireless broadband networking infrastructures, such as 3G and 3.5G, WLAN and WLAN-mesh, and WiMAX, makes available multimedia (audio and video) information and entertainment ("infotainment") in our lives anytime, anywhere, on any device. However, wireless multimedia delivery faces several challenges, such as a high error rate, bandwidth variation and limitation, battery power limitation, and so on. Take, for example, the voice over IP (VoIP) and video streaming applications, which are quite mature in wireline infrastructure. At the same time, wireless broadband based on WLAN and WiMAX is also becoming widespread. While these wireless networks were not designed with real-time multimedia communication services in mind, their widespread availability and low cost makes them an inviting solution for adding mobility to these communication services. The major issue is how to achieve a wireless broadband system which can deliver real-time interactive multimedia smoothly and still satisfy the QoS metrics typically used to define the quality of a VoIP or video conferencing session, e.g., the one-way delay, jitter, packet loss rate, and throughput (see Section 7.2).

Advances in media coding over wireless networks are governed by two dominant rules [1]. One is the well-known Moore's law, which states that computing power doubles every 18 months. Moore's law certainly applies to media codec evolution, and there have been huge advances in technology in the ten years since the adoption of MPEG-2. The second governing principle is the huge bandwidth gap (one or two orders of magnitude) between wireless and wired networks. This bandwidth gap demands that coding technologies must achieve efficient compact representation of media data over wireless networks.

Even though computing power is increasing exponentially, the power control of hardware still remains a challenging issue. In addition to media coding efficiency and power dissipation, error resilience is an important issue, since wireless networks generally cannot guarantee error-free communication during fading periods. According to [1], there are four layers of error control for wireless video communication (see Figure 10.1): the layers 1 and 2 (layer-1/2) transport, which corresponds to the physical and media access control (MAC) layers; an end-to-end transport layer such as TCP/IP or RTP/UDP/IP; an error-resilience tool and network adaptation layer such as H.264 NAL [2]; and a source coder layer. To provide error robustness, a number of approaches have to be taken on the basis of one specific layer of error control or a combination of several layers [3].

In Chapter 9, we discussed extensively layer-1/2 error control for wireless networks. More specifically, in the MAC layer, the link adaptation of WLAN, the QoS provisioning provided by 802.11e, and WiMAX service classes differentiation were all discussed. In the PHY layer, the adaptive modulation and coding schemes (e.g., the adaptive combination of BPSK, QPSK, 16QAM, and 64QAM with different amounts of error correction coding redundancy for various wireless channel conditions) used in WLAN and WiMAX, along

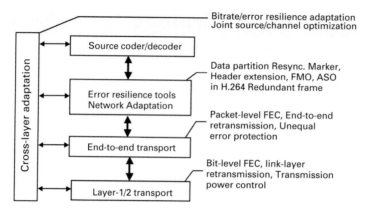

Figure 10.1 The four layers of error control for wireless video communication [1] (© IEEE 2005).

with retransmission, are also popular techniques. Since we discussed extensively QoS mechanisms for end-to-end transport layers such as TCP/IP or RTP/UDP/IP (layers 2 and 3) in Chapter 7 for video streaming in general, in Section 10.1 we will discuss mainly end-to-end transport techniques for wireless error control. Many techniques discussed in Chapter 8 regarding QoS provisioning for streamlining video over wireline networks are still applicable, such as the error control, discussed in Section 8.2.2.3. In Section 10.2 we will discuss the error resilience techniques offered in the source coding layer for better network adaptation; many video coding techniques that are useful in QoS provisioning were referred to in Chapter 5. Some examples of QoS provisioning for VoIP and scalable video over wireless networks will be also presented in this chapter.

10.1   End-to-end transport error control

Most multimedia data are transmitted over Internet-based UDP transport, which further uses the real-time transport protocol (RTP) and its control protocol, RTCP, to control the rate on the basis of information about losses in a certain period. However, reliance on information about losses alone may lead to erroneous control. In a wireless network environment, common channel errors due to multipath fading, shadowing, and attenuation may cause bit errors and packet loss; this is quite different from the packet loss caused by network congestion. In congestion control, packet loss information can serve as an index of network congestion for effective rate adjustment; therefore wireless packet loss can mistakenly guide congestion control and lead to dramatic performance degradation. More specifically, for a network topology containing wireless links, packet loss can be caused by either congestion loss or wireless channel errors resulting from multipath fading, shadowing, or attenuation. Packet loss due to wireless channel errors will result in an improper reduction in sending rate and dramatically throttle the throughput. There have been studies to improve the TCP and UDP over wireless networks. The Snoop protocol [4] uses a retransmission mechanism for lost packets at the base station to improve the TCP throughput. Westwood TCP [5] exploited the use of TCP-acknowledged packets to enable the sender to estimate the bandwidth and adequately adjust the slow-start threshold and congestion window. Lee et al. [6] explored the linear relationship between the probability of packet loss and the packet size, under the assumption of uniformly

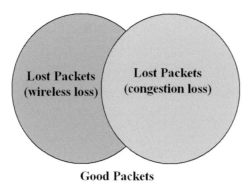

Figure 10.2 Two packet-loss classes, wireless loss and congestion loss.

distributed wireless errors. Elaarag [7] surveyed various techniques for TCPs to improve performance in wireless networks. Another approach is to perform packet loss classification so that congestion control algorithms can adapt the sending rate more effectively. In general, two kinds of packet loss, congestion loss and wireless loss, as shown in Figure 10.2, are commonly categorized. Note that the intersection of wireless loss and congestion loss, which means that the network suffers from both network congestion and severe wireless fading, is regarded as congestion loss. Packet-loss classification (PLC) algorithms depend on analysis of the statistical behavior of some observed values, such as the packet timestamp and packet serial number in the packet header. The successful classification of packet loss, as either due to congestion or due to wireless transmission errors, can allow the sender to respond appropriately.

10.1.1 Packet loss classification (PLC)

Biaz and Vaidya [8] suggested using the inter-arrival gap time at the receiver to discriminate congestion loss from wireless loss, so that a sender can respond appropriately. The *spike-train* method, [9], provides two predefined thresholds on the relative one-way trip time (ROTT) to classify packet loss. Cen [10] further extended the inter-arrival gap and spike-train packet loss classification (PLC) methods and proposed a Zigzag scheme which uses different threshold values based on the mean and deviation of ROTT for different numbers of lost packets. It was reported in [10] that none of the above PLC methods could perform well for different network topologies and a switching algorithm based on the inter-arrival gap, spike train and Zigzag, depending on various values of ROTT, was suggested. The main drawback of using thresholds on either packet inter-arrival time or delay time is that it may cause a misclassification of packet loss, because it is difficult to conclude that congestion loss and wireless loss will exhibit distinct boundaries on either packet inter-arrival time or packet ROTT time. Another PLC algorithm based on the trend of the ROTT was also proposed [11], to assist PLC in the ambiguous area of ROTT distribution. All these PLC algorithms can also assist other congestion control protocols that might lead to unnecessary bandwidth reduction in the presence of wireless packet loss.

10.1.1.1 Classification via inter-arrival gap
Even though most multimedia data are transmitted over the Internet using UDP transport, it is possible to transmit using TCP transport. Therefore, to improve the performance of TCP

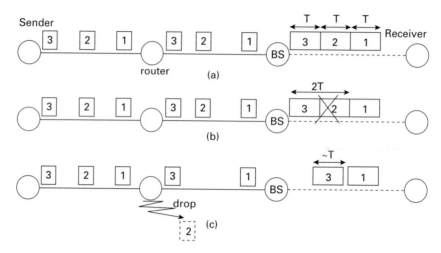

Figure 10.3 The inter-arrival gap analysis proposed in [8] (© IEEE 1998).

traffic over wireless networks, a PLC algorithm based on analyzing the inter-arrival gaps of transmitted packets on the receivers was proposed [8].

Transmission control protocol connections may traverse wireless links in several scenarios; let us consider a popular scenario where the sender is on the wired network and the receiver is connected via the wireless link, which is also assumed to be the bottleneck for the connection. Thus the packets tend to queue up at the access point (AP) or the base station (BS). Therefore, most packets are sent back-to-back on the wireless link. This last characteristic is the key to the simple heuristic scheme developed in [8] to distinguish packet losses due to transmission errors from congestion losses on the wired network.

As shown in Figure 10.3 [8], there are three scenarios that may occur when the sender sends packets 1, 2, and 3 to the receiver. In Figure 10.3(a), since none of the packets is lost the receiver receives all the packets. In this case, the "packet inter-arrival gap," or the time between the *arrivals* of consecutive packets, is approximately equal to the time T required to transmit one packet on the wireless link. In practice, the inter-arrival gap can vary owing to other factors, including queuing delay. The time of arrival of a packet is the time when all bits belonging to the packet have been received by the receiver. Note that, for most constant bitrate (CBR) multimedia data, the sender sends packets fairly regularly. At the router, the delay between packets may be modified by cross traffic and the queuing policy of the router. However, at the base station the packets are sent back-to-back because the wireless link is assumed to be the slowest link (the bottleneck) along the path for the connection. Figure 10.3(b) shows a scenario where packet 2 is lost when being transmitted over the wireless link. In this case, the time between the *arrivals* of the two packets at the receiver (i.e., packets 1 and 3) is 2T, since packet 2 (lost due to wireless transmission) uses the wireless link for T time units. Now consider a third scenario, shown in Figure 10.3(c), where packet 2 is lost due to queue overflow (congestion) at the intermediate router. In this case, the inter-arrival gap between packets 1 and 3 will be comparable to T (this holds true if packet 3 arrives at the base station either before, or just after, the base station has transmitted packet 1). When packet 2 has been lost and packet 3 arrives at the receiver, packet 3 is called an *out-of-order packet*. On the basis of the above observations, an heuristic PLC rule was developed [8]. Let P_0 denote an out-of-order packet received by the receiver, let P_i denote the last in-sequence packet received before P_0, let T_g denote the time between the arrivals of packets P_0 and P_i, and let T_{min} denote the minimum inter-arrival time observed so far by the receiver during the connection. Finally, let the number of packets missing between P_i and P_0 be n (assuming that all the packets are of the same size). If

$$(n+1)T_{min} \leq T_g < (n+2)T_{min} \qquad (10.1)$$

then the n missing packets are assumed to be lost by wireless transmission errors. Otherwise, the n missing packets are assumed to be lost by congestion.

The concept here is that, on the basis of the arrival time of P_i, if P_{i+n+1} arrives at around the time that it was supposed to arrive, we can assume that the missing packets were properly transmitted and were then lost to wireless errors. If P_{i+n+1} arrives much earlier than it expected then at least some packets ahead of it, $P_{i+1}, \ldots, P_{i+n}$, were probably dropped at a buffer, and if it arrives much later than expected then it is likely that queuing times at buffers have increased. Either way, we can attribute the loss to congestion. The inter-arrival gap scheme works best when the last link is both the wireless link and the bottleneck link of the connection and is not shared by other connections competing for the link [10]. Note that the condition for identifying a packet loss as a wireless loss is quite restrictive. The reason is that it is preferable, for the sake of satisfactory operation of the network, to mistake a wireless loss for a congestion loss rather than vice versa [8].

10.1.1.2 Classification via spike-train analysis

As discussed in Section 7.5.4, the relative one-way trip time (ROTT) can be a useful indication of end-to-end congestion. As previously defined, ROTT is measured by the receiver as the time difference between the receiving time and the packet-sending time-stamp, recorded in the field of the multimedia UDP packet header plus a fixed bias. Tobe et al. [9] observed that the ROTT values increase sharply in the case of congestion, i.e., a plot of the ROTT versus time exhibits spike trains, as shown in Figure 10.4. They found that sequences of these spikes, or spike trains, are related only to congestion losses and not to random losses such as wireless losses.

The spike-train scheme was initially derived to differentiate between degrees of congestion but it did not explicitly differentiate wireless loss from congestion loss. The

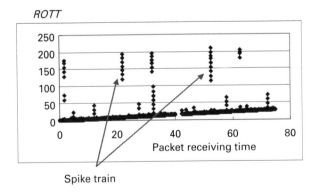

Figure 10.4 The ROTT values increase sharply when there is congestion, i.e., a successive ROTT plot vs. time exhibits spike trains [9] (© IEEE 2000).

ROTT was used to identify the state of the current connection: if the connection is in the *spike state*, losses are assumed to be due to congestion; otherwise, losses are assumed to be wireless.

The spike state, which derives its name from the fact that plots of ROTT versus time tend to show spikes during periods of congestion, is determined as follows. On receipt of a packet with sequence number i, if the connection is currently not in the spike state and the ROTT for packet i exceeds the threshold $B_{spikestart}$ then the algorithm enters the spike state. Otherwise, if the connection is currently in the spike state and the ROTT for packet i is less than a second threshold $B_{spikeend}$ then the algorithm leaves the spike state. When the receiver detects a loss because of a gap in the sequence number of received packets, it classifies the loss on the basis of the current state (see Figure 10.5 [10]).

The threshold values $B_{spikestart}$ and $B_{spikeend}$ were empirically determined and hard-coded to be $ROTT_{min} + 20$ ms and $ROTT_{min} + 5$ ms, respectively, in the original publication [9], and were slightly reformulated in [10] to make them more practical:

$$B_{spikestart} = ROTT_{min} + a \times (ROTT_{max} - ROTT_{min}),$$
$$B_{spikeend} = ROTT_{min} + \beta \times (ROTT_{max} - ROTT_{min}), \tag{10.2}$$

where $ROTT_{max}$ and $ROTT_{min}$ are the maximum and minimum ROTT values observed so far and $a \geq \beta$. It was found empirically that the values $a = 1/2$, $\beta = 1/3$ can result in a good tradeoff between low congestion-loss misclassification and reasonable wireless-loss misclassification in the wireless last-hop topology; overall performance in the wireless backbone topology is relatively insensitive to the choice of these two parameters.

10.1.1.3 Classification via Zigzag scheme

Further extending the classification schemes based on the inter-arrival gap [8] and spike-train analysis [9], a new scheme called Zigzag was proposed in [10]. Zigzag classifies packet losses as wireless on the basis of the number of losses n and on the differences between the current value of the relative one-way trip time $ROTT$ and its mean $ROTT_{mean}$ or deviation $ROTT_{dev}$; these are recursively calculated using the exponential average:

$$ROTT_{mean} = (1 - a) \times ROTT_{mean} + a \times ROTT,$$
$$ROTT_{dev} = (1 - 2a) \times ROTT_{dev} + 2a \times |ROTT - ROTT_{mean}|. \tag{10.3}$$

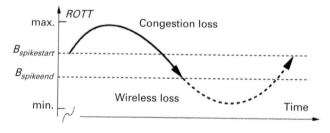

Figure 10.5 The spike state is determined as follows. On receipt of a packet with sequence number i, if the connection is currently not in the spike state and the ROTT for packet i exceeds the threshold $B_{spikestart}$ then the algorithm enters the spike state [10] (© IEEE 2003).

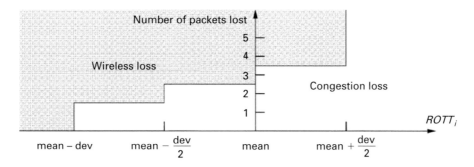

Figure 10.6: The packet loss classification rule of Zigzag [10] (© IEEE 2003).

In the Zigzag classification scheme, a loss is classified as wireless if

$$(n = 1 \text{ AND } ROTT < ROTT_{mean} - ROTT_{dev}) \text{ OR}$$
$$(n = 2 \text{ AND } ROTT < ROTT_{mean} - ROTT_{dev}/2) \text{ OR}$$
$$(n = 3 \text{ AND } ROTT < ROTT_{mean}) \text{ OR} \qquad (10.4)$$
$$(n > 3 \text{ AND } ROTT < ROTT_{mean} - ROTT_{dev});$$

otherwise, the loss is classified as congestion loss (see Figure 10.6).

Since congestion loss usually comes with a higher delay, according to Eq. (10.4) larger values of the ROTT would be classified correctly as congestion loss. The reasoning behind increasing the threshold for the number of losses encountered is that a more severe loss is associated with higher congestion and with higher ROTTs. In this way, a loss event containing four or more packets would be classified as congestion loss only when relatively large ROTT values are observed [10].

The insight behind the ROTT classification in Eq. (10.4) is that with the additive increase and multiplicative decrease (AIMD) rules used in TCP or TCP-friendly UDP traffic, the ROTT often exhibits a sawtooth pattern, i.e., the instantaneous ROTT tends to be less than its mean after a multiplicative decrease action taken against congestion, and the probability that the instantaneous ROTT is greater than its mean increases linearly with increasing window size. This pattern is characteristic of AIMD congestion control regardless of other network parameters. Therefore, the misclassification rate of Zigzag is rather insensitive to changes in network topology.

10.1.1.4 Classification via delay-trend index

As discussed in Section 7.5.4, the delay-trend analysis of ROTTs also provides an effective indication of end-to-end congestion, which provides further evidence of packet loss caused by congestion. Taking advantage of this observation, an ROTT-based delay trend analysis was also proposed for packet loss classification [11]. The basic concept is that when a packet loss is observed at time t, it should be considered as a congestion loss if the current ROTT is in an ascending phase; otherwise, it should be classified as a wireless loss.

On the basis of this concept and the spike-train algorithm, a packet loss classification algorithm that uses delay trend detection is proposed, mainly for packet loss in the gray zone of the ROTT. As shown in Figure 10.7, the gray zone of the ROTT is defined to be the interval between TG^{up} and TG^{low}, where TG^{up} denotes the upper bound of the gray zone and TG^{low} is the lower bound of the gray zone. When the receiver observes packet loss from information about the packet serial number and the ROTT value of the received packet is

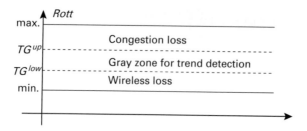

Figure 10.7 Delay-trend-analysis-based packet loss classification [11] (© IEEE 2005).

greater than TG^{up}, the packet loss will be classified as congestion loss. If the ROTT value is smaller than TG^{low} then the packet loss will be classified as wireless loss. For ROTT values falling in the gray zone, a delay-trend detection process is performed to classify the packet loss. In this method, the moving average of the delay-trend index S_f is defined as

$$<S_f> = (1 - \gamma) \times S_f + \gamma \times I(ROTT_i > ROTT_{i-1}), \qquad (10.5)$$

where $I(X)$ is defined as 1 if X is valid and 0 otherwise; $ROTT_i$ is the ROTT value of the ith packet and γ is the smoothing factor of S_f and is empirically set to 1/30 to achieve the best results. The delay-trend index S_f can take values from 0 to 1; it will be around 0.5 if $ROTT_i$ is randomly distributed without an increasing trend. If there is a strong increasing trend, however, S_f will approach unity. A threshold $S_{f,th}$ is used so that if $S_{f,th} < S_f$ then the packet loss will be classified as congestion loss. For packet loss classification in a congestion control protocol, the relatively conservative value $S_{f,th} = 0.4$ is chosen for better congestion control stability.

As in the modified spike-train scheme [10], the upper and lower bounds of the gray zone of ROTT values are defined as follows:

$$TG^{up} = ROTT_{min} + a(ROTT_{max} - ROTT_{min}),$$
$$TG^{low} = ROTT_{min} + \beta(ROTT_{max} - ROTT_{min}), \qquad (10.6)$$

where a and β control the range of the gray zone and are empirically chosen to be 0.8 and 0.3 respectively. The values $ROTT_{min}$ and $ROTT_{max}$ still denote the minimum and maximum ROTTs observed so far.

10.1.2   Forward error correction (FEC) via block erasure codes

The conventional approach to disseminating packets end-to-end over IP networks relies on the retransmission on demand of lost packets. One such implementation is the automatic repeat request (ARQ) of lost packets, which is based on TCP transport. Since most multimedia data over IP networks are based on UDP transport, a different approach based on the use of forward error correction (FEC) or hybrid FEC plus ARQ techniques is widely used. Pure FEC relies on the transmission of redundant data (generated by a suitable encoder) that allows the receiver to reconstruct the original message even in the presence of some communication errors. Hybrid FEC plus ARQ is typically implemented by reducing the initial amount of redundancy and sending repair packets only on demand, as in ARQ-based protocols. This feature can make a feedback channel unnecessary, and this is what makes FEC attractive for sending multimedia over IP systems.

In wireless networks, transmission errors are generally handled at a low level in the protocol stack, and the corrupted data is either corrected or completely removed. Examples are the symbol-level FEC used in the adaptive modulation and coding scheme (MCS) of the PHY layer, the retransmission adopted in the MAC layer, and the parity check done in the network layer. Thus upper protocol layers, such as the application layer, mainly have to deal with *erasures*, i.e., missing packets in *known* locations. The computations necessary for erasure recovery are slightly simpler than those for full error recovery. An (n, k) *block erasure code* (or, the corresponding encoder) takes k source packets and produces n encoded packets $(n > k)$ in such a way that any subset of k encoded packets allows the reconstruction of the k source packets in the decoder.

Most wireless networks are modeled as a lossy channel characterized by an error probability ε, i.e., a fraction ε of data symbols (e.g., each symbol could be 8-bit) will be corrupted. This is also referred to as a channel of capacity $1 - \varepsilon$. From information theory, it is known that a message of size k symbols can be encoded into $n = k/(1 - \varepsilon)$ transmission symbols, so that any k (out of n) of the transmission symbols can be used to decode the original message (see Figure 10.8) [12]. One famous example of such coding is the Reed–Solomon code [77], which is also a systematic code whose output symbols include the unmodified input symbols. Systematic codes are much cheaper to decode when only a few erasures are expected; besides, they might allow the partial reconstruction of data even when fewer than k packets are available. A rather simplified systematic $(n, k) = (5, 3)$ erasure code is given as follows:

$$
\begin{aligned}
O_1 &= S_1, \\
O_2 &= S_2, \\
O_3 &= S_3, \\
O_4 &= S_1 + S_2 + 2S_3, \\
O_5 &= 2S_1 + S_2 + 3S_3,
\end{aligned}
\tag{10.7}
$$

where the three input symbols $\{S_1, S_2, S_3\}$ are erasure coded to generate five output symbols

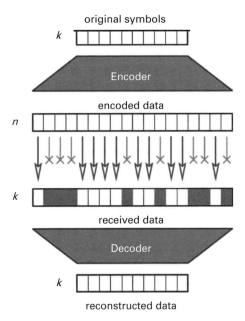

original symbols

encoded data

received data

reconstructed data

Figure 10.8 An (n, k) block erasure code (or the corresponding encoder) takes k source packets and produces n encoded packets $(n > k)$ in such a way that any subset of k encoded packets allows the reconstruction of the source packets in the decoder [12] (© IEEE 1997).

$\{O_1, O_2, O_3, O_4, O_5\}$ for transmission. Suppose that the symbols $\{O_1, O_3\}$ are lost or corrupted and that the symbols $\{O_2, O_4, O_5\}$ are successfully received and can be used to recover fully the original input symbols. Then, from Eq. (10.7),

$$S_1 = O_2 - 3O_4 + 2O_5,$$
$$S_2 = O_2, \qquad\qquad (10.8)$$
$$S_3 = -O_2 + 2O_4 - O_5.$$

One efficient implementation of an erasure code for small values of k is called the Vandermonde code [13]; it interprets k source symbols as the coefficients of a polynomial P of degree $k - 1$. As the polynomial is fully characterized by its values at k different points, we can produce the desired amount of redundancy by evaluating P at n different points. Reconstruction of the original data (the coefficients of P) is possible if k of these values are available. In practice, the encoding process requires the original data to be multiplied by an $n \times k$ encoding matrix G, which happens to be a Vandermonde matrix. The decoding process requires the inversion of a $k \times k$ submatrix G' taken from G. By simple algebraic manipulation G can be transformed to make its top k rows constitute the identity matrix, thus making the code systematic. Because the actual implementation of the encoding and decoding computations is made for a finite field, or *Galois field* [14], basic erasure encoding and decoding operations can often be implemented efficiently using XOR and table lookups and solving linear systems.

A block erasure code is a straightforward extension of the erasure code concept. An *(n, k) block erasure code* takes k source packets and produces n encoded packets in such a way that any subset of k encoded packets allows the reconstruction of the source packets in the decoder. Normally, in block erasure codes the values *(n, k)* are relatively small compared with those in symbol-based erasure codes, to avoid long delays and large buffering. Figure 10.9 shows how $n - k$ redundant packets are created by applying the consecutive erasure encoding processes to k separate symbols from k packets [15]. This is particularly important for multimedia networking over IP networks, where the minimum unit of transmission is a packet, i.e., the whole packet will be erased if one or more corrupted bits in the packet cannot be corrected when received by the lower level of the IP stack of the receiver.

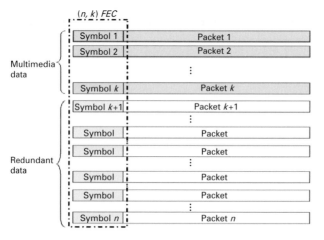

Figure 10.9 The creation of $n-k$ redundant packets by applying the consecutive erasure encoding process to k separate symbols from k packets [15] (© IEEE 2005).

10.1.3 Wireless channel model

To decide the block erasure code parameters, i.e., n and k, the error probabilities of the wireless channels should be appropriately estimated. The error probabilities can stochastically represent loss rates at bit level, symbol level, or packet level, depending on the granularity and the protocol layers used in the modeling. The Gilbert–Elliot two-state Markov chain model [16] is the most popular model for simulating the bursty nature of multipath wireless fading. As shown in Figure 10.10, the model assumes two states corresponding to good and bad channels, with the following transition matrix:

$$\begin{pmatrix} 1-q & p \\ q & 1-p \end{pmatrix}. \tag{10.9}$$

It is beneficial to parameterize the transition probabilities p and q (see below) of the model by simply counting "real data," collected from temporal measurements at various levels (bit, symbol, or packet) whenever available. The two states "good" and "bad" are, more specifically, error-free bursts and error bursts respectively. The idea behind this is to capture the "bursty" nature of the channel, as observed in low-level measurements (e.g., bits or symbols) or high-level measurements (e.g., packets). Conceptually, after transmission of every transmitted bit, symbol, or packet the new channel state can be determined from the reception status using the discrete two-state Markov chain, and the corresponding transition probabilities can be updated.

The average bit, symbol, or packet error rate P_b can thus be expressed as

$$P_b = \frac{q}{p+q}. \tag{10.10}$$

It has been reported that the Gilbert–Elliot model is suitable for short-term rather then long-term error correlation [17]. For long-term bursty errors, block interleaving is usually adopted to remove the long-term bursty phenomenon. Besides the interleaving technique, channel coding, such as block coding and convolutional coding, is also an important component to reduce errors induced by impaired channels.

It is always highly preferable to use the Gilbert–Elliot model for simulating the bit error patterns of a wireless channel if bit-level data can be collected; the packet error patterns can then be derived from bit-wise simulation. Using the Gilbert–Elliot model for packet-level error patterns cannot be as accurate, even though this is the only option for application-layer data analysis, since a reliable packet-level channel model is difficult to achieve owing to the many factors involved. These include the bit error behavior of the channel, the coding and

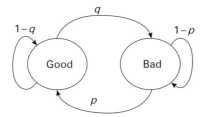

Figure 10.10 Gilbert–Elliot two-state Markov chain model.

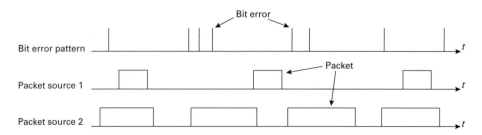

Figure 10.11 The resulting packet error characteristics, estimated from different packetization and transmission behaviors using the Gilbert–Elliot model, can be quite different under the same bit error patterns [18].

packet formats, and the source characteristics (e.g., the packet-length distribution) [18]. As shown in Figure 10.11, the resulting packet error characteristics, estimated from different packetization and transmission behaviors using the Gilbert–Elliot model, can be quite different under the same bit error patterns if the packet arrival patterns are different. More specifically, the longer packets in source #2 are more likely to be hit by a bit error than the shorter packets of source #1.

10.1.4   From channel model (p, q) to *FEC* (n, k)

The Gilbert–Elliot model is used to simulate the packet error rate on the basis of a two-state Markov chain with parameters p and q representing the transition probabilities from (packet) loss state to received state and from received state to loss state respectively. During the process, the target decoding error rates $e_{FEC(n,k)}$ can be derived accordingly [19]. More specifically, k is first decided according to the amount of allowed delays and buffering; then an estimated (p, q) value representing the wireless channel conditions can be used to determine the minimum n value so as to satisfy

$$e_{FEC(n,k)} < e_{FEC}, \qquad (10.11)$$

where e_{FEC} is the preset loss requirement for multimedia applications. If the block erasure code (see Section 10.1.2) is used, $e_{FEC(n,k)}$ can be derived using the following steps [19]. First the steady-state average packet loss rate P_L is given as

$$P_L = \frac{q}{p+q}, \qquad (10.12)$$

and then we have

$$e_{FEC,(n,k)} = \sum_{m=n-k+1}^{n} P(m,n) \qquad (10.13)$$

where

$$P(m,n) = \sum_{s=1}^{n-m+1} P_L G(s) H(m, n-s+1) \qquad (10.14)$$

with

$$G(s) = \begin{cases} 1 & \text{for} \quad s = 1, \\ p(1-q)^{s-2} & \text{for} \quad s > 1. \end{cases} \tag{10.15}$$

$$H(x,y) = \begin{cases} G(y) & \text{for} \quad x = 1, \\ \sum_{s=1}^{y-x+1} g(s)H(x-1,y-s) & \text{for} \quad 2 \leq x \leq y, \end{cases} \tag{10.16}$$

$$g(s) = \begin{cases} 1-p & \text{for} \quad s = 1, \\ p(1-q)^{s-2} & \text{for} \quad s > 1, \end{cases} \tag{10.17}$$

where $P(m, n)$ is the probability that m packets will be lost out of n consecutive packets. In conclusion, the number n of FEC packets needed for the given k data packets can thus be decided using the available (p, q) values estimated at receivers for a known FEC scheme.

10.2 Error resilience and power control at the source coding layer

In addition to the error control techniques created for the various layers of multimedia-over-wireless-network protocols, there have been a great deal of efforts to provide error control and recovery technologies at the source coding layer. As discussed in Section 8.2.2.3, many error-resilient and concealment techniques have been developed to enhance the robustness of compressed video to packet loss. The standardized error-resilient encoding schemes include resynchronization marking, data partitioning, and data recovery. All these techniques are perfectly suited for wireless video transmission too. However, error concealment decoding is performed by the receiver when packet loss has already occurred. There are two basic approaches for error concealment, spatial and temporal interpolation. In this section, we will discuss mainly the additional error-resilient features offered by H.264 through its network abstraction layer (NAL) of the coding process.

10.2.1 The H.264 network abstraction layer for network adaptation

The H.264 standard provides additional functionalities to support network transport as well as error control. It has two layers in its coding process. One is the video coding layer (VCL), which contains the signal processing functionality (transform, quantization, motion search and compensation, deblocking filter, and so on). The other is the network abstraction layer (NAL), which encapsulates the slice output of the VCL into NAL units suitable for transmission over packet networks. The network abstraction in H.264 is something like a network adaptation. The H.264 error-resilience structure is based on the flexible network adaptation structure provided in NAL and elegantly separates the H.264 source coder tasks from error-resilience tasks such as data partitioning. Moreover, this error-resilience structure is based on the assumption that bit-erroneous packets have been discarded by the receiver, and this design concept is perfectly suited for IP-based content delivery. According to [20], the following features of NAL create robustness with respect to data errors and losses and flexibility of operation over a variety of network environments.

(1) *Parameter set structure*   The H.264 coding parameter set (e.g., picture size, display window, macroblock allocation map, and so on) is commonly exchanged in advance and reliably by out-of-band signaling, for example, using the session description protocol (SDP).

(2) *Flexible macroblock ordering (FMO)*   This allows the transmission of a macroblock in non-raster scan order so that spatial concealment can be more effectively applied.

(3) *Arbitrary slice ordering (ASO)*   This diminishes the end-to-end delay in real-time applications, particularly when used on best-effort Internet with frequent out-of-order delivery behavior.

(4) *Redundant pictures*   Redundant slices and spare macroblocks can provide additional protection against the propagation of errors via interprediction.

(5) *Data partitioning*   Like the MPEG-4 data partitioning concept, this produces better visual quality if the underlying network provides unequal error protection using the priority of the packet.

(6) *SPSI synchronization and picture switching*   This is a feature unique to H.264 that allows the switching of a decoder between sequences of the same video content, having different data rates. This feature contributes to error recovery in the case of insufficient bandwidth using graceful degradation.

The use of NAL unit syntax structure allows greater customization in carrying the video content to the transport layer, and thus represents a major advance from the error-resilience structure of other video coding standards such as MPEG-4. A typical example is the packetization of an RTP payload [21]. For example, the flexible syntax with arbitrary slice ordering (ASO) enables robust packet scheduling, where important packets are sent earlier and lost packets are retransmitted so that the transmission reliability is improved for important pictures. The H.264 NAL structure together with underlying transport technologies such as RTP can support most of the packet-loss resilience schemes. Of the error-resilience tools newly introduced in H.264, slice interleaving enabled by flexible macroblock ordering (FMO) and ASO is a key feature. However, there is some concern that handling FMO may complicate the implementation of decoder buffer management significantly while providing benefits only in the case when slices are randomly and moderately lost. Switching between FMO and non-FMO coding may be desirable, depending on the link condition, if the complexity allows.

10.2.2 Power control with rate–distortion optimization

Since most wireless multimedia are encoded and decoded with portable devices powered by batteries, power consumption issues become increasingly important. For example, in a video transmission scheme the total power consumed in the portable devices consists of the transmission power and the processing power. The latter is mainly determined by the computational costs for source coding (data compression) and channel coding (error detection and correction). The transmission power, however, depends on the bit energy and total bitrate to be delivered. Both the transmission power and the processing power need to be controlled so as to adapt to changing channel conditions. Reducing the transmission power of a user will lengthen its battery lifetime and lower the interference with neighbors but will increase the bit error rate in its own transmission. More specifically, if the media coding power dissipation increases beyond a modest 100 mW, corresponding to an 802.11 WLAN receiver's power consumption, it will be hard to implement the media application in portable devices. There have been some efforts to tackle these power control issues for wireless multimedia through transmission power optimization together with the rate–distortion (R–D) optimization of media coding [22] [23] [24]. According to current mobile network operations and implementations, layer-1/2 transport control is isolated from the upper layers, except through the predefined QoS class designation. Future networks might provide a programmable function for layer-1/2, which would involve the cross-layer optimization of the associated power consumption. In addition, if necessary, other key parameters would include the spreading factor, the number of multiplexing spreading codes, the coding rate of the error correction code, the level of modulation, etc.

Most power control research for wireless systems aims to predetermine the signal-to-noise ratio (SNR) for the terminals at the base station while minimizing the transmission power. Wireless multimedia, however, has characteristics that necessitate a different type of power control. First of all, a chief concern in multimedia is to keep the end-to-end distortion D of the signal at a constant value. The distortion occurs because of lossy source compression and channel errors, and it depends on both the channel SNR and the video encoder parameters such as the complexity and bitrate. Also, multimedia signal compression consumes an amount of power comparable with the transmission power, which necessitates joint optimization of the source encoder and transmitter in order to minimize the total power consumption. This was investigated for a single-user system in [25]. In a multi-user system each user experiences interference with others, and the power allocation becomes a multivariate problem depending on the source encoder and transmission parameters of all users. In centralized power control schemes, a central controller collects the operating parameters of all users in the network, such as the channel conditions, the required end-to-end distortion, etc. and finds the optimal operating powers for all users by jointly optimizing the system [26]. However, centralized schemes have a complexity that increases exponentially with the number of users in the network and they also have large delays since the optimization has to be redone each time users enter or leave the network. An iterative power control algorithm for the uplink of multiple users that transmit H.263 video in a CDMA cell system was proposed in [23], in which the total power consumed, including the compression and transmission power, is minimized subject to a predetermined end-to-end distortion at each terminal. This algorithm updates the compression parameters (complexity, rate) and transmission power of each user simply on the basis of the total interference plus noise level of that user and then iterates between the users. One of the advantages of an iterative scheme is that it adapts easily to changing conditions such as the link SNRs and the number of users.

Another interesting effort along this line of research combines the power-control and joint source-channel coding approaches to support QoS for robust video communication over wireless networks [27]. The joint source-channel coding is performed by introducing redundancy to combat the transmission errors and so minimize the distortion. By simultaneously controlling the transmission power, source rate, and error protection level (using an unequal error protection scheme based on the Reed–Solomon codes to protect the compressed video bitstream) to minimize the power consumption of mobile stations, the desired QoS for wireless video transmission can be better maintained in a multi-user environment.

10.3   Multimedia over wireless mesh

As discussed in Section 9.3.3, to build a small-to-large-scale wireless distribution system based on a WLAN infrastructure, the IEEE 802.11s extended service set (ESS) was proposed for the installation, configuration, and operation of WLAN multi-hop mesh. The 802.11-based wireless mesh networks (WMNs) have emerged as a key technology for a variety of new multimedia services that require flexible network support [28]. An 802.11-based wireless mesh network can be regarded as a dynamic collection of backhaul routers. It is similar to an ad hoc network but the topology, composed of backhaul routers, is static. The conventional 802.11 WLAN standards use the CSMA/CA access mechanism in the MAC-layer protocol, which is based on medium sharing and single-hop transmission and is not suitable for multi-hop communication such as is used in wireless mesh networks. Owing to the backhaul networking requirement for wireless mesh, a new MAC mechanism, which can guarantee throughput, capacity, latency, reach, the QoS capability, is critically needed. Therefore, in addition to the 802.11s standardization efforts, some proprietary radio technologies are used instead of IEEE 802.11-based radio technology [29], such as the products made by Radiant and MeshNetwork; their systems are, however, incompatible with others [30]. Several wireless mesh network research prototypes have also been proposed, such as the MIT Roofnet [31] and Microsoft Research [32]. Roofnet, which provides broadband Internet access to users in Cambridge MA, is an experimental 802.11b/g mesh network with 20 active nodes in development at the MIT Computer Science and Artificial Intelligence Laboratory (CSAIL). Microsoft Research makes a similar testbed, called the mesh connectivity layer (MCL) [32], which consists of IEEE 802.11 interfaces in ad hoc mode and ad hoc routing and link quality measurement in a module. However, these two projects consider only the backbone network, not mobility management [33].

In a typical WMN, the network configuration maintains the ad hoc communication structure but with two architectural levels, mesh routers and mesh clients [28] (see Figure 10.12). The mesh routers form the infrastructure of a mesh backbone for mesh clients. In general, mesh routers have minimal mobility and operate just like a network of fixed routers, except that they are connected by wireless links using wireless technologies such as IEEE 802.11. As shown in Figure 10.12, the WMN can access the Internet through a gateway mesh router connected to the IP core network by wires. In WMNs, every mesh router is equipped with a traffic aggregation device (similar to an 802.11 access point) that interacts with individual mesh clients. The mesh router relays the aggregated data traffic of mesh clients to and from the IP core network. Typically, a mesh router has multiple wireless interfaces for communication with other mesh routers, and each wireless interface, corresponding to one wireless channel, has different characteristics, owing to the inherent features of a wireless environment. In practice, wireless interfaces are usually running on different frequencies and are built on either the same or different wireless access technologies such as IEEE 802.11a/b/g/n.

Since WMNs can serve as indoor or outdoor networks, multimedia networking services can benefit greatly from this new kind of infrastructure, with greater bandwidth at lower cost than 3G cellular networks. With the strict QoS requirement in multimedia communications, channel availability and network latency problems become quite important in WMN, especially when the size and complexity of multiple-hop mesh networks increase. If the delay, bandwidth, or packet loss rate are too constraining then the delivery of voice or video packets may be unsuccessful. Hence, the adaptive and scalable coding and streaming

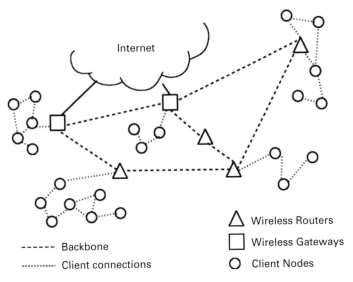

Figure 10.12 In a typical WMN, the network configuration maintains the ad hoc communication structure but consists of two architectural levels, mesh routers and mesh clients [28] (© IEEE 2008).

methods discussed in Chapter 8 become crucial for effective multimedia communication applications in the emerging wireless mesh networks.

10.3.1 Capacity problems with wireless mesh

In a mesh network, a packet destined to reach a certain node in the network may hop through multiple nodes to reach its destination. Analysis of the capacity of such networks shows that they suffer from scalability problems, i.e., as the size of the network increases their capacity degrades significantly with the increasing number of nodes [34]. Moreover, research has demonstrated that a node should only communicate with nearby nodes to maximize the network performance in terms of bandwidth [35]. This requires a large number of consecutive hops to deliver the packets and may severely limit the QoS experienced by real-time multimedia applications, especially with regard to performance metrics such as the end-to-end delay, jitter, and packet loss ratio. A latency of several ms per hop due to processing or transmission delay may preclude delay-intolerant applications such as voice and real-time interactive video after a few hops only. This problem is mainly due to the single-radio-channel nature of early generation WMNs, where each node operates in half-duplex mode and shares the same radio frequency, i.e., all radios on the same channel must remain silent until the packet completes its hops within the same collision domain. The use of more sophisticated (and more expensive) multiple-radio mesh networks can significantly increase the system's scalability [36]. More specifically, a common signaling channel takes charge of the multi-radio mesh communication. It assigns multiple radio channels to multiple wireless terminals and provides an effective solution to the problems of request to send (RTS) and clear to send (CTS) in a multi-hop scenario [37]. Five control packets are used in [38]: the first RTS/CTS handshake is used to contact the appropriate communication partner while the probing packets on the dedicated channel are used to probe the actual conditions of the data channels. The last RTS/CTS exchange is needed to set the transmission parameters properly and is used to inform the neighboring network nodes of the upcoming communication. It is claimed that the backoff scheme in the standard protocol should be modified for the reserved signaling channel. However, if the sending node does not receive the CTS, the size of congestion window (CW) would not be increased in the modified version. Because the common channel is used only for signaling, there will be no evidence of congestion on the dedicated channels from the common channel. Thus it makes no sense to double the CW even if there is a collision on the common signaling channel.

Using the multi-radio channel scheme the control of resources is made more efficient; not only can the hidden-node problem be lessened but also the exposed-node problem (see Section 9.2.2) can be reduced in the wireless mesh environment. In such an environment the system throughput would degrade dramatically, owing to the inefficient utilization of channels. To increase concurrent transmission by better spatial reuse [37], power control algorithms and directional antennas can be used. Another approach is to intelligently modify the MAC itself to allow neighboring nodes to synchronize their reception periods so that, at explicitly defined instants, one-hop neighbors can agree to switch their roles between transmitting and receiving. The problem of packet collisions can thus be avoided. This modified MAC performs a distributed coordination function (DCF) handshake by intro-ducing a variable "control gap" between the RTS/CTS exchange and the DATA/ACK phases; this gap gives the neighboring nodes an opportunity to synchronize their DATA/ACK phases.

Before 802.11s was standardized, the use of the IEEE 802.11e for wireless mesh in multimedia networking applications was appealing. As discussed in Chapter 9, the 802.11e, which is a QoS-enhanced MAC protocol, introduces two additional MAC modes, the enhanced distributed coordination function (EDCF) and the hybrid coordination function (HCF). Since the IEEE 802.11e was also designed for a single-hop environment, the per-formance of ad hoc networks running EDCF is not optimal: the EDCF parameters cannot be adapted to the network conditions [39] [40]. The collision rate will increase very fast in ad hoc networks when the number of contentions to access the shared medium is high. This significantly affects the goodput and latency and thus decreases the performance of delay-bounded traffic. In [39] a modified EDCF, called EDCF dual-measurement (EDCF-DM), was proposed to increase channel utilization by dynamically modifying the contention-window size according to the current collision rate and state of each traffic category. A similar method [40], named adaptive EDCF, adjusts the size of the contention window of each traffic class by taking into account both application requirements and network conditions [37].

10.3.2   Routing in wireless mesh

While excessive hop counts can sometimes be minimized by proper network architecture design, this is not the case for spontaneous, unstructured, ad hoc WMNs that require support at the application level to mitigate the effect of excessive delay and jitter. Routing protocols maintain information on the topology of the network in order to calculate routes for packet forwarding. A crucial metric in the design of efficient routing strategies is the number of hops that user traffic must make to reach its destination. Some other useful parameters can be considered, such as the network topology, the length of the links, and the wireless technology. A considerable amount of research has addressed routing problems specific to WMNs [41]. The routing mechanism may use information about the underlying topology of the network to collect the hop count or the distances between nodes or to determine how nodes are connected to each other. Some proposals utilize the shortest-hop-count metric as the path selection metric. This metric has been shown to result in poor network throughput because it favors long low-bandwidth links over short high-bandwidth links [42]. More recent proposals aim instead to improve routing performance by utilizing route-selection metrics [43], which take into account not only the throughput but also the contributions of both bandwidth and delay. A combination of several metrics, along with their contributions to the end-to-end distortion of the media application, is however necessary for optimal routing and rate allocation strategies.

Naturally, a mesh topology also enables the definition of multiple routes between two end points. Such routes may be utilized by multipath routing techniques to increase the QoS robustness in multimedia transmission. If the current path becomes unusable, the traffic flow can then quickly switch to an alternative path without waiting for a new routing path to be set up. The existence of multiple paths can also help to reduce the chance that the service will be interrupted by node mobility [44]. Even better, data partitioning over multiple paths can reduce short-term correlations in real-time traffic and therefore improve the performance of multimedia streaming applications, since burst losses in general cause important degradations in the video stream quality [45]. Clearly, the monitoring of several available paths is necessary to ensure a sustained quality of service.

10.3.3 Handoff in wireless mesh

Finally, a typical problem of WMNs with mobile peers is handoff management. As shown in Figure 10.12, one of the main characteristics of mesh networks is that they have only a few wireless gateways connected to a wired network while the wireless routers (WRs) provide network access to mobile clients (i.e., they act as APs to the clients). The client may move freely within the range of a given WR. But as it moves away from one WR and gets closer to another WR, it should hand all its open connections over to the new one in order to preserve network connectivity. Ideally the handoff should be completely transparent to mobile clients, with no interruption, loss of connectivity, or transmission "hiccups." In cellular data and voice systems the handoff problem is typically coordinated by the network itself using signaling embedded in the low-level protocols, which are able to leverage (obtain) considerable information about the network topology and client proximity. In contrast, 802.11 networks currently lack efficient and transparent handoff solutions. Consequently, as a mobile 802.11 client reaches the limits of its current coverage region inside the mesh, it must abandon its current WR, actively probe the network to discover alternatives, and then reconnect to a new best WR. Such delays may be disastrous for streaming applications with strict timing constraints. Similarly, as one cannot know in advance whether the necessary QoS resources are available at a new AP, a transition can lead to poor application performance. Moreover, forcing an additional session at the new AP may even result in the degradation of ongoing connections.

Handoff is identified as a major obstacle to the deployment of large-scale voice over IP (VoIP) using wireless mesh. Since the distance covered by an AP is, approximately, only 100 m (300 ft), handoff is inevitable during a voice conversation in which the speaker moves. In a typical VoIP session it is observed that during handoffs the packet loss rate is high and delay increases. Three important findings were reported in [46] about handoffs. First, the handoff duration depends on the environment. Second, every single handoff lasts more than one second, which is beyond the acceptable range for VoIP application. Finally, the handoff duration varies greatly from instance to instance. This undesirable behavior of handoffs is due to the design of the inter-access-point protocol (IAPP), defined in the IEEE 802.11f standard, which dictates that the mobile unit should conduct the handoff without help from the APs. In addition, for the sake of simplicity and security, the IAPP forces any mobile unit to create a unique association with several APs in an extended service set (ESS). On the basis of this design, the handoff consists of a sequential process composed of four steps, scanning, authentication, association, and re-association. Of these steps, scanning is the most time-consuming phase; it can amount to 90% of the entire handoff time and can take several seconds [47]. There have been a number of efforts to optimize the IEEE 802.11 handoff process, mainly in the scanning phase [48]. It is worth noting that the handoff problem is a well-known weakness of IEEE 802.11 and is currently under redevelopment in the IEEE 802.11r standard.

To enhance the capabilities of admission and congestion control, the 802.11e standard uses the hybrid coordination function (HCF) in the MAC layer to improve QoS for wireless multimedia. The improvements continue in the developing standard 802.11n. Multihop ad hoc networking, high traffic loads, and the lack of coordination among nodes and of facilities for route reservation or clustering still remain the key challenging issues in building a real-time multimedia communications WMN infrastructure.

10.3.4   SIP-based VoIP over wireless mesh

With the growing importance of wireless mesh networks (WMN) in the research community as a quick and affordable solution for broadband wireless access [36], the session initiation protocol (SIP) has been considered as a feasible signaling solution for VoIP applications over WMNs since it was selected as the call control protocol for third generation (3G) IP-based mobile networks [49]. Owing to the combination of wireless infrastructure, user mobility, and heterogeneous network computing needed, to deploy SIP in a WMN, two important issues have to be resolved in a SIP architecture for WMNs, one is the mobility support to allow the users in WMNs freedom to move anywhere at any time in WMNs, and the other is the requirement to adapt to the varying end-to-end available bandwidth encountered in wireless channel environments.

To overcome these issues, an enhanced SIP proxy server was proposed in [49]. This enhanced SIP proxy server employs a common open policy service (COPS) protocol [50] to dynamically reserve the access bandwidth in the IP core network for all SIP terminals in a WMN. The COPS protocol, as defined by the IETF as a part of the IP suite, specifies a simple client–server model for supporting policy control over QoS signaling protocols. Moreover, the enhanced SIP proxy server contains two special modules to deal with traffic-prediction and call admission control (CAC) problems.

A CAC mechanism must be employed when the predicted and reserved access bandwidths are different from the actual bandwidth. This mechanism is used to accept or reject connection requests on the basis of the QoS requirements of these connections and the system state information. The CAC mechanism prevents the oversubscription of VoIP networks and is a concept that applies only to real-time media traffic, not to data traffic. A CAC mechanism complements the capabilities of QoS tools in protecting audio and video traffic from the negative effects of other such traffic and in keeping excessive audio and video traffic away from the network. Call admission control can also help WMNs to provide different types of traffic load with different priorities by manipulating their blocking probabilities.

10.4 Wireless VoIP and scalable IPTV video

Voice over IP (VoIP) applications in wireless networks have gained increasing popularity in recent years [51] [52]. It has been reported that the 802.11 standard based on the CSMA/CA MAC access mechanism is inefficient in meeting the QoS requirement for VoIP applications. To provide satisfactory QoS support for delay-sensitive real-time VoIP applications, a VoIP flow is usually given higher priority in accessing the shared wireless channel than delay-insensitive non-real-time flows, such as in the use of a higher access category based on the 802.11e MAC layer. For end-to-end VoIP based on standard WLAN MACs, application-layer QoS support is thus needed. Moreover, the advances in IEEE 802.16e wireless broadband and scalable video technologies have also made it possible for Internet protocol television (IPTV) to become the next "killer" application for modern Internet carriers in metropolitan areas with mobility support. It is a very strategic but challenging leverage for a carrier to glimpse the potential of IPTV using WiMAX as the access network. Challenges are posed for IPTV over WiMAX as this means multicasting under a diversity of fading conditions.

In this section we will provide two end-to-end wireless multimedia networking systems; one is for VoIP and the other is for scalable video communication. Both systems take into account congestion and error control strategies discussed earlier to combat specifically the deficiencies created by the wireless broadband infrastructure.

10.4.1 End-to-end WLAN-based VoIP system

An end-to-end wireless VoIP system with QoS support is illustrated in Figure 10.13 [53], where the last-mile connection is assumed to be a WLAN. The system is built on top of

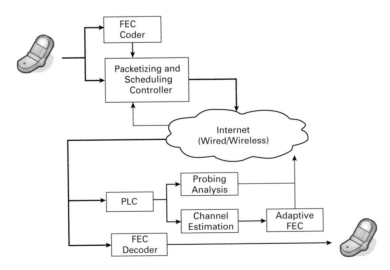

Figure 10.13 An end-to-end wireless VoIP system with QoS support [53] (© IEEE 2006).

UDP and RTP for real-time support and RTCP for periodic feedback. The process starts with voice encoding at the sender side. The encoded frames then pass through the packet formation and scheduling controller for transmission jobs in the application layer. This controller determines (1) how many frames there should be in a packet, (2) how much packet loss protection to provide, and (3) the time to send out the next packet, which depends on RTCP feedback from the receiver. There are several framing choices and two modes, normal and probing, for scheduling. At the receiver side, received packets first go through the packet loss classification (PLC) unit to categorize the packet losses from congestion and wireless separately. The receiver is responsible for making a conclusion about the congestion status and sending the message back to the sender using embedded probing mechanisms. The feedback includes an (n, k) combination from the analysis of the adaptive FEC unit.

10.4.1.1 Bandwidth inference via embedded probing

In wireless VoIP, the voice data are commonly encoded at a constant bitrate without using scalable coding. Therefore receiver-driven layered multicast techniques for available bandwidth probing, such as the BIC technique discussed in Section 7.5, cannot be applied directly. To estimate the available bandwidth for efficient congestion and error control in wireless VoIP, an embedded probing technique is thus adopted.

Figure 10.14 shows examples of packet scheduling and packetizing using embedded probing for bandwidth inference. Assuming that the receiver observes the wireless loss percentage above some preset quality threshold, it has to decide whether to add extra protection (i.e., filled block erasure FEC packets, as shown in Figure 10.14(c)), which will increase the data rate from $r1$ to $r2$. At such a moment, source packet transmission can be rescheduled as in Figure 10.14(b) from the regular scheduling in Figure 10.14(a). Note that, during this short packet train rescheduling, the rate is temporarily set as $r2$, while maintaining an overall

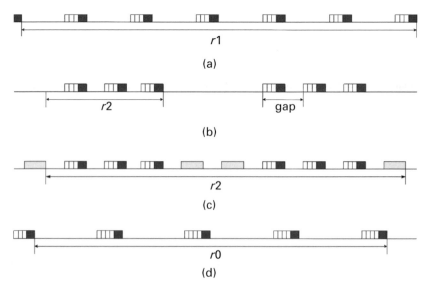

Figure 10.14 Packetizing and scheduling examples. The Internet overhead, the length of each frame, and the FEC packets are indicated [53] (© IEEE 2006).

Table 10.1 With an IP/UDP/RTP header, the 8 kbps G.729 can give a 40 kbps rate after being packetized using one frame per packet

Frames per packet	n	k	Packet/sec	BW (kbps)	Delay (ms)
1	1	1	100	40	10
	9	6	150	60	60
2	1	1	50	25.6	20
	7	4	87.5	44.8	80
3	1	1	33.5	18.6	30
	5	3	55.5	31	90
4	1	1	25	16	40
	4	2	50	32	80
	5	3	41.6	26.7	120

transmission rate $r1$. More rate options are available that depend on packetizing and involve unavoidable Internet overhead with a 40 bytes per packet header under the IP/UPD/RTP structure [53].

In this wireless VoIP system, the speech codec used is the multiple-pulse-coding-based G.729, which is also referred to as the conjugate structure – algebraic code excited linear prediction (CS-ACELP, see Section 2.4) [54]. The 8 kbps G.729 has been shown to achieve a toll quality similar to that of 32 kbps G.726 adaptive delta pulse-code modulation (ADPCM) and has been adopted in several Internet-based VoIP or session initiated protocol (SIP) phones. The G.729 codec uses 10 ms frames, with 5 ms (40 sample) subframes for excitation signal representation. For example, as shown in Table 10.1, with an IP/UDP/RTP header, the 8 kbps G.729 can give a rate of 40 kbps after being packetized using one frame per packet. However, packetizing G.729 frames using three frames per packet yields 18.6 kbps. To achieve an even lower transmission rate, four frames per packet will reduce it to 16 kbps, as indicated in Figure 10.14(d). With the extra saving in packaging more frames per packet, it is possible to trade off the longer delay thereby introduced with a lower packet loss (by adding more FEC packets), under the same bandwidth budget. Note that any packet formation is required to meet an overall delay constraint, i.e., that the one-way trip time is less than 400 ms.

In order to reduce the delay either by using fewer frames per packet or by inserting FEC packets, both of which actions require a higher available bandwidth, it is important to determine whether on the basis of the embedded probing there is enough bandwidth for the required target rate, say $r2$ as shown in Figure 10.14(b). According to the traffic model, the equal event EQE for the sum of the gaps (see Figure 10.14(b)) is defined as

$$EQE = I\left(\frac{|G_{sum,sender} - G_{sum,receiver}|}{\max(G_{sum,sender}, G_{sum,receiver})} < \delta\right) \tag{10.18}$$

where $I(x)$ is a logic function with value 1 if the inequality inside the bracket is true or 0 if not; G is the sum of the gap times within a train of sender or receiver data, and δ is the threshold suggested in [55]. The EQE is taken for each packet train. The sending gap and receiving gap are regarded as being equal, which indicates the available bandwidth being identified, if the EQE is 1. To avoid a wrong conclusion being drawn from these relatively short packet trains, a supplementary index S_{EQE} is also continuously calculated to facilitate the decision. The supplementary index at the arrival of the nth packet train is defined as

$$S_{EQE,n} = (1 - a) \times S_{EQE,n-1} + a \times EQE_n. \tag{10.19}$$

To make the decision whether the probing bandwidth is indeed the available bandwidth, not only do the sending gap and receiving gap need to be more or less equal but also two additional conditions have to be satisfied. First, S_{EQE} must rise to a high enough value quickly (within a time equal to *period 1*), and, second, it must stay at a reasonably high value for a certain time period (*period 2*). More specifically, if S_{EQE} stays at high values then this implies that there exist a number of consecutive EQE values equaling 1, and the channel is less likely to be congested. Obviously, two thresholds, $Th_{S_{EQE}}$ and $Th_{S_{EQE\_high}}$, and two time periods need to be specified. Both these time periods are set equal, *period 1 = period 2*, as the time interval for receiving 150 packets. The reason why 150 packets was chosen is based on the experience that generally 50 packets per train are required to get a good cross sample [56], therefore three times this value compensates well for any short packet trains. Taking nine packets per train for example, there are at most 16 EQEs and S_{EQE} samples per period. A positive decision can be resulted from reaching $Th_{S_{EQE\_high}}$ within 150 packets and not dropping below $Th_{S_{EQE}}$ for another 150 packets.

10.4.1.2 Performance improvement

To monitor the quality, a receiver always keeps four observation values: the congestion loss rate, the wireless loss rate, and the estimated Gilbert–Elliot two-state transition probabilities p, q shown in Figure 10.10. For the delay, the overall effect when choosing (n, k) for FEC is approximately as below:

$$Total\_Delay \approx t_f \times F \times k + Internet\_OTT, \tag{10.20}$$

where the parameter t_f represents the per-frame interval of the voice codec, F is the number of frames per packet, and k is one of the FEC parameters. The Internet one-way trip time (OTT) is around 100–150 ms in real-world ISP environments [57]. Therefore, in order to improve transmission efficiency (higher F) or achieve a particular error protection level with lower redundancy (higher k), the overall delay is also increased. In the proposed algorithm, Eq. (10.20) is checked every time a rate change is performed. A total OTT delay constraint of 250 ms is commonly set but a hard limit of 400 ms is adopted to ensure delay quality even under heavily congested conditions, when more frames per packet need to be used.

For the case where both loss rates are lower than the corresponding thresholds, no action should be taken. When there is no congestion at the current rate, bandwidth estimation based on embedded probing is performed first and then additional FEC packets can be inserted. When the channel is congested, the system will reduce its rate to relieve the congestion by removing FEC packets or packaging more frames in a packet (i.e., relaxing the delay constraint). To support the same quality at different packetizing conditions, the number of frames per packet can be increased one by one and a search made, for every (n, k), for the most suitable FEC protection. Also, even under a longer delay tolerance the system will keep looking for a chance to provide services with less delay. It also needs to be noticed that the congestion loss threshold is set slightly higher than the wireless loss because FEC, since it is the main mechanism for recovering wireless loss packets, needs to be more aggressively added.

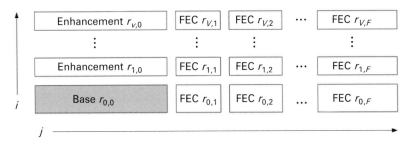

Figure 10.15 Video data and error protection codes are both formatted in layers [58].

10.4.1.3 Extension to FEC-based scalable video over WLAN

Scalable video and FEC layers Using embedded VoIP packets to infer the available bandwidth and using the Gilbert–Elliot model to estimate the wireless-channel-condition p,q values and so derive the necessary FEC (n, k) packet insertion, end-to-end scalable video can also be disseminated over WLAN [58]. Scalable video can be created by a scalable extension of H.264/AVC [59] [60], which is an emerging compression technology with coding efficiency comparable with the original H.264/AVC standard (as discussed in Section 5.7). To provide maximum flexibility in both quality and protection, the video data and error protection codes are both formatted in layers as shown in Figure 10.15, with rate $r_{i,j}$; i and j are the indexes of the video and FEC layers respectively. Layers with $j = 0$ contain only video streams; otherwise, j indicates the level of protection for a specific i. There are V enhancement video layers and F FEC protection layers.

In other words, by modeling the network conditions using F sets of channel parameters (p, q), the corresponding $n_{i,j}, j = 1 \sim F$, in ascending order for each video layer i, to generate F protection layers, can be derived (see Figure 10.16). In order to judge how many FEC layers a receiver should subscribe to after the whole scenario has been constructed, a video data layer i is chosen, the newly estimated (p, q) set is inserted, and the minimum n_{new} is found using the same process as that given in Section 10.1.4; then the value of j with $n_{i,j}$ closest to but larger than n_{new} is chosen.

Regarding the interlayer dependency and protection levels for every layer, the overall data rate is

$$R_{v,f} = \sum_{i=0}^{v} \sum_{j=0}^{f} r_{i,j} \tag{10.21}$$

where v is the number of enhancement layers *subscribed* to and f, which is the same for every i, is the number of FEC layers requested. Unequal protection for different video layers is supported by different $e_{FEC,i}$ values, not by f. For instance, under the channel condition $(p, q) = (0.8, 0.2)$, the values $e_{FEC,0} = 0.001$, $e_{FEC,5} = 0.01$, and $k_0 = k_5 = 8$ are predetermined and result in $n_{0,j} = 17$ and $n_{5,j} = 15$ with block-erasure-code-based FEC. This also implies that one video layer requires all lower layers to be available for decoding (owing to the cumulative layer structure of the adopted scalable codec), although not every sublayer FEC is needed.

Embedded layered probing and join decision In order to increase the data rate, either for more video data or more loss recovery, available-bandwidth estimation has to be performed

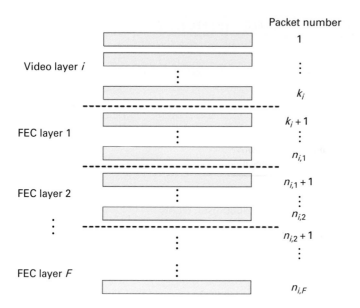

Figure 10.16 Layered FEC structure for the appropriate level of protection [58].

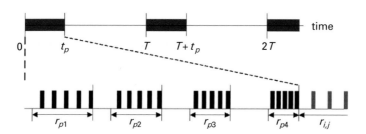

Figure 10.17: Embedded probing mechanism for the available bandwidth [58].

in advance to prevent congestion. The embedded probing can be again used. The probing streams use the regular video data through the effective scheduling of packet transmission and take advantage of the fact that streaming systems usually have decoding buffers to tolerate some amount of delay [58].

As shown in Figure 10.17, in embedded probing the stream is periodically separated into alternate probing and regular intervals with period T. The length of the probing interval, t_p, is further divided into certain uniform probing regions according to the number of possible layers to be probed, e.g., r_{p1} to r_{p4}. In each region, previously generated packets are delayed in transmission, creating a temporarily higher sending rate within it. At the receiver side, the interval and region to which a packet belongs is distinguished by its RTP timestamp. Full-search delay-trend detection, as shown in Eq. (7.5), is then applied to packets in time slots at objective rates. The duration of each probing region is set to be the time taken to send 50 packets at regular intervals in base layer $(i, j) = (0, 0)$ and remains the same for every layer.

Table 10.2 Adaptation rules for all possible cases and reactions in the proposed architecture where five action paths are allowed in the system [58]

Events	Channel estimation	Probing result	Congestion loss	Action path
Increase FEC	worse	positive	—	I
		negative	(yes)	II
Decrease FEC	better	—	—	III
Change quality	same	positive	—	IV
		negative	yes	V

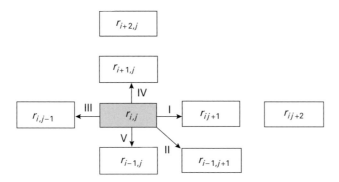

Figure 10.18 A partial view of Figure 10.15 showing switching between five action paths [58].

Other important facts are that the probing rate for which we are looking is an aggregation rate $R_{v,f}$, not $r_{i,j}$. The aggregation rates of all possible target subscriptions should be included in a probing interval. Given that the layered streams are ready at the server, the decision about which layers to subscribe to is based on three types of information, the channel estimation (p, q), probing results for the available bandwidth, and the observed packet loss rates. On the basis of the packet loss classification (PLC), the estimated wireless channel model $(\hat{p}, \hat{q})$ (updated every T seconds, depending on the number of samples) and the congestion packet loss (updated every second) can be continuously monitored. Table 10.2 addresses all possible cases and reactions in the proposed architecture where five action paths are allowed in the system. A dash means that the possibility corresponding to this table entry is irrelevant because the rate is getting lower or because low congestion loss is necessary for a positive probing. After every period T, the receivers yield a (p, q) pair and map it to demand that the FEC value, j, for the current video layer i satisfies the condition in (Eq. 10.11). If the new j is larger than the current j, it is categorized as a "worse" wireless condition. "Better" and "the same" are defined analogously. For these three conditions, the corresponding responses are to increase the FEC, decrease the FEC, or change the video quality accordingly, taking observations into account. For example, if the wireless channel is "worse" and probing for more FEC in the same video layer has failed then it should follow the second path with more FEC but less quality; if the channel is "the same" and probing for a higher rate at the same protection level is positive then it is time to subscribe to more video content. Figure 10.18 shows a partial view of Figure 10.15 with five action paths, I–V. The aggregation rates of the probing regions will match target rates. Depending on the channel quality, end users select the regions corresponding to r_{p1} and r_{p2} or r_{p3} and r_{p4} in Figure 10.17 for delay-trend detection and make a decision to either stay or take one of the five action paths. The only information needed for feedback is the resulting change in layer subscription.

10.4.2 Scalable IPTV video over WiMAX

With the capability for point-to-multipoint and multicast of IEEE 802.16d/e WiMAX, broadband wireless access can be extended to IPTV services. Nonetheless, there are challenges posed for IPTV over WiMAX relating to multicasting under a diversity of fading conditions. A cross-layer design framework based on a novel two-level superposition coded multicasting (SCM) scheme was introduced in [61], see below.

10.4.2.1 Why use WiMAX for IPTV?

WiMAX offers ease of deployment and affordable cost similar to other wireless technologies but with a larger service coverage and more bandwidth. It can capture the maximum number of IPTV subscribers under the same infrastructure and provide even better accessibility to the same pool of video content for mobile users in the future. Therefore, WiMAX is considered the best candidate to offer quadruple play (i.e., Internet, phone, TV based on wireless access) services. Launching IPTV over WiMAX can further achieve economy of scale in terms of more services and better service availability under a common infrastructure.

WiMAX is also an emerging technology that can adequately support the QoS requirement of IPTV. The QoS features of WiMAX best suited for such support include its reservation-based bandwidth allocation, cost-effective and infrastructure-free deployment, and stringent QoS support for five types of service, unsolicited granted service (UGS), real-time polling service (rtPS), extended real-time polling service (ertPS) non-real-time polling service (nrtPS), and best-effort (BE) traffic [62]. Enabling rtPS in wireless broadband access can support perfectly the bandwidth requirements of the managed content of IPTV service providers, especially for paid high-definition and standard-definition TV (HDTV and SDTV).

10.4.2.2 Multicast of scalable IPTV in WiMAX

To maximize the efficiency of large-scale IPTV over WiMAX, the IPTV system must take advantage of the fading-channel diversity and multicast capabilities of WiMAX. Note that in a single-user communication scheme, there are many QoS adaptation approaches for optimizing data throughput along a time-varying channel. These approaches include adaptive coding, modulation, power control schemes, and so on in response to variations in the channel condition. However, in the multicast scenario of WiMAX, where a single multicast transmission of a video packet is sent from a BS to multiple SSs using the multicast and broadcast service (MBS), the video packet can be received by all the subscribed SSs but each SS is subject to a different bit error rate owing to the associated heterogeneous channel conditions. This makes centralized QoS adaptation impossible since different SSs have different channel conditions. Take for example a scenario in which an IPTV program could be multicast-disseminated by WiMAX either with the more efficient 16-QAM modulations or with the less efficient QPSK modulations. When a multicast signal based on 16-QAM is sent from the BS, only those SSs with a good channel state, which will support 16-QAM well, can decode the received multicast signals. However, if QPSK is used for the video multicast then it could result in resource underutilization of those SSs with good channel states. Minimizing bandwidth consumption by the use of more efficient modulations in the content delivery of IPTV becomes critical for scaling the system to support more TV channels simultaneously.

10.4.2.3 Two-level superposition coded multicast

To satisfy the heterogeneous quality requests of IPTV from different SSs, a two-level superposition coded multicasting (SCM) scheme was proposed in [61]. The two-level SCM approach can serve as a framework for applying superposition coding at the physical (PHY) layer in WiMAX to modulate or demodulate multicast signals. Superposition coding was originally introduced to increase the overall user capacity of a wireless communication system by exploiting the spatial or temporal power disparities perceived by multiple users of common broadcast signals [63] [64]. An example of superposition coding is shown in Figure 10.19 [61], where the nodes are indexed in an increasing order according to their distance from the BS. Assume the SNR experienced by these nodes decreases linearly as the distance increases. When the BS transmits signals to $M3$ with the targeted signal-to-noise ratio (SNR) level for $M3$, the SNR of this transmitted signal experienced by both $M1$ and $M2$ is much greater than the targeted SNR levels (by $A + B$ decibels and C decibels respectively). Similarly, when the BS transmits signal to $M2$, $M1$ receives an additional A decibels of power above the targeted SNR level. This implies that $M1$ has a more than sufficient SNR to decode the messages intended for both $M2$ and $M3$, and $M2$ has a more than sufficient SNR to decode the messages intended for $M3$. The power disparities at nodes $M1$, $M2$, and $M3$ suggest that the information for $M1$ could be included in the transmission to $M2$ or $M3$ through the adoption of superposition coding. Similarly, the information for $M2$ can be included while transmitting information to $M3$. The dotted line in Figure 10.19 indicates that by employing superposition coding the BS transmits information to $M2$ while transmitting to $M3$ at the targeted SNR level.

Using the principle of superposition coding, the two-level SCM scheme is applied to WiMAX video multicasting as follows. The video data is first encoded using scalable video coding (SVC) technology, which encodes the video at the highest resolution and allows the

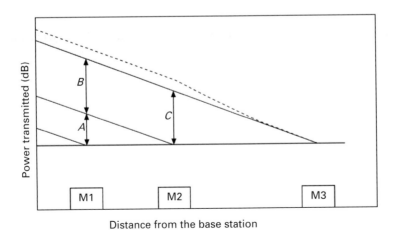

Distance from the base station

Figure 10.19 An example of superposition coding. The nodes are indexed in an increasing order according to their distance from the BS [61] (© IEEE 2007).

bitstream to be adapted to provide various lower resolutions. To achieve scalability, the video data is encoded into several layers. The lower layers contain the lower-resolution data. This data is more important because it provides basic video quality at a low bitrate. The higher layers contain the refinement data. This refines the lower-resolution data to provide higher-resolution video. The refinement data is less important and can be removed when the bandwidth or decoding capability is not sufficient. With the use of a layered structure, which puts data of different importance into different layers, unequal erasure protection (UEP) can be easily incorporated with SVC to provide more protection for the important data. With such features, the SVC bitstream is more suitable than the non-scalable bitstream when the video packets are transmitted over an error-prone channel with fluctuating bandwidth. In the proposed superposition coded multicast (SCM) scheme, the video is compressed into two scalable layers, the base-quality layer and the enhancement-quality layer. Instead of using one modulation scheme at a time, multicast signals in a time slot of WiMAX are generated by performing the superposition of data modulated by QPSK and data modulated by 16-QAM in the physical layer. It is assumed that all the SSs can support QPSK even in their worst channel conditions. For ensuring video quality, the base-quality layer of each video frame is modulated by QPSK, and the enhancement-quality layer is modulated by 16-QAM. With such superposition coded multicast signals, each SS can at least decode and obtain the base video quality layer of an IPTV channel modulated with QPSK when it is in a bad channel state, or achieve the full video quality with all the combined base and enhancement layers modulated by QPSK and 16-QAM separately when it is in a good channel state. Even in the presence of fluctuations in channel conditions between good and bad in all the SSs from time to time, an acceptable video quality of IPTV channel for every SS can be adaptively achieved.

10.4.3 Cross-layer congestion control for video over WLAN

In spite of the proliferation of wireless technology for a variety of multimedia services, e.g., IPTV streaming (downlink), real-time gaming, video surveillance (uplink), and conferencing (bidirectional), the wireless-enabling technology IEEE 802.11 WLAN can unfortunately be easily demonstrated to fail when subjected to the congested conditions associated with many of these real applications. These effects are a serious hurdle for the deployment and advancement of wireless technologies and therefore a congestion-control remedy is needed. As discussed in Chapter 9, there are some issues related to the use of WLAN for real-time multimedia communication.

(1) *Performance anomaly* Conventional use of the 802.11 DCF results in a system that achieves throughput fairness, but at the cost of decreased aggregate system throughput and increased congestion in a multirate WLAN environment [65].
(2) *Imperfect link adaptation (LA)* At the MAC level, devices implement LA intelligence [66] to select modulation and channel coding schemes according to the estimated channel conditions. However, today's LA algorithms can be misled by collision errors – thus lowering the PHY rate even though the SNR is good, which in turn leads to an even higher collision probability. This "Snowball Effect" [67] quickly leads the system into catastrophic failure.
(3) *Available bandwidth estimation* It is often desirable for congestion control algorithms to estimate the available bandwidth. At the application (APP) layer, systems suffer from slow

and inaccurate capabilities for estimating the end-to-end available bandwidth. A measure of the available bandwidth cannot be very accurate in the absence of MAC-layer information.

Given these difficulties of the single-layer solution, a distributed cross-layer adaptation system for video transmission is highly desirable for WLAN. One such solution is proposed in [68], this addresses the aforementioned issues and adaptively ensures wireless video transmission of acceptable quality in an uplink, multi-camera, topology. This cross-layer solution greatly improves the performance in video over WLAN in terms of goodput and packet loss. The central idea is that airtime fairness is preferred to throughput fairness, i.e., the use of airtime fairness combined with video source rate adaptation increases the aggregate goodput and decreases the packet loss rate (PLR). The solution uses a module (which runs within each wireless IP camera) that handles two tasks: (1) it induces airtime fairness, and (2) it perform video source rate adaptation. Implementing these tasks involves:

(1) *a fair airtime throughput estimation (FATE)* formulation to upper bound the video encoder's target bitrate. The FATE formulation computes the maximum throughput that the encoder is allowed to use if system-wide airtime fairnes is to be guaranteed;
(2) *a source rate adaptation (SRA)* scheme to scale the video encoder's target bitrate in inverse proportionality to the camera's observed packet loss rate;
(3) *a contention window adaptation (CWA)* scheme to adapt the 802.11 contention window in order to control access to the channel in a way that guarantees system-wide airtime fairness on a quick timescale.

The idea is that the APP offers powerful control loop gain on a slower timescale while the MAC offers feedback and fine tuning on a faster timescale. By exploiting and harmonizing these properties, a distributed congestion-control system that has bottom-up robustness on a frame-by-frame scale can be created.

10.4.3.1 Why airtime fairness?

It is well known that per-station throughput in a multirate WLAN BSS tends to move quickly toward the rate of the lowest rate station in the basic service set (BSS) [65]. Channel access among competing stations is managed in such a way that the stations divide the available throughput according to CSMA/CA, which implements a throughput fairness policy. As shown in Figure 10.20, all 10 stations are initially at 11 Mbps. At $t = 15$, one of the 10 stations switches from 11 Mbps to 2 Mbps. Throughput fairness results in: decreased aggregate system throughput (from about 5.5 to 4 Mbps in Figure 10.20); no throughput guarantees above a minimum level; and a possibility of congestion with the added danger that some video flows will be dropped completely. Consequently, if a station switches its PHY rate then we want to adapt that station's channel share and video source rate accordingly, in order to protect the system from catastrophic behavior. Therefore a balanced allocation of airtime is needed such that a slow station does not consume more airtime than a fast station; this principle is referred to as airtime fairness. The fairness issue in IEEE 802.11 WLAN has received considerable attention in the literature. Self-clocked fair queuing (SCFQ) [69] has been adopted by the distributed fair scheduling (DFS) protocol [70] for weighted fairness. This applies the systems virtual clock in a fully distributed fashion so as to emulate centralized fairness. Other algorithms model the distributed coordination function (DCF) or enhanced distributed channel access (EDCA) using a Markovian structure and then allow for tuning of the CSMA/CA parameters including the

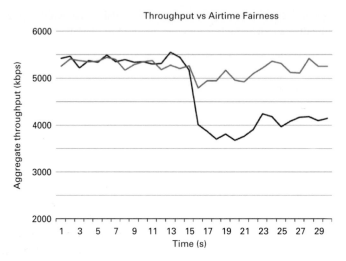

Figure 10.20 Throughput vs. airtime fairness. Note that the aggregate system throughput (lighter curve) is reduced in the case of throughput fairness (darker curve) [68] (© IEEE 2008).

contention-window size, frame size, retry limit, and AIFS number to achieve the desired fairness in relation to airtime [68].

10.4.3.2 Controlling airtime

As discussed in Section 9.2, there are two common approaches to controlling access to the WLAN medium and therefore to the allocation of airtime, as follows.

(1) *MAC-layer approach*   The contention window is adapted. If two stations are contending for the channel, the one with a lower CW_{min} is the one that is most likely to win the contention.

(2) *APP-layer approach*   The video bitrate is adapted. A station that transmits video at a 400 kbps bitrate will not contend as aggressively for the channel as a station that transmits video at a 800 kbps bitrate.

Although airtime control can be performed from either the APP layer or the MAC layer, it is found that the best solution is to take advantage of both layers simultaneously. Each of the above approaches has its own drawbacks, but controlling airtime simultaneously from the MAC and the APP allows the benefits of each approach to cancel out the other's drawbacks.

MAC-layer airtime control   As discussed in Section 9.2.3, the initial contention-window size CW_{min} is an effective parameter for controlling airtime share. An intuitive explanation of contention-window adaptation (CWA) is as follows. Assume a set of stations $S = \{1, \ldots, N\}$. In a time interval $[t1, t2]$ an arbitrary station i under static channel conditions consumes data airtime A_i in n_i transmissions:

$$A_i = \frac{1}{r_i}\sum_{k=1}^{n_i} f_k = n_i\frac{\bar{f}}{r_i}, \tag{10.22}$$

where f_k is the MAC service data unit (MSDU) size for the kth frame transmission and r_i is the PHY rate applied. From the CSMA/CA process, the expected value of the backoff time B is

$$E\left[\sum_{k=1}^{n_i} B_k\right] = n_i \times E[B] \approx n_i \times \frac{CW_{\min}}{2} \times SlotTime, \tag{10.23}$$

assuming that collisions and wireless errors are both low owing to appropriate CWA and link adaptation. Since all stations perform backoff simultaneously, it can be approximated that, for stations i and j,

$$\sum_{k=1}^{n_i} B_k^i \approx \sum_{k=1}^{n_j} B_k^j \tag{10.24}$$

if and only if

$$\frac{n_i CW_{\min}^i}{n_j CW_{\min}^j} = 1.$$

Airtime fairness is achieved if $A_i = A_j$ (see Eq. 10.22). If this equality is to hold then, since $\bar{f}$ is constant, airtime fairness is achieved if $n_i/n_j = r_i/r_j$. Substituting this into the above equation, we have airtime fairness in a multirate environment if

$$r_i CW_{\min}^i = r_j CW_{\min}^j. \tag{10.25}$$

Thus, by tuning the $CW_{\min}$ ratio, a fair airtime share can be realized.

The advantage of MAC-layer adaptation is that it can react on a fast frame-by-frame timescale, changing $CW_{\min}$ instantaneously to modify the STA's channel access intensity. The disadvantage is that it can lead to high packet loss, as shown in the following example. Consider two STAs, camera A and camera B, both transmitting video encoded at a bitrate of 800 kbps. Camera A uses an 11 Mbps PHY rate and camera B uses a 2 Mbps PHY rate. Under airtime fairness, the CWA algorithm penalizes (lowers) camera B's channel access probability by a factor proportional to its reduced PHY rate. This reduction in access probability combined with the lower PHY rate puts a theoretical limit on the maximum throughput that camera B can attain. If this limit is less than 800 kbps (approximately the minimum throughput needed to sustain the encoded video), then the received video will suffer high packet loss.

APP-layer airtime control The upper bound on throughput imposed by contention-window adaptation (CWA) guarantees system-wide airtime fairness. This airtime fairness can also be achieved by changing the video encoding rate in the application (APP) layer, which can implicitly induce airtime fairness without using CWA. The module responsible for this task is called cross-layer fair airtime throughput estimation (FATE). The advantage of controlling airtime using bitrate adaptation in APP is that it solves the problem of high packet loss that plagues the CWA solution. More specifically, if CWA puts a 500 kbps upper bound on camera B's throughput, the camera still tries to transmit video encoded at an 800 kbps target bitrate, resulting in high packet loss. However, when the APP-layer approach is used, the encoder's target bitrate is physically changed from 800 kbps to 500 kbps, thus greatly reducing packet loss compared to CWA. The disadvantage of bitrate adaptation is that it operates on a slow timescale because video encoders take several seconds to fully reach a newly set target bitrate.

Reducing packet loss   To supplement the concept of airtime fairness, a smart source rate adaptation (SRA) technique is also applied to reduce packet loss further, so that the target bitrate can be adjusted *within* the boundaries set by FATE. By applying CWA, FATE, and SRA, it is now conceptually clear that simultaneously using APP-layer and MAC-layer airtime control yields the best combination of low packet loss (due to APP) and quick reactivity and stability (due to MAC).

10.4.3.3 Distributed cross-layer congestion control

The cross-layer congestion control (CLC) system is shown in Figure 10.21 for a single station (camera). Since the system is distributed, each camera has its own instantiation. All instantiations share the number N of active cameras in the basic service set (BSS). The CLC module operates as follows on each camera: the SRA module constantly interacts with the MAC to track packet loss. If the packet loss rate is too high then the SRA module incrementally reduces the target bitrate. If the packet loss rate stabilizes at a low value then the SRA module incrementally increases the target bitrate as long as the new target bitrate does not exceed the maximum set by the FATE module. The FATE module takes frame error rate (FER) information from the MAC and the value of N from the application server to compute the maximum throughput that a station can attain to maintain system-wide airtime fairness. During every adjustment to the target bitrate, CWA is used to instantly provide airtime fairness while allowing the relatively slow encoder to catch up with the new target bitrate. The specific way that each component uses its inputs and calculates its outputs is detailed below.

Cross-layer fair airtime throughput estimation: There are several ways to evaluate per-station system throughput in IEEE 802.11 networks. Using the framing and timing details in the standards, theoretical maximum throughput can be calculated assuming no channel or collision errors [71]. With errors, the Markovian model can be applied to model the throughput in a multirate WLAN deployment [72]. Given information such as the wireless error rate, transmission rate, and backoff parameters for all stations, one can solve non-linear equations for transmission probabilities in order to compute system and per-station throughput. However, in a multi-camera setting, this information may not all be available, nor the computation power for accurate modeling and equation solving. Therefore, a cost efficient scheme is adopted for per-station throughput evaluation. Any device which is capable of calculating its FER and receiving information about the total number of stations in its BSS can use the proposed formula.

Assume that G^i is the theoretical system throughput as seen by station i, computed according to [71] depending on the average MSDU_size $\bar{f}$ and r_i (the PHY rate of station i); then

$$G^i = \frac{MSDU\_Size}{MSDU\_Time} = \frac{\bar{f}}{k + \bar{f}/r_i},$$ (10.26)

where $k = T_{DIFS} + T_{SIFS} + T_{BACKOFF} + T_{ACK}$ and $\bar{f}/r_i = T_{DATA}$.

This simplified approximation implies that station i will "perceive" the maximum theoretical system throughput as if r_i were the PHY rate of the system (in reality, each station may use a different PHY rate). It was shown [68] that this approximation does not have a significant effect on accuracy, and it greatly improves the performance of the algorithm by precluding the need for stations to exchange detailed configuration information and solve complex equations. From the theoretical maximum system throughput, the maximum per-station fair throughput as seen by station i can be approximated as $G_e^i \approx \frac{1}{N} \times G^i$.

The next step is to compensate for the effect of errors due to the observed FER at each station, p_f^i, where all unacknowledged frames are considered errors. The idea is to estimate the portion of frames successfully transmitted in a given amount of airtime share, so we discount the throughput estimate by p_f^i, i.e.,

$$G_e^i \approx \frac{1}{N} G^i \times (1 - p_f^i). \qquad (10.27)$$

Note that frame errors can be classified into "collisions" and "channel errors," the former having probability p_c and the latter probability p_e. In other words $p_f = p_c + p_e - p_c\, p_e$. When an error is due to a poor channel, only one station is involved (the station that transmits the packet), but when an error is due to a collision it involves two or more stations so, in a steady-state average sense, the airtime wasted in collisions is uniformly divided across all stations involved in the collision. Thus the term $1 - p_f$ overdiscounts the maximum per-station throughput, so a correction factor is needed. Assuming that $p_{c,3+} \ll p_c$, where $p_{c,3+}$ is the probability that at least three stations collide, the estimated fair airtime maximum throughput for station i is [68]

$$G_e^i \approx \frac{1}{N} G^i \times (1 - p_f^i) \times \left(1 + \frac{p_c^i}{2}\right), \qquad (10.28)$$

where p_f^i is observed at each station (for example, from the driver), p_e^i is estimated, for example, using received signal strength indication (RSSI) mapping, and

$$p_c^i = \frac{p_f^i - p_e^i}{1 - p_e^i}.$$

The necessary observations, f, r, p_f, and p_e, are evaluated every 100 ms and used to recompute G_e^i. The result is then passed to the source rate adaptation (SRA) module, as indicated by the arrow labeled "Max Throughput" in Figure 10.21; the condition on G_e^i is shown in the SRA pseudo-code (see Figure 10.22). Consequently, FATE enables airtime fairness via the SRA process and is a sufficient solution under a static channel (without PHY rate changes). As mentioned earlier, the 5–10 second delay required by the encoder to catch up with the new target bitrate is too long. If there is a change in the PHY rate, the system could easily suffer catastrophic failure during the transition period. Thus CWA, detailed below, is utilized during these transition periods.

Contention-window adaptation   The CWA module computes the appropriate CW_{min} for each data frame immediately before the frame enters the backoff process. The computation uses the PHY rate as an input and computes its output using an abstraction of Eq. (10.25). Namely, let the right-hand side of Eq. (10.25) be equal to a constant, $r_{max}\, CW_{min}^0$, where r_{max} and CW_{min}^0 are the maximum PHY rate and the initial CW_{min} value respectively. Then, the left-hand side is divided by r_i to obtain CW_{min} for the ith camera. Thus we have

$$CW_{min}^i = min\{\lfloor c_i\, CW_{min}^0 \rfloor, CW_{max}\} \qquad (10.29)$$

where

$$c_i = \frac{r_{max}}{r_i}. \qquad (10.30)$$

The value of CW_{min} is set as inversely proportional to the PHY rate. Thus, stations using low channel rates due to degraded signal quality have a lower probability of transmitting than

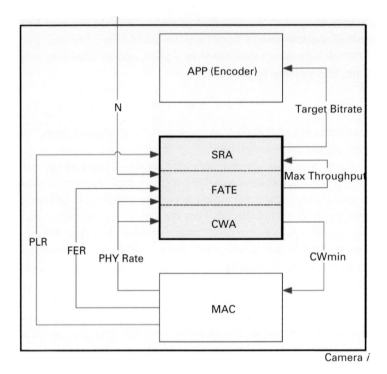

Camera *i*

Figure 10.21 Cross-layer congestion control system diagram. The number of active cameras is N [68] (© IEEE 2008).

```
while(1)
    Compute PLR
    if PLR > η1
        Bitrate = Bitrate − δ
    else if PLR < η2 for t consecutive seconds
        if(Bitrate + δ < Ge)
            Bitrate = Bitrate + δ
    sleep ε
```

Figure 10.22 The pseudo-code for source rate adaptation (SRA) [68] (© IEEE 2008).

high-rate stations. Statistically, equal-data airtime share among stations is maintained under this adaptation. This implementation guarantees QoS according to a set of requirements and also releases the congestion on channel utilization time.

10.4.4 Cross-layer scalable video over WiMAX

For wireless mobile communication, scale video coding (SVC) has several advantages over non-scalable video coding. For a single user, the transmission bandwidth is time varying due to the mobility and fluctuation of the available resources. Besides, the users are located at different positions; the varying signal quality leads to varying transmission bandwidth. It is difficult to support all users with a single non-scalable bitstream. Moreover, there is no priority in the non-scalable bitstream. This leads to inefficient error protection because both the more important data and the less important data have the same quality of service in an error-prone channel. Scalable video coding provides simple solutions to these problems. The SVC bitstream can be easily adapted to the varying bandwidth and various receivers, and the layered data structure allows the more important data to obtain more protection easily. As a result, these features provide more efficient video transmission over an error-prone channel with fluctuating bandwidth. In Section 10.4.2 we discussed a cross-layer (MAC+PHY) design framework based on a two-level superposition coded multicasting (SCM) scheme for the delivery of two-layer SVC using the multicast capability of WiMAX. In the present section we present another cross-layer design, based on end-to-end unicast connections of WiMAX, for the delivery of several layers of H.264/SVC video.

As discussed in Section 9.4, the IEEE 802.16 WiMAX defines several burst profiles that are a combination of a modulation and coding scheme in each PHY configuration. With link adaptation, the system can decide the proper modulation and coding scheme according to the current SNR value. In IEEE 802.16e, the SNR of each mobile station may change with time. At the beginning of the frame the base station decides the burst profile of each downlink (DL) and uplink (UL) data burst. For a DL data burst, the base station can decide the burst profile of the data burst according to the feedback DL channel condition in the UL fast feedback channel (the channel quality indicator, CQI).

To provide better QoS for scalable streaming services in mobile WiMAX systems, a cross-layer design between the streaming server and mobile WiMAX base station was proposed [73]. In this cross-layer design, multiple connections with feedback information on the available transmission bandwidth are implemented for supporting scalable video streaming in which transmission packets can be further separated into multiple levels of importance.

An end-to-end transmission of a streaming video in the mobile WiMAX system is shown in Figure 10.23, where the last-mile transmission system is IEEE 802.16e/Mobile WiMAX and consists of base stations (BSs) and mobile stations (MSs). The streaming server and WiMAX subsystem are interconnected by an IP-based backhaul network. The streaming service in this study is encoded by the scalable extension of H.264/AVC. In the proposed cross-layer design, the BS periodically reports the average bandwidth, with regard to the residual BS capacity and RF condition, to the streaming server during the *report period*. The main functional duties of the streaming server and base station MAC are described in detail as follows.

10.4.4.1 The controls in the streaming server

During each report period, the BS will request a target bitrate from the SVC streaming server, which then analyzes the bitstream at the group of pictures (GOPs) that correspond to the current report period, as shown in Figure 10.24. In the SVC bitstream, the data at lower spatial-SNR resolution is more important. And in each spatial-SNR layer, the lower temporal layer is more important. Therefore, according to the requested bitrate from the BS, the lower-spatial-SNR layers are extracted first. At the spatial-SNR layer where the bitrate cannot cover all the data, only the data at the lower temporal layers are extracted. With fine granular scalability (FGS) support, the SVC data at the same temporal layers can be

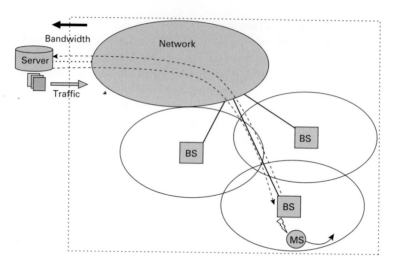

Figure 10.23 End-to-end transmission of a streaming video in the mobile WiMAX system, where the last-mile transmission system is IEEE 802.16e/Mobile WiMAX, and consists of base stations (BSs) and mobile stations (MSs) [73].

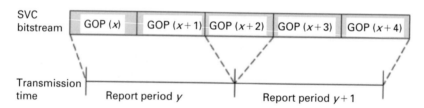

Figure 10.24 GOP (*x*), GOP (*x* + 1), (*x*+2) and GOP (*x* + 2), GOP (*x* + 3), GOP (*x* + 4) will be analyzed and transmitted during report period *y* and report period *y* + 1 respectively [73] (© IEEE 2007).

truncated at any position to provide exactly the requested bitrate. The extracted data are then sent from the streaming server to the BS.

If a GOP lies between two report periods, as does the GOP(*x* + 2) in Figure 10.24, this GOP, which was already sent in the first report period, will not be sent again. However, if the second report period allows a higher bandwidth, the remaining data in this GOP can still be transmitted to create smoother video quality when the bandwidth changes frequently.

To support multiple connections between the BS and MS, the data sent to the BS from the streaming server are assigned to one of two connections according to their importance level, as shown in Figure 10.25. The more important data (e.g., 80% of the total data) are allocated to the first connection, which has a higher transmission priority with MAC retransmission; the remaining data are then allocated to the second connection. This means that the BS need only retransmit the more important data when the actual transmission bandwidth is smaller than the expected bandwidth.

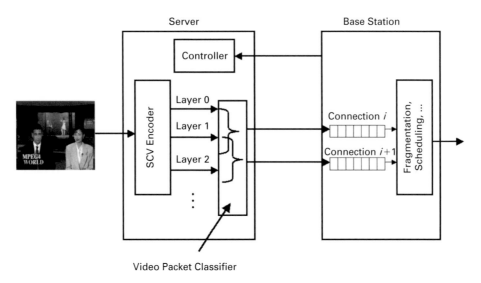

Figure 10.25 To support multiple connections between the BS and MS, the data sent to the BS from the streaming server are allocated into several connections according to their importance levels [73] (© IEEE 2007).

10.4.4.2 The controls in the base station

Upon receiving the packets from the streaming server, the WiMAX base station stores those packets as MAC service data units (MSDUs) in the queues. These MSDUs have been categorized into different levels of importance in the streaming server, therefore they are treated differently depending on the service types and MAC controls. The base station is responsible for collecting the downlink (DL) channel condition and translating the information to the available bandwidth resources. Besides, the base station (BS) also supports multiple connections, with flexibility in the provision of different QoS and priorities to different connections. Some important issues associated with the base station are summarized below.

(1) *RF bandwidth estimation and bandwidth report* Depending on the buffer size in the mobile stations (MSs) and BS, the BS should request the target streaming bitrate for MSs from the server and report the average bandwidth to the server during the report period, which is set to be 0.5 second. More specifically, for each WiMAX frame, the BS can collect the reported DL channel information (e.g., in terms of SNR) from the channel quality indicator (CQI) in the uplink subframe. Then the BS can decide the proper modulation and coding scheme for each DL burst according to the adaptive modulation and coding (AMC) mapping of the OFDMA slot, as defined by the 802.16e standard [74]. By combining the information from the AMC scheme and the remaining capacity of the burst profile, the BS can calculate the available bandwidth for the streaming service. Finally, the BS can calculate the average bandwidth available for the streaming service over the duration of a report period and then report it to the server.
(2) *Multiple connections* In a WiMAX system, the service flow is mapped to a connection in the MAC layer and each MAC connection relates to a set of QoS parameters

according to its service type. The 802.16 MAC is flexible and allows multiple connections for an MS at a given time. Because the streaming server adopts scalable video coding and can decide which video packet is important and which is not, the BS can thus establish two or more connections with the server. As shown in Figure 10.25, a two-connection scenario is adopted: the first connection is for the more important video packets and is given higher priority.

(3) *ARQ and MAC retransmission*   Medium access control retransmission can be used to increase robustness and hence decrease the video packet loss rate. Since the first connection is for the more important video packets, it should be managed as an automatic-repeat-request (ARQ)-enabled connection, while the other connections can be optional in support of MAC retransmission, depending on the tradeoff between robustness and resource efficiency.

(4) *Handover control*   Handover control needs to be considered carefully since 802.16e supports mobility. Several available enhanced MAC handover algorithms for 802.16e [78] [79] have been proposed, mainly to reduce the handover gap and so avoid packet loss.

10.4.4.3 The tasks of the mobile station

The mobile station (MS) monitors the downlink subframe and reports the downlink channel condition to the BS through the CQI channel feedback in the uplink subframe. It also feeds back the ACK information for each received ARQ block.

10.4.5   Scalable video multicast over WiMAX

The IEEE 802.16e standard and the associated WiMAX forum are expected to provide a promising last-mile wireless broadband technology. Owing to the dynamically changing quality of the signal received by a WiMAX subscriber station (SS), an intelligent approach to managing the wireless infrastructure multicast resource is the first key step to successful delivery.

A utility-based resource allocation scheme is adopted for efficient scalable video multicasting [75] [76]. Given N subscriber stations $\{SS_n, n = 1, \ldots, N\}$, the goal is to adjust dynamically the burst profile for each multicast layer, corresponding to each scalable video layer, in order to maximize the total utility $U = \sum_n U_n$ subject to the total resource B pre-allocated to the MBS at the BS. For each multicast data layer, a utility $u_{c,l}$ is assigned to the cth video and lth layer. Therefore U_n equals $\sum_{c(n),l(n)} u_{c,l}$, the sum of the the utilities received by all the layers at SS_n. A representative example of utility assignment for layered map or traffic information and surveillance videos used in the intelligent transportation system (ITS) is shown in Table 10.3, which suggests that the lower layers are more important to users. The ITS process first selects proper burst profiles serving base layers to *all* SSs subscribing the channel. For the enhancement layers, we assign their burst profiles according to the marginal utility $\Delta U / \Delta B_c$, the ratio of the total utility gain and the extra resources consumed. Depending on the number of SSs in good and poor channel conditions, ΔU can vary and the best value among all layers will be chosen and served until B is fully used up. This effective resource allocation should be operated at every frame for best channel usage. After the process, the resources allocated for traffic multicast service are optimized at the WiMAX BS, as illustrated in Figure 10.26 for a downlink subframe. The

Table 10.3 A representative example of utility assignment for layered map or traffic information and surveillance videos used in the intelligent transportation system (ITS)

Type	Layers	Contents	Utility
	Base layer (TB)	Coarse map with summarized traffic conditions on specific locations	1
Type I: map and traffic information	Enhancement layer 1 (TE1)	Detailed map, buildings, and detailed traffic parameters	0.75
	Enhancement layer 2 (TE2)	Local information service (hotels, restaurants, etc.)	0.5
Type II: surveillance videos	Base layer (VB)		1
	Enhancement layer (VE)	Layer-encoded videos	0.5

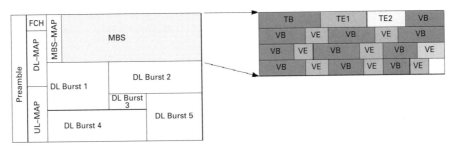

Figure 10.26 On the left, a WiMAX downlink subframe and on the right a detailed MBS zone. The MBS zone is further divided into burst profiles for each layer.

multicast and broadcast service (MBS) zone, shown on the right, is allocated for the multicast contents. The base layers of the traffic and video (TB, VB) are usually encoded in more robust modulations and cost more resources; the enhancement layers of the traffic and video (TE1, TE2, and VE) are intended for faster modulations. Therefore, at the client side, all SSs receive at least the base layer of the channels subscribed while some with good channel quality can receive more enhancement layers.

References

[1] M. Etoh and T. Yoshimura, "Advances in wireless video delivery," *Proc.* IEEE, 93(1): 111–122, January 2005.

[2] T. Stockhammer, M. M. Hannuksela, and T. Wiegand, "H.264/AVC in wireless environments," *IEEE Trans. Circuits Syst. Video Technol.*, 13(7): 657–673, July 2003.

[3] Y. Wang and Q.-F. Zhu, "Error control and concealment for video communication: a review," *Proc. IEEE*, 86(5): 974–997, May 1998.

[4] R. Yavatkar and N. Bhagawat, "Improving end-to-end performance of TCP over mobile internetworks," in *Proc. IEEE Workshop on Mobile Computing Systems and Applications*, pp. 146–152, December 1994.

[5] S. Mascolo, C. Casetti, M. Gerla, M. Y. Sanadidi, and R. Wang, "TCP Westwood: bandwidth estimation for enhanced transport over wireless links," in *Proc. ACM Mobicom 2001*, Rome, July 16–21.

[6] C.-L. Lee, C.-F. Liu, and Y.-C. Chen, "On the use of loss history for performance improvement of TCP over wireless networks," *IEICE Trans. Commun.*, 2457–2467, November 2002.

[7] H. Elaarag, "Improving TCP performance over mobile networks," *ACM Computing Surveys*, 34 (3): 357–374, September 2002.

[8] S. Biaz and N. Vaidya, "Discriminating congestion losses from wireless losses using inter-arrival times at the receiver," in *Proc. IEEE Symp. on Application-Specific Systems and Software Engineering and Technology (ASSET)*, pp. 10–17, March 1999.

[9] Y. Tobe, Y. Tamura, A. Molano, S. Ghost, and H. Tokuda, "Achieving moderate fairness for UDP flows by path-status classification," in *Proc. IEEE LCN2000*, Tampa FL, pp. 252–261 November 2000.

[10] S. Cen, P. C. Cosman, and G. M. Voelker, "End-to-end differentiation of congestion and wireless losses," *IEEE Trans. on Networking*, 11(5): 703–717, October 2003.

[11] H.-F. Hsiao, A. Chindapol, J. Ritcey, Y.-C. Chen, and J.-N. Hwang, "A new multimedia packet loss classification algorithm for congestion control over wired/wireless channels," in *Proc. ICASSP IEEE*, March 2005.

[12] L. Rizzo and L. Vicisano, "A reliable multicast data distribution protocol based on software FEC techniques," in *Proc. IEEE Workshop on High-Performance Communication Systems (HPCS'97)*, pp. 116–125, 1997.

[13] L. Rizzo, "Effective erasure codes for reliable computer communication protocols," *ACM Computer Commun. Rev.*, 27(2): 24–36, April 1997.

[14] R. E. Blahut, *Theory and Practice of Error Control Codes*, Addison Wesley, 1984.

[15] H.-F. Hsiao and J.-N. Hwang, "The dynamics and stabilities of layered congestion control for multimedia streaming," in *Proc. IEEE ISCAS 2005*, Kobe, May 2005.

[16] E. Elliot, "Estimates of error rates for codes on burst-noise channels," Bell System Technical Journal, 1963.

[17] A. Willi, "A new class of packet- and bit-level models for wireless channels," in *Proc. IEEE Symp. on Personal, Indoor and Mobile Radio Commun.*, September 2002.

[18] J.-P. Ebert and A. Willig, "A Gilbert–Elliot bit error model and the efficient use in packet level simulation," TKN Technical Report TKN-99-002, Technical University of Berlin, Telecommunication Networks Group, March 1999.

[19] Q. Zhang, G. Wang, Z. Xiong, J. Zhou, and W. Zhu, "Error robust scalable audio streaming over wireless IP networks," *IEEE Trans. Multimedia*, 6(6): 897–909, December 2004.

[20] T. Wiegand, G. Sullivan, G. Bjontegaard, and A. Luthra, "Overview of the H.264/ AVC video coding standard," *IEEE Trans. Circuits Syst. Video Technol.*, 13(7): 560–576, July 2003.

[21] S. Wenger, M. Hannuksela, T. Stockhammer, M. Westerlund, and D. Signer, "RTP payload format for H.264 video," Internet Eng. Task Force, IETF draft-ietf-avt-rtp-h264-03.txt, October 2003.

[22] Q. Zhang, Z. Ji, W. Zhu, and Y.-Q. Zhang, "Power-minimized bit allocation for video communication over wireless channels," *IEEE Trans. Circuits Syst. Video Technol.*, 12(6): 398–410, June 2002.

[23] Y. Eisenberg, C. Luna, T. Pappas, R. Berry, and A. Katsaggelos, "Joint source coding and transmission power management for energy efficient wireless video communications," *IEEE Trans. Circuits Syst. Video Technol.*, 12(6): 411–424, June 2002.

[24] I.-M. Kim, H.-M. Kim, and D. Sachs, "Power-distortion optimized mode selection for transmission of VBR videos in CDMA systems," *IEEE Trans. Commun.*, 51(4): 525–529, April 2003.

[25] Y. Wang, G. Wen, S. Wenger, and A. K. Katsaggelos, "Error resilient video coding techniques," *IEEE Signal Process. Mag.*, 17, 61–82, July 2000.

[26] W. M. Lam, A. R. Reibman, and B. Liu, "Recovery of lost or erroneously received motion vectors," in *Proc. ICASSP*, Minneapolis, MN, Vol. 15, pp. 417–420, April 1993.

[27] Z. Ji, Q. Zhang, W. Zhu, J. Lu, and Y.-Q. Zhang, "Joint power control and source-channel coding for video communication over wireless networks," in *Proc. IEEE Vehicular Technology Conf.* (VTC 2001), Vol. 3, pp. 1658–1662, October 2001.

[28] P. Frossard, J. C. De Martin, M. Reha Civanlar, "Media streaming with network diversity," *Proc. IEEE*, 96(1): 39–53, January 2008.

[29] H. Song, B. C. Kim, J. Y. Lee, and H. S. Lee, "IEEE 802.11-based wireless mesh network testbed," in *Proc. IEEE Mobile and Wireless Communications Summit*, pp. 1–5, July 2007.

[30] R. Bruno, M. Conti, and E. Gregori, "Mesh networks: commodity multihop ad hoc networks," *IEEE Commun. Mag.*, 371: 123–131, March 2005.

[31] "MIT Roofnet," http://pdos.csail.mit.edu/roofnet/.

[32] Networking Research Group, Microsoft Research, http://research.microsoft.com/mesh/.

[33] V. Navda, A. Kashyap, and S. R. Das, "Design and evaluation of iMesh: an infrastructure-mode wireless mesh network," in *Proc. 6th IEEE Int. Symp. on the World of Wireless Mobile and Multimedia Networks*, pp. 164–170, June 2005.

[34] P. Gupta and P. Kumar, "The capacity of wireless networks," *IEEE Trans. Inf. Theory*, 46: 377–404, March 2000.

[35] J. Li, C. Blake, D. D. Couto, H. Lee, and R. Morris, "Capacity of ad hoc wireless networks," in *Proc. 7th Int. Conf. on Mobile Comput. Network.*, Rome, pp. 61–69, 2001.

[36] I. Akyildiz, X. Wang, and W. Wang, "Wireless mesh networks: a survey," *Comput. Networks*, 47: 445–487, 2005.

[37] T.-J. Tsai and J.-W. Chen, "IEEE 802.11 MAC protocol over wireless mesh networks: problems and perspectives," in *Proc. IEEE Int Conf. on Advanced Information Networking and Applications (AINA'05)*, Vol. 2, pp. 60–63, March 2005.

[38] A. Acharya, A. Misra, and S. Bansal "MACA-P: a MAC for concurrent transmissions in multi-hop wireless networks," in *Proc. IEEE PerCom'03*, March 2003.

[39] H. Zhu, G. Cao, A. Yener, A. D. Mathias, A.D.; "EDCF-DM: a novel enhanced distributed coordination function for wireless ad hoc networks," in *Proc. IEEE Int. Conf. on Communications*, Vol. 7, pp. 3886–3890, June 2004.

[40] L. Romdhani, N. Qiang, T. Turletti, "Adaptive EDCF: enhanced service differentiation for IEEE 802.11 wireless ad-hoc networks," in *Proc. IEEE Conf. on Wireless Communications and Networking*, (WCNC 2003), Vol. 2, pp. 1373–1378, March 2003.

[41] S. Gray, D. Kotz, C. Newport *et al.* "Outdoor experimental comparison of four ad hoc routing algorithms," in *Proc. 7th, ACM Int. Symp. Model., Anal. Simul. Wireless Mobile Syst. (MSWiM)*, Venice, pp. 220–229, December 2004.

[42] D. D. Couto, S. Aguayo, B. Chambers, and R. Morris, "Performance of multihop wireless networks: shortest path is not enough," *ACM SIGCOMM Comput. Commun. Rev.*, 33: 83–88, January 2003.

[43] R. Draves, J. Padhye, and B. Zill, "Comparison of routing metrics for static multi-hop wireless networks," in *Proc. ACM Annual Conf. on Special Interest Group Data Commun. (SIGCOMM)*, pp. 133–144, August 2004.

[44] K. Rojviboonchai, F. Yang, Q. Zhang, H. Aida, and W. Zhu, "AMTP: a multipath multimedia streaming protocol for mobile ad hoc networks," in *Proc. IEEE ICC*, Vol. 2, pp. 1246–1250, May 2005.

[45] Y. J. Liang, J. G. Apostolopoulos, and B. Girod, "Analysis of packet loss for compressed video: does burst-length matter?" in *Proc. IEEE ICASSP*, Vol. 5, pp. 684–687, April 2003.

[46] A. F. da Conceiqdo, J. Lit, D. A. Florenciot, and F. Kon, "Is IEEE 802.11 ready for VoIP?" in *Proc. IEEE Multimedia Signal processing Workshop*, Victoria BC, pp. 108–113, October 2006.

[47] H. Velayos and G. Karlsson, "Techniques to reduce IEEE 802.1lb handoff time," in *Proc. IEEE Int. Conf. on Communications (ICC)*, June 2004.

[48] I. Ramani and S. Savage, "SyncScan: practical fast handoff for 802.11 infrastructure networks," in *Proc. IEEE INFOCOM*, Miami FL, March 2005.

[49] B. Rong, Y. Qian, and H.-H. Chen, "An enhanced SIP proxy server for wireless VoIP in wireless mesh networks," *IEEE Commun. Mag.*, 46(1): 108–113, January 2008.

[50] "Common open policy service (COPS) protocol," IETF RFC 2748, http://tools.ietf.org/html/rfc2748.

[51] F. A. Tobagi, "Voice over IP: the challenges behind the vision," in *Proc. Asilomar Conf. on Signals, Systems and Computers*, November 2004.

[52] X. Ling, Y. Cheng, X. (Sherman) Shen, and J. W. Mark, "Voice capacity analysis of WLANs with channel access prioritizing mechanisms," *IEEE Commun. Mag.*, 46(1): 82–89, January 2008.

[53] C.-W. Huang, S. Sukittanon, J. A. Ritcey, A. Chindapol, and J.-N. Hwang, "An embedded packet train and adaptive FEC scheme for VoIP over wired/wireless IP networks," in *Proc. IEEE Int Conf. on ASSP*, Toulouse, May 2006.

[54] "Coding of speech at 8 kbit/s using conjugate-structure algebraic-code-excited linear prediction (CS-ACELP)," ITU-T Recommendation G.729, http://www.itu.int/rec/T-REC-G.729/e.

[55] N. Hu and P. Steenkiste, "Evaluation and characterization of available bandwidth probing techniques," *IEEE J. Selected Areas in Commun. (JSAC)*, 21(6): 879 – 894, August 2003.

[56] Q. Liu, J. Yoo, B.-T. Jang, K. Choi and J.-N. Hwang, "A scalable videoGIS system for GPS-guided vehicles," *Signal Process.: Image Commun.*, 20(3): 205–208, March 2005.

[57] "ICFA SCIC network monitoring report," Standing Committee on Inter-Regional Connectivity (SCIC), of the International Committee for Future Accelerators (ICFA), January 2004, http://www.slac.stanford.edu/xorg/icfa/icfa-net-paper-jan04/.

[58] C.-W. Huang and J.-N. Hwang, "An embedded packet train and adaptive FEC scheme for effective video adaptation over wireless broadband networks," in *Proc. 15th Int. Packet Video Workshop*, Hangzhou, April 2006.

[59] H. Schwarz, D. Marpe, and T. Wiegand, "Overview of the scalable video coding extension of the H.264/AVC standard," *IEEE Trans. Circuits Syst. for Video Technol.*, 17(9): 1103–1120, September 2007.

[60] "Advanced video coding for generic audiovisual services," ITU-T Rec. H.264 and ISO/IEC 14496–10 (MPEG-4 AVC), ITU-T and ISO/IEC JTC 1, Version 8 (including SVC extension), consented to in July 2007.

[61] J. She, F. Hou, P.-H. Ho, and L.-L. Xie, "IPTV over WiMAX: key success factors, challenges, and solutions," *IEEE Commun. Mag.*, 45(8): 87–93, August 2007.

[62] C. Cicconetti *et al.*, "Quality of service support in IEEE 802.16 networks," *IEEE Network*, 20 (2): 50–55, March 2006.

[63] T. M. Cover, "Broadcast channels," *IEEE Trans. Info. Theory*, 18: 2–14, January 1972.

[64] S. Bopping and J. M. Shea, "Superposition coding in the downlink of CDMA cellular systems," in *Proc. IEEE Wireless Communications and Networking Conf.*, Vol. 4 pp. 1978–83, April 2006.

[65] M. Heusse, F. Rousseau, G. Berger-Sabbatel, and A. Duda, "Performance anomaly of 802.11b," in *Proc. IEEE Conf. on Computer Communications (Infocom)*, 2003.

[66] A. Kamerman and L. Monteban, "WaveLAN II: a high-performance wireless LAN for the unlicensed band," *Bell Laboratories Technical Journal*, 2(3): 118–133, 1997.

[67] M. Loiacono, J. Rosca, and W. Trappe, "The snowball effect: detailing performance anomalies of 802.11 rate adaptation," in *Proc. IEEE Global Telecommunications Conf. (GLOBECOM)*, November 2007.

[68] C.-W. Huang, M. Loiacono, J. Rosca, and J.-N. Hwang, "Distributed cross layer congestion control for real-time video over WLAN," in *Proc. IEEE Int. Conf. on Communications (ICC'08)*, Beijing, May 2008.

[69] S. Golestani, " A self-clocked fair queueing scheme for broadband applications," in *Proc. of the IEEE Conf. on Computer Communications (Infocom)*, 1994.

[70] N. Vaidya, A. Dugar, S. Gupta, and P. Bahl, "Distributed fair scheduling in a wireless LAN," *IEEE Trans. Mobile Computing*, 4(6): 616–629, 2005.

[71] J. Jun, P. Peddabachagari, and M. Sichitiu, "Theoretical maximum throughput of IEEE 802.11 and its applications," in *Proc. IEEE Int. Symp. on Network Computing and Applications*, pp. 249–256, 2003.

[72] D.-Y. Yang, T.-J. Lee, K. Jang, J.-B. Chang, and S. Choi, "Performance enhancement of multirate IEEE 802.11 WLANs with geographically scattered stations," *IEEE Trans. Mobile Computing*, 5, 906–919, 2006.

[73] H.-H. Juan, H.-C. Huang, C. Huang, and T. Chiang, "Cross-layer system designs for scalable video streaming over mobile WiMAX," in *Proc. IEEE WCNC 2007*, March 2007.

[74] "IEEE standard for local and metropolitan area networks, Part 16: Air interface for fixed and mobile broadband wireless access systems, Amendment 2: Physical and medium access control layers for combined fixed and mobile operation in licensed bands, and corrigendum 1," IEEE Standard 802.16e–2005 and IEEE Standard 802.16–2004/Cor 1–2005 (Amendment and Corrigendum to IEEE Standard 802.16–2004), 2006.

[75] J. Kim, J. Cho, and H. Shin, "Resource allocation for scalable video broadcast in wireless cellular networks," in *Proc. IEEE Int. Conf. on Wireless and Mobile Computing, Networking and Communications*, vol. 2, pp. 174–180, August 2005.

[76] W. H. Kuo, T. Liu, and W. Liao, "Utility-based resource allocation for layer-encoded IPTV multicast in IEEE 802.16 (WiMAX) wireless networks,"in *Proc. IEEE Int. Conf. on Communications*, pp. 1754–1759, June 2007.

[77] I. S. Reed and G. Solomon, "Polynomial codes over certain finite fields," *SIAM J. Applied Math.*, 8, pp. 300–304, 1960.

[78] S. Choi, G. H. Hwang, T. Kwon, A. R. Lim, and D. H. Cho, "Fast handover scheme for real-time downlink services in IEEE 802.16e BWA system," VTC 2005, IEEE 61st, Spring 2005.

[79] D. H. Lee, K. Kyamakya, and J. P. Umondi, "Fast handover algorithm for IEEE 802.16e broadband wireless access system," Wireless Pervasive Computing, 2006.

11 Digital rights management of multimedia

Owing to the proliferation of digitized media applications, such as e-Book, streaming videos, web images, shared music, etc., there is a growing need to protect the intellectual property rights of digital media and prevent illegal copying and falsification. This explains the strong demand for digital rights management (DRM), which is an access control technology that protects and enforces the rights associated with the use of digital content, such as multimedia data. The most important functions of DRM are to prevent unauthorized access and the creation of unauthorized copies of digital content, and moreover to provide a mechanism by which copies can be detected and traced (content tracking). Digital rights management is the most critical component of the intellectual property management and protection (IPMP) protocol widely promoted in the MPEG standards. Under the IPMP's scope, intellectual property (IP) is anything whose use owes the inventors or the owners some form of compensation. This could be a particular media application or it could be the technology used by the media. The management of IP involves storage and serving, appropriate authorization of use, and correct billing and tracking. The protection of IP prevents unauthorized use or misuse of the IP and can make legitimate use easy. According to [1], an effective DRM system should have the following four requirements.

(1) The DRM system must package the content to be protected in a secure manner.
(2) The DRM system must obtain the access conditions (license) specified by the owner of the protected content.
(3) The DRM system must determine whether the access conditions have been fulfilled.
(4) The DRM system must be tamper-proof to prevent or deter attempts to circumvent, modify, or reverse-engineer the security protocols used by the DRM system.

Secure packaging can ensure that all access to the protected content is governed by the DRM system. Secure packaging is usually accomplished by encryption [2] [3], where the content is scrambled and rendered unintelligible unless a decryption key is known. The DRM system provides the decryption key to unseal the package only when all the access conditions specified by the content owner are satisfied.

The access conditions, commonly called a "license," for DRM have three important components. The first is a rights expression language (REL) or protocol for the content owner to express flexibly the access conditions or rules. The second is a mechanism to associate or bind the access conditions to the content. This typically uses metadata or watermarks. Metadata is information that is stored alongside (but is separate from) the content, while watermarks are embedded directly into the content itself. The third is security, to prevent users from circumventing DRM by modifying the access conditions.

Having obtained the access conditions for protected content, the DRM system also requires a secure means for determining whether the access conditions have been fulfilled. The ease of satisfying this requirement depends greatly on the flexibility of the access conditions supported by the DRM system. Some representative access conditions are user restriction, expiration time, access counts, etc.

Security is essential to prevent users from circumventing the DRM by supplying false credentials. For DRM, it is essential that media delivery mechanisms prevent unauthorized access from the source to the consumption device. This is sometimes referred to as "end-to-end" security. One way to achieve "end-to-end" security is for devices to authenticate themselves prior to sending or receiving the media. Another important task for ensuring the security of the DRM system is to protect the media encryption keys during the key delivery process. One of the most widely used approaches for this mission is the public key infra-structure (PKI).

In this chapter, most techniques developed to satisfy the requirements (1)–(4) are first addressed. Then efforts towards the standardization of these techniques based on MPEG-21 are further discussed. The latter, formally referred to as the ISO/IEC 21000 multimedia framework, aims to address the issues of secure and interoperable end-to-end solutions for multimedia networking by standardizing interfaces and tools to facilitate the exchange of multimedia resources across heterogeneous devices, networks, and users.

11.1   A generic DRM architecture

Digital rights management systems are typically set up as client–server systems where a system receives requests and provides services to clients. Figure 11.1 shows a generic DRM architecture, which consists of three modules: client, license server, and content server. A user uses the client device to obtain, either by downloading in advance or live streaming online,

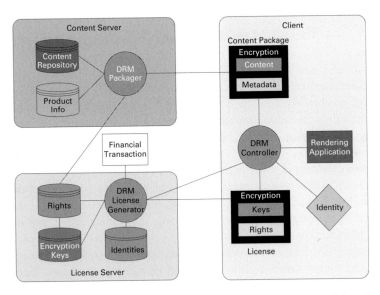

Figure 11.1 A generic DRM architecture, which consists of three modules, client, license server, and content server.

multimedia content packages from the content server, and then requests operation (e.g., view, play) of these contents. The DRM controller residing in the client device starts to collect information, such as the content ID, the user ID, and the requested rights, etc. When the necessary information has been collected, the DRM controller contacts the license generator residing in the license server; the license generator validates (authorizes) the forwarded content ID and user ID, and calls for the requested rights from the client device to be imposed on the downloaded or streamed multimedia content. The information being forwarded from client to DRM server requires a secure transaction, which is commonly carried out by a public key encryption (to be discussed later). The license generator goes on to extract the data encryption keys (commonly based on secret key encryption technologies such as DES or AES, to be discussed later) from the key repository corresponding to the downloaded or streamed multimedia content, then creates and sends the license to the client device; this is followed by the generation of a financial transaction, if necessary. The sending of the license from DRM server to client again calls for the help of public key encryption, i.e., sending a digitally signed license. The DRM controller at the client side, after receiving and opening the license from the DRM server, will extract data encryption keys from the received license to decrypt the downloaded or live streamed multimedia contents and will generate a financial transaction on the client side if necessary. The DRM controller then hands the decrypted content to the rendering (i.e., media decoding) application to be played back.

11.1.1   DRM content server

A content server is a computer system that provides content or media to devices that are connected to a communication system. The main function of a content server is to receive and process requests for media content, to set up a connection to the requesting device, and to manage media transfer during the communication session. As shown in Figure 11.1, the content server consists of three main functional modules: the content repository, the product information and the DRM packager.

The *content repository* includes three further components: the file server, the content management system (CMS), and the digital asset management (DAM) system. The contents stored in the file server include computer files, audio files, image and video files, electronic documents, and web content. The CMS system is used to manage the content stored in the file servers through the web and is deployed primarily for interactive use by a potentially large number of users. The DAM system is responsible for tasks and decisions on metadata annotating, metadata cataloging, and the storage and retrieval of digital assets such as images, animations, videos, and music. The DAM is also responsible for the protocol for downloading, renaming, backing up, grouping, archiving, optimizing, maintaining, and exporting of those digital assets.

The *product information* module contains the digital rights and product metadata associated with the digital assets, which denote contents, resources, and services in the digital domain. Digital rights refers to privileges for creating, distributing, using, and managing digital assets. Metadata, or so-called meta-tags, are commonly included in media files to describe the attributes of the media content. These attributes typically include the title of the media and media format details (media length and encoding formats) and may include additional descriptive information such as media category, actors, or related programs. With the advent of multimedia communication over IP networks, metadata standards have been developed for the management of all types of digital resources and for a variety of purposes, such as resource discovery, information retrieval, record keeping, rights management and copyright, etc.

A *DRM packager* has the task of combining the multimedia content (digital audio and/or video) and the associated metadata and encrypting the multimedia content with the keys provided by the license server to a media format or file that is sent from a content provider to a user or viewer of the content.

11.1.2 DRM license server

A license server is a system that maintains a list of license holders and their associated permissions to access licensed content. The main function of a license server is to confirm or provide the necessary codes or information elements to users or systems with the ability to provide access to licensed content. The license server may provide to client devices an encryption key, or other information, that enables a license holder to access the information they have requested. The encryption key is generated from an *encryption key server*, which is a system that can create, manage, and assign key values for an encryption system. A key is a word, algorithm, or program used to encrypt or decrypt a message that is created in a way that does not allow a person or system to discover the process used to create the keys.

License servers use licensing *rights* to determine whether users or devices have authorization to access data or media. Licensing rights are the processes and/or restrictions that are to be followed as part of a licensing agreement. The identity information of users or devices are stored in the *identity database*, which contains the critical information for authorization or authentication purposes during the user operation requests and license generation processes.

A *DRM license generator* is the coordinator of software and/or hardware that generates the license which allows users to access content through a DRM system. The DRM license generator receives requests to access digital content, obtains the necessary information elements (e.g., user ID, content ID, and requested rights), performs authentication (if requested) and retrieves the necessary encryption keys that allow for the decoding of digital media.

11.1.3 DRM client

A DRM client is either a hardware device or a software program that is configured to receive from the network the *content package* from the content server and request a DRM license from license server. The DRM *Controller* residing on the client side is the nerve center of the operation of DRM services. The controller is responsible for communicating with the license server for validating and authenticating the *identity* of the user and requesting the license from the license server so as to extract the data encryption keys for decrypting the downloaded or live streamed content packages, as well as to extract the license *rights* to enforce the usage conditions. Finally, the decrypted multimedia content can then be passed over to the rendering application for media data decoding and playback.

11.1.4 Separating content from license

In the common architecture for most modern DRM systems, the content and license are handled by the content server and license server separately, as shown in Figure 11.1. This strategy enables the content provider to separate rights clearance from content packaging and delivery. There are several reasons to support this separate server design, in which the license is not included in the same package as the content. First, there are normally multiple sets of rights for a given content, since there are multiple types of users with different access requirements. Second, one set of rights can be applicable to multiple content items, especially with regard to subscription to a library of content. Third, contents such as the live streaming media from networks may not reside on a user's device; this would make the license delivery with content more difficult.

11.2   Encryption

The encryption process uses a cryptographic algorithm scrambling confidential data (called plaintext) to an unintelligible form (called ciphertext) so as to keep it safe from external "eyes" and thus ensure a high level of security. In multimedia networking applications, the plaintext is normally referred as a block (say 128 bits) of compressed audio or video bitstream data. There are two major types of encryption: one is symmetric encryption (secret key cryptography, SKC) and the other is asymmetric encryption (public key cryptography, PKC). As shown in Figure 11.2 and Eq. (11.1) below, producing the ciphertext c requires the use of an encryption algorithm E operating on the plaintext m with the encryption key $k1$ and the inverse (decryption, D) process of recovering the original plaintext m from the ciphertext c using the decryption key $k2$ [2] [3]:

$$c = E(m, k1),$$
$$m = D(c, k2) = D(E(m, k1), k2) \tag{11.1}$$

For SKC, both keys ($k1$ and $k2$) are identical (i.e., $k1 = k2$) and are called the secret key, i.e., the same key that scrambles the message is used to unscramble the message. Therefore, in SKC both parties have to arrange in advance and communicate which key to use. One of the most difficult tasks in SKC is protection of the key from disclosure during the key exchange process. However, for PKC two different keys, which are mathematically related, are used ($k1 \neq k2$); one is called the public key and the other is called the private key. In PKC, if one key (either $k1$ or $k2$) is used for encryption of the plaintext in an application (to be discussed later), then the other key has to be used for decryption. It is also assumed that only the owner of the private key knows the value of

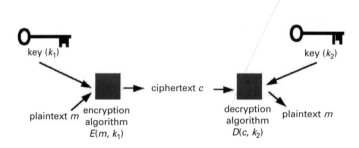

Figure 11.2 A cipher algorithm produces the ciphertext using an encryption key ($k1$), and the inverse process of recovering the original plaintext from the ciphertext is based on the decryption key ($k2$).

that key and that the private key is always safely stored in the owner's computers or devices without being exchanged or delivered outside.

11.2.1   Secret key cryptography (SKC)

Since only a single key is used for both encryption and decryption, secret key cryptography (SKC) is called symmetric encryption. In SKC the sender uses the key to encrypt the plaintext and sends the ciphertext to the receiver. The receiver applies the same key to decrypt the cyphertext and recover the plaintext. There are many SKC algorithms in use today. The *data encryption standard (DES)* is the most common SKC scheme used so far. This standard was designed by IBM in the 1970s and adopted by the National Bureau of Standards (NBS) of the USA, now the National Institute for Standards and Technology (NIST), in 1977 for commercial and unclassified government applications. The data encryption standard is a block-cipher approach that employs a 56-bit key on a 64-bit plaintext block to yield a 64-bit ciphertext. It is defined in American National Standard X3.92 [4] and three Federal Information Processing Standards (FIPS), FIPS 46-3, FIPS 74, and FIPS 81. To strengthen the protection capability of DES, an important variant, Triple-DES (3DES), was proposed. The 3DES variant, which is also defined in FIPS 46-3, employs up to three different 56-bit keys and makes three encryption or decryption passes over the plaintext block.

The DES standerd helped to focus and unify the cryptographic research community and is still a widely used cryptosystem. Owing to the increases in computing power, the protection capabilities offered by DES have been weakened considerably. More specifically, the Electronic Frontier Foundation in 1998 built a special-purpose machine, called Deep Crack, which consists of 27 boards each containing 64 chips and is capable of testing 90 billion keys a second, it can recover the key for a message encrypted with DES in about four days. Therefore, in 1997, NIST initiated a very public and long process to develop a new secure cryptosystem for US government applications. The result, the advanced encryption standard (AES) [5], became the official successor to DES in December 2001. The advanced encryption standard adopts the Rijndael SKC scheme, which is a block-cipher algorithm designed by the Belgian cryptographers Joan Daemen and Vincent Rijmen that can use a variable block length and key length; the latest specification allows any combination of key lengths of 128, 192, or 256 bits and blocks of length 128, 192, or 256 bits. For example, FIPS 197 describes a 128-bit AES block cipher employing a 128-bit, 192-bit, or 256-bit key.

In addition to the DES, 3DES, and AES standards, there are some other cryptography algorithms available internationally. CAST (IETF RFC 2144 [6]), named after its developers, Carlisle Adams and Stafford Tavares, is a DES-like substitution–permutation cryptography algorithm. CAST-128 employs a 128-bit key operating on a 64-bit block and CAST-256 is an extension of CAST-128, using a 128-bit block size and a variable-length (128-, 160-, 192-, 224-, or 256-bit) key. CAST is available internationally and CAST-256 was one of the round-1 algorithms in the advanced encryption standard (AES) development process.

The International Data Encryption Algorithm (IDEA) is another secret key cryptogram, designed by Xuejia Lai and James Massey of ETH Zurich in 1991 [7] . It is a 64-bit block cipher using a 128-bit key. This algorithm is patented in the USA and in most of the

European countries, the patent being held by Ascom-Tech. It is freely available for non-commercial use; it was used in Pretty Good Privacy (PGP) v2.0, which is a high-security encryption application for MS-DOS, Unix, VAX/VMS, and other computers.

Rivest Ciphers (RCs), named after Ron Rivest, is a series of SKC algorithms. More specifically, RC2 (IETF RFC 2268) is a 64-bit block cipher using variable-sized keys designed to replace DES. The code used in RC2 has never been made public although many companies have licensed RC2 for use in their products. Another block cipher supporting a variety of block sizes, key sizes, and number of encryption passes over the data is RC5 (IETF RFC 2040); RC6 [9] is an improvement on RC5 and was one of the five finalists submitted to the AES standard selection.

11.2.1.1 DES encryption/decryption algorithm

The data encryption standard (DES) algorithm transforms a 64-bit block of plaintext, using a 56-bit DES key K_0, into a 64-bit block of ciphertext. In this process each 64-bit block of plaintext goes through an initial permutation IP and this is followed by 16 rounds of key-dependent substitution and transposition operations based on 16 48-bit subkeys $\{K_i, i = 1, \ldots, 16\}$ derived from the initial 56-bit DES key K_0. Finally, the process ends by taking the inverse of the initial permutation (IP^{-1}) as shown in Figure 11.3.

Initial permutation and inverse of initial permutation The 64 bits of the input plaintext block to be enciphered are first subjected to the above-mentioned initial permutation IP, which is defined by the permutation matrix shown in Figure 11.4. For example, the permuted input has bit 58 of the input as its first bit, bit 50 as its second bit, and so on, with bit 7 as its last bit. The permuted input block is then the input to a 16-round key-dependent cipher computation to be described later. The output of the 16-round computation, called the preoutput, is subjected to a final permutation IP^{-1}, which is the inverse of the initial permutation IP, as defined by the permutation matrix shown in Figure 11.5. For example, the output of the algorithm has bit 40 of the preoutput block as its first bit, bit 8 as its second bit, and so on until we come to bit 25 of the preoutput block, which is the last bit of the output.

One round of key-dependent cipher operation Each round of key-dependent computation can be simply defined in terms of a cipher function F and a key-schedule function $K_i = KS(i, C_{i-1}, D_{i-1})$ as shown in Figure 11.6. In each round of cipher computation, the 64-bit matrix derived from the $(i-1)$th round is split into two blocks L_{i-1} and R_{i-1} having 32 bits each, L_{i-1} representing the first 32 bits and R_{i-1} representing the remaining 32 bits. From L_{i-1} and R_{i-1}, the new round of ciphered blocks L_i and R_i (also 32 bits each) can be computed, using Eq. (11.2):

$$L_i = R_{i-1},$$
$$R_i = L_{i-1} \oplus F(R_{i-1}, K_i), \tag{11.2}$$

where

$$K_i = KS(i, C_{i-1}, D_{i-1}),$$
$$C_i = Left\_Shift(i, C_{i-1}); \quad D_i = Left\_Shift(i, D_{i-1})$$

and $\oplus$ denotes the bitwise XOR operation.

The initial input blocks L_0 and R_0 are derived by splitting the output after the initial permutation given in Figure 11.4. The KS function in each round, however, takes a 56-bit transformed key (represented as a block pair $C_{i-1}D_{i-1}$) from the $(i-1)$th round to derive

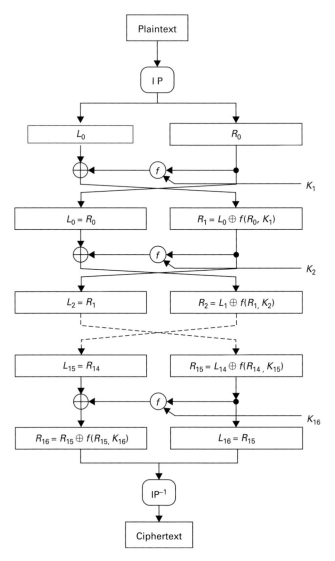

Figure 11.3 The DES algorithm transforms a 64-bit block of plaintext, using a 56-bit DES key, into a 64-bit block of ciphertext [4].

58	50	42	34	26	18	10	2
60	52	44	36	28	20	12	4
62	54	46	38	30	22	14	6
64	56	48	40	32	24	16	8
57	49	41	33	25	17	9	1
59	51	43	35	27	19	11	3
61	53	45	37	29	21	13	5
63	55	47	39	31	23	15	7

Figure 11.4 The initial permutation matrix IP shown in Figure 11.3.

40	8	48	16	56	24	64	32
39	7	47	15	55	23	63	31
38	6	46	14	54	22	62	30
37	5	45	13	53	21	61	29
36	4	44	12	52	20	60	28
35	3	43	11	51	19	59	27
34	2	42	10	50	18	58	26
33	1	41	9	49	17	57	25

Figure 11.5 The permutation matrix used in the final permutation IP^{-1} shown in Figure 11.3.

the 48-bit subkey K_i (see also Eq. 11.2), each block C_{i-1} and D_{i-1} being 28 bits each. From C_{i-1} and D_{i-1} the new round of transformed key C_i and D_i blocks (also 28 bits each) can be derived easily by (circularly) left-shifting each block separately, as shown in Figure 11.6. The initial transformed key block pair $C_0 D_0$ is equal to the originally chosen 56-bit DES key. The number of left shifts for each round is given in Table 11.1. For example, C_3 and D_3 are obtained from C_2 and D_2, respectively, by two left shifts and C_{16} and D_{16} are obtained from C_{15} and D_{15}, respectively, by one left shift.

Note that the resulting ciphertext is obtained by performing the inverse IP of the initial permutation, as defined in Figure 11.5, on the preoutput block $R_{16}L_{16}$. Several operations need to be carried out in the cipher F and key scheduling KS functions. More specifically, as seen in Figure 11.6, the expansion/permutation function defined in the cipher function takes a block of 32 bits as input and yields a block of 48 bits as output. This is achieved by using a permutation based on the E table shown in Figure 11.7: the expansion is done by repeating some bits during the permutation process. For example, the permuted result has bit 32 of the input as its first bit and also its 47th bit.

Similarly, as seen in Figure 11.6, the permutation/contraction function defined in the KS function takes a block of 56 bits as input and yields a block of 48 bits as output. This is achieved by using a permutation based on the PC-2 table shown in Figure 11.8, where the contraction is done by skipping some bits during the permutation process. For example, bit 9 of the input is never used in the permutation and is skipped in the output.

The S-box (selection function) used in the cipher function F takes a number m of input bits and transforms them into a number n of output bits; an $m \times n$ S-box is implemented as a (fixed) lookup table, e.g., in DES a 6×4 S-box is used. More specifically, given an S-box function $S(\cdot)$ defined in DES and a block B of 6 bits, then $S(B)$ can be determined as follows: The first and last bits of B represent in base 2 a number in the range 0 to 3; let that number be i. The middle four bits of B represent in base 2 a number in the range 0 to 15; let that number be j. Look up in the table the 4-bit number in the ith row and jth column. It is a number in the range 0 to 15 and is the output $S(B)$ of this S-box in response to the input B. For example, given the S-box S_5 used in DES (see Figure 11.9), an input 011011 has outer bits **01** and inner bits 1101; the corresponding output would be 0101, because for input 011011 the row is 01, i.e., row 1 and the column is determined by 1101, i.e., it is column 13. In row 1 column 13 we have 5, so that the output is 0101.

As shown in Figure 11.6, the 48-bit output from the expansion/permutation operation is XORed with the 48-bit subkey to produce a different 48-bit data output to be processed by the S-box. Since a 6×4 S-box is used in DES, we need eight separate S-boxes to process these 48 bits, resulting in a 32-bit output which is further permuted using another

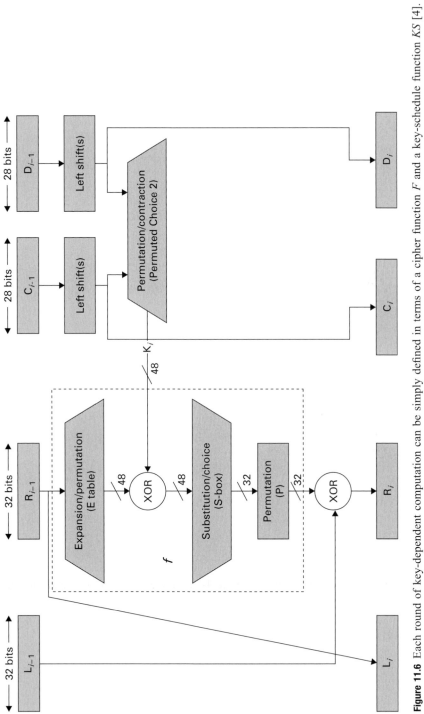

Figure 11.6 Each round of key-dependent computation can be simply defined in terms of a cipher function F and a key-schedule function KS [4].

Table 11.1 The number of left shifts of bits for
each round (see Figure 11.6)

Iteration number	Number of left shifts
1	1
2	1
3	2
4	2
5	2
6	2
7	2
8	2
9	1
10	2
11	2
12	2
13	2
14	2
15	2
16	1

32	1	2	3	4	5
4	5	6	7	8	9
8	9	10	11	12	13
12	13	14	15	16	17
16	17	18	19	20	21
20	21	22	23	24	25
24	25	26	27	28	29
28	29	30	31	32	1

Figure 11.7 The expansion/permutation function used in DES takes a block of 32 bits as input and
yields a block of 48 bits as output.

14	17	11	24	1	5
3	28	15	6	21	10
23	19	12	4	26	8
16	7	27	20	13	2
41	52	31	37	47	55
30	40	51	45	33	48
44	49	39	56	34	53
46	42	50	36	29	32

Figure 11.8 The PC-2 table for DES.

Row	Column Number															
No.	0	1	2	3	4	5	6	7	8	9	10	11	12	13	14	15
0	14	4	13	1	2	15	11	8	3	10	6	12	5	9	0	7
1	0	15	7	4	14	2	13	1	10	6	12	11	9	5	3	8
2	4	1	14	8	13	6	2	11	15	12	9	7	3	10	5	0
3	15	12	8	2	4	9	1	7	5	11	3	14	10	0	6	13

Figure 11.9 The S-box S_5 used in DES.

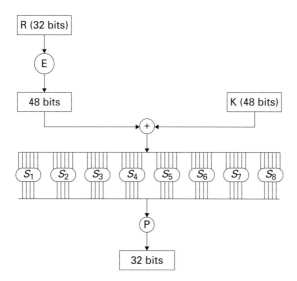

Figure 11.10 Since a 6×4 S-box is used in DES, eight separate S-boxes $\{S_1 - S_8\}$ are needed to process these 48 bits, resulting in a 32-bit output, which is further permuted using another permutation matrix, P [4].

permutation matrix, P, as shown in Figure 11.10. These eight S-boxes $\{S_1 - S_8\}$ are provided in Figure 11.11. The permutation matrix P is given in Figure 11.12.

DES decryption To decrypt a 64-bit ciphertext, exactly the same procedures as defined for encryption are used, as shown in Figure 11.3. The input is now the ciphertext and the 16 subkeys $\{K_i\}$ are used in the reverse order. That is, instead of applying K_1 for the first round apply K_{16}, then K_{15} for the second round, and so on, down to K_1. The final output will be the decrypted plaintext.

11.2.1.2 Triple DES

Let $E_K(I)$ and $D_K(I)$ represent the DES encryption and decryption of I using the DES key K. The triple DES encryption/decryption operation as defined in FIPS 46-3 is a compound operation of DES encryption and decryption operations based on a set of three keys $K1$, $K2$, $K3$. More specifically, in a triple DES encryption operation, the transformation of a 64-bit block I into a 64-bit block O is defined as follows [4] (see Figure 11.13):

$$O = E_{K3}(D_{K2}(E_{K1}(I))). \tag{11.3}$$

Similarly, in the triple DEA decryption operation, the transformation of a 64-bit block I into a 64-bit block O is defined as follows

$$O = D_{K1}(E_{K2}(D_{K3}(I))). \tag{11.4}$$

The triple DES standard specifies several keying options for the bundle $K1$, $K2$, $K3$:

(1) keying option 1 $K1$, $K2$, and $K3$ are independent keys;
(2) keying option 2 $K1$ and $K2$ are independent keys and $K3 = K1$;
(3) keying option 3 $K1 = K2 = K3$.

S_1

14	4	13	1	2	15	11	8	3	10	6	12	5	9	0	7
0	15	7	4	14	2	13	1	10	6	12	11	9	5	3	8
4	1	14	8	13	6	2	11	15	12	9	7	3	10	5	0
15	12	8	2	4	9	1	7	5	11	3	14	10	0	6	13

S_2

15	1	8	14	6	11	3	4	9	7	2	13	12	0	5	10
3	13	4	7	15	2	8	14	12	0	1	10	6	9	11	5
0	14	7	11	10	4	13	1	5	8	12	6	9	3	2	15
13	8	10	1	3	15	4	2	11	6	7	12	0	5	14	9

S_3

10	0	9	14	6	3	15	5	1	13	12	7	11	4	2	8
13	7	0	9	3	4	6	10	2	8	5	14	12	11	15	1
13	6	4	9	8	15	3	0	11	1	2	12	5	10	14	7
1	10	13	0	6	9	8	7	4	15	14	3	11	5	2	12

S_4

7	13	14	3	0	6	9	10	1	2	8	5	11	12	4	15
13	8	11	5	6	15	0	3	4	7	2	12	1	10	14	9
10	6	9	0	12	11	7	13	15	1	3	14	5	2	8	4
3	15	0	6	10	1	13	8	9	4	5	11	12	7	2	14

S_5

2	12	4	1	7	10	11	6	8	5	3	15	13	0	14	9
14	11	2	12	4	7	13	1	5	0	15	10	3	9	8	6
4	2	1	11	10	13	7	8	15	9	12	5	6	3	0	14
11	8	12	7	1	14	2	13	6	15	0	9	10	4	5	3

S_6

12	1	10	15	9	2	6	8	0	13	3	4	14	7	5	11
10	15	4	2	7	12	9	5	6	1	13	14	0	11	3	8
9	14	15	5	2	8	12	3	7	0	4	10	1	13	11	6
4	3	2	12	9	5	15	10	11	14	1	7	6	0	8	13

S_7

4	11	2	14	15	0	8	13	3	12	9	7	5	10	6	1
13	0	11	7	4	9	1	10	14	3	5	12	2	15	8	6
1	4	11	13	12	3	7	14	10	15	6	8	0	5	9	2
6	11	13	8	1	4	10	7	9	5	0	15	14	2	3	12

S_8

13	2	8	4	6	15	11	1	10	9	3	14	5	0	12	7
1	15	13	8	10	3	7	4	12	5	6	11	0	14	9	2
7	11	4	1	9	12	14	2	0	6	10	13	15	3	5	8
2	1	14	7	4	10	8	13	15	12	9	0	3	5	6	11

Figure 11.11 The eight separate S-boxes used in DES.

16	7	20	21
29	12	28	17
1	15	23	26
5	18	31	10
2	8	24	14
32	27	3	9
19	13	30	6
22	11	4	25

Figure 11.12 The permutation matrix P used after the S-box operations.

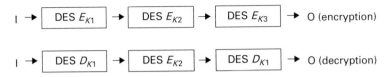

Figure 11.13 The encryption steps of the triple DES.

Input bit sequence	0	1	2	3	4	5	6	7	8	9	10	11	12	13	14	15	16	17	18	19	20	21	22	23	- - -
Byte number	0								1								2								- - -
Bit number in byte	7	6	5	4	3	2	1	0	7	6	5	4	3	2	1	0	7	6	5	4	3	2	1	0	- - -

Figure 11.14 A bit sequence in AES is first converted into a byte sequence with specific ordering [5].

11.2.1.3 Advanced encryption standard

The advanced encryption standard (AES) is also a block cipher and is based on the Rijndael algorithm. Unlike the bit-wise operations used in DES, the basic unit for processing in the AES algorithm is a byte, i.e., a sequence of eight bits treated as a single entity. The input, output, and cipher key bit sequences are all processed as arrays of bytes formed by dividing these sequences into groups of eight contiguous bits to form arrays of bytes. Figure 11.14 illustrates how a bit sequence is converted into a byte sequence and how the bytes and the bit ordering within bytes are defined for AES manipulation [5].

Byte-based mathematical operations All bytes in the AES algorithm are interpreted as finite-field (also called Galois-field GF(2^8)) elements, which can be added and multiplied, these operations being defined over a finite field and different from those used for numbers. For example, the addition of two elements in a finite field is achieved by "adding" the coefficients for the corresponding powers in the polynomials for the two elements. The addition is performed with the XOR operation denoted by $\oplus$, i.e., modulo 2. Consequently, the subtraction of two bytes over a finite field is identical to the addition of two bytes. Alternatively, the addition of two finite-field elements can be described as the modulo 2 addition of corresponding bits in the byte. For example,

$$(a_7a_6a_5a_4a_3a_2a_1a_0) \oplus (b_7b_6b_5b_4b_3b_2b_1b_0) = (c_7c_6c_5c_4c_3c_2c_1c_0),$$
$$(a_7x^7 + a_6x^6 + a_5x^5 + a_4x^4 + a_3x^3 + a_2x^2 + a_1x^1 + a_0)$$
$$\oplus (b_7x^7 + b_6x^6 + b_5x^5 + b_4x^4 + b_3x^3 + b_2x^2 + b_1x^1 + b_0)$$
$$= (c_7x^7 + c_6x^6 + c_5x^5 + c_4x^4 + c_3x^3 + c_2x^2 + c_1x^1 + c_0), \qquad (11.5)$$

where $c_7 = a_7 \oplus b_7$, $c_6 = a_6 \oplus b_6$, $c_5 = a_5 \oplus b_5$, $\cdots$, $c_0 = a_0 \oplus b_0$.

In a polynomial representation, multiplication in GF(2^8), denoted as $\otimes$, corresponds to the pre-multiplication of polynomials each coefficient in the product polynomial being accumulated using the $\oplus$ operation; then the resulting binary coefficient polynomial is taken modulo an irreducible polynomial of degree 8. A polynomial is irreducible if its only divisors are one and itself. For the AES algorithm, this irreducible polynomial is

$$m(x) = x^8 + x^4 + x^3 + x + 1. \qquad (11.6)$$

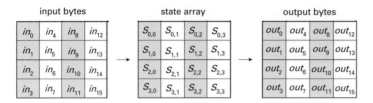

Figure 11.15 Internally, the AES algorithm's operations are performed on a two-dimensional array of bytes called the state [5].

Take, for example, the multiplication of two bytes, $87 = (01010111)$ and $131 = (10000011)$. In $GF(2^8)$ this can be performed as follows:

$$87 \otimes 131 = (x^6 + x^4 + x^2 + x + 1) \otimes (x^7 + x + 1)$$
$$= x^{13} + x^{11} + x^9 + x^8 + x^7 + x^7 + x^5 + x^3 + x^2 + x + x^6 + x^4 + x^2 + x + 1$$
$$= x^{13} + x^{11} + x^9 + x^8 + x^6 + x^5 + x^4 + x^3 + 1 \bmod (x^8 + x^4 + x^3 + x + 1)$$
$$= x^7 + x^6 + 1$$
$$= (11\,000\,001)$$
$$= 193.$$

The AES encryption algorithm Internally, the AES algorithm's operations are performed on a two-dimensional array of bytes called the state. The state consists of four rows of bytes, each containing Nb bytes, where Nb is the block length (in bits) divided by 32. For example, for a 128-bit block $Nb = 4$ and for a 192-bit block $Nb = 6$, etc. In the state array denoted by the symbol s each individual byte has two indices, a row number r in the range $0 \le r < 4$ and a column number c in the range $0 \le c < Nb$. This allows an individual byte of the state to be referred to as $s_{r,c}$. An input block of 128 bits, say, corresponds to an input array of 16 bytes arranged as a 4×4 array. This input array is copied into the state array. The cipher or inverse cipher operations are then conducted on this two-dimensional indexed state array, after which its final value is copied to the output array, as shown in Figure 11.15.

For the AES algorithm, the lengths of the input block, the output block, and the state are always 128 bits. This is represented by $Nb = 4$, which reflects the number of 32-bit words (the number of columns) in the state. However, the length of the cipher key K can be 128, 192, or 256 bits (corresponding to AES-128, AES-192, and AES-256). The key length is represented by $Nk = 4$, 6, or 8, which also reflects the number of 32-bit words (number of columns) in the cipher key. The number of rounds to be performed during execution of the algorithm is dependent on the key size. More specifically, the number of rounds is represented by Nr, where $Nr = 10$ when $Nk = 4$, $Nr = 12$ when $Nk = 6$, and $Nr = 14$ when $Nk = 8$.

At the start of the AES cipher, the 128-bit block input is copied to the state array as shown in Figure 11.15. After an initial AddRoundKey (ARK) operation, the state array is transformed by implementing a round function 10, 12, or 14 times depending on the key length. Each round includes Substitute Bytes (SubBytes), ShiftRows, MixColumns, and AddRoundKey operations; the final (10th) round differs slightly from the first $Nr - 1 (= 9)$ rounds (without including the MixColumns operation), as shown in Figure 11.16 for AES-128. The final state is then copied to the output.

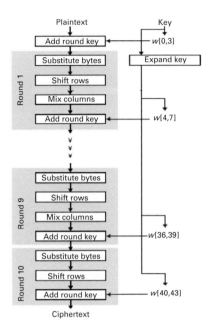

Figure 11.16 Each round of AES includes Substitute Bytes (SubBytes), ShiftRows, MixColumns, and AddRoundKey operations.

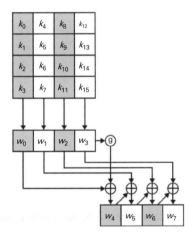

Figure 11.17 The four functions included in the round function are parameterized using a (Rijndael) key schedule that consists of a one-dimensional array of 4-byte words derived using the key expansion procedure.

Taking for example the case of AES-128, the four functions included in the round function are parameterized using a (Rijndael) key schedule that consists of a one-dimensional array of 4-byte words derived using the key expansion procedure shown in Figure 11.17. More specifically, the original 128-bit (four 32-bit words) AES key is taken and is expanded into an array of 44 32-bit words (i.e., there are 10 subkeys for each 128

Table 11.2 The 8-bit (Rijndael) S-box used to replace each input byte with an 8-bit substitution box S for which $s'_{ij} = S(s_{ij})$

										y						
(hex)	0	1	2	3	4	5	6	7	8	9	a	b	c	d	e	f
0	63	7c	77	7b	f2	6b	6f	c5	30	01	67	2b	fe	d7	ab	76
1	ca	82	c9	7d	fa	59	47	f0	ad	d4	a2	af	9c	a4	72	c0
2	b7	fd	93	26	36	3f	f7	cc	34	a5	e5	f1	71	d8	31	15
3	04	c7	23	c3	18	96	05	9a	07	12	80	e2	eb	27	b2	75
4	09	83	2c	1a	1b	6e	5a	a0	52	3b	d6	b3	29	e3	2f	84
5	53	d1	00	ed	20	fc	b1	5b	6a	cb	be	39	4a	4c	58	cf
x 6	d0	ef	aa	fb	43	4d	33	85	45	f9	02	7f	50	3c	9f	a8
7	51	a3	40	8f	92	9d	38	f5	bc	b6	da	21	10	ff	f3	d2
8	cd	0c	13	ec	5f	97	44	17	c4	a7	7e	3d	64	5d	19	73
9	60	81	4f	dc	22	2a	90	88	46	ee	b8	14	de	5e	0b	db
a	e0	32	3a	0a	49	06	24	5c	c2	d3	ac	62	91	95	e4	79
b	e7	c8	37	6d	8d	d5	4e	a9	6c	56	f4	ea	65	7a	ae	08
c	ba	78	25	2e	1c	a6	b4	c6	e8	dd	74	1f	4b	bd	8b	8a
d	70	3e	b5	66	48	03	f6	0e	61	35	57	b9	86	c1	1d	9e
e	e1	f8	98	11	69	d9	8e	94	9b	1e	87	e9	ce	55	28	df
f	8c	a1	89	0d	bf	e6	42	68	41	99	2d	0f	B0	54	bb	16

bits or each four 32-bit word). First, the original 128-bit key is copied into the first four words ($Subkey_0 = (w_0, w_1, w_2, w_3)$) then additional words are created with a loop. Every four new words constitute a new *Subkey* set. To expand a new word, say w_i, we need the values of the previous word w_{i-1} belonging to the same subkey and of the word four places back, i.e., w_{i-4}, in the previous subkey. Three out of 4 new words belonging to the same subkey can simply use XOR to combine these two words together, i.e., $w_i = w_{i-1} \oplus w_{i-4}$. The first word of each subkey should be created using $w_i = g(w_{i-1}) \oplus w_{i-4}$, where g represents a sequence of the following operations on w_{i-1}:

(1) a circular left shift of one byte;
(2) an 8-bit S-box *substitution* $b_{ij} = S(a_{ij})$ according to Table 11.2;
(3) an XOR with a predefined value from a round constant table, see Table 11.3, which depends on the round number;
(4) an XOR with w_{i-4}.

In operation (2), assuming that a_{ij} is represented as xy in hexadecimal format, the substitution value is determined by the intersection of the row in Table 11.2 with index x and the column with index y. For example, $S(01010011) = S(53) = ed = 11101101$. The S-box used in AES is derived from the multiplicative inverse of $GF(2^8)$, which is known to have good non-linearity properties.

The four round functions defined in an AES can be summarized as follows.

(1) *SubBytes* This is a non-linear substitution step where each byte is replaced with another according to a lookup table.
(2) *ShiftRows* This is a transposition step where each row of the state is shifted cyclically a certain number of steps.
(3) *MixColumns* This is a mixing operation, which operates on the columns of the state, combining the four bytes in each column.

Table 11.3 A round constant table in the subkey expansion of AES

Round	Round constant (hex)
1	01 00 00 00
2	02 00 00 00
3	04 00 00 00
4	08 00 00 00
5	10 00 00 00
6	20 00 00 00
7	40 00 00 00
8	80 00 00 00
9	1b 00 00 00
final	36 00 00 00

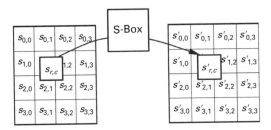

Figure 11.18 In the SubBytes step, each byte in the array is updated using the 8-bit substitution box S, where $s'_{ij} = S(s_{ij})$ [5].

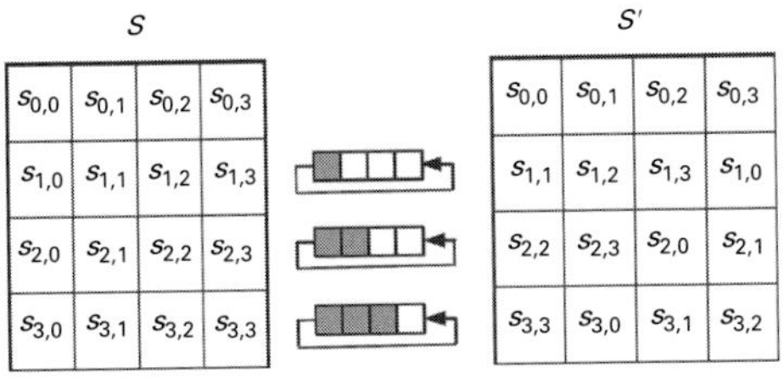

$$
\begin{array}{cc}
S & S' \\
\end{array}
$$

$s_{0,0}$	$s_{0,1}$	$s_{0,2}$	$s_{0,3}$		$s_{0,0}$	$s_{0,1}$	$s_{0,2}$	$s_{0,3}$
$s_{1,0}$	$s_{1,1}$	$s_{1,2}$	$s_{1,3}$		$s_{1,1}$	$s_{1,2}$	$s_{1,3}$	$s_{1,0}$
$s_{2,0}$	$s_{2,1}$	$s_{2,2}$	$s_{2,3}$		$s_{2,2}$	$s_{2,3}$	$s_{2,0}$	$s_{2,1}$
$s_{3,0}$	$s_{3,1}$	$s_{3,2}$	$s_{3,3}$		$s_{3,3}$	$s_{3,0}$	$s_{3,1}$	$s_{3,2}$

Figure 11.19 In the ShiftRows step, bytes in each row of the state are shifted cyclically to the left. The number of places each byte is shifted differs for each row [5].

(4) *AddRoundKey* In this operation, each byte of the state is combined using a round key; each round key is derived from the cipher key using a key schedule.

As shown in Figure 11.18, in the SubBytes step each byte in the array is updated using the 8-bit substitution S-box to give $s'_{ij} = S(s_{ij})$ using Table 11.2.

In the ShiftRows step, bytes in each row of the state are shifted cyclically to the left. The number of places each byte is shifted differs for each row as shown in Figure 11.19. More

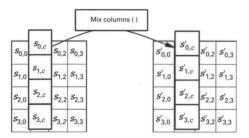

Figure 11.20 In the MixColumns step, each column is treated as a polynomial over GF (2^8) and is then multiplied with a fixed polynomial $c(x)$ [5].

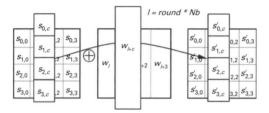

Figure 11.21 In the AddRoundKey step, each byte of the state is combined with a byte of the subkey of the current round using the XOR operation [5].

specifically, the first row is left unchanged; each byte of the second row is shifted one to the left; similarly, the third and fourth rows are shifted by offsets of two and three respectively.

In the MixColumns step (see Figure 11.20), each column is treated as a polynomial over GF(2^8) and is then multiplied by a fixed polynomial $c(x)$. The MixColumns step can also be viewed as multiplication by the following maximum distance separable (MDS) matrix in Rijndael's finite field, i.e.,

$$\begin{bmatrix} s'_0 \\ s'_1 \\ s'_2 \\ s'_3 \end{bmatrix} = \begin{bmatrix} 2 & 3 & 1 & 1 \\ 1 & 2 & 3 & 1 \\ 1 & 1 & 2 & 3 \\ 3 & 1 & 1 & 2 \end{bmatrix} \begin{bmatrix} s_0 \\ s_1 \\ s_2 \\ s_3 \end{bmatrix}$$

or (11.7)

$$s'_0 = 2s_0 + 3s_1 + s_2 + s_3,$$
$$s'_1 = s_0 + 2s_1 + 3s_2 + s_3,$$
$$s'_2 = s_0 + s_1 + 2s_2 + 3s_3,$$
$$s'_3 = 3s_0 + s_1 + s_2 + 2s_3.$$

Since this multiplication is done in Rijndael's Galois field, the addition and multiplication above need to follow byte-based mathematical manipulation under a finite field, discussed previously.

In the AddRoundKey step, each byte of the state is combined with a byte of the subkey of the current round using the XOR operation, as shown in Figure 11.21. For each round, a subkey $(w_i, w_{i+1}, w_{i+2}, w_{i+3})$ is derived from the main key using Rijndael's key expansion schedule (see Figure 11.17). The subkey then is added by combining each byte of the state with the

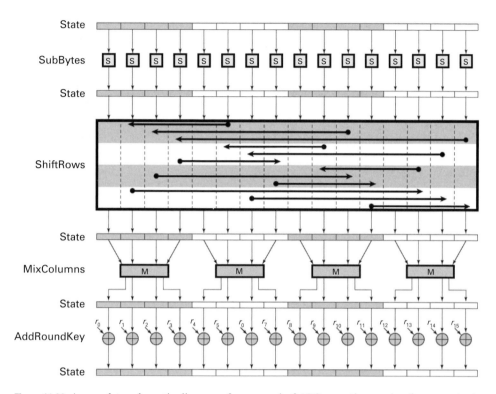

Figure 11.22 A complete schematic diagram of one round of AES operations, going from one 4×4 state to the next 4×4 state.

corresponding byte of the subkey using bit-wise XOR. A complete schematic diagram of one round of AES operations going from one 4×4 state to the next 4×4 state is shown in Figure 11.22.

The AES decryption algorithm The AES decryption algorithm can be implemented using an algorithm having the same structure as the encryption algorithm shown in Figure 11.16, i.e., it has the same sequence of transformations as the encryption algorithm but with each round operation replaced by its corresponding inverse, such as InvSubBytes, InvShiftRows, and InvMixColumns, as shown in Figure 11.23. Note that the AddRoundKey operation is the same for both encryption and decryption since it only involves the XOR operations with the subkeys. More specifically, the inverse S-box used in InvSubBytes is provided in Table 11.4.

The InvShiftRows operation is the inverse of the ShiftRows transformation. The bytes in the last three rows of the state are cyclically shifted to the right over different numbers of bytes (offsets) but the first row is not shifted as shown in Figure 11.24.

Finally, the InvMixColumns operation is the inverse of the MixColumns transformation, and this can be written as a matrix multiplication, i.e.,

$$
\begin{bmatrix} s_0' \\ s_1' \\ s_2' \\ s_3' \end{bmatrix} = \begin{bmatrix} 2 & 3 & 1 & 1 \\ 1 & 2 & 3 & 1 \\ 1 & 1 & 2 & 3 \\ 3 & 1 & 1 & 2 \end{bmatrix}^{-1} \begin{bmatrix} s_0 \\ s_1 \\ s_2 \\ s_3 \end{bmatrix} = \begin{bmatrix} 0e & 0b & 0d & 09 \\ 09 & 0e & 0b & 0d \\ 0d & 09 & 0e & 0e \\ 0b & 0d & 09 & 0e \end{bmatrix} \begin{bmatrix} s_0 \\ s_1 \\ s_2 \\ s_3 \end{bmatrix}, \qquad (11.8)
$$

where all the matrix elements are represented in hexadecimal format.

Table 11.4 The inverse S-box used in InvSubBytes [5]

										y						
(hex)	0	1	2	3	4	5	6	7	8	9	a	b	c	d	e	f
0	52	09	6a	d5	30	36	a5	38	bf	40	a3	9e	81	f3	d7	fb
1	7c	e3	39	82	9b	2f	ff	87	34	8e	43	44	c4	de	e9	cb
2	54	7b	94	32	a6	c2	23	3d	ee	4c	95	0b	42	fa	c3	4e
3	08	2e	a1	66	28	d9	24	b2	76	5b	a2	49	6d	8b	d1	25
4	72	f8	f6	64	86	68	98	16	d4	a4	5c	cc	5d	65	b6	92
5	6c	70	48	50	fd	ed	b9	da	5e	15	46	57	a7	8d	9d	84
x 6	90	d8	ab	00	8c	bc	d3	0a	f7	e4	58	05	b8	b3	45	06
7	d0	2c	1e	8f	ca	3f	0f	02	c1	af	bd	03	01	13	8a	6b
8	3a	91	11	41	4f	67	dc	ea	97	f2	cf	ce	f0	b4	e6	73
9	96	ac	74	22	e7	ad	35	85	e2	f9	37	e8	1c	75	df	6e
a	47	f1	1a	71	1d	29	c5	89	6f	b7	62	0e	aa	18	be	1b
b	fc	56	3e	4b	c6	d2	79	20	9a	db	c0	fe	78	cd	5a	f4
c	1f	dd	a8	33	88	07	c7	31	b1	12	10	59	27	80	ec	5f
d	60	51	7f	a9	19	b5	4a	0d	2d	e5	7a	9f	93	c9	9c	ef
e	a0	e0	3b	4d	ae	2a	f5	b0	c8	eb	bb	3c	83	53	99	61
f	17	2b	04	7e	ba	77	d6	26	e1	69	14	63	55	21	0c	7d

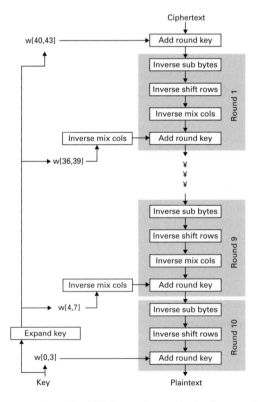

Figure 11.23 The AES decryption can be implemented using an algorithm that has the same structure as the encryption algorithm [5].

S S'

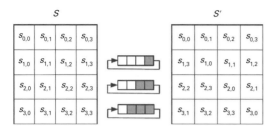

Figure 11.24 InvShiftRows is the inverse of the ShiftRows transformation.

11.2.2 Public key cryptography

With the great advances in computer CPU power, the computational cost of data encryption
has greatly reduced. One of the most challenging issues of secret key cryptography (SKC) is
the key exchange (or key distribution) problem, i.e., the problem of securely transmitting keys
to the users who need them. The classical, ineffective, solution to key distribution is based on
the use of a secure channel, which is not used for direct transmission of the plaintext or the
ciphertext messages because it is too slow or expensive. Diffie and Hellman in 1976 [10] and
independently Merkle in 1978 [11] proposed a radically different approach, called public key
cryptography (PKC) [12], to resolve the key distribution problem in SKC. Furthermore, PKC
also resolves the problem of digital signature, which would provide the recipient of a purely
digital electronic message with a way of demonstrating to other people that the message had
come from a particular person, just as a written signature on a letter allows the recipient to
trust the authenticity of the author and the content.

Public key encryption is based upon a key pair: one is called the public key (*Kpub*) and
the other private key (*Kpriv*). Moreover, of these two keys one is used for encryption and the
other for decryption. Even though these two related keys represent inverse operations, there
must be no easily computed method of deriving the private key from the public key. The
public key can then be made public without compromising the private key.

The RSA scheme, jointly contributed by the three researchers R. L. Rivest, A. Shamir,
and L. Adleman in 1978 [13], was the first and the most important contribution towards
making PKC a practical realization. The RSA implementation is based on the fact that it is
easy to generate two large primes and multiply them together but much more difficult to
factor the result. The product can therefore be made public as part of the encryption key
without compromising the factors which effectively constitute the decryption key. By
making each of the factors 100 digits long, the multiplication can be done in a fraction of a
second but factoring would require many years using the best-known algorithm.

11.2.2.1 The RSA public key cryptography
The RSA PKC design starts by finding two large prime numbers p and q, each about 100
decimal digits long. Let

$$n = pq \quad \text{and} \quad z = (p-1)(q-1), \tag{11.9}$$

where z denotes the number of integers between 1 and n that have no common factor with n.
Note that it is easy to compute z if the factorization of n is known, but computing z directly
from n is equivalent in difficulty to factoring n.

Choose a random integer e between 3 and z that has no common factors with z. Then find an integer d, which is co-prime with z and is the "inverse" of e modulo z, i.e.,

$$(de) \text{ modulo } z = 1 \tag{11.10}$$

The public key is equal to e with side information n available, and the private key is equal to d with side information n available [12].

RSA encryption Given a plaintext message m that is an integer between 0 and $n-1$ and the public key e, the ciphertext integer c can be formed:

$$c = m^e \text{ modulo } n. \tag{11.11}$$

In other words, raise m to the power e, divide the result by n, and let c be the remainder.

RSA decryption Using the secret key d, the plaintext m can be recovered as

$$m = c^d \text{ modulo } n. \tag{11.12}$$

Note that RSA PKC encryption and decryption both involve an exponentiation in modular arithmetic and that this can be accomplished in at most $2(\log 2n)$ multiplications mod n. A practical way to do this computation is given in [14]: to evaluate $y = a^x$, the exponent x is represented in binary form, the base a is raised to the first, second, fourth, eighth, etc. powers (each step involving only one squaring or multiplication), and an appropriate set of these is multiplied together to form y.

Cryptanalysis In order to determine the private key d, a cryptanalyst must factor the roughly 200-digit number n. This task would take many years with the best algorithm known today, using today's computers.

A simple RSA example To illustrate RSA PKC encryption and decryption, suppose that $p = 11$ and $q = 19$. Then $n = p \times q = 209$ and $z = (p-1) \times (q-1) = 180$. If the public key is picked as $e = 23$ $(23 < 180)$ then the private key is determined as $d = 47$ where d is co-prime with z, i.e., $\gcd(47,180) = 1$ and 47×23 modulo $180 = 1$. Given a plaintext message $m = 17$, we have

$$\text{encryption} \quad c = m^e \text{ modulo } n = 17^{23} \text{ modulo } 209 = 139,$$
$$\text{decryption} \quad m = c^d \text{ modulo } n = 139^{47} \text{ modulo } 209 = 17.$$

11.2.2.2 PKC for encryption and authentication

As previously discussed, the public and private keys e, d are inverse to each other and can be applied in two different ways: one way is called *public key encryption* and uses the public key to encrypt the plaintext, whose ciphertext can be later decrypted by the corresponding private key; the other way is called *public key authentication* and uses the private key to encrypt the plaintext, whose ciphertext can be later decrypted by the corresponding public key.

Public key encryption Public key encryption is used mainly in applications which allow a person to send a private message to someone and simply encrypt it with the recipient's public key, which is assumed to be available to the public. The received ciphertext can only be decrypted by the recipient's private key, which is assumed to be hidden securely the whole time (see Figure 11.25). Therefore even if this transmitted ciphertext is intercepted during delivery, no one can accurately decrypt it except the intended recipient, who owns the corresponding

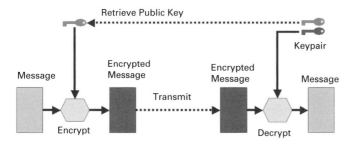

Figure 11.25 Public key encryption encrypts the message with the recipient's public key. The received ciphertext can only be decrypted by the recipient's private key [15].

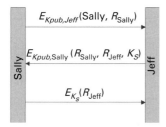

Figure 11.26 A secret-key-distribution protocol based on public key encryption.

private key. The mathematical formulation for this public key encryption is given as follows:

$$D(E(m, e), d) = m, \tag{11.13}$$

where the message m is normally not multimedia data; it can be the secret key we need to distribute to the receiver end or it can be a credit card number or your password to be transmitted and verified over the network.

A major reason why we do not use public key encryption for multimedia data is the computational cost associated with encryption and decryption, which is about two orders of magnitude higher than that of secret key encryption. Therefore multimedia data are mainly encrypted by SKC, while the key distribution of SKC is done by PKC. Figure 11.26 shows a more complete secret-key-distribution protocol based on public key encryption. To be identified and establish multimedia communication with Jeff, Sally first sends her identity, including also a challenge message, R_{Sally}, to Jeff using public key encryption with Jeff's public key. The challenge message is commonly in terms of a unique challenge cookie based on a local secret to prevent guessing. It is suggested that the cookie be a hashed local secret and a timestamp to provide uniqueness. Jeff decrypts the received request and sends back Sally's challenge message along with his own challenge message, R_{Jeff} and the secret key K_s to be shared by both sides, using public key encryption with Sally's public key. Sally decrypts the returned message and authenticates Jeff's identity on the basis of the returned challenge message R_{Sally}, that she originally sent to Jeff. Sally now sends back Jeff's challenge message using the attached secret key K_s. Jeff is now free to send all kinds of multimedia data to Sally using the securely distributed secret key K_s.

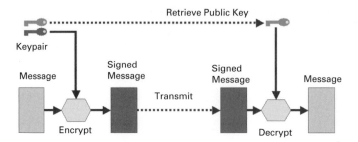

Figure 11.27 Whoever receives the ciphertext (the authenticated message) may verify the signature using the sender's public key, which is assumed to be publicly available [15].

Public key authentication Public key authentication is used mainly in applications which allow a person to sign his or her own message by encrypting the plaintext with his or her own private key. Whoever receives the ciphertext (the authenticated message) may verify the signature using the sender's public key, which is assumed to be available to the public (see Figure 11.27). The aim of public key authentication is to ensure that the received message was indeed sent by the sender who claims who he/she is and has not been "spoofed" by others. The mathematical formulation for this public key authentication is given as follows:

$$D(E(m, d), e) = m. \tag{11.14}$$

Since PKC requires high computational cost and the message (such as a long legal document) to be digitally signed may be quite long, a slightly modified scheme, based on public key authentication, called digital signature can be applied.

Digital signature A digital signature is the electronic analog of a written signature in that the digital signature can be used by a third party to determine that the entity named by the signature did in fact sign the information. In contrast with handwritten signatures, a digital signature also indicates proof that the information has not been changed since the signing. A digital signature is created by first creating a hash of the information, then encrypting the hash with the private key, and finally affixing the encrypted hash to the information.

Digital signatures are different from public key authentication, where the whole message is encrypted and is unintelligible to recipients unless they have the corresponding public key with which to decrypt it, from which authenticity can certainly be verified. However, the concept in digital signature is to retain the intelligibility of the message while attaching it to an additional signature to verify authenticity. Digital signatures have an advantage over written signatures because written signatures are the same whatever the message, whereas a digital signature of an electronic message is context dependent. However, there is a disadvantage associated with it, i.e., if a private key is stolen it could easily be used for forgery, while a human written signature cannot be easily stolen.

The fundamental idea behind digital signatures is shown in Figures 11.28 and 11.29. Instead of using a private key to encrypt the whole message, we encrypt only a very condensed fixed-length version of the whole message, called a *message digest h(m)*; this is created by applying a one-way hash function to the (full) message (see Eq. (11.15) below). The hash function is an efficient transformation of an arbitrary message to a hash value of fixed length, e.g., 128 bits or higher. Normally the hash value is much smaller than the original input, and it is difficult to reverse a one-way hash function. According to the collision freeness principle of hash functions, it is very difficult to find two messages that

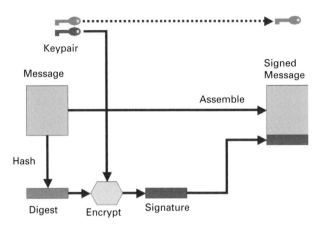

Figure 11.28 For a digital signature, the public key authentication using the private key is applied only to the hashed message digest, and the encrypted message digest is attached to the original message to form a signed message [15].

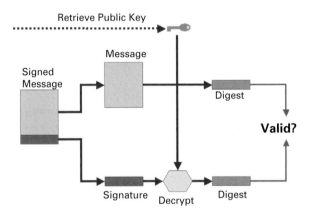

Figure 11.29 The recipient can apply the same hash function to the received (original) message to create a new digest and then compare this with the decrypted message digest attached to the signed message to verify the signed message [15].

result in the same hash values. Examples of widely used hash functions are message digest 5 (MD5, 128 bits) [16] and the secure hash algorithm (SHA, 160 bits) [17].

As shown in Figure 11.28, public key authentication using the private key is applied only to the hashed message digest, and the encrypted message digest is attached to the original message to form a signed message which still contains the original message. This is similar to the human written signature attached to the end of a letter or a legal document. Everybody can read the original message and the hashed message once decrypted with the sender's public key. If the recipient is doubtful about the authenticity of the original message, he or she can apply the same hash function to the received (original) message to create a new digest and compare it with the decrypted message digest attached to the signed message. Authenticity is proved if both message digests match, i.e., nothing changes in the original message (see Figure 11.29):

$$D(E(h(m),d),e) = h(m). \tag{11.15}$$

11.2.2.3 Public key infrastructure (PKI)

Even with the powerful capability of PKC in solving the issues of secret key distribution and the digital signatures, another challenge still exists to the large-scale deployment of PKC technologies, i.e., how to build a framework or services that can provide effective use of public keys. For example, the key generation and issuance system must be protected against disclosure and corruption, the public key must be available from a reliable and trusted source, how to ensure that the individual generating the keys is who he or she claims to be, and, moreover, there must be provision for the termination of disclosed or corrupted keys. This calls for the use of digital certificates. A public key digital certificate, called an X.509 certificate, is a data file that binds the identity of an entity to a public key. A digital certificate is like an electronic credit card that establishes your credentials, when you are doing business or other transactions, through a certification authority (CA). Certificates contain the name of the entity (the subscriber), the validity period start and end dates, the public key, the name of the CA that issued the certificate, and an identifier that links the certificate to the certificate policy that describes the system under which the certificate was issued. The information contained in the certificate is digitally signed by the issuing CA and the signature is considered part of the certificate. A CA is an entity trusted by one or more users to create and assign certificates. Certification authorities are responsible for issuing certificates, publishing certificates, and revoking certificates by placing them on certificate revocation lists.

A public key infrastructure (PKI) is the framework and services that provide for the generation, production, distribution, control, accounting, and destruction of public key digital certificates. A PKI integrates digital certificates, public key cryptography, and certification authorities into a total, enterprise-wide, network security architecture. A typical enterprise PKI encompasses: the issuance of digital certificates to individual users and servers; end user enrollment software; integration with certificate directories; tools for managing, renewing, and revoking certificates; and related services and support. Figure 11.30 shows a

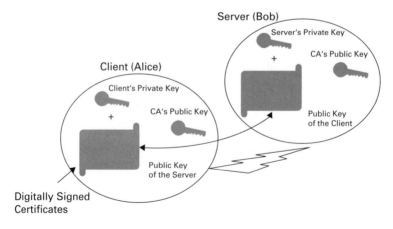

Figure 11.30 A simple example of how Alice and Bob (client and server) establish secure communication under PKI framework.

simple example of how, say, Alice and Bob (client and server) establish secure communication under PKI framework. The procedure is as follows.

(1) Alice creates a public and private key pair locally, e.g., via OpenSSL or MS Web Server Certificate Wizard. She then safeguards the created private key.
(2) Alice sends a certificate creation request (CSR) to a trusted CA, and then sends her public key to CA.
(3) The certification authority signs and issues a digital certificate and sends this back to Alice. Alice can share this digital certificate with the public.
(4) Bob downloads this trusted CA's signed digital certificate, which includes a public key. With this public key Bob can download Alice's digital certificate.
(5) Alice's digital certificate has been signed by the trusted CA (using the CA's private key); this can be verified only by the CA's corresponding public key, obtained earlier by Bob.
(6) In a similar way Bob sends his digital certificate signed by the CA to Alice to be verified by Alice (or Alice herself downloads Bob's certificate from the CA).
(7) Once all this has been agreed, Alice can use Bob's public key to encrypt data and send it to Bob, and only Bob can decipher it as he is the only owner of the corresponding private key. In the same way, Bob can send data to Alice.
(8) With this setup Joe cannot spoof Bob's entity since Alice can always verify (using the trusted CA's certificate) the public key that Joe is presenting in the form of a certificate. So if Joe tried to present fake certificates not signed by a trusted CA then Alice can discard them.

11.3 Digital watermarking

Encryption is very effective in restricting access to data; however, once the encrypted data has been decrypted, encryption techniques cannot offer any protection at all. This is a significant limitation and encryption alone may not be sufficient for digital rights management (DRM). Another limitation of encryption is the need for customized hardware and software to perform the cryptography, while many existing multimedia appliances (e.g., CD players and DVD players) cannot be used if the data are encrypted. Digital watermarking has thus been proposed as a complementary means (rather than a replacement) for content protection even after data has been decrypted or when using the existing multimedia appliances [1]. Digital watermarking is defined as the *imperceptible* insertion or embedding of information into multimedia data, i.e., the digital data is modified in an imperceptible way so as to avoid degradation of the host data or easy perceptual identification of the embedded information.

Digital watermarking, being a technique for embedding information into digital content, requires no decryption for playing back the digital content unless an attempt has been made to extract the embedded watermark. This technique is not the same as the use of digital signatures, which encrypt a hashed file. Figure 11.31 shows a simple example where an image is imperceptibly embedded with a binary logo watermark. Depending on the application, a watermark may be a string of characters, a number, an image (e.g., logo), a piece of sound, etc., which represents a copyright message, a serial number, plain text, a control signal, etc. The digital content embedded by a watermark can be further compressed, encrypted, and protected by error correcting codes. The amount of information that can be stored in a watermark depends on the application. Taking for example the audio

| Original Image | Watermarked Image | Extracted logo |

Figure 11.31 An image imperceptibly embedded with a binary logo watermark.

watermarking technology defined by the International Federation for the Phonographic Industry (IFPI), the minimum payload for an audio watermark should be 20 bits per second, independently of the signal level and music type. For the protection of intellectual property rights, it seems reasonable to assume that one would want to embed an amount of information similar to that used for the International Standard Book Number (ISBN, roughly 10 digits) or, better the International Standard Recording Code (ISRC, roughly 12 alphanumeric letters). On top of this, one should also add the year of copyright, the permissions granted on the work, and the rating for it. This means that about 60 or 70 bits of information should be embedded in the host data, which can be an image, video frame, or audio fragment [18]. Enlarging the watermark to this extent may not be desirable. For digital videos, one second of video is considered to be the smallest entity that can be copyrighted. Therefore, the 60–70 bits of watermark information has to be embedded in a less than one-second fragment of the video stream (approximately 25 frames).

The watermarking of videos for fingerprinting or authentication requires watermarks that depend on each video frame. Indeed, if only one watermark pattern were inserted into each frame this would lead to a very vulnerable watermarking scheme with a serious security gap. It has been shown that by processing the video frames it is possible to statistically recover a good approximation to the watermark pattern. Therefore, the requirement of reliable watermarking for video is either that the watermark depends on the frame index or that it is determined by the video frame content itself. Obviously, the latter case leads to more versatile schemes.

11.3.1   Watermarking applications

Digital watermarking can be used in several applications to assist a DRM system to overcome intellectual property management and protection (IPMP) issues [1], as follows.

(1) *Identification of legal ownership* An embedded watermark can be used to identify the ownership of the digital content. The watermark provides a proof of ownership if the copyright notice has been altered or removed (see Figure 11.32).
(2) *Usage restriction* A watermark can encode the number of times that the content may be (legally) used. A compliant device checks the watermark and determines whether an additional usage is allowed. Each time a usage is made, the watermarked content is modified (renewed) to decrement the count of allowable usages (see Figure 11.33).
(3) *Fingerprinting or content tracking* A watermark can be used to encode the identification of the user or recipient of the content (e.g., credit card information about the buyer of a

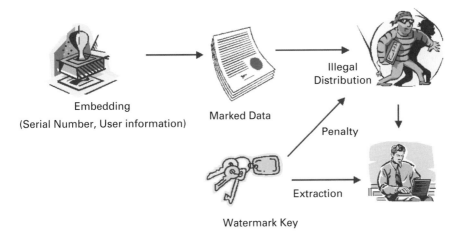

Figure 11.32 A watermark provides a proof of ownership if the copyright notice has been altered or removed.

Figure 11.33 A watermark can encode the number of times the content may be (legally) used.

music CD). If a copy of the watermark is found in an illegal copy, the embedded watermark can identify the source of the suspect copies (see Figure 11.34). This will prevent users from making illegal copies to give to others.

(4) *Authenticity checking* A fragile watermark, which is easily removed or destroyed by content modification, can be designed to detect any slight change to the watermarked data with a high probability (see Figure 11.35). The main application of fragile watermarks is in content authentication, since failure to detect the watermark proves that the host data has been modified and is no longer authentic.

11.3.2 Components of digital watermarking

As shown in Figure 11.36, there are three main components within the digital watermarking framework [19]:

(1) *watermark embedding* embeds the watermarks into the original data imperceptibly, by taking advantage of the human visual and auditory systems;

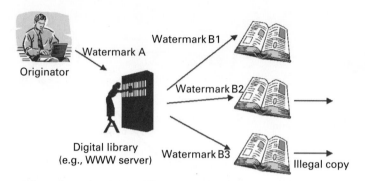

Figure 11.34 A watermark can be used to encode the identification of the user or recipient of the content.

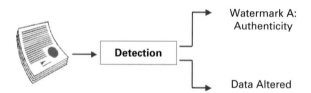

Figure 11.35 A fragile watermark can be designed to detect any slight change to the watermarked data with high probability.

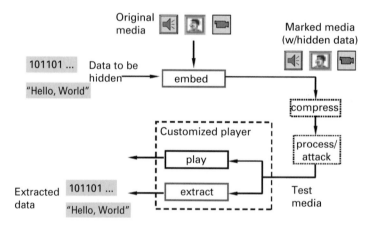

Figure 11.36 Three main components within the digital watermarking framework: watermark embedding, watermark attack and watermark detection [19].

(2) *watermark attack* intentionally or unintentionally obliterates (removes or forges) watermarks or makes them undetectable;

(3) *watermark detection* detects whether the signal contains a watermark and extracts any message carried by the watermark.

These three components will now be discussed in turn.

11.3.2.1 Watermark embedding

In watermark embedding, the watermark is created and inserted into the original digital content (text, speech, audio, image and video, etc.) to produce the watermarked content. There are two main groups of watermark embedding technologies [20]: coefficient-based and system-based. Coefficient-based approaches are the most obvious since the embedding process is performed by a direct modification of pixel values or transform coefficient values. The simplest coefficient-based method for watermark insertion is additive (as in Figure 11.31); the watermark is added to the original media signal in an analogous way to additive noise. Other embedding methods include multiplicative embedding and quantization embedding. The watermark may be inserted directly into the spatial domain (i.e., the pixels of the image or video) or after the image or video has been transformed. Common transformations include the discrete Fourier transform (DFT), the discrete cosine transform (DCT), and the wavelet transform (WT) [1]. The watermark should not be placed in perceptually insignificant regions of the image (or its spectrum), since many common signal and geometric processes affect these components. For example, a watermark placed in the high-frequency spectrum of an image can be easily eliminated with little degradation to the image by any process that directly or indirectly performs low-pass filtering. Therefore it is common practice to insert a watermark into the most perceptually significant regions of the spectrum in a fidelity-preserving fashion.

In the second group of watermark embedding technologies, the system-based group, the embedding process is performed by slightly changing an existing processing system. One example is fractal image watermarking, where the algorithm constrains the search areas of the range blocks to embed the information. Another example is the embedding of a sinusoidal wave into the average interword distance of a binary text document, i.e., changing the spacing structure of the text without modifying the gray levels.

Spread-spectrum modulation Advantage can be taken of the successful application of spread-spectrum technologies in communications, where one transmits a narrowband signal over a much larger bandwidth such that the signal energy present in any single frequency is undetectable. Therefore if we view the frequency domain of the image or audio as a communication channel, then the watermark can be viewed as a signal to be transmitted through it. The use of a direct-sequence spread spectrum to modulate the watermark message can increase immunity to noise caused by attacks and unintentional signal distortions.

In the spread-spectrum scheme, for each bit b_j of an n-bit watermark bit string $b_0 b_1 \ldots b_{n-1}$, a different stochastically independent binary-valued pseudo-random pattern RP_i is generated to replace b_j, using the simple rule that we take $+RP_i$ if b_j represents a 0 and $-RP_i$ if b_j represents a 1. Prior to adding the watermark to an image or video content, we can tile the one-dimensional spread-spectrum-modulated watermark to a two-dimensional watermark and scale the watermark by a gain factor or limit it to a certain small range. Spreading the watermark throughout the spectrum of the media content ensures a large measure of security against unintentional or intentional attacks. First, the location of the watermark is now spread over a large region of the overall transform domain and becomes less obvious to attackers. Second, a watermark that is well placed in the frequency domain of an image or a sound track will be practically impossible to perceive if the energy in the watermark is sufficiently small in any single frequency coefficient. Moreover, it is possible to take advantage of the human perceptual-masking effect, where information in certain regions of an image or a sound is occluded by perceptually more prominent

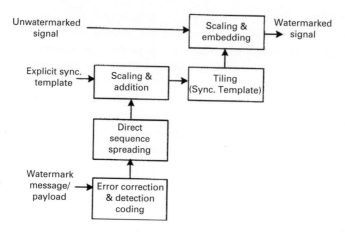

Figure 11.37 A typical spread-spectrum-based watermark generation mechanism, where forward error correction (FEC) coding can be also incorporated before the spread spectrum is applied [47] (© IEEE 2003).

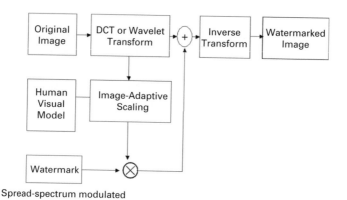

Figure 11.38 The general procedure for frequency-domain watermarking.

information in another part of the scene, or by the sound, to increase the watermark energy. Figure 11.37 illustrates a typical spread-spectrum-based watermark generation mechanism, where forward error correction (FEC) coding can be also incorporated before the spread spectrum is applied. The tiling is needed since the spread spectrum is carried out in one dimension and the frequency domain of an image frame is two dimensional [47].

Figure 11.38 illustrates the general procedure for frequency-domain watermarking. Upon applying a frequency transformation (DCT or wavelet) to the data, a human perceptual model is computed that highlights perceptually significant regions in the spectrum that can support the watermark without affecting perceptual fidelity. This will also help to determine the appropriate amount of content adaptive scaling that should multiply the spread-spectrum-modulated watermark. The scaled watermark signal can then be inserted into these regions. In practice, in order to place a lengthy watermark into an image or

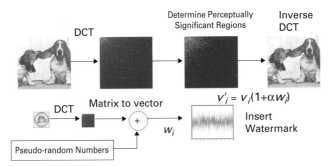

Figure 11.39 A secure spread-spectrum-based watermarking system for images [21] (© IEEE 1997).

audio the watermark is placed in the highest magnitude coefficients, excluding the DC component, of the transform matrix of the DCT of the image or audio data. A secure spread-spectrum-based watermarking system for images was proposed in [21]; its procedures are illustrated in Figure 11.39.

11.3.2.2 Watermark attacks

The watermarked media content may be subjected to attack before it can be extracted by the watermark detector. An attack is an intentional or unintentional process which may remove the embedded watermark, increase the difficulty in detecting the watermark, or subvert the security of the watermark. Attacks on digital watermarking schemes create two side effects: they either reduce the effective channel capacity or fully disable detection of the embedded watermark. Some attacks arise from the processing of the watermarked data by users without malicious intent. Of course, if there is a motivation for hackers or pirate users to remove a watermark or render it undetectable in the watermarked contents then the benefits and protection that watermarking offers in the DRM system are lost.

Four different groups of attacks can be identified [21]: removal attacks, geometrical attacks, cryptographic attacks, and protocol attacks. *Removal attacks* try to remove the embedded watermark and are commonly based on denoising or watermark prediction and removal. The effect of removal attacks is, in general, a decrease in the effective channel capacity. A special form of unauthorized removal is called *collusion attack* [22] [23]; here the attacker obtains several copies of a given work, each with a different fingerprint watermark, and combines them to produce a copy with no watermark. It is generally believed that a fairly small number of copies is sufficient to make a successful collusion attack. *Geometrical attacks* do not attempt to remove the embedded watermark but aim to distort it so that detection is impaired. This kind of attack may be considered as a process similar to the desynchronizing of signals in a communication system. *Cryptographic attacks* are used to remove or destroy the embedded watermark and are based on concepts such as a brute force search of the key space, the statistical averaging of several watermarked images, or the collusion of several watermarked images. Finally, *protocol attacks* take a more global approach by identifying weaknesses on a system level and revealing that a given watermarking method is not secure. An example of a protocol attack is a copy attack, which can be very effective in cases where a watermark is used to authenticate the image on an identification document such as a passport.

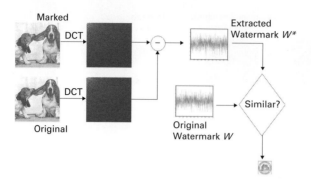

Figure 11.40 A type-2 non-oblivious watermarking detector is used to detect the existence of the embedded (spread-spectrum-modulated) watermark from a watermarked image [21] (© IEEE 1997).

11.3.2.3 Watermark detection

A watermark detector is used to determine whether a watermark is present in the host data. From the watermark detection perspective, watermarking can be classified into either non-oblivious or oblivious watermarking. Type-1, non-oblivious, watermarking requires the use of the original unwatermarked data to determine where the watermark is. Type-2, non-oblivious, watermarking uses both the original unmarked data and a copy of the watermark to find an embedded watermark. Since the original unwatermarked data is needed for watermark extraction in non-oblivious watermarking, which is more robust to attacks, the necessity of having the original unwatermarked data is clearly a disadvantage that severely limits the applicability of non-oblivious techniques. Non-oblivious watermarking techniques are commonly used in applications such as copyright protection and data monitoring. However, all that an oblivious watermarking system requires for watermark detection and extraction is a hint that tells where to find the watermark from the watermarked or attacked data. Oblivious watermarking is usually considered to be the less robust method but can be used in many applications, e.g., copy protection and indexing, where a watermark detector does not have access to the original unwatermarked data.

In some watermarking techniques, one must have access at least to a hash of the unwatermarked data in order to recreate the watermark sequence at the receiving end and so be able to correlate the received watermark with the true watermark. Such techniques are not truly oblivious because the hash needs to be exchanged prior to watermark detection. For many DRM applications, the watermark detection is usually oblivious, which means that the detector does not have access to the original unmarked data. Some DRM applications, such as content tracking, and other watermarking applications, may use non-oblivious watermark detection, where the original unmarked data is assumed to be available to the detector.

Figure 11.40 illustrates the use of a type-2, non-oblivious, watermarking detector to detect the existence of the embedded (spread-spectrum-modulated) watermark from the watermarked image shown in Figure 11.39. It is assumed that the unwatermarked image and the original watermark message are all available to the detector. Note that it is highly unlikely that the extracted modulated watermark W^* will be identical to the original modulated watermark W owing to all kinds of intentional or unintentional attacks. Even the act of requantizing the watermarked data for delivery will cause W^* to deviate from W. The similarity measure between W^* and W is defined as follows [21]:

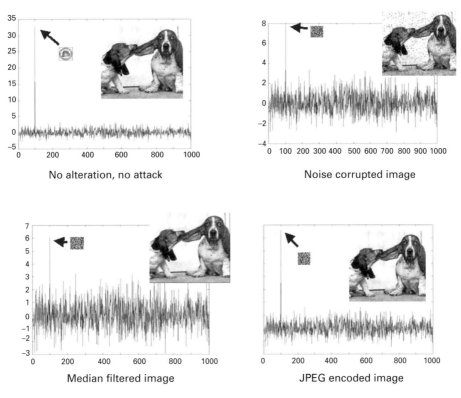

Figure 11.41 Several simulation results using the non-oblivious detection scheme based on a similarity measure [21]: watermark detector response is plotted vs. series of random watermarks (© IEEE 1997).

$$sim(W, W^*) = \frac{WW^*}{\sqrt{W^*W^*}} \qquad (11.16)$$

Many other measures are possible, including the standard correlation coefficient. To decide whether W^* and W match, one can determine whether $sim(W,W^*) > T$, where T is a predefined threshold and can be set to minimize both the rate of false negatives (missed detections) and false positives (false alarms) [21].

Figure 11.41 shows several simulation results using the non-oblivious detection scheme based on the similarity measure defined in Eq. (11.16). It can be seen that when there is no alteration (i.e., no attack) of the watermarked image, the resulting normalized correlation $sim(W,W^*)$ has a very strong value. This similarity measure also exhibits strong responses, which indicate the presence of the desired watermark and the robustness of the spread-spectrum watermarking scheme in the cases of salt-and-pepper-noise corrupted, median filtered, or JPEG compressed watermarked images.

11.4   MPEG-21

Thanks to the standardization efforts of the MPEG committee in their pioneering work in creating the MPEG-1, -2, -4, and -7 standards, it is widely believed that today's

Table 11.5 The various parts of the MPEG-21 standard [24] (© IEEE 2005)

Part	Title	Approx. publication date
1	Vision, Technologies and Strategy, 2nd edition	Nov. 2004
2	Digital Item Declaration, 2nd edition	April 2005
3	Digital Item Identification	April 2003
4	Intellectual Property Management and Protection Components	Jan. 2006
5	Rights Expression Language	April 2004
6	Rights Data Dictionary	May 2004
7	Digital Item Adaptation	Oct. 2004
8	Reference Software	July 2005
9	File Format	April 2005
10	Digital Item Processing	July 2005
11	Evaluation Methods for Persistent Association Tools	Nov. 2004
12	Test Bed for MPEG-21 Resource Delivery	Feb. 2005
13	EMPTY	—
14	Conformance	April 2006
15	Event Reporting	Jan. 2006
16	Binary Format	July 2005
17	Fragment Identification for MPEG Resources	April 2006

communication infrastructures have evolved to enable access of multimedia services any-time and anywhere. These standards define an extensive set of tools for audio-visual compression and transport as well as metadata for multimedia content description. However, secure and interoperable end-to-end solutions are still not available for broad ranges of users. MPEG-21, formally referred to as ISO/IEC 21000 multimedia framework, was thus proposed to address the above problem by standardizing interfaces and tools to facilitate the exchange of multimedia resources across heterogeneous devices, networks, and users. More specifically, MPEG-21 standardizes the requisite elements for packaging, identifying, adapting, and processing these resources as well as managing their usage rights. This framework will benefit the entire consumption chain, from creators and rights holders to service providers and consumers [24][25].

The basic unit of transaction in the MPEG-21 multimedia framework is the *digital item* (*DI*), which packages resources along with identifiers, metadata, licenses, and methods that enable interaction with the DIs. Another key concept in the MPEG-21 multimedia framework is that of a user, which stands for any entity that interacts in the MPEG-21 environment or makes use of DIs. Such users include individuals, consumers, communities, organizations, corporations, consortia, and governments. At the most basic level, MPEG-21 can be seen as providing a framework in which one user interacts with another user, the object of that interaction being a digital item. Some examples of interactions include content creation, management, protection, archiving, adaptation, delivery, and consumption. Table 11.5 provides a current list of the parts, along with their approximate date of publication, of the MPEG-21 standard.

The goal of MPEG-21 is to define the technologies needed to support users in exchanging, accessing, consuming, trading, and otherwise manipulating digital items (DIs) in an efficient, transparent, and interoperable way. MPEG first produced a technical report, "Vision, technologies, and strategy," which was later adopted as MPEG-21 Part 1. Different parts of MPEG-21 were then completed one by one, through iterations of proposals based upon requirements issued by MPEG. Several parts of MPEG-21 deal with DRM issues. More specifically, MPEG-21 Parts 2 and 3 deal with the declaration and identification of DIs and Parts 4, 5, and 6 deal with issues related to intellectual property management and protection (IPMP). Part 7, digital item adaptation (DIA), has particular relevance to universal multimedia access (UMA), and Part 10, digital item processing (DIP), enables the incorporation of interoperable descriptions of programmability related to multimedia experience. The following sections will discuss these parts in more detail.

11.4.1 Digital item declaration

A *digital item* (DI) is defined as a structured digital object with a standard representation, identification, and metadata. A DI, as defined specifically by MPEG in MPEG-21 [26] [27] consists of resources, metadata, and structure. A resource denotes an individual asset, e.g., an MPEG-2 video; a metadata denotes descriptive information, e.g., an MPEG-7 description of the media; and a structure denotes the relationship between parts of the digital item. An example of a DI could be an "MPEG-21 music album," which comprises a series of media resources [28]:

(1) 10 audio files, representing the "tracks" that form the basis of the album;
(2) two text files, representing the lyrics of two tracks;
(3) two images, representing the cover photograph and other artwork of the album;
(4) a text file, representing the introductory text for the album.

The relationship between these resources and how they relate to the DI itself is expressed in a digital item declaration (DID), which is a document that specifies the makeup, structure, and organization of a DI. More specifically, the DID formally expresses and identifies the resources (e.g., MP3 files) and metadata (e.g., descriptions of the singer and lyrics) which are considered by the author to be the constituents of the DI. Further, the DID binds together individual resources and metadata and groups of resources and metadata. This is further extended by the capability to allow metadata to be anchored to certain fragments in a media resource. It should be noted that the DID model is an abstract model of building blocks useful for declaring DIs. The MPEG-21 DID system divides the representation of DIs into three distinct parts [27].

(1) *Model* This describes a set of abstract terms and concepts that form a useful model for defining DIs. Within this model, a DI is a structured representation of a digital object, and, as such, it is the entity that is acted upon (managed, described, exchanged, collected, etc.) within the model. The declaration of a DI compliant with an abstract model is referred to as a digital item declaration (DID).
(2) *Representation* This describes the syntax and semantics of each DID element, as represented in extensible markup language (XML). This XML-based syntax is referred to as the MPEG-21 digital item declaration language (DIDL). A DID represented according to the MPEG-21 DIDL syntax is referred to as a DIDL document.
(3) *Schema* This contains the entire XML syntax and grammar for the structure of DIDL documents.

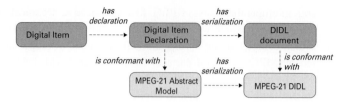

Figure 11.42 The relationship between an MPEG-21 abstract model and the MPEG-21 DIDL [29].

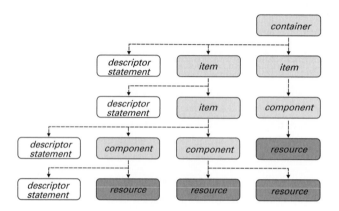

Figure 11.43 A *descriptor–statement* construct introduces an extensible mechanism that can be used to associate information with other entities of the abstract model [29].

The relationship between an MPEG-21 abstract model and the MPEG-21 DIDL is shown in Figure 11.42 [29].

The MPEG-21 abstract model defines several constituent entities of a DID. These entities make up the backbone of a DID and are presented in a bottom-to-top approach for declaring digital items (see Figure 11.43) [29], as follows.

- *A resource* denotes an individually identifiable data stream such as a video file, image, audio clip, or a textual asset.
- *A component* denotes one or more equivalent *resources* of a (set of) *descriptor–statement* construct(s) which contain secondary information related to all the resource entities bound by the component.
- *An item* denotes one or more *items* and/or *components* of a (set of) *descriptor–statement* construct(s), which contain information about the particular item. In the abstract model, an item entity is equivalent to a digital item and is the first point of entry to the content for a user. If items contain other items then the outermost item represents the composite item and the inner items represent the individual items that make up the composite.
- *A container* denotes one or more items and/or containers of a (set of) *descriptor–statement* construct(s) which contain information about the particular container. Containers can be used to form groupings of digital items.

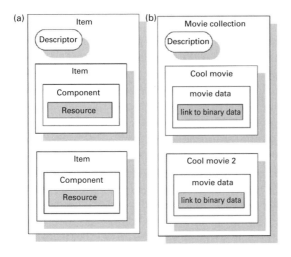

```
<?xml version="1.0" encoding="UTF-8"?>
<DIDL xmlns="urn:mpeg:mpeg21:2002:02-DIDL-NS">
 <Item>
  <Descriptor>
   <Statement mimeType="text/plain">
   My movie collection
   </Statement>
  </Descriptor>
  <Item>
   <Component>
    <Resource mimeType="video/mpeg" ref="cool_movie.mpg"/>
   </Component>
  </Item>
  <Item>
   <Component>
    <Resource mimeType="video/mpeg" ref="cool_movie_2.mpg"/>
   </Component>
  </Item>
 </Item>
</DIDL>
```

Figure 11.44 An example of DID for a digital movie collection. (a) A graphical representation of the DID. (b) The corresponding digital item declaration language (DIDL) elements based on XML [30] (© IEEE 2005).

- *A descriptor–statement construct* associates a "statement," i.e., textual secondary information with its enclosing entity, e.g., information attached as a "child" entity to an item (or a container) and providing secondary information about that item (or container). Examples of likely statements include information supporting discovery, digital preservation, and rights expressions.

A DID is in fact a static extensible markup language (XML) declaration containing resources and metadata (either by reference or by inclusion in the XML document). An example of a DID for a digital movie collection is shown in Figure 11.44. A graphical representation of the DID is shown in Figure 11.44(a) and the corresponding digital item declaration language (DIDL) elements based on XML are given in Figure 11.44(b).

11.4.2 Digital item identification

In the analog world, we use identifier schemes to identify physical objects by number, such as the International Standard Recording Code (ISRC) for music/video recording and the International Standard Book Number (ISBN) for commercial books. It is also highly desirable to have a similar identifier system for DIs. Since there are already many identification schemes available, the digital item identification (DII) part of MPEG-21 [31] simply integrates existing identification schemes for various application space into the MPEG-21 framework, instead of creating a new identification scheme. More specifically, DII provides a normative way to express how this identification can be associated with DIs, containers, components, and/or fragments so as to be included in a specific place in the DID [27]. Digital items and their parts within the MPEG-21 multimedia framework are identified by encapsulating uniform resource identifiers (URIs), as specified in IETF RFC 2396.

DII elements Two XML elements are introduced in DII: identifiers and relatedidentifiers. Both these elements are intended to contain URIs. The difference between them is subtle but important as it results in clear differences in usage and implementation. The identifier contains a URI that identifies a DI, container, component, or fragment. In contrast, the relatedidentifier carries identifiers that are related to the DI. One example is the identifying of an abstraction of a work (e.g., a composition as an abstraction of a sound recording). Note that the identifiers covered by the DII specification can be associated with digital items by including them in a specific place in the statement element of the digital item declaration. Examples of likely statements include descriptive, control, revision tracking, and/or identifying information. Figure 11.45 shows the relationship between DID and DII. The shaded boxes are subject to the DII specification while the empty boxes are defined in the DID specification. Figure 11.46 shows a music album example, where the DII identifier is included in the statement element of the DID.

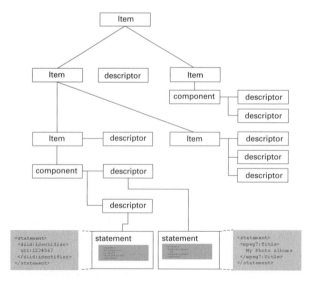

Figure 11.45 The relationship between DID and DII. The shaded boxes are subject to the DII specification while the empty boxes are defined in the DID specification.

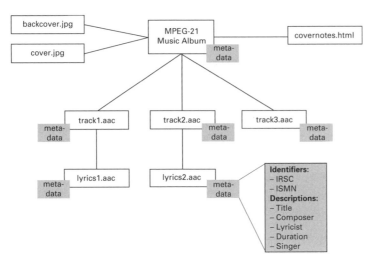

Figure 11.46 A music album example whose DII identifier is included in the statement element of the DID [25].

```
<?xml version="1.0"?>
<DIDL xmlns="urn:mpeg:mpeg21:2002:01-DIDL-NS"
    xmlns:dii="urn:mpeg:mpeg21:2002:01-DII-NS">
  <Item>
    <Component>
    <Descriptor>
      <Statement mimeType="text/xml">
<dii:Identifier>urn:mpegRA:mpeg21:dii:isrc:US-ZO3-99-
32476</dii:Identifier>
    <!-- ISRC identifying the sound recording -->
      </Statement>
    </Descriptor>
    <Descriptor>
      <Statement mimeType="text/xml">
<dii:RelatedIdentifier>urn:mpegRA:mpeg21:dii:iswc:T-
034.524.680-1</dii:RelatedIdentifier>
    <!-- ISWC of the underlying musical work -->
      </Statement>
    </Descriptor>
      <Resource ref="Track01.mp3" MimeType= "audio/mp3"/>
    </Component>
  </item>
</DIDL>
```

Figure 11.47 A sound recording in the form of an MP3 file is identified using the identifier element and an ISRC [27] (© IEEE 2005).

A DII example As shown in Figure 11.47 [27], a sound recording in the form of an MP3 file is identified using the identifier element and an ISRC. Note that the relatedidentifier is included in the form of a URI version of the International Standard Musical Work Code, which identifies the underlying musical work recorded in the MP3. It can be seen that the DII elements are included in descriptor statement DIDL combinations, which are bound to the resource using a component. It is important to note, however, that an application will need to be DII aware to find and act upon these identifiers in an intelligent and meaningful manner. The DID is valid regardless of whether a given application understands the DII part of the MPEG-21 standard [31].

Registering identifiers in DII As discussed previously, the DII elements must contain valid URIs. While many internationally used identifiers are now available in this form and possess their own uniform resource namespace (URN), it is possible that users of MPEG-21 may wish to use other identifiers not available in this form. To allow this to happen with relative ease, MPEG has created a registration authority mechanism that allows identifiers intended for use in the MPEG-21 framework to be recognizable. This provides the ability for identifiers to be expressed in the form *urn:mpegRA:mpeg21:dii:sss:nnn*, where the string *sss* denotes the identifier for an identification system and *nnn* denotes a unique identifier within that identification system.

11.4.3 Intellectual property management and protection (IPMP)

MPEG-21 Part 4 defines an interoperable framework for intellectual property management and protection (IPMP), which is a much more interoperable extension of MPEG-4 IPMP. The project includes standardized ways of retrieving IPMP tools from remote locations and exchanging messages between IPMP tools and between these tools and the terminal (see Figure 11.48). It also addresses the authentication of IPMP tools and has provisions for integrating rights expressions according to the rights expression language (REL) and the rights data dictionary (RDD) [32].

The *rights expression language* [33] defined in MPEG-21 is a machine-interpretable language intended to provide flexible interoperable mechanisms to support the transparent and augmented use of digital resources in the publishing, distributing, and consuming of electronic books and in broadcasting, digital movies, digital music, interactive games, computer software, and other creations in digital form in a way that protects digital content and honors the rights, conditions, and fees specified for digital contents. It is also intended to support the specification of access and use controls for digital content in cases where

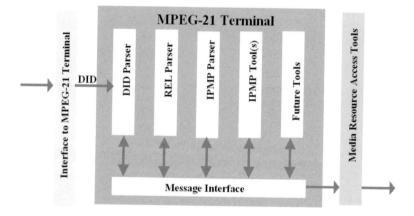

Figure 11.48 MPEG-21 IPMP includes standardized ways of retrieving IPMP tools from remote locations and exchanging messages between IPMP tools and between these tools and the terminal [25].

financial exchange is not part of the terms of use and to support the exchange of sensitive or private digital content. The REL is also intended to provide flexible interoperable mechanisms to ensure that personal data is processed in accordance with individual rights and to meet the requirement for users to be able to express their rights and interests in a way that addresses issues of privacy and the use of personal data.

The *rights data dictionary* (RDD) [34] defined in MPEG-21 is a dictionary of key terms required to describe rights of those who control DIs, including intellectual property rights and the permissions they grant, that can be unambiguously expressed using a standard syntactic convention and can be applied across all domains in which rights and permissions need to be expressed. The RDD comprises a set of clear, consistent, structured, integrated, and uniquely identified terms to support the MPEG-21 REL. It specifies the structure and core of the dictionary and specifies how further terms may be defined under the governance of a registration authority.

11.4.3.1 MPEG-21 rights expression language

The MPEG-21 REL is an XML-based declarative language, derived from the extensible rights markup language (XrML) initially developed by Xerox Palo Alto Research Center (PARC, http://www.parc.com/) and now owned by ContentGuard (http://www.contentguard.com/), which can be used to specify rights and conditions for the authorized distribution and use of any content, resources, or services. Its goals are as follows [32]:

(1) to define the syntax and semantics of a machine interpretable language that can be used to specify rights unambiguously;
(2) to provide an authorization model to determine whether an authorization or access control request can be granted according to a set of rights expressions in the above language;
(3) to support the many usage models in the end-to-end distribution value chain.

REL data model As shown in Figure 11.49, the MPEG REL data model starts with a license which contains one or more grants and an issuer, whereas a grant contains four basic entities (*principal, right, resource,* and *condition*). The relationship between these four entities is shown in Figure 11.50 [32], and the quantities are described as follows.

(1) A *license* contains grants, issuers, and some other related information. Conceptually, a license is a collection of grants issued by one or more issuers.

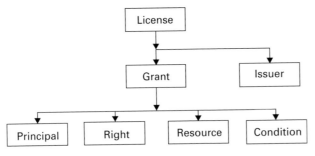

Figure 11.49 The MPEG REL data model starts with a license which contains one or more grants and an issuer, whereas a grant contains four basic entities (principal, right, resource, and condition) [32] (© IEEE 2005).

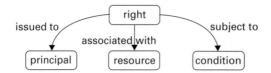

Figure 11.50 The relationship between the four entities of REL [25].

(2) An *issuer* is an element within the license that identifies a principal who issues the license. The issuer can also contain a digital signature of the license signed by the principal to signify that the principal does indeed bestow the grants contained in the license and to facilitate reliable establishment of the trustworthiness of the license by others.

(3) A *grant* is the element within the license that essentially grants a principal with a right over a resource, possibly subject to a condition.

(4) A *principal* within a grant encapsulates the identification of a party to whom a right is granted. A principal denotes the party that it identifies by information unique to that party. For example, the MPEG REL supports the concept of a key-holder, who is someone possessing the private key of a public and private key pair.

(5) A *right* is the action that a principal can be granted to exercise against some resource under some condition. The MPEG REL provides a set of commonly used specific rights, such as play, print, and adapt, as well as rights relating to other rights, such as obtain, issue, and revoke.

(6) A *resource* is the "object" to which a principal can be granted a right. A resource can be a digital work (such as an ebook, an audio or video file, or an image), a service (such as an email service, or B2B transaction service), or even a piece of information that can be owned by a principal (such as a name, an email address, a role, or any other property or attribute). A resource can also be a grant, and in this case a right about that grant (as a meta right) would be specified within its parent grant.

(7) A *condition* specifies the terms, conditions, and obligations under which rights can be exercised. For example, a simple condition is a time interval within which a right can be exercised; a slightly more complicated condition may require the existence of a valid, prerequisite, right that has been issued by some trusted entity. Moreover, a condition can be the conjunction of several other conditions.

On the basis of MPEG-21 REL, anyone owning or distributing digital resources can identify principals (such as users, groups, devices, and systems) authorized to use those resources, the specific rights accorded to those principals, and the terms and conditions under which those rights may be exercised.

For example, consider a song "When the thistle blooms," distributed by a label "PDQ Records," to Alice's MP3player. A typical REL expression might make the statement, under the authority of PDQ Records, that Alice is granted the right to play "When the thistle blooms" in December 2003. Figure 11.51 shows the REL expression of this license and the associated grants. In the MPEG REL terminology, "Alice" is considered a *principal*; "play," a *right*; "When the thistle blooms," a *resource*; "in December 2003," a *condition*; and "PDQ Records," an *issuer* of the right. This example should capture the essence of an MPEG REL expression, which is a statement of the following form: an issuer states that a *principal* has some *right* to a *resource* under some *condition*. The right-granting portion of this statement (e.g., Alice is granted the right to play "When the thistle blooms" in December 2003) is

```
<license>
<grant>
<keyHolder licensePartID="Alice">
<info⨯dsig:KeyValue⨯dsig:RSAKeyvalue⨯dsig:Modulus>oRUTUiTQk
...</dsig:Modulus>
    <dsig:Exponent>AQABAA==</dsig:Exponent/dsig:RSAKeyValue⨯/dsig:keyValue/info>
    </keyHolder>
    <mx:play/>
    <mx:diReference>
    <mx:identifier>urn:PDQRecords:song:WhenTheThistleBlooms.mp3</mx:identifier>
    </mx:diReference>
    <validityInterval>
    <notBedore>2003-12-01T00:00:00</notBefore>
    <notAfter>2003-12-31T23:59:59</notAfter)
    </validityInterval>
    </grant>
    <issuer licensePartID="Bob">
    <dsig:Signature>
    <dsig:SignatureValue>zlRYax15EX...</dsig:SignatureValue>
    <dsig:SinatureValue>zlRYax15EX... /dsig:SignatureValue>
    <dsig:KeyInfo⨯dsig:keyValuedsig:RSAKeyValue⨯dsig:Modulus>yQ==...
    </dsig:Modulus>
    <dsig:Exponent>AQAB==</dsig:Exponent></dsig:RSAKeyvalue⨯/dsig:KeyValue⨯/dsig:KeyInf
    0>
    <dsig:Signature>
    <issuer>
    </license>
```

Figure 11.51 An REL expression for a license by which Alice is granted the right to play "When the thistle blooms" in December 2003 [32] (© IEEE 2005).

called a *grant* and the entire statement is called a *license*, which in this case consists of the grant and the issuer, PDQ Records.

11.4.3.2 MPEG-21 rights data dictionary

The MPEG-21 rights data dictionary (RDD), now including more than 2000 terms, was originally based on the <indecs> project [35], which set out an event-based approach to the interoperability of metadata. Building on this approach, the MPEG-21 RDD is based on a logical structure, referred to as the context model, which is used to construct a natural language ontology for the terms used in rights management. Terms are derived from verbs, the first term being "act," and are organized into families according to the roles they play in contexts. Contexts come in two varieties, i.e., events (through which change happens) and situations (which are unchanging), which conform to the rules of the MPEG-21 RDD context model [32]. The 14 ActTypes listed in Table 11.6 cover the most common expected actions that a user might wish to undertake in REL grants with respect to digital content.

The RDD context model (see Figure 11.52) defines a group of five terms, which form a "basic term set" – context, agent, resource, time, and place [32]. These terms and their associated classes and relators form the core semantic architecture of the MPEG-21 RDD; that is, the mechanisms by which meanings are derived from one another. Using the context model, terms can be introduced in a logical manner, ensuring that the structure of the dictionary is not broken. For instance, when a new act (say, modify) is defined, a full range of related terms embodying some aspect of the act can also be defined, and each new term is placed in its logical node of the dictionary.

In order to provide more complex right-expression terms in the future for use in the REL, more granular definitions may be required and managed through the process of specialization, using the methodology of the context model to suit the particular circumstances.

Table 11.6 The most common 14 expected actions a user might wish to undertake in the REL grants with respect to digital content [32] (© IEEE 2005)

Act type	Definition
Adapt	To change transiently an existing resource to derive a new resource
Delete	To destroy a digital resource
Diminish	To derive a new resource which is smaller than its source
Embed	To put a resource into another resource
Enhance	To derive a new resource which is larger than its source
Enlarge	To modify a resource by adding to it
Execute	To execute a digital resource
Install	To follow the instruction provided by an installing resource
Modify	To change a resource, preserving the alterations made
Move	To relocate a resource from one place to another
Play	To derive a transient and directly perceivable representation of a resource
Print	To derive a fixed and directly perceivable representation of a resource
Reduce	To modify a resource by taking away from it
Uninstall	To follow the instruction provided by an uninstalling resource

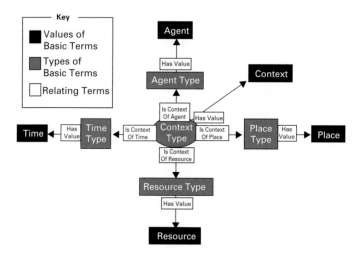

Figure 11.52 The RDD Context model defines a group of five terms, which form the basic term set – context, agent, resource, time and place [32] (© IEEE 2005).

The relationship between REL and RDD is shown in Figure 11.53; the MPEG-21 RDD is designed to handle many different metadata schemes that already exist and to support the mapping of terms from different namespaces. Such mapping will enable the transformation of metadata from the terminology of one namespace (or authority) into that of another namespace (or authority). This is an essential prerequisite to semantic interoperability of rights expressions. The mapping will be the responsibility of the registration authority, which will be established under ISO/IEC Joint Technical Committee (JTC) 1 directives.

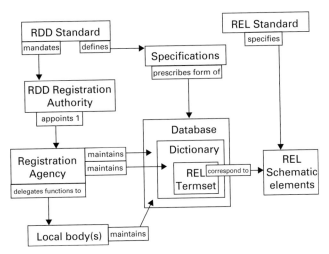

Figure 11.53 The relationship between REL and RDD [36] (© IEEE 2002).

11.4.4 Digital item adaptation (DIA)

Owing to the growing diversity of networks, terminals, and media formats, it is getting more difficult to achieve universal media access. More specifically, the diversity is in bandwidths, display and audio capabilities, processing power, user preferences, and various kinds of non-compatible audio, image, and video formats etc. It is widely believed that efficient adaptation is needed in the delivery route and that metadata can give such efficiency. Therefore, the standardization of metadata interfacing to adaptation is highly desired. The MPEG-21 digital item adaptation (DIA) [37] was thus proposed, to achieve interoperable transparent access to distributed multimedia content by shielding users from network and terminal installation, management, and implementation issues. This should enable the provision of network and terminal resources on demand, to form user communities where multimedia content can be created and shared, always with the agreed contracted quality, reliability and flexibility, and so allowing multimedia applications to connect diverse sets of users in such a way that the quality of user experience will be guaranteed. As shown in Figure 11.54, which illustrates the conceptual architecture of DIA, a digital item is subject

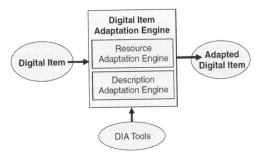

Figure 11.54 The conceptual architecture of DIA [28].

to a resource adaptation engine and a descriptor adaptation engine, which produce together the adapted digital item. Even though the adaptation engines themselves are non-normative tools of DIA, the descriptions and format-independent tools that provide support for DIA in terms of resource adaptation, descriptor adaptation, and/or QoS management are within the scope of these requirements. Among several DIA tools, the adaptation QoS tool is an important resource adaptation tool for guaranteeing the QoS of a network and a terminal. It describes relations between resource constraints (e.g., bandwidth), adaptation operations, and the corresponding qualities. Feasible adaptation operators for various types and formats of content are provided in the classification schemes of MPEG-21 DIA.

DIA requirements MPEG-21 DIA addresses requirements that are specific to the usage environment descriptions, which include the following [28]:

(1) *usage environment* in terms of terminal, network, delivery, user, and natural environment capabilities;
(2) *terminal capabilities* including acquisition properties, device types (e.g., encoder, decoder, gateway, router, camera), profiles, output properties, hardware properties (e.g., processor speed, power consumption, memory architecture), software properties, system properties (e.g., processing modules, interconnection of components, configuration options, ability to scale resource quality, manage multiple resources), and IPMP related capabilities;
(3) *network capabilities* including delay characteristics (e.g., end-to-end delay, one-way delay, delay variation), error characteristics (e.g., bit-error rate, packet loss, burstiness), and bandwidth characteristics (e.g., amount of bandwidth, variation in bandwidth);
(4) *delivery capabilities* including the types of transport protocols supported and the types of connections supported (e.g., broadcast, unicast, multicast);
(5) *user characteristics* including user preferences and demographic information (note that this requirement is partially addressed by the existing MPEG-7 description schemes)
(6) *natural environment characteristics* including location, type of location (e.g., indoor, outdoor, public place, home, office), available access networks in a given area, velocity of user or terminal, and illumination properties affecting user or terminal;
(7) *service capabilities* including user roles and types of services;
(8) *interactions and relations among users* including the relation between various dimensions of the usage environment description.

DIA adaptation engine As shown in Figure 11.54, only the tools used to guide the adaptation engine are specified by the standard. The adaptation engines themselves are left open to various implementations. A generic DIA adaptation engine is illustrated in Figure 11.55, where the inputs to an adaptation engine include the bitstream resource, resource description, and adaptation constraints; the outputs from the adaptation engine include the adapted bitstream resource and adapted resource description. The DIA engines can be placed on the media server side of a server–client delivery architecture, or they can be placed in the edge server closer to the clients. Moreover, they could be placed in the relay server in the CDN-based application-level multicast architecture. The adaptation operation in an adaptation engine can be broken up conceptually into two functional modules, as shown in Figure 11.56:

(1) The adaptation decision taking engine (ADTE);
(2) The bitstream and description adaptation engine (BDAE).

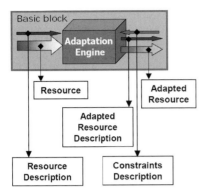

Figure 11.55 A generic DIA adaptation engine [42].

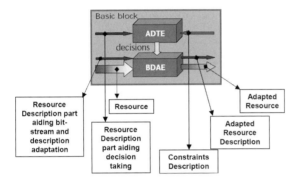

Figure 11.56 The adaptation operation in an adaptation engine can be broken up conceptually into two functional modules, ADTE and BDAE [42].

In an MPEG-21 DIA terminal, user characteristics (user information, usage preference, history, accessibility, location), terminal capabilities (codec capabilities, device properties, I/O characteristics), and network characteristics (capabilities, conditions) are represented using a usage environment description (UED) [37] in terms of a fixed XML grammar for various items. Then this information is transmitted to a server having an adaptation decision taking engine (ADTE), which converts these descriptions to constraints under certain assumptions (this conversion is not defined in MPEG-21).

Towards the goal of universal accessibility many different ways for video adaptation have been developed, and they can be categorized into two approaches: transcoding-based methods [38] [39] and scalable bit-stream-based methods [40]. Transcoding is the technology used to reformat video content so that it can be viewed on various different types of device with different characteristics (see Section 8.1.2). Otherwise video content would have to be developed separately for display on different platforms and available transmission bandwidths. For example, requantization, frame dropping, and DCT-coefficient dropping are commonly used in transcoding for adjusting the bitrate of the compressed video. These kinds of method usually have high complexity since the video content needs to be partially or fully decoded and re-encoded. However, adaptation methods based on scalable video coding (SVC) in standards like H.263 and MPEG-4 Visual have also been proposed (see Section 5.7). These adaptation methods are quite simple, involving just the truncation of a number of embedded coded layers. Even though providing very low complexity, they are unlikely to be used owing to their very low coding efficiency.

A great advance in coding efficiency has been offered by SVC standardization based on the scalable extension of H.264 AVC, which provides scalability at a bitstream level with coding efficiency similar to that of H.264 AVC. The scalability allows parts of the SVC bitstream to be removed in the spatial, temporal, and SNR domains. A dynamic adaptation method of SVC using MPEG-21 digital item adaptation (DIA) that can provide an optimally adapted video stream over heterogeneous networks was proposed in [40]. This proposed SVC adaptation operator for MPEG-21 DIA was also adopted in the ISO/IEC JTC1 standard [41].

Taking the usage environment description (UED) fully into account, the ADTE module in the DIA engine proposed in [40] finds an optimal adaptation point along the axes of the spatial, temporal, and SNR dimensions. In a dynamic extractor, SVC network abstraction layer (NAL) units can be cropped or dropped to produce the desired scalable layers corresponding to the resulting adaptation point, so that the picture size, frame rate and/or bitrate of a transmitting video stream are adapted in real time. More specifically, as shown in Figure 11.57, a server is composed of a parser for collecting usage environment information from the received UED, a streamer for transmitting packets, an FEC block for reducing transmission errors, an I/O block, an ADTE, and a dynamic extractor. A client consists of a UED authoring tool, a network monitoring block for measuring the available network conditions, an SVC decoder, a rendering block for displaying video, and an I/O block for communicating via RTP, RTCP, and RTSP.

More specifically, by taking account of adaptation QoS, UED, and network characteristic information the ADTE determines the optimal adaptation operation along the axes of spatial, temporal, and quality resolution. It can appropriately determine the best adaptation operation by setting different values on the following three SVC control parameters [40]:

(1) *SpatialLayers* indicates the number of enhancement layers for which spatial resolution should be truncated from the input bitstream. It is assumed that the highest enhancement layer is truncated first.

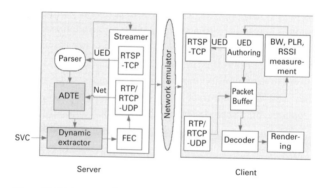

Figure 11.57 A server is composed of a parser, a streamer, an FEC block, an I/O block, an ADTE, and a dynamic extractor [40] (© IEEE 2007).

(2) *TemporalLevels* indicates the number of enhancement layers for which temporal resolution should be truncated from the input bitstream. It is assumed that the highest enhancement layer is truncated first.

(3) *QualityReduction* indicates the SNR enhancement fraction to be truncated from the input bitstream. It is assumed that the quality reduction value for no truncation is "0" and for full truncation is "1."

11.4.5 Digital item processing

Multimedia applications and services commonly involve the combined consumption of audio, image, video, and/or text. Examples of such interactions could be adding or removing tracks from a digital music album, requesting adaptations for resources, configuring digital items according to user preferences, etc. Scene composition descriptions for MPEG-4 and the synchronized multimedia integration language (SMIL) [43] were proposed to provide descriptive information on how individual media resources may be combined into a global presentation based on the description of temporal relationships between individual resources. To incorporate more interoperable descriptions of programmability related to multimedia experience and to further allow interactions between a user (the term includes both human end users and machines, as defined in MPEG-21), a digital item (DI), and the other parts of MPEG-21, MPEG-21 Part 10, digital item processing (DIP, ISO/IEC 21000-10) [44] was proposed.

For a quick introduction to the MPEG-21 DIP technology and its functionality, one could think of a DI. To make the presentation of the DI more attractive to an end user it is possible to use SMIL or MPEG-4 as a presentation language for the DI. Since MPEG-21 DIP interacts with DIs, it is necessary to look at the MPEG-21 DID [26] to understand the motivation behind the development of DIP. Take for example the DID shown in Figure 11.44, which corresponds to a digital movie collection using DIDL elements. How should we process this DID using a DIP engine?

11.4.5.1 Structure of DIP information

Before answering the above question, how a DIP engine could be constructed to process the information provided by the DID, let us define some information terms and structures used in DIP [30].

Digital item method (DIM) A DIM expresses the interaction of an MPEG-21 user with a DI. It contains calls to digital item base operations (DIBOs) and describes the possible interactions of an MPEG-21 user (e.g., a human consumer) with the DI. A DIBO is the smallest possible, elementary, action that can be performed on a DI.

Digital item method language (DIML) The DIML is the language in which DIMs are expressed. It provides the syntax and structure for authoring a DIM using DIBOs and digital item extension operations (DIXOs). Instead of defining a new DIML, MPEG decided to use ECMAScript [45] as the DIM language. The selection of ECMAScript was made to ensure a lightweight DIML in terms of memory, footprint, and processing power. A DIM object is an object representation of an element in the DIDL and can be accessed in the DIML. Note that, in the DIML, the set of DIBOs is treated as a library of functions.

Digital item extension operation (DIXO) A DIXO is an action allowing non-standardized operations to be invoked by a DIM. The DIXO mechanism realizes the extension of the set of standardized DIBOs in an interoperable way. Apart from being a mechanism to access native functionality, DIBOs also serve to abstract out complicated operations and provide a high-level interface to DIMs. The language in which DIXOs are implemented is not standardized in the MPEG-21 DIP specification, while a Java version called J-DIXO was provided in [46].

Object type and map An object type is a type of argument, i.e., variable, term, or other expression, passed to a DIM. It allows the argument to be processed within the DIM according to its semantics. An object map provides the relationship between the DID elements of the object types. The object type and the object map are created to allow DID elements to act as arguments for the DIMs. The object type information realizes the coupling of a type to a DID element.

DIP engine This engine is the processing unit responsible for generating an object map and executing the DIMs. The execution of DIMs includes the calling of DIBOs, loading of DIXOs, execution of DIXOs, and interpreting the DIML code. As a result, this DIP engine is a vital part of an MPEG-21 DIP-enabled peer.

11.4.5.2 Types of DIP information

Two types of MPEG-21 DIP information are included in a DID, allowing the processing of DIs, DIM information and object type information (see Figure 11.58 [30]).

DIM information The DIM information can be split into two different parts, the method information and the DIM implementation. The former contains all the information necessary to allow a DIP engine to call a DIM. The latter contains the actual implementation of the DIM, i.e., the ECMAScript code with the calls to the DIBOs and DIXOs.

Object type information When creating DIs, the DIDL is used to declare the different parts of the DI. The resulting DID is a composition of different DID elements. To allow the usage of those DID elements as arguments for DIMs, the object types and object map are defined in the DIP specification.

A walkthrough of DIP operations Now it is time to give an answer to the question "What happens when a DID arrives at an MPEG-21 DIP-enabled peer?" A walkthrough of the relevant steps is given in Figure 11.59 [30] and below. It should be noted that this diagram is not a normative part of the MPEG-21 DIP specification, but it can be used as a guideline when implementing DIP on a terminal.

(1) As a first step, a DID arrives at an MPEG-21 peer, which recognizes the received DID by looking at the namespace in which the XML code is started and transfers control to the DID engine.
(2) The DID engine opens the DID and searches through the DID to find DIP information.
(3) The DID engine transfers the DIP information to the DIP engine.
(4) The DIP engine builds a list of DIMs and creates an object map.
(5) If there is a DIM with the name "main" that requires no arguments in the DID, this DIM is selected automatically. The walkthrough then continues from step (8).
(6) The list of DIMs is presented to the MPEG-21 user. This list can be seen as a menu of possible actions that the MPEG-21 user can request from the DIP engine.

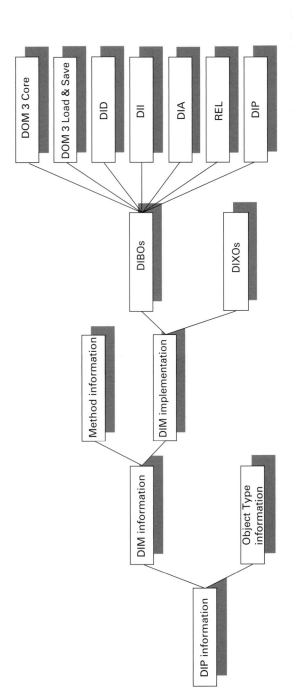

Figure 11.58 Two types of MPEG-21 DIP information are included in a DID, allowing processing of DIs: (1) DIM information and (2) object-type information [30] (© IEEE 2005).

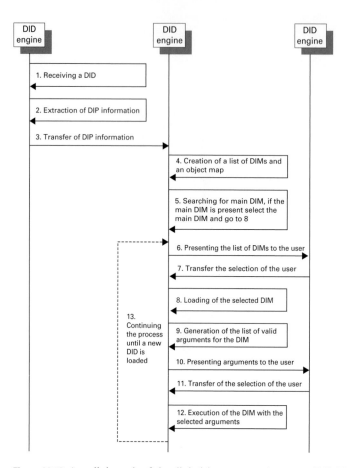

Figure 11.59 A walkthrough of the digital item processing steps [30] (© IEEE 2005).

(7) The MPEG-21 user selects a DIM and his or her selection is transferred to the DIP engine.

(8) The DIP engine ensures that the DIM implementation is loaded and available for execution in the ECMAScript environment. If some DIXOs have been used in the DIM, those DIXOs are downloaded and initialized.

(9) Using the information present in the object map, the DIP engine generates a list of arguments valid for the selected DIM. If there are no arguments for this DIM then the walkthrough continues from step (12).

(10) The list of valid arguments is presented to the MPEG-21 user. This list can be seen as a menu of possible elements on which the selected DIM can be performed.

(11) The MPEG-21 user selects an argument and notifies the DIP engine of the selection.

(12) The DIM is now executed by calling the DIBOs and DIXOs listed in the DIM implementation. Note that the execution of the DIBOs and DIXOs can trigger several rights checks. For example, when a PlayResource DIBO is executed, the implementation of the DIBO will need to check whether the conditions for playing the resource are met.

(13) After execution of the DIM, the DIP engine returns to the state in which it presents the list of DIMs to the MPEG-21 user.

References

[1] E. T. Lin, A. M. Eskicioglu, R. L. Lagdendijk, and E. J. Delp, "Advances in digital video content protection," *Proc. IEEE*, 93(1): 171–183, January 2005.

[2] A. J. Menezes, P. C. van Oorschot, and S. A. Vanstone, *Handbook of Applied Cryptography*, p. 265, CRC Press, 1997.

[3] B. Schneier, *Applied Cryptography*, second edition, pp. 319–325, John Wiley & Sons, 1996.

[4] "Data encryption standard (DES)," American National Standard X3.92, Federal Information Processing Standards (FIPS), FIPS 46–3 FIPS 74 and FIPS 81, http://csrc.nist.gov/publications/fips/fips46-3/fips46-3.pdf.

[5] "Advanced encryption standard (AES)," FIPS 197, http://csrc.nist.gov/publications/fips/fips197/fips-197.pdf.

[6] CAST, IETF RFC 2144, http://www.rfc-editor.org/rfc/rfc2144.txt.

[7] X. Lai, *On the Design and Security of Block Ciphers*, ETH Series in Information Processing, J. L. Massey (ed.), Vol. 1, Hartung-Gorre Verlag Konstanz, Technische Hochschule (Zurich), 1992.

[8] "Pretty Good Privacy (PGP)" v2.0, http://www.pgpi.org/.

[9] Rivest Ciphers (RCs), RC2 (IETF RFC 2268), RC5 (IETF RFC 2040), RC6, http://people.csail.mit.edu/rivest/Rc6.pdf.

[10] W. Diffie and M. E. Hellman, "New directions in cryptography," *IEEE Trans. Information Theory*, IT-22, 644–54, November 1976.

[11] R. C. Merkle, "Secure communication over an insecure channel," *Commun. Ass. Comp. Mach.*, 21, 294–99, April 1978.

[12] Martin E. Hellman, "An overview of public key cryptography," *IEEE Communi. Mag., 50th Anniversary Commemorative Issue*, 42–49, May 2002.

[13] R. L. Rivest, A. Shamir, and L. Adleman, "On digital signatures and public key cryptosystems," *Commun. Ass. Comp. Mach.*, 21, 120–26, February 1978.

[14] R. J. McEliece, "A public key system based on algebraic coding theory," JPL DSN Progress Report, 1978.

[15] APNIC Certificate Authority Status Report, Special Interest Group, March 1st, 2000, Seoul.

[16] R. Rivest, "The MD5 message-digest algorithm," IETF RFC 1321, http://tools.ietf.org/html/rfc1321.

[17] D. Eastlake III and T. Hansen, "US secure hash algorithms (SHA and HMAC-SHA), IETF RFC 4634," http://tools.ietf.org/html/rfc4634.

[18] G. C. Langelaar, I. Setyawan, and R. L. Lagendijk, "Watermark in digital image and video data," *IEEE Signal Processing Mag.*, 17(5): 20–46, September 2000.

[19] M. Wu and B. Liu, *Multimedia Data Hiding*, Springer-Verlag, 2003.

[20] J. R. Hernandez Martin and M. Kutter, "Information retrieval in digital watermarking," *IEEE Communi. Mag.*, 38(8): 111–116, August 2000.

[21] I. J. Cox, J. Kilian, F. Thomson Leighton, and T. Shamoon, "Secure spread spectrum watermarking for Multimedia", *IEEE T-IP*, 6(12): 1673–1687, December 1997.

[22] D. Boneh and J. Shaw, "Collusion-secure fingerprinting for digital data," *IEEE Trans. Inf. Theory*, 44: 1897–1905, September 1998.

[23] W. Trappe, M. Wu, and K. J. R. Liu, "Anti-collusion codes: multiuser and multimedia perspectives," in *Proc. IEEE Int. Conf. Image Processing*, vol. 2, pp. 149–152, 2002.

[24] F. Pereira, J. R. Smith, and A. Vetro, "Introduction to special section on MPEG-21," *IEEE Trans. Multimedia*, 7(3): 397–399, June 2005.

[25] Jan Bormans, Keith Hill, MPEG-21 Multimedia Framework, ISO/IEC JTC1/SC29/WG11/N5231, October 2002. http://www.chiariglione.org/mpeg/standards/mpeg-21/mpeg-21.htm.

[26] "Information technology – multimedia framework (MPEG-21) – Part 2: Digital item declaration," ISO/IEC 21000–2:2003, March 2003.

[27] I. S. Burnett, S. J. Davis, and G. M. Drury, "MPEG-21 digital item declaration and identification – principles and compression," *IEEE Trans. Multimedia*, 7(3): 400–407, June 2005.

[28] J. Bormans, J. Gelissen, and A. Perkis, "MPEG-21: the 21st century multimedia framework," *IEEE Signal Process. Mag.*, pp. 53–62, March 2003.

[29] J. Bekaert and H. Van de Sompel, "Representing digital assets using MPEG-21 digital item declaration," Technical Report LA-UR-05–6633, Los Alamos National Laboratory.

[30] F. De Keukelaere, S. De Zutter, and R. Van de Walle, "MPEG-21 digital item processing," *IEEE Trans. Multimedia*, 7(3): 427–434, June 2005.

[31] "Information technology – multimedia framework (MPEG-21) – Part 3: Digital item identification," ISO/IEC 21000-3:2003, March 2003.

[32] X. Wang, T. DeMartini, B. Wragg, M. Paramasivam, and C. Barlas, "The MPEG-21 rights expression language and rights data dictionary," *IEEE Trans. Multimedia*, 7(3): 408–417, June 2005.

[33] "Information technology – multimedia framework – Part 5: Rights expression," ISO/IEC 21000-5:2004, 2004.

[34] Information technology – multimedia framework – Part 6: Rights data dictionary, ISO/IEC 21000-6:2004, 2004.

[35] G. Rust and B. Bide. "<indecs> final report," 2000, http://www.indecs.org/project.htm#finalDoc.

[36] D. Borses, "MPEG-21 multimedia framework," presentation to the IEEE Computer Society, OCCS, November 25, 2002.

[37] "Information technology – multimedia framework (MPEG-21) – Part 7: Digital item adaptation," ISO/IEC IS 21000-7:2004, 2004.

[38] A. Vetro, C. Christopoulos, and H. Sun, "Video transcoding architectures and techniques: an overview," *IEEE Signal Process. Mag.*, 20(2): 18–29, March 2003.

[39] I. Ahmad, X. Wei, Y. Sun, and Y.-Q. Zhang, "Video transcoding: an overview of various techniques and research issues," *IEEE Trans. Multimedia*, 7(5): 793–804, October 2005.

[40] H. Choi, J. W. Kang, and J.-G. Kim, "Dynamic and interoperable adaptation of SVC for QoS-enabled streaming," *IEEE Trans. Consumer Electron.*, 53(2): 384–389, May 2007.

[41] T. C. Thang, Y. S. Kim, J. W. Kang, Y. M. Ro, and J.-G. Kim, "SVC video adaptation with MPEG-21 DIA adaptation QoS," ISO/IEC JTC1/SC29/WG11, doc. m12638, Nice, October 2005.

[42] D. Mukherjee, "MPEG-21 DIA: objectives and concepts," Class report, ECE 289J, UC Davis, 2004. This report also appeared in D. Mukherjee, G. Kuo, A. Said, G. Beretta, S. Liu, and S. Hsian, "Proposals for end-to-end digital item adaptation using structured scalable meta-formats (SSM)," ISO/IEC JTC1/SC 29/WG 11 MPEG2002/M8898.1, Shanghai, October 2002.

[43] World Wide Web Consortium Ebrahimi, "Synchronized multimedia integration language (SMIL 2.0)," 2001, http://www.w3.org/TR/2001/REC-smil20-20010807/.

[44] "Information technology – multimedia framework, – Part 10: FCD digital item processing," ISO/IEC 21 000-10 (MPEG-21), N6780, October 2003.

[45] "Standard ECMA-262 ECMAScript language specification third edition," ECMA, 1999, http://www.ecma-international.org/publications/standards/Ecma-262.htm.

[46] "Java programming language", Sun Microsystems, http://java.sun.com/.

[47] A. M. Alattar, E. T. Lin, M. U. Celik, "Digital watermarking of low bit-rate advanced simple profile MPEG-4 compressed video," *IEEE Trans. Circuits Syst. Video Technol.*, 13(8): 787–800, August 2003.

12 Implementations of multimedia networking

12.1 Speech and audio compression module

This speech and audio compression module, which (as well as all the software associated with this chapter) can be downloaded from http://allison.ee.washington.edu/index_files/ page565.htm, will allow you to perform speech and audio compression. There is a restriction on the input speech files. When we select a speech file for compression, the file must be *.wav. After compression we can either detect the effect on different tones or discern the increasing improvement in speech compression through distinct encoding methods.

12.1.1 Introducing the graphical user interface (GUI)

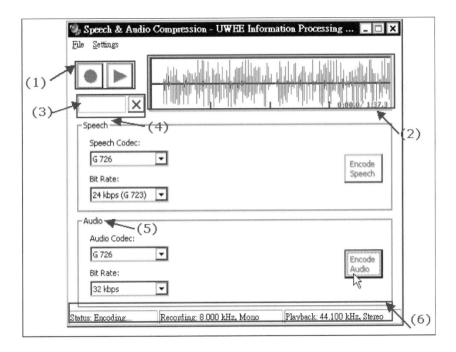

(1) **The Record and Play buttons** allow us to record or play our speech signals and audio files.

(2) **The waveform window** shows the signal waveform on which we shall be operating.

(3) **The Progress bar** informs us of the compressing progress by the filling of the bar. If we want to stop encoding, we just push the Stop button. (This will only be visible during encoding.)

(4) **Using the speech codec block** we can choose our speech codec, bitrate, and other options we want to use during encoding.

(5) **Using the audio codec block** we can choose our audio codec, bitrate, and other options we want, and then begin encoding from this block.

(6) **The Status bar** shows the recording sampling rate and the playback sampling rate.

12.1.2 Quick start guide

12.1.2.1 Speech codecs

(1) Step 1   Plug the headphones into your computer.

(2) Step 2   Start Sound Recorder from the Start menu.

(3) Step 3   Now record some speech, saying what you like.

(4) Step 4   Save the file in any directory.

(5) Step 5   Start Speech & Audio compression.

The Speech & Audio compression window will show up.

(6) Step 6   Open a *.wav file that you have recorded and listen to the speech signal by clicking Play. We will use the file speech_2.wav in this example.

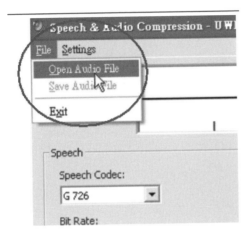

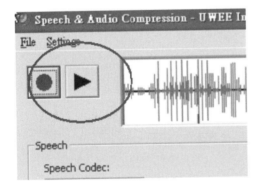

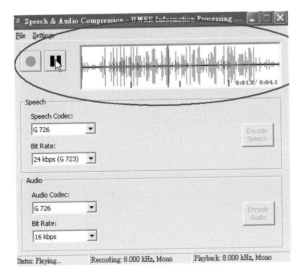

(7) Step 7   Select an encoding option. Here, we choose the standard G.723.1 with 5.3 kbps bitrate.

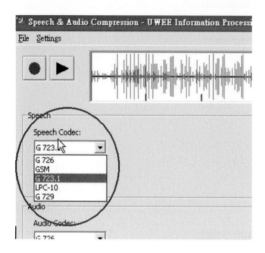

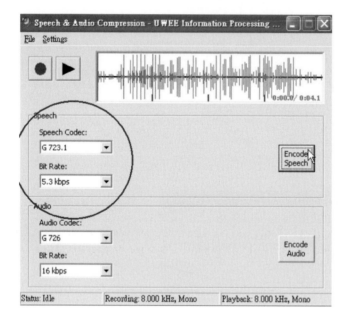

(8) Step 8   When compression has been completed, we can see that the compression factor is 24.110 : 1. To hear the difference, we can save the compressed file and then compare it with the uncompressed file.

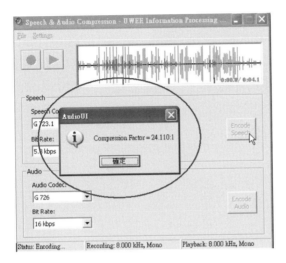

Note that if we use Save Audio File to save the compressed file, which is in fact already decompressed from the encoded bitstream, we should find out whether the size of this saved file is the same as the original speech/audio file. To obtain a copy of the real compressed bitsream, which indeed has a smaller size but requires a special player to play it, we can select Save Bitstream under Settings to get the corresponding compressed bitsream. Please verify the file size using the compression factor. These bit-streams are the actual data to be saved in the storage or packetized for networking purposes.

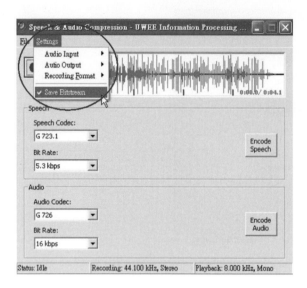

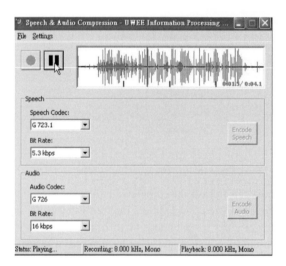

12.1.2.2 Audio codecs

(1) Step 1   Before starting, please prepare a *.wav file sampled at various sampling frequencies (e.g., 44.1 kHz). These can be pre-recorded or live captured audio. If we are compressing files using MPEG-4 AAC, we can also select a file sampled at 96 kHz.

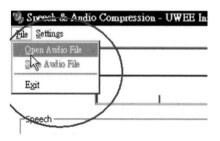

Play the music to make sure that is the file you want. We will use the file audio_example1.wav in this example.

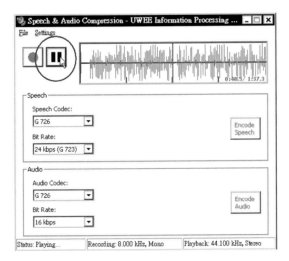

(2) Step 2 Select the audio codec. Here, we have selected the G.726 (ADPCM) standard with a 32 kbps bitrate.

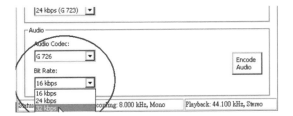

(3) Step 3   Now push the Encode Audio button to start compressing the file.

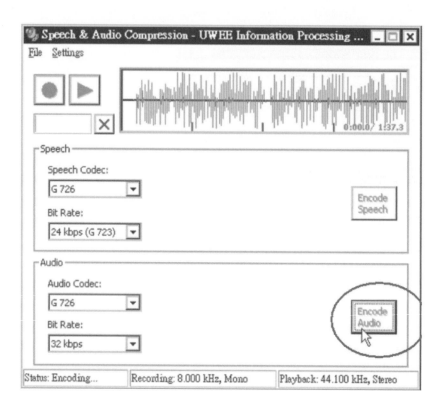

A Progress bar will show up under the Play button to tell us how the compression is progressing. To cancel it, just push the red Stop button.

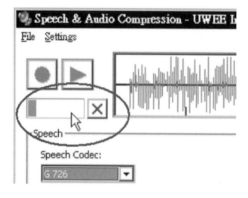

An error dialog will show up.

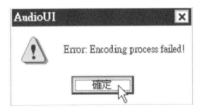

After compressing the file, there will be a message box showing the compression factor.

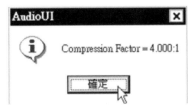

(4) Step 4   Finish.

We can now hear the encoded-then-decoded file and compare the quality with the uncompressed file.

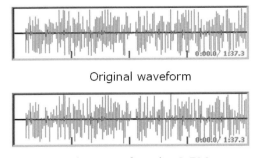

Original waveform

Encoding waveform by G.726

Also, we can save the encoded-then-decoded file, which again should have the same size as the original audio file. We can also save the compressed (encoded) bitstream alone for reference purpose.

12.1.3 Advanced usage

12.1.3.1 Speech coding

Before recording, we can choose either mono or stereo in accordance with our preferences and select our desired sampling rate.

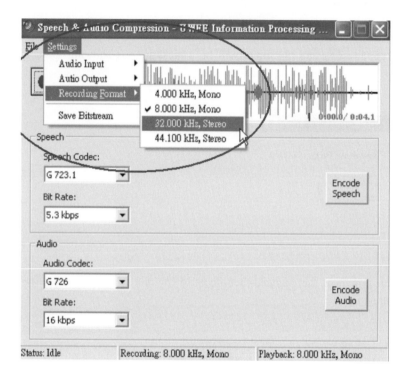

12.1.3.2 Audio coding
MPEG layer 3

For MP3 we can decide to use a variable bitrate (VBR) and also choose a quality factor.

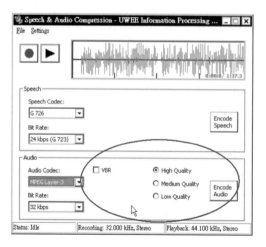

Now, let us compare a file without using VBR and the same file when VBR is used. First, without VBR:

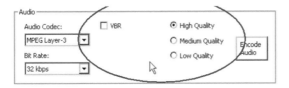

The compression factor is 44.113 : 1.

Second, with VBR (the chosen bitrate will not be enforced):

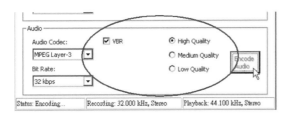

The compression factor is 6.884 : 1. If we save the compressed bitstream with file extension .mp3, this saved file infact can be played back by most commercially available MP3 players, such as the Microsoft WMA player.

Finally, compare the quality of the original audio and the compressed audio:

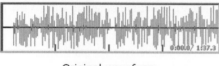

Original waveform

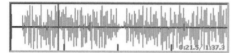

Encoding waveform by MPEG layer3 without VBR

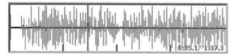

Encoding waveform by MPEG layer3 with VBR

MPEG-4 AAC

There are two additional coding options for MPEG-4 AAC that will generate different compression ratios.

First, without selecting any coding option:

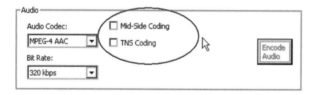

The compression factor is 16.891 : 1.

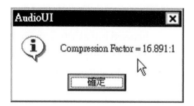

Second, with both coding options chosen:

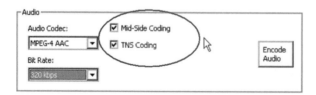

The compression factor is 17.231 : 1.

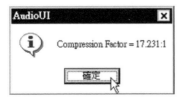

Finally, observe the quality of the original audio and the compressed audio. The compression factor is almost the same. This shows that the right and left sound channels of our example file are quite different:

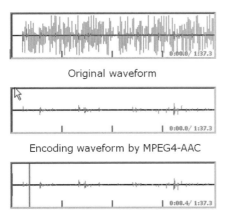

Original waveform

Encoding waveform by MPEG4-AAC

Encoding waveform by MPEG4-AAC with Mid-side coding and TNS coding

12.2   Image and video compression module

This image and video compression module, which can be downloaded from http://allison. washington.edu/index_files/Page565.htm, will allow you to perform image and video compression. The restriction on image compression is that the uncompressed file must be a *.bmp. The restriction on video compression is that files must be a *.avi without any prior compression.

12.2.1　Introducing the graphical user interface (GUI)

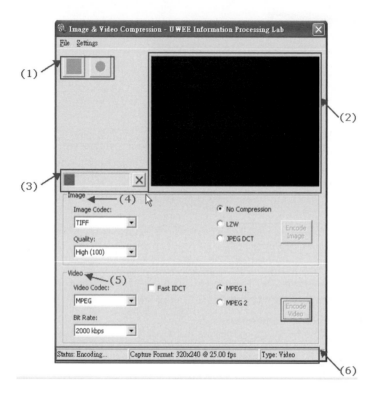

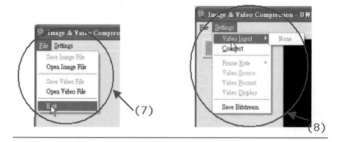

(1) **The Play and Record buttons** allow us to play or record image or video files.

(2) **The display window** displays the images or video files on which we are operating.

(3) **The Progress bar** informs us of the compressing progress by the filling of the bar. If we want to stop encoding, we just push the Stop button. (This will only be visible during encoding.)

(4) **Using the image codec block**, we can choose our image codec, bitrate, and other options. When ready we can begin encoding.

(5) **Using the video codec block**, we can choose our video codec along with the bitrate and other options before we begin encoding.

(6) **The Status bar** shows the capture format and the type of file that you are using at the moment.
(7) **The File menu** lists all file operations such as save and open.
(8) **The Settings menu** has three sections: choosing a camera, setting the video capture format, and saving the compressed bitstream file. Bitstream saving will be mentioned below in Section 12.2.3 on advanced usage.

12.2.2 Quick start

12.2.2.1 Image codecs

This software will help us encode our *.bmp files into various compression formats such as TIFF, JPEG, JPEG2000, PNG, and GIF. See whether your eyes can tell the difference!

(1) Step 1 First, prepare a *.bmp file (or capture it directly using the Record button of this software. Please refer to Section 12.2.3.3)
(2) Step 2 Open the Image & Video Compression program and then open a saved *. bmp file. In this example, we are using the "leaf.bmp" file shown below.
(3) Step 3 View the original *.bmp file.

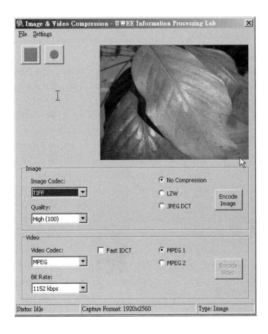

(4) Step 4 From the list, choose the coding scheme that you prefer. In this example, we will compress the image using baseline JPEG with low quality.

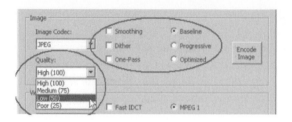

(5) Step 5   Encode the image. The Progress bar will show how the compression is proceeding.

(6) Step 6   The compression factor dialog will show up when finished, and we can also get the compressed bitstream if Save Bitstream is checked (with *.jpg extension). Again, this saved .jpg bitstream can be opened by most standard JPG viewers embedded in the computer browsers.

(7) Step 7   Save the compressed then decompressed image and compare the results.

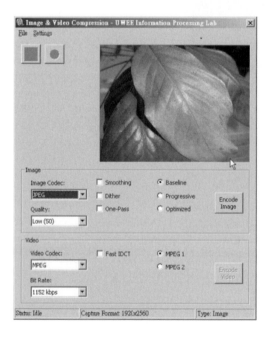

(8) Step 8   We can try other files with different encoding schemes by following the same procedure as in steps (1)–(7).

12.2.2.2 Video codecs

With this software we can compress the video to reduce the file size. We can also observe the distortion and tradeoff when using different video compression schemes. We have MPEG-1, MPEG-2, H263+, and H264 codecs available for use.

(1) Step 1   Prepare a *.AVI file. We could use the container.avi file for instance.

(2) Step 2   Open the Image & Video compression program and the saved *.AVI file.

(3) Step 3   First view the original file by clicking Play. We can see the file's format by looking at the status bar shown in the bottom of the software window. The capture format for this video is CIF (352*288), displayed at 15 frames per second.

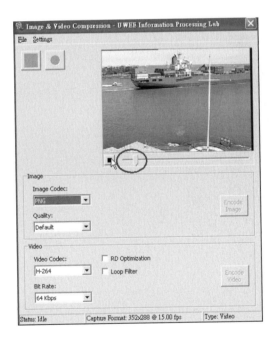

(4) Step 4   Choose from the list a coding scheme according to your preference. In this example, we have selected MPEG-1 at 2000 kbps and have used the fast IDCT (a

fast algorithm to speed up the computation without changing the compression results) for compression.

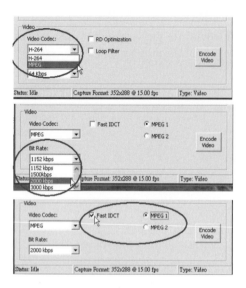

(5) Step 5   Run the coding scheme. The Progress bar will show how the compression is proceeding.

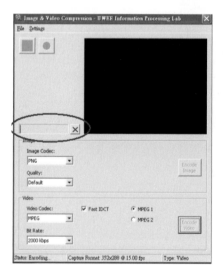

(6) Step 6   The compression factor dialog will pop up to show us that the compression factor is 15.203 : 1, which can be justified if the Save Bitstream option is enabled. The saved .mpg file can be played (decoded) by most standard MPEG 1/2 players, including Microsoft WMV players.

(7) Step 7   Save the file and observe the results.

(8) Step 8   We can try other video files using different codecs by following the same procedures as in steps (1)–(7).

12.2.3 Advanced usage

12.2.3.1 Image coding

Save bitstream   Before we start to encode a file, we can mark Save Bitstream to save the bitstream of the compressed file for future viewing.

First, we check the size of leaf.bmp uncompressed (14 MB).
Second, after compression, we can check the file size of Capture.jpg (362 kB).

Did the file size decrease? If yes, what is the compression ratio corresponding to the ratio of these two files? Note that files compressed by the JPEG2000 codec can be viewed by another, more sophisticated, viewer, kdu_show.exe (also available in the software download directory).

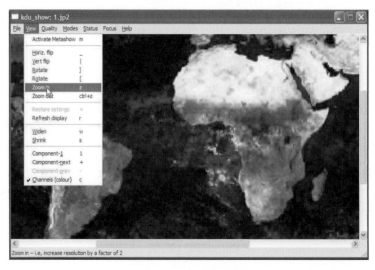

More specifically, this viewer allows us to zoom in and zoom out to or from a more detailed picture, and it also allows us to view various layers (resolutions) of the compressed image (for scalable purposes).

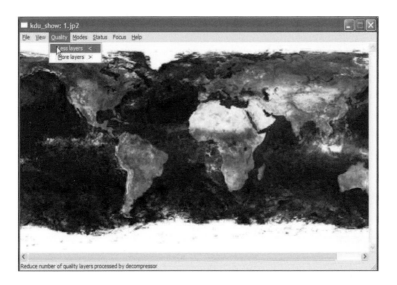

12.2.3.2 Video coding

Save bitstream   Before we start to encode a file, we can mark Save Bitstream to save the bitstream of the compressed file for future viewing.

First we check the size of container.avi before compression (43.5 MB).

Second, after compressing, we can check the size of capture.mpg (2.86 MB).

The file size has indeed decreased significantly! The compression ratio of these two files is 15.023 : 1, the same as step 6 in Section 12.2.2.2

Note that this compressed file can be opened by real player, windows media player, etc., as long as it supports the MPEG compression scheme.

12.2.3.3 Connect to camera

We may connect a camera to the computer to capture images or record live videos directly from our software before using the compression engines.

(1) Step 1   Connect a camera device to your computer, and choose at the bottom Connect.

(2) Step 2   Choose the Frame Rate to fix the number of frames per second at which you want to record.

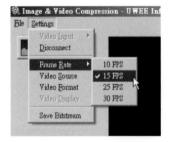

(3) Step 3   Press Video Source, which will set up the camera. You will see several options that can modify how the camera is used.

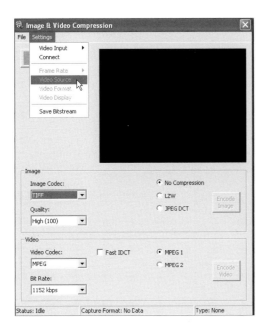

(4) Step 4   Press Video Format, which sets up video capture and enables several options that can modify how the capturing is done. In our example, we will use 640*480 resolution for the digital video format.

(5) Step 5   Now we can record a video clip or simply capture a snapshot of the image.

(6) Step 6   We can push the Play button to observe the results.

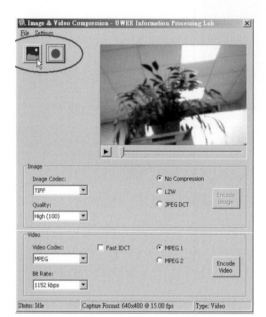

12.3   IP networking module

There are four layers in a TCP/IP stack, the application layer, the transport layer, the network layer, and the data-link and physical layer. To facilitate the communications between two computers within a network, we can use sockets. Sockets are used by an application to send or receive packets. Using a socket is similar to opening a door and only allowing computers that have the same type of protocol and port to pass through; also, they must know each other's IP address. The protocols supported by the socket's transport layer are the TCP and UDP.

The IP networking module, which can be downloaded from http://allison.ee.washington. edu/index_files/Page565.htm. can be used on a single computer as long as there are two network sockets and packets applications activated on your PC (one as the server and the other as the client). It is better, however, to experiment with this application on two separate computers.

12.3.1 Introducing the GUI

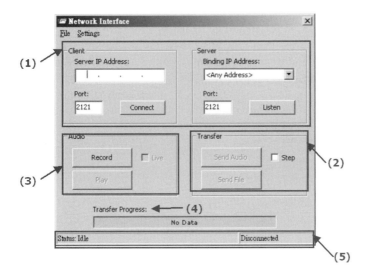

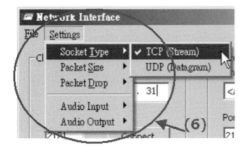

(1) From the connection block you can set the client and server options. In the next section, we will talk about how to set up a connection.

(2) Using the transfer window, we can send data to our destination, from either the client or server by pushing these buttons. If you check step you will divide the data into smaller packets that you can see being sent and whose size can be set in the setting menu. You only send one packet size of data each time you push the Send Audio button or the Send File button.

(3) The audio block is used to select options if you want to record an audio file. The details will be discussed in the following section.

(4) The Transfer Progress bar only fills up when the Send Audio button or the Send File button has been pushed. From this we can see the transmitting progress from source to destination.

(5) The Status bar shows the program status and the status of the network.

(6) The Settings menu has four parts: the protocol type, the packet settings, the Internet simulation settings, and the audio settings.

12.3.2 Quick start

(1) Step 1　Open the Net & Socket program on two different computers or open it twice on the same PC. One serves as the client, the other as the server. In this example, we are using the software on the same PC, so we have to open the program twice.

(2) Step 2　Now, decide the socket type to be used, the packet size, and the percentage of packets that will be lost in transmission. Make sure that both the client and the server have the same socket type.

Note: UDP is typically used, instead of TCP, when sending audio.

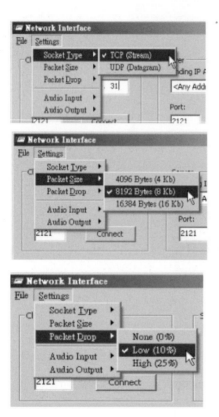

(3) Step 3　On the server side, we need to specify the IP address and port number of our server. On the client side, we need to type in the server's IP address and port number.

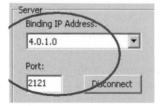

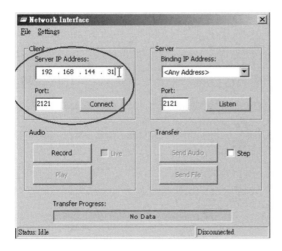

(4) Step 4   We now press Listen on the server side. The client can start by pressing the Connect button. Remember that the server must listen before the client connects, since all IP connections have to be initiated by clients.

(5) Step 5   The message Connected will show up in the Status bar when the network connection is set up, and the block for transferring will be activated. For instance, on the server side we see:

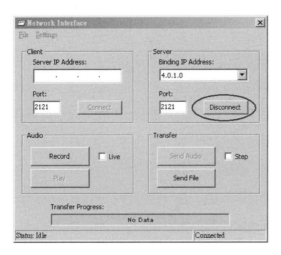

(6) Step 6   Send a file from client to server. We could also send a file from server to client. Both connections, however, are only one-way transmissions.

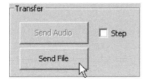

When transmission begins, there will be a Transfer Progress bar to let us know that the packets are being forwarded. When the transmission is completed, a success message will show up in the Status bar.

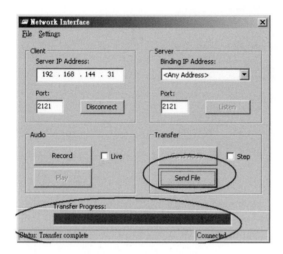

If you mark Step then the Transfer Progress bar will behave like this:

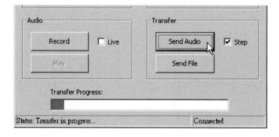

Each time you push the Send File or Send Audio buttons, one packet will be sent.

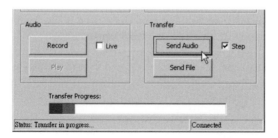

(7) Step 7 Disconnection: If either the client or server wants to break the connection, click on Disconnect.

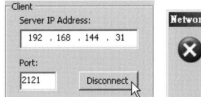 

(8) Step 8 Change computers and try steps (1)–(7) again. Is there a difference in performance?

12.3.3 Advanced usage

12.3.3.1 Send audio by marking Live

Following steps 1–5 in Section 12.3.2, we can also send audio files live. Either the client or server will send audio and the other will just receive audio.

If the sender presses the Record button their voice will be sent straight away, and the receiver will hear the audio immediately unless the sender pushes the Stop button.

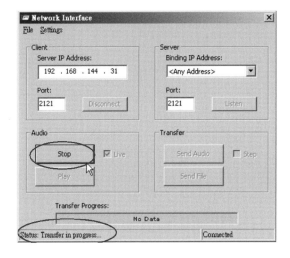

12.3.3.2 Send audio without marking Live

Follow steps1–5 in Section 12.3.2. Press Record, and when the sender has finished recording their voice press Stop. When the sender presses the Send Audio button, the Server will then receive the complete audio.

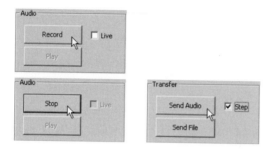

The receiver can press Play after receiving the audio to listen to the recording.

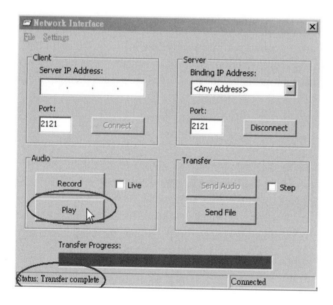

Note that you cannot choose the Stop button if the audio is still sending; in the screenshot it had actually finished.

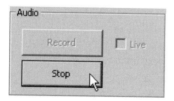

If you choose the Packet Drop option, selecting either a 10% or a 25% packet loss, different outcomes will occur after the communication between the sender and the receiver. For example,

(1) if the sender has set the packet drop to Low (10%) Packet Drop, then this dialog will appear:

(2) If the receiver has set the packet size to High (25%) Packet Drop, then this dialog will appear:

12.4   Audio and video capturing and displaying

In this section, we will discuss how to create Microsoft Foundation Class (MFC) applications for video/audio capture and playback. We will demonstrate how to set up and choose the options for an MFC application. The source codes provided, which can also be downloaded from the same website, have to be copied into the projects and then built and run. These steps are basic skills in programming and creating your own multimedia applications. The objectives of this section are to

• create a GUI application using MFC
• create an application to capture and display video
• create an application to capture and playback audio

Creating an application that can support video capability is an easy task if you are using Microsoft's libraries. Although Video for Windows (VFW) is not the most advanced technology, when compared with the more extensible DirectShow, it can support many video related operations such as reading and displaying a video file and capturing and saving a video file. We will take advantage of VFW's ease of use and simple model. Here we give an introduction to the use of Visual Studio 2005 that is by no means exhaustive. However, it should serve as a primer to aid your understanding of the examples presented in the coming few sections. In this section we will demonstrate how to create a new MFC application to capture and playback video and audio. Our first, detailed, example shows clearly the steps in creating the MFC application. The following two examples follow the same idea and concepts as the first example.

12.4.1 Creating a new MFC project

12.4.1.1 Choosing the project type

To create a new application, start Visual Studio 2005, go to the menu, and click File -> New -> Project. Our project type is C++, the subtype is MFC, and we are creating an MFC Application. You will need to enter a name for your project. For the first example, you need to enter VideoCapture as the name. Click OK when everything is ready.

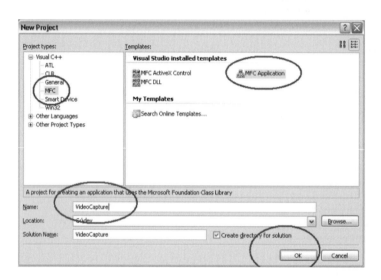

The next dialog will give you an introduction to the MFC Application Wizard. You have a choice to click Finish or Next. Do **NOT** click Finish because we have a few more options to decide. Clicking Finish will only set the default options.

12.4.1.2 Choosing the application type

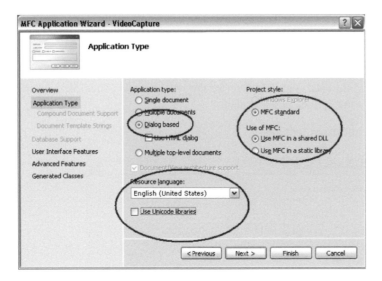

The next dialog will show the Application Type options. Since we are not creating any documents such as a word or power point document we will not choose Single or Multiple documents but instead will choose Dialog based for our application. For the language choose English (United States), which in any case would be selected by default. For our project style choose MFC standard. Choose Use MFC in a static library for the option *Use of MFC*. Note that the above does *not* reflect this choice for Use of MFC. Use the specified choice mentioned above. The last option to choose is to uncheck the Use Unicode libraries box. This will simplify our string manipulation functions. Click Next when all the options are set for the application type.

12.4.1.3 Choosing the interface features

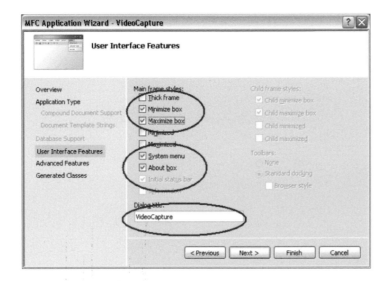

The next dialog window is User Interface Features. These options allow you to choose how the user interface is displayed. Make sure that the following boxes are checked:

Minimize box
Maximize box
System menu
About box

For the dialog title enter VideoCapture for the first example. Click Next when you have finished setting the options in this dialog.

12.4.1.4 Choosing the advanced features

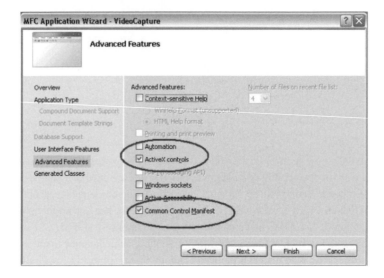

The next dialog will let you choose advanced features. By default ActiveX controls and Common Control Manifest are checked. You may leave them as they are and click Next.

12.4.1.5 Generated classes

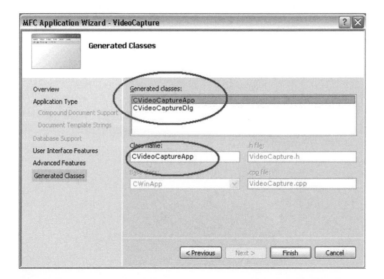

The next dialog lets you choose the name of the generated class. By default, the wizard will name the class in the format CxxxxxxxxApp and CxxxxxxxxxDlg. You do not need to change these names. Just click Finish and you have completed the setting up of a new MFC application.

12.4.2 Navigating Visual Studio 2005

The following sections will show how to explore the Visual Studio 2005 resources such as the code, graphics, and user interfaces so that you will become familiar with how to obtain these items for modification.

12.4.2.1 Browsing the solution tree

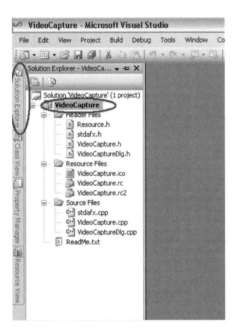

On the right-hand side of the development environment you should see four tabs each having a tree that lets you expand the branches. The Solutions tab will allow you to look at the source documents in the project.

12.4.2.2 Browsing the Class View tab

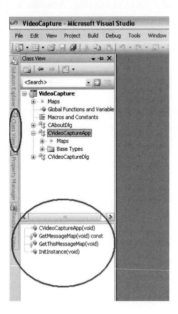

The Class View tab will show you the classes and their member variables and functions.

12.4.2.3 Browsing the Resource View tab

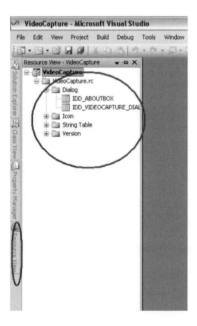

The Resource View tab will show you the multimedia resources that are used in the project.

12.4.3 Adding files in the project solution

On the Solutions tab, click on Source Files. Check that the following files are there:

- stdafx.cpp
- VideoCapture.cpp
- VideoCaptureDlg.cpp

On the tree, click on Header Files and you should see the following files:

- Resource.h
- stdafx.h
- VideoCapture.h
- VideoCaptureDlg.h

In order to simplify your work, the following files have been prepared to add to the project.

- Header Files:
    VideoCapture.h
    VideoCaptureDlg.h
    VideoIn.h
    VideoOut.h
- Source Files:
    VideoCapture.cpp
    VideoCaptureDlg.cpp
    VideoIn.cpp
    VideoOut.cpp
- APS Files:
    VideoCapture.aps
- Resource Files:
    resource.rc
- Resource Folder:
    res

You will need all the software files provided. There are three folders, one each for VideoCapture, VideoPlayback, and AudioUI. For this example copy all of the files in VideoCapture and place them in your project folder where you find similar codes. For example, if in your project directory you find .h and .cpp files, you should copy the provided source file into this directory.

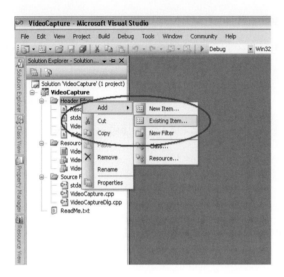

Now go back to the project. On the solutions tree right-click the Header Files branch and click Add Existing Files. Select all the header files in your project folder. Then right-click on the Source Files branch and click Add Existing Files; select all the CPP files and click OK.

12.4.4 Adding external dependencies

We are now almost ready to compile and test the project. Open the stdafx.h file and add the following two lines after the lines at the end of the document if they are not already there:

```
#include "vfw.h"
#pragma comment(lib, "vfw32")
```

Now you should have the complete file needed for creating the video capture example and will be ready to compile and build the project.

12.4.5 Building the application

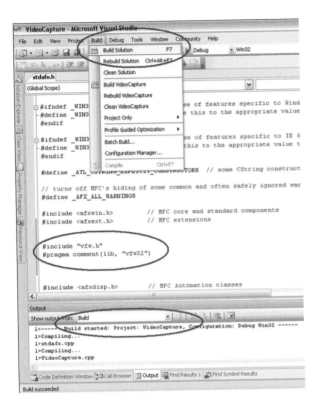

To build the projects go to the menu and click Build -> Build Solution. The project output window at the bottom of Visual Studio will start showing outputs. To take a shortcut you can press the F7 key to start the build process.

12.4.6 Building a solution versus building a project

Here is a short description to clarify the differences between building and rebuilding a solution and between building a solution as opposed to a project.

The difference between Build and Rebuild is that Build will compile and link only if you have changed the source code and the changed parts have been compiled and linked. For Rebuild it will compile and link all files afresh regardless of any changes.

The difference between building a solution and building a project is that in a solution you may have more than one project. Building a solution will build all the projects in the solution. Building a project will only build the project and leave other projects untouched unless the project is dependent upon another project.

12.4.7 Running the application

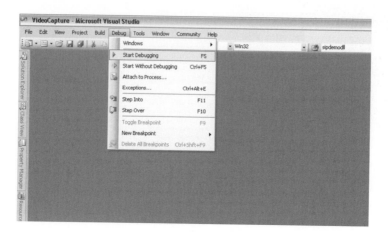

To run the application go to the menu and click Debug -> Start Debugging and/or press the F5 shortcut key to start the debugging process.

12.4.8 Connecting to the camera to start recording

The following will describe how to run the application.

Above is a screenshot of the application when you start. In order to start recording a video you must first connect to the video capturing device. On the menu click Settings -> Connect Camera.

After the capturing device is connected you may start capturing with the device by clicking the Record button •, which was grayed out before it was connected.

To stop recording, simply click the Stop button ■.

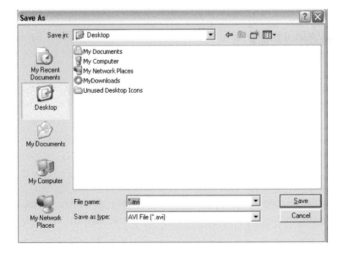

To save a file, on the menu bar click File -> Save. This will let you save the captured video to a file.

12.4.9 Creating and running the VideoPlayback applications

Now that we have an application to capture video we can create another application to display the video. Using the steps in Section 12.4.1, start another project named "VideoPlayback" with the same options as those used for creating the MFC application.

Your next task will be to copy the files in the folder VideoPlayback into your own project folder just as for the last example.

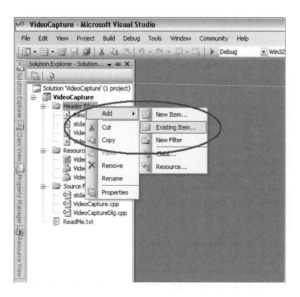

Now you can add all the files within the project in Solution Explorer. On the Solutions tree right-click the Header Files branch and click Add -> Existing Item. Select all the header files in your project folder. Then right-click on the Source Files branch and click Add -> Existing Item, then select all the CPP files and click OK.

We are now almost ready to compile and test the project. Open the stdafx.h file and add the following two lines after the lines at the end of the document if they are not already there:

```
#include "vfw.h"
#pragma comment(lib, "vfw32")
```

Now you should have all the files necessary. Build and run the application as you did for the VideoCapture example. To open up a file simply go to the menu and click File -> Open. The VideoPlayback application only plays back files in .avi extension and format.

12.4.10 Creating and running the Audio Rec/Playback applications

We are now ready to build the audio capture and playback application. Again, use the same method to start a new project in Visual Studio 2005 with the same settings as for the previous two projects.

Copy all the files in the AudioUI folder into your own project directory. Now go back to Visual Studio 2005 and add the files from the solution menu. Be sure to add all the header files into the Include directory and all the CPP source files into the Source directory.

After all the files have been added, also add the following four lines to the end of stdafx.h file if they are not already there:

```
// Multimedia support
#include <mmsystem.h>
#include <mmreg.h>
#pragma comment(lib, "winmm.lib")
```

The next step is to build the application and run it. Build the project and see whether there is any error. If not you are ready to test the application. The figure below shows the application.

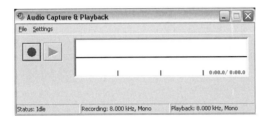

You can begin capturing audio by clicking the Record button ●.

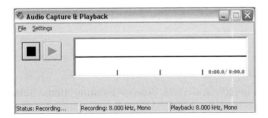

When the application is recording the Record button ● will change to the Stop button ■ and the status bar will indicate that it is recording.

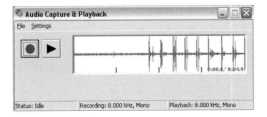

After you click Stop the waveform of the audio just recorded will show and you can either click Record to record another audio clip or click the Play button ▶ to hear the recording.

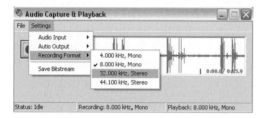

To set the sampling frequency of the recorded audio click Settings -> Record Format and choose the desired frequency.

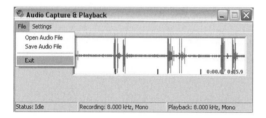

After you have recorded the audio clip you can save it and exit the application.

12.5 Encoding and decoding of video or audio

Our previous example of video capture and playback allowed us to record and play video and audio. Our next set of examples will demonstrate how to encode and decode these video and audio files based on the video or audio encoders. These examples demonstrate how to write an application to control the encoding and decoding of video and audio with the MPEG or H264 video codecs and the advanced audio coding (AAC) codec. The video codecs are packaged in either dynamic or static libraries. The main task will be to create a console application to call the encoding and decoding functions contained in the libraries.

12.5.1 MPEG video encoding and decoding

12.5.1.1 Starting a new console project

The first example will be MPEG encoding and decoding. First, create a console application with Visual Studio. In order to create a console application, on the menu click File -> New -> project.

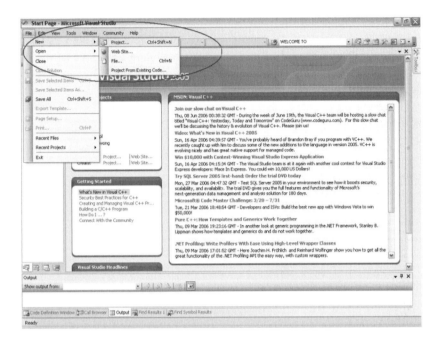

12.5.1.2 Choosing the project options

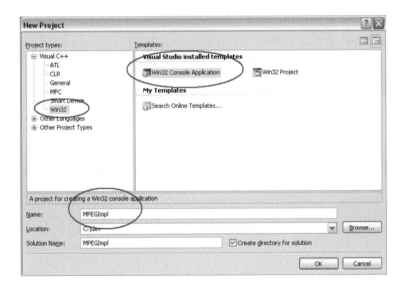

Now choose *Wind32* as the project type and *Win32 Console Application* as the template type. Lastly, name the project *MPEGImpl*. (Note that you must name the project exactly because the example code uses the same name.)

12.5.1.3 Setting project Properties

There are a few project settings that we want to set before moving on. First, on the menu click Project -> Project Property. The dialog below will show up. On the left-hand side click Configuration Property and, on the right-hand side, for the field *Character Set* choose the option Use Multi-Byte Character Set.

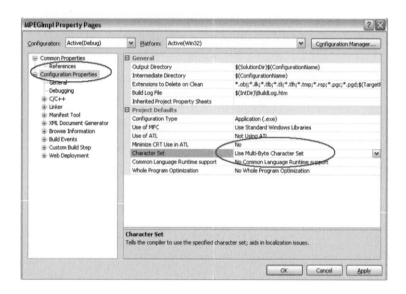

In the same window go to Configuration Properties -> General -> C/C++ -> Additional Include Directories and set it to ..\Include. This will add another folder allowing the compiler to search for header files.

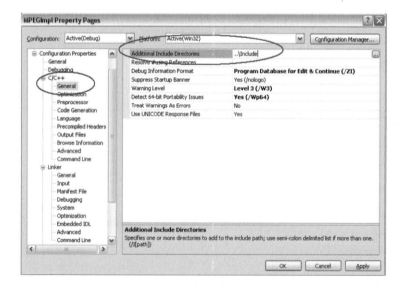

Also in the same window as shown below, on the left-hand side click Linker -> General. Then on the right-hand side set the field Additional Library Directories to ..\Lib. This will add another folder for the linker to search for libraries. Then on the left-hand side again click Input. On the right-hand side, for the field Ignore Specific Library enter LIBCMTD. This library will then be ignored; including it would cause redefinition errors.

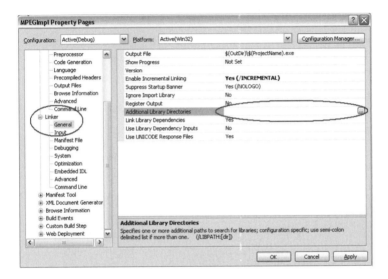

Note that the above steps are required if you change the project configuration from Debug to Release or any other type of configuration. If you desire to apply the above settings to all configurations you must choose All Configurations before you make the changes.

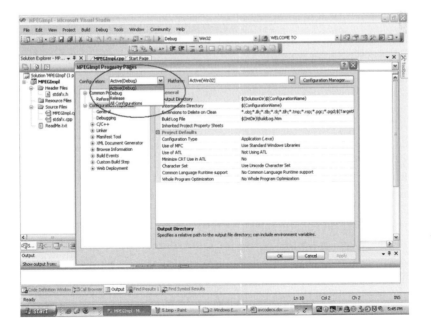

12.5.1.4 Adding external dependencies and building the project

Once you have set the project options, you are ready to copy the source files provided into your own project folder. First, copy the two folders Include and Lib into your own project folder. Then copy the files from the MPEGImpl folder into the same folder in your project directory. The operating system allows you, if you want, to replace the files that are already in this folder. Choose YES.

Another step is needed before you can build and run the MPEG codec. Open the file stdafx.h and add the following lines if they are not already there:

```
#pragma comment(lib, "vfw32")
#pragma comment(lib, "aviconvd")
#pragma comment(lib, "mpeg2enc")
#pragma comment(lib, "mpeg2dec")
```

The lines above add a reference to the libraries required in this project.

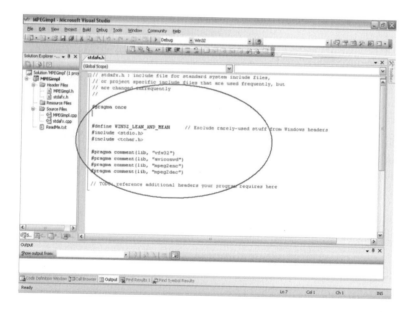

Once you have finished these steps you are ready to build the project. When the project is successfully built, start the console by clicking Start -> Run ... In the text box type cmd and click Enter. This will start the console window. In the command prompt, change to the directory of the project and go into the debug folder.

12.5.1.5 Encoding a video file with an MPEG encoder

First, use the VideoCapture application you captured from Section 12.4 to record and save a video. Copy this video file into the debug directory of your project. Now run the application by typing MPEGImple.exe in the command prompt. The application will ask you a few questions in order to set the parameter for the codec. First, it will ask for a source file. Enter the name of the test video file that you have just copied into your own project. The temporary file is used for the intermediary file. You may name the file as you want but it must have an .mpg extension. The application will then ask you for the destination file. This will be the file reconstructed after the source file was encoded into the temporary file. If you open the temporary file with a media player such as Windows Media Player you will be able to see the encoded video.

12.5.2 Encoding H.264 video

In the next example we will build a console application to run the H264 codec. Again we will encode the video, create an intermediary file, and also reconstruct the encoded video file. In this example we will use a dynamically linked library as opposed to the statically linked library we used for the MPEG example.

12.5.2.1 Starting a new console project

First create another console application project in exactly the same way as in the MPEG example. Make sure that you also set the project settings in the same way, for example, setting the paths for Additional Include Directory and Additional Library Directory. Once you have a console application you can add the necessary source, header, and library files provided in the H264 directory and copy all the files including the folders into your own project directory.

12.5.2.2 Adding external dependencies and building the project

Adding external dependencies can be done by adding the following codes to the end of the stdafx.h file if they are not already there:

```
#pragma comment(lib, "aviconvd")
#pragma comment(lib, "h264")
```

12.5.2.3 Building and running the application

Now that all the necessary files have been added and the project properties set up, you should be able to compile the project without any problem. Check to see whether the debug folder has been created in your main project folder. Copy the H264.dll file into the debug folder. You are now ready to execute the H264 codec. Copy the test video file that you created for the MPEG example into the debug folder of the H264 project. Now start up a console and change to the directory in which the application resides. Now type the name of the program and click Enter. The application will ask you a sequence of questions and will let you choose the options. Submit your choices accordingly. Note that the temporary file will be the encoded file and the destination file will be the reconstructed file after decompression.

12.5.3 Encoding AAC audio

In the next example we will build a console application to run the AAC audio codec. Like the previous two video encoding examples, we will first encode the uncompressed file and then create an intermediary file and also reconstruct the compressed file. In this example we will use statically linked libraries again. Also, we will be building the libraries from the source code of the codec.

12.5.3.1 Starting a new console project

First, create another console application project with exactly the same options as in the previous two examples. Make sure that you also set the project settings in the same way, for example, setting the paths for Additional Include Directory and Additional Library Directory. There is only one additional option you need to set that is different from the other two examples. On the project properties dialog on the left-hand side, click Configuration Properties -> General and set the field Use of MFC to Use MFC in Shared DLL. Once you have a console application you can add the necessary source and header files from the provided FAAC directory. Copy all the files including the folders from the provided directory into your own project directory. Add all the .h files and .cpp files in the FAACImpl solution. Do this by right-clicking Header Files -> Add -> Existing Item and also Source Files -> Add -> Existing Item.

12.5.3.2 Creating the library projects

There are two separate libraries with the AAC codec. The source code for the encoder is in the libfaac directory and the source code for the decoder is in the libfaad directory. In order to build these two libraries we need to create two new Win32 projects. Within the same project as the FAAC project add a new project by right-clicking the Solution menu and clicking Add -> New Project.

The familiar Add New Project dialog will appear.

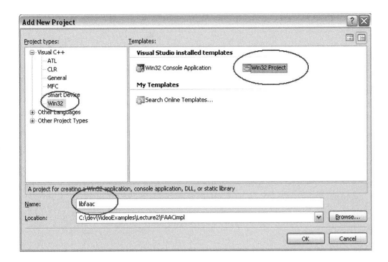

For the project type choose Win32 and for the template choose Win32 Project. Enter the name libfaac for the project name and click OK.

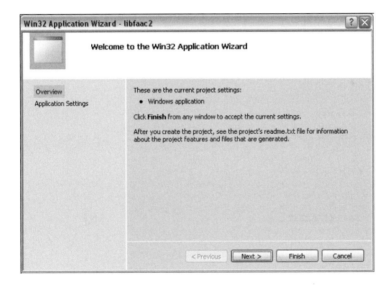

Above is the next dialog that will show up. Just click Next.

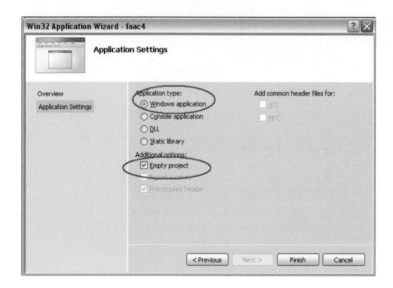

For the Application setting choose Windows application and choose Empty project for Additional options. Click Finish when you are done.

Before you add the source files for the codec first copy the files from the provided libfaac source into the project folder for your new libfaac.

After you have added the source files into your own project directory, add the header files from the solution browser by right-clicking on the Header files branch for the libfaac project (shown above). Then on the menu click Add -> Existing Item. Add all the .h files from your project directory.

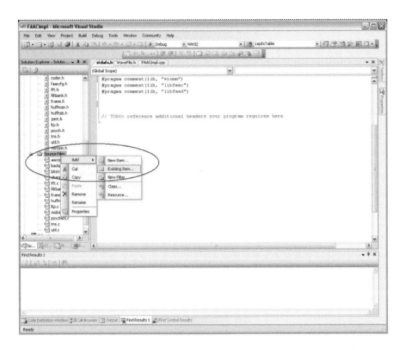

Also add the .c files to your project.

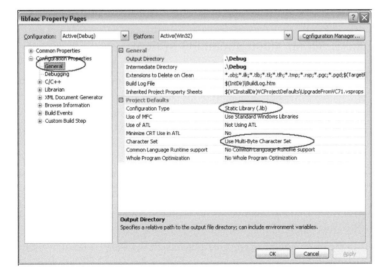

When you have finished adding all the files needed, set up the project properties. On the menu click Project -> Properties and it will bring you to the dialog above. Then Click on Configuration Properties -> General. On the right-hand side you will see settings for Project Defaults. Change the field Configuration Type to Static Library (.Lib) and also change the field Character Set to Use Multi-Byte Character Set.

Build the libfaac project now and you should be able to find the .lib file in the debug directory. Repeat for the libfaad source files the process used for the libfaac source files and project settings. After you have built the libraries for both libfaac and libfaad, copy the two .lib files into the project directory for the FAACImpl\Lib project.

12.5.4 Adding external dependencies and building the project

Add external dependencies by adding the following lines of code to the end of the stdafx.h file for the FAACImple project, if they are not already there:

```
#pragma comment(lib, "winmm")
#pragma comment(lib, "libfaac")
#pragma comment(lib, "libfaad")
```

Now that all the necessary files have been added and the project properties set up you should be able to compile the project without any problem. Check to see whether the debug folder has been created in your main project folder. You are now ready to execute the codec with a .wav file created with the audio capture and playback application from the last section. Copy a test audio file into the debug folder. Now start up a console and change to the directory of where an application resides. Now type the name of the program and click Enter. An easier way to start the application is just to click the Play button ▶ located at the top center of Visual Studio. The application will ask you a sequence of questions and will let you choose the options. Submit the choices accordingly. Note that the temporary file will be the encoded bitstream file and the destination file will be the reconstructed file after decompression.

12.6   Building a client–server video streaming system

We have explained how to capture, playback, encode, and decode video and audio data. We will now add one more component to the previous capture and playback application in order to provide network capability. Our main objective in this section is to enable the sending of video data from the server machine to the client machine. More specifically, we use the following procedure.

- Add a socket class to the video capture application to make a video server.
- Add a socket class to the video playback application to make a video client.

The basic algorithm for a video client or server is quite similar to that for video capture and playback. The difference here is that instead of saving the video to a file we first encode the video file so that it can be compressed, then pocketsize the video data and send it through the network. The client will receive the packet, extract the encoded data from the packet, decode the file, and display it on the screen.

12.6.1 Creating a video server

For the next two examples, we will be modifying the applications from Section 12.4 to create two new applications that are network capable. More specifically, to create a video server, several modifications have to be made to the VideoCapture application.

12.6.1.1 Adding the UDPSocket class

In order to add network capability to the VideoCapture application we need to add a socket class that will support the sending and receiving of data. Multimedia content is usually sent through the network using UDP packets since using TCP may cause delay in the case of retransmission. Our socket class will therefore be based on a UDP socket. This class will support the sending and receiving of data. To add the socket class to the VideoCapture application, we need to add UDPSocket .h and .cpp files to the VideoCapture project. To avoid getting compiling errors, please make sure to add the header file for Windows socket support, i.e., go to the stdafx.h file and add the following line, after the line:

```
#include "vfw.h"
#include "winsock2.h"
```

12.6.1.2 Adding SocketThreadProc

After adding the socket class, a new thread needs to be added into our application to handle packets that are received. We will need to add a few variables and functions to the VideoCaptureDlg class. First of all, add the line #include "UDPSocket.h" to VideoCaptureDlg.h. Adding this line will allow you to use the class variables and functions declared in that header file. Now add a class member variable of the type CUDPSocket called m_socket to the VideoCaptureDlg class. To do this, add the line CUDPSocket m_socket under Private Members.

Now add a declaration for a thread function that will handle the receiving of data from the network. Add the line

```
static UINT SockThreadProc(LPVOID lpParameter);
```

under Protected Members of the VideoCaptureDlg class. We need now to add a socket thread function by appending the following text to the end of VideoCaptureDlg. cpp file:

```
UINT CVideoCaptureDlg::SockThreadProc(LPVOID
lpParameter)
{
   CVideoCaptureDlg* p_VCDlg = (CVideoCaptureDlg*)
   lpParameter;
   fd_set rfs;
   struct timeval tv;
   SOCKADDR_IN sRecv;
   int iRc = 0;
   int iLen = 0;

   CUDPSocket udpSocket = p_VCDlg->m_socket;
   char* pBuf = new char[MAX_UDP_SIZE];

   while ( udpSocket.m_Socket )
   {
```

```
      FD_ZERO( &rfs );
      FD_SET( udpSocket.m_Socket, &rfs );
      tv.tv_sec = 1;
      tv.tv_usec = 0;

      iRc = select( 1, &rfs, NULL, NULL, &tv );
      if ( iRc != 1 )
        continue;

      iLen = udpSocket.RecvFrom(pBuf, MAX_UDP_SIZE,
    (SOCKADDR*)&sRecv);

        if (iLen == SOCKET_ERROR)
          continue;

        ((COMMON_HEADER*)pBuf)->sockSrc = sRecv;
        switch(((COMMON_HEADER*)pBuf)->flag)
        {
        case FLAG_CMD:
            p_VCDlg->OnRemoteCommand(iLen, pBuf);
          break;
        }
      }
      // Return Zero is the same as end thread
      // AfxEndThread(0, true);
      return 0;
  }
```

Here is a description of the socket thread procedure. There are some declarations just before the "while loop," The "while loop" checks whether the socket member of the VideoCaptureDlg class is still available. Then, it will check for activities for the socket. If there are activities then it will attempt to receive data from the network. The length and the data source will be recorded as well. The packet received should have a recognizable header that specifies the packet type. Currently the server receives command packets that are sent from clients to request video streams. When the member socket of VideoCaptureDlg is released, this thread will terminate.

12.6.1.3 Adding the function to handle commands

We need to now add a member function to handle commands. Again, we need to declare the member function and also add the function definition to the VideoCaptureDlg class. Add the member function declaration to the private members of the VideoCaptureDlg class declaration. Append the following definition code to the end of the VideoCaptureDlg.cpp file.

Declaration:

```
void OnRemoteCommand(int iLen, char* pBuf);
```

Definition:

```
void CVideoCaptureDlg::OnRemoteCommand(int iLen, char* pBuf)
{
```

```
CMD_HEADER *pRemoteCommand = (CMD_HEADER* )pBuf;

switch(pRemoteCommand->cmd)
{
case CMD_REQUEST_VIDEO:
   m_ClientAddr = pRemoteCommand->sockSrc;
   //m_ClientAddr.sin_port = htons(CLIENT_PORT);
   break;

case CMD_STOP_VIDEO:
   m_ClientAddr.sin_port = 0;
      break;
   }
   return;
}
```

12.6.1.4 Modifying the video capture class

Now that we have networking capability we need to modify the video capture class to send the video to the encoder and then to the network instead of saving the video data to the file. We need to modify the CVideoIn class so that when the video begins capturing, it will activate a callback function on the VideoCaptureDlg class. That function will then be responsible for encoding the video frame and sending it to the remote client.

Declaration:

```
bool BeginVideoCapture(BITMAPINFO bmi, LPVOID
callBackProc, DWORD IUser);
```

Definition:

```
bool CVideoIn::BeginVideoCapture(BITMAPINFO bmi,
LPVOID callBackProc, DWORD IUser)
{
   // Check for connection
   if (!IsConnected())
      return false;

   // Get settings
   CAPTUREPARMS captureParms;
   if (!capCaptureGetSetup(m_hWnd, &captureParms, sizeof
(CAPTUREPARMS)))
      return false;

   captureParms.vKeyAbort    = 0;
   captureParms.fCaptureAudio = false;
   captureParms.fYield    = true;
   captureParms.fAbortLeftMouse = false;
   captureParms.fAbortRightMouse = false;
   captureParms.wPercentDropForError = 100;
   captureParms.dwRequestMicroSecPerFrame = (DWORD)
(1.0e6 / 20.0);

   if (!capCaptureSetSetup(m_hWnd, &captureParms, sizeof
```

```
(CAPTUREPARMS)))
     return false;
  if (!capSetVideoFormat(m_hWnd, &bmi, sizeof
  (BITMAPINFO)))
     return false;

  if (!capSetCallbackOnVideoStream(m_hWnd,
  callBackProc))
     return false;

  if (!capSetUserData(m_hWnd, IUser))
     return false;

  if (!capCaptureSequenceNoFile(m_hWnd))
     return false;

  return true;
}
```

12.6.1.5 Adding a polymorphic function

In C++, it is easy to add a function that has the same name but behaves differently when different types of parameter are passed through. Here we have defined two implementations of BeginVideoCapture, one that saves file after capturing and one that just passes a video frame after capturing.

With the function added to CVideoIn we can now modify the callback function on the VideoCaptureDlg class to take care of coding the captured frame (VideoStream-CallbackProc).

Declaration:

```
static LRESULT CALLBACK VideoStreamCallbackProc
(HWND hWnd, LPVIDEOHDR lpVHdr );
```

Definition:

```
LRESULT CALLBACK CVideoCaptureDlg::
VideoStreamCallbackProc(HWND hWnd, LPVIDEOHDR lpVHdr)
{
   CVideoCaptureDlg* p_VCDlg =
(CVideoCaptureDlg*)capGetUserData(hWnd);
   p_VCDlg->OnEncodeVideoData((char*)lpVHdr->lpData,
lpVHdr->dwBytesUsed);
   return 0;
}
```

12.6.1.6 Adding a function for encoding video data

One more function, On EncodeVideoData, was also added to the VideoCaptureDlg class for encoding videodata.

Declaration:

```
void OnEncodeVideoData( char* pVid, int len);
```

Definition:

```
void CVideoCaptureDlg::OnEncodeVideoData(char* pVid, int len )
{
#ifdef TESTING
   int rlen;
   if ( m_VideoCodec.EncodeVideoData(pVid, len,
m_VideoPacket + sizeof(VIDEO_HEADER), &rlen,
&(((VIDEO_HEADER*)m_VideoPacket)->key)))
   {
      char* pout = new char [100000];
      int lenr;
      if (m_VideoCodec.DecodeVideoData(m_VideoPacket +
sizeof(VIDEO_HEADER), pout, &lenr, NULL))
      {
         videoOut.DrawVideoFrame(pout);

         // Need to add code to stop wait until the video stops
decompressing
         return;
      }
   }
#else
   int rlen;
   VIDEO_HEADER* VideoHeader = (VIDEO_HEADER*)
   m_VideoPacket;
   int HeaderSize = sizeof(VIDEO_HEADER);

   VideoHeader->flag = FLAG_VIDEO;

   if ( ! m_VideoCodec.EncodeVideoData(pVid, len,
m_VideoPacket + HeaderSize, &rlen, &VideoHeader->key))
      return;

   if (m_ClientAddr.sin_port == 0)
      return;
   if ( rlen > FRAG_SIZE )
   {
      int count = 1;
      int PacketSize;
      int RemainSize;
      int Offset;

      VideoHeader->id                = m_idVideo++;
      VideoHeader->data_size_total   = rlen;
      RemainSize                     = rlen;
      Offset                         = HeaderSize;
      do
      {
         PacketSize = RemainSize > FRAG_SIZE ? FRAG_SIZE :
RemainSize;
         RemainSize -= FRAG_SIZE;
```

```
        VideoHeader->data_size  = (unsigned short)
PacketSize;
        VideoHeader->subid    = count;

        memcpy(m_VideoPacket + HeaderSize, m_VideoPacket +
Offset, PacketSize);
        m_socket.SendTo(m_VideoPacket,   PacketSize  +  Head-
erSize, (SOCKADDR*)&m_ClientAddr);
          Offset += PacketSize;
          Sleep (1);
          count ++;
      } while (RemainSize > 0);
    }
    else
    {
      VideoHeader->data_size  = (unsigned short) rlen;
      VideoHeader->data_size  = rlen;
      VideoHeader->id    = m_idVideo++;
      VideoHeader->subid    = 0;

      m_socket.SendTo(m_VideoPacket, rlen + HeaderSize,
(SOCKADDR*)&m_ClientAddr);
          // Send packet to all others
    }
#endif
}
```

With all the codes added to the VideoCapture application you should have an application
that can be used as a server to send live captured video. Now build and run the project.

12.6.2 Creating a video client

Our previous example showed how to add the socket class to the video capture application
in order to make it a server; we will now add the socket class to the video playback
application and make it a client. Please refer to the previous example for instruction on how
to add a UDP socket class (see Section 12.6.1.1).

12.6.2.1 Adding SockThreadProc to the VideoPlayback application

After adding the socket class we also need to add the socket thread procedure just as we did
for the server but now, instead of receiving commands, this new thread will be responsible
for receiving video packets.
Declaration:

```
static UINT SockThreadProc( LPVOID lpParameter );
```

Definition:

```
UINT CVideoPlaybackDlg::SockThreadProc( LPVOID
lpParameter )
{
   CVideoPlaybackDlg* p_VPDlg = (CVideoPlaybackDlg*)
   lpParameter;
```

```
fd_set rfs;
struct timeval tv;
SOCKADDR_IN sRecv;
int  iRc   = 0;
int iLen = 0;

CUDPSocket udpSocket = p_VPDlg->m_socket;
char* pBuf = new char[MAX_UDP_SIZE];
char* pFraBuf = new char[MAX_FRAG_SIZE];


int count = 400;
while( count )
{
    FD_ZERO( &rfs );
    FD_SET( udpSocket.m_Socket, &rfs );
    tv.tv_sec  = 1;
    tv.tv_usec  = 0;

    iRc = select( 1, &rfs, NULL, NULL, &tv );
    if( iRc != 1 )
      continue;
    iLen = udpSocket.RecvFrom( pBuf, MAX_UDP_SIZE, (
SOCKADDR* )&sRecv );
      if( iLen == SOCKET_ERROR )
       continue;
      if( iLen == 0 )
       continue;

    // Capture the source of the packet
    //( ( COMMON_HEADER* )pBuf )->sockSrc= sRecv;
    // Process the packet
    switch( ( ( COMMON_HEADER* )pBuf )->flag ) //check the,
flag
    {
        case FLAG_VIDEO:
          p_VPDlg->OnVideoPacket( iLen, pBuf, pFraBuf);
          //count--;
        break;
    }
}

AfxEndThread( 0, true);
return 0;
}
```

12.6.2.2 Adding the OnVideoPacket function

Upon receivng a video packet the socket thread procedure will call the OnVideoPacket function in the VideoPlaybackDlg class to handle the decoding and the displaying of the video.

Declaration:

```
void OnVideoPacket(int iLen, char* pBuf, char*
pFraBuf);
```

Definition:

```
void CVideoPlaybackDlg::OnVideoPacket( int iLen, char* pBuf,
char* pFraBuf )
{
    int HeaderSize = sizeof(VIDEO_HEADER);
    if( iLen < HeaderSize )   // Victor: set a threshold for
the received buffer size
        return;

    if ( m_VideoQueue.count > MAX_QUEUESIZE - 1 )
        m_VideoQueue.Remove();
    // Check to see if it's a fragment
    if ( ( ( VIDEO_HEADER* )pBuf )->subid != 0 )
    {
        int FraLen;
        ProcessVideoFragment(pBuf, pFraBuf, &FraLen);
        //char* buf = new char[MAX_FRAG_SIZE];
        //memcpy(buf, pFraBuf, FraLen);
        //m_VideoQueue.Add( buf, FraLen );
        PlayVideoFrame(pFraBuf);
    }
    else
    {
        //char* buf = new char[MAX_FRAG_SIZE];
        //memcpy(buf, pBuf + HeaderSize, iLen - HeaderSize);
        //m_VideoQueue.Add( buf, iLen - HeaderSize );
        PlayVideoFrame(pBuf + HeaderSize);

    }
}
```

The above function will receive video fragments if the video frame is too big and cannot be contained in one video packet. It will run through a loop until all the video fragments associated with a single video frame have been received.

12.6.2.3 Adding the PlayVideoFrame function

PlayVideoFrame is responsible for decoding and displaying the video frame.
Declaration:

```
void PlayVideoFrame(char* pIn);
```

Definition:

```
void CVideoPlaybackDlg::PlayVideoFrame(char* pIn)
{
    if( m_VideoCodec.DecodeVideoData(pIn, m_VideoBuf, 0,
0))
        DrawVideoFrame(m_VideoBuf);
}
```

12.6.2.4 Adding the DrawVideoFrame function

In order to display the decoded video frame the OnVideoPacket function needs to call the DrawVideoFrame function.

Declaration:

```
bool DrawVideoFrame(char* VidFrame);
```

Definition:

```
bool CVideoPlaybackDlg::DrawVideoFrame(char* VidFrame)
{
CRect frameRect;
   this->GetClientRect(frameRect);
   BITMAPINFOHEADER m_biHeader;
   m_biHeader.biWidth       = 176;//frameRect.Width();
   m_biHeader.biHeight      = 144;//frameRect.Height();
   m_biHeader.biSize        = sizeof( BITMAPINFOHEADER );
   m_biHeader.biCompression = BI_RGB;
   m_biHeader.biPlanes      = 1;
   m_biHeader.biBitCount    = 24;
   m_biHeader.biSizeImage   = 3 * m_biHeader.biWidth *
m_biHeader.biHeight;

   BITMAPINFO bmi;
   bmi.bmiHeader = m_biHeader;
   bmi.bmiColors[0].rgbBlue = 0;
   bmi.bmiColors[0].rgbGreen = 0;
   bmi.bmiColors[0].rgbRed = 0;
   bmi.bmiColors[0].rgbReserved = 0;

   ShowWindow( SW_SHOWNORMAL );
   if ( VidFrame )
   {
      /*
      HDC hdc,   // handle to DC
      int XDest, // x-coord of destination upper-left corner
      int YDest, // y-coord of destination
upper-left corner
      int nDestWidth, // width of destination rectangle
      int nDestHeight, // height of destination rectangle
      int XSrc, // x-coord of source upper-left corner
      int YSrc, // y-coord of source upper-left corner
      int nSrcWidth, // width of source rectangle
      int nSrcHeight, // height of source rectangle
      CONST VOID *lpBits, // bitmap bits
      CONST BITMAPINFO *lpBitsInfo, // bitmap data
      UINT iUsage, // usage options
      DWORD dwRop // raster operation code
      */
      HDC m_hdc = c_VideoFrame.GetDC()->m_hDC;
      CRect frameRect;
      c_VideoFrame.GetClientRect(frameRect);
```

```
    SetStretchBltMode( m_hdc, COLORONCOLOR );
    int Res = StretchDIBits( m_hdc, 0, 0, frameRect.Width
(), frameRect.Height(), 0, 0, m_biHeader.biWidth,
m_biHeader.biHeight, VidFrame, &bmi, DIB_RGB_COLORS,
SRCCOPY );
    if ( Res ==GDI_ERROR )
    {
      return false;
    }
  }
  return true;
}
```

12.6.3 Building the server and client applications

Now that you have created the server and client applications, we will put these applications
to the test. First build and start the Video Server application. Since the video codec used in
the client–server video streaming system is based on the Windows Media Video 9 (WMV9)
codec (see Section 5.7), you will have to also download and install the WMV9 video
compression manager (VCM) from our software download site (or the Windows Media
Services 9 series SDK from the download center of Microsoft, which is http://www.
microsoft.com/windows/windowsmedia/download/AllDownloads.aspx).

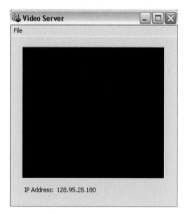

Note that on the bottom of the application the IP address of the server is indicated. This IP
address needs to be typed in at the client application in order to connect to the server.Now
start the server by clicking File -> Start Server.

The above snapshot shows the application after the server has started.

Now start the client application: go to the menu and click File -> Open Stream.

The above dialog will show. Enter the IP address shown on the server application and click OK.

The above figure shows the client application after it starts running. To stop the server, go to menu and click File -> Stop Streaming.

12.7   Creating a small P2P video conferencing system

The objective of this section is to demonstrate how to create a simple peer-to-peer (P2P) video conferencing system, without the use of a centralized streaming server, based on the Microsoft Windows Media Video 9 (WMV9) codec.

12.7.1   Installing the WMV9 codec and SDK

You have to again download and install the WMV9 video compression manager (VCM) from our software download site (or the Windows Media Services 9 series SDK from the download center of Microsoft, i.e., http://www.microsoft.com/windows/windowsmedia/download/AllDownloads.aspx).

Once you have the WMV9 SDK installed you should be able to build the project. Some warnings may be given in the linking process but that is fine; as long as the project builds without compiling errors your application will still run correctly.

12.7.2   Starting the MCU

A multipoint control unit (MCU) is used to perform user registration, admission control, IP address forwarding, etc. To run the MCU, you have to extract the MCU project from the downloaded software, open the vmcu solution, and build the project. You may start this project by using the project Play button directly from Visual Studio; you may also start it by opening the .exe file in the debug folder. Note that the icq_user.xml and uw_icq_user.xml files must be located in the same directory as vmcu.exe.

The client application dictates the use of the MCU to locate other peers and to reach them with an invitation to start a video conference. Below is an illustration of the MCU application.

Start up the MCU (Multiple Control Unit) server and click Start. This will start the server that listens to the connection. When a connection is established, the client application will download the list of buddies that are peers of the current clients.

12.7.3 Building the MCU presence server

Go to the directory where you have extracted the Video Conference project. Then go into the "prj" folder. Open up the project solution by double-clicking the .sln file. Once you have opened up the solution, make sure you can build the project.

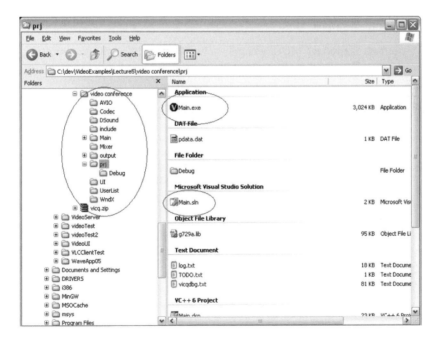

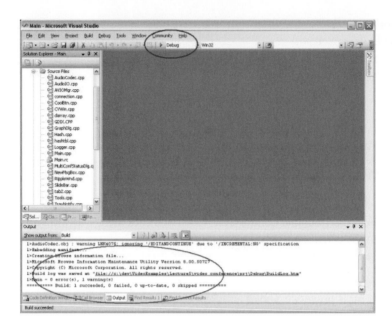

You may start this project by using the project Play button indicated at the top center of the above picture within Visual Studio.

12.7.4 Logging onto the MCU

Once you have started the MCU, you may now sign in with the client application. Click the Sign In button and the Sign In dialog will appear.

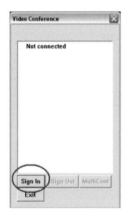

There are six accounts, User ID "a" to "e," on the MCU presence server. Without loss of generality, each account is given the same password as its User ID. For example, if the User ID is "a" then the password for that account will be "a" also. The server IP has the same IP address as the host of the MCU application that has been activated. You must find out this IP address and enter it into the text box to connect to the MCU server. If you do not know the

IP address of the server simply open up a console application and type "ipconfig" in the command prompt. You will then find out the IP address of the host.

12.7.5 Starting a video conference

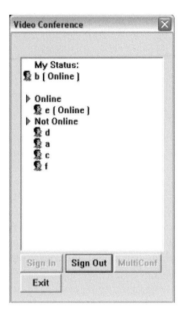

After a successful connection to the MCU, the client application will download the list of buddies and will display it on the user list dialog shown above.

There are two ways to start a conference and each depends on the number of participants you want to have. You may have a one-to-one video conversation with someone by simply clicking their User ID on the user list. If you want to start a conference with more than one client then you need to press the control key while selecting different User ID's on the user list. Then, once you have selected your desired peers, just click the Multiconf button to send an invitation to all those peers.

12.7.6 Multiconference status and canceling a call

Once an invitation is sent, the multiconf dialog will appear. This dialog indicates the status of the invitation. Whenever you see "connecting ... " it means that the invitation has not reached the desired peer. Whenever you see "connected ... " it means that the invitation has been received and we are waiting for the other party to respond with an accept or decline message. At this point you may cancel a call by simply clicking the Cancel Call button at the bottom of the Multiconf dialog.

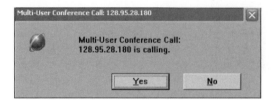

The above illustration shows the dialog at the receiving end of the invitation. This peer has the choice of accepting or declining the invitation.

Once any peer accepts your invitation a video conference starts. First, your own video window will show up, and then the video window of the peers will also show up.

Below there are two video windows. The one on the left shows your own video and the one on the right shows the video of one of the peers.

It may take a while for the video to start showing on your own and/or the peer video windows. As more peers accept an invitation, more peer video windows will pop up. If none of the peers that received an invitation accepts then neither your own video window nor a peer video window will show.

Whenever a video conference is in progress you can hang up and leave it. A multi-conference will remain as long as there are still two peers chatting with each other. Whenever there is only one peer left on the conference it will terminate.

Index

.